PUBLICATION P201

HANDBOOK OF STRUCTURAL STEELWORK
3rd Edition

Jointly published by:

The British Constructional
Steelwork Association Ltd
4 Whitehall Court
London SW1A 2ES

Tel: 020 7839 8566
Fax: 020 7976 1634

The Steel Construction Institute
Silwood Park
Ascot
SL5 7QN

Tel: 01344 623345
Fax: 01344 622944

© The British Constructional Steelwork Association Ltd and The Steel Construction Institute, 2002

© The British Constructional Steelwork Association Ltd, 1990, 1991.

Apart from any fair dealing for the purposes of research or private study or criticism or review, as permitted under the Copyright Designs and Patents Act, 1988, this publication may not be reproduced, stored, or transmitted, in any form or by any means, without the prior permission in writing of the publishers, or in the case of reprographic reproduction only in accordance with the terms of the licences issued by the UK Copyright Licensing Agency, or in accordance with the terms of licences issued by the appropriate Reproduction Rights Organisation outside the UK.

Enquiries concerning reproduction outside the terms stated here should be sent to the publishers, at the addresses given on the title page.

Although care has been taken to ensure, to the best of our knowledge, that all data and information contained herein are accurate to the extent that they relate to either matters of fact or accepted practice or matters of opinion at the time of publication, The British Constructional Steelwork Association Limited and The Steel Construction Institute assume no responsibility for any errors in or misinterpretations of such data and/or information or any loss or damage arising from or related to their use.

Publications supplied to the Members of SCI and BCSA at a discount are not for resale by them.

Publication Number: P201   ISBN 1 85942 133 4

(ISBN 0 85073 023 6, Second Edition, 1991)

(ISBN 0 85073 023 6, First Edition, 1990)

British Library Cataloguing-in-Publication Data.

A catalogue record for this book is available from the British Library.

FOREWORD

The objective of this publication is to present a practical guide to the design of structural steel elements for buildings. The document comprises three principal Sections: general guidance, design data, and design tables.

The guidance is in accordance with BS 5950-1:2000, *Structural use of steelwork in building – Code of practice for design. Rolled and welded section.* Worked examples are presented where appropriate. No attempt has been made to consider complete structures, and it is to be noted therefore that certain important design matters are not dealt with – those for instance of overall stability, of interaction between components, and of the overall analysis of a building.

Section on General Design Data includes bending moment diagrams, shear force diagrams and expressions for deflection calculations. A variety of beams and cantilevers with different loading and support conditions are covered. Expressions for properties of geometrical figures are also given, together with useful mathematical solutions and metric conversion factors.

The design tables also include section property, member capacity and ultimate load tables calculated according to BS 5950-1:2000. The tables are preceded by a comprehensive set of explanatory notes. Section ranges listed are those that were readily available at the time of printing. In addition, both hot finished and cold formed structural hollow sections are included in the 'Tables of Dimensions and Section Properties'.

A list of references is given at the end of the explanatory notes to the design tables.

ACKNOWLEDGEMENTS

This publication is jointly published by the BCSA and the SCI. The preparation of this publication was carried out under the guidance of a steering group consisting of the following members:

Mr D Brown   The Steel Construction Institute
Dr P Kirby   University of Sheffield
Mr A Way   The Steel Construction Institute
Mr P Williams The British Constructional Steelwork Association

Dr P Kirby wrote Chapters 1 to 5 of the publication.

The section property and member capacity tables were produced by Mr A Way.

Valuable comments were also received from:

Mr A Malik   The Steel Construction Institute
Mr A Rathbone CSC (UK) Ltd.

The publication has been jointly funded by the BCSA and the SCI.

Contents

		Page No.
FOREWORD		iii
ACKNOWLEDGEMENTS		v
CHAPTER 1	**GENERAL DESIGN CONSIDERATIONS**	1
1.1	Design aims	1
1.2	Methods of design	1
1.3	Loadings	3
1.4	Limit state design	4
1.5	Stability limit state	8
1.6	Design strengths	11
CHAPTER 2	**LOCAL RESISTANCE OF CROSS-SECTIONS**	13
2.1	Local buckling	13
2.2	Classification	14
2.3	Example – Section classification	20
2.4	General Guidance	22
CHAPTER 3	**BEAMS**	23
3.1	Design considerations	23
3.2	Moment and shear capacities	25
3.3	Design of beams without full lateral restraint	25
3.4	Equivalent slenderness	27
3.5	Effective length	27
3.6	Equivalent uniform moment factor, m_{LT}	29
3.7	Calculation of bending resistance for beams without full restraint	30
3.8	Calculation of bending resistance – a simpler approach	30
3.9	Example – Beam with full lateral restraint	32
3.10	Example – Unrestrained beams	33
3.11	Web bearing capacity and web buckling resistance	35
3.12	Web stiffeners	39
3.13	Example – Web bearing and buckling	41
3.14	Example – Web stiffeners	43
CHAPTER 4	**MEMBERS IN TENSION AND COMPRESSION**	46
4.1	Introduction	46
4.2	Ties	46
4.3	Simple tension members	47
4.4	Tension members also subjected to moments	48

4.5	Struts	48
4.6	Columns in simple construction	60
4.7	Compression members with moments	61
4.8	Example – Angle section used as a tie	63
4.9	Example – Axially loaded strut 1	64
4.10	Example – Axially loaded strut 2	65
4.11	Example – Column in simple construction	66
4.12	Example – Column under axial load and moment	68

CHAPTER 5 TRUSSES 72
5.1	Introduction	72
5.2	Typical uses	72
5.3	Design concept	74

GENERAL DESIGN DATA 79
Bending moment and deflection formulae for beams 80
Moving loads 91
Fixed end moments 94
Trigonometrical formulae 95
Solution of Triangles 96
Properties of geometrical figures 98
Metric conversions 106

EXPLANATORY NOTES 107
General 108
Dimensions of sections 109
Section properties 110
Capacity and resistance tables 121
Bending tables 122
Web bearing and buckling tables 124
Tension tables 128
Compression tables 129
Axial and bending tables 136
Bolts and welds 139

REFERENCES 143

Yellow Pages

TABLES OF DIMENSIONS AND GROSS SECTION PROPERTIES 147

Universal beams	148
Universal columns	154
Joists	158
Parallel flange channels	162
ASB (Asymmetric Beams)	166
Equal angles	169
Unequal angles	170
Equal angles back to back	172
Unequal angles back to back	173
Tees cut from universal beams	174
Tees cut from universal columns	178
Hot-finished circular hollow sections	180
Hot-finished square hollow sections	182
Hot-finished rectangular hollow sections	184
Cold-formed circular hollow sections	186
Cold-formed square hollow sections	189
Cold-formed rectangular hollow sections	191

	Pink Pages	Green Pages
MEMBER CAPACITIES	**S275**	**S355**
Universal beams subject to bending	196	280
Universal columns subject to bending	199	283
Joists subject to bending	200	284
Parallel flange channels subject to bending	201	285
Universal beams web bearing and buckling	202	286
Universal columns web bearing and buckling	205	289
Joists web bearing and buckling	206	290
Parallel flange channels web bearing and buckling	207	291
Equal angles subject to tension	208	292
Equal angles back to back subject to tension	211	295
Unequal angles subject to tension	214	298
Unequal angels back to back subject to tension	217	301

	S275	S355
MEMBER CAPACITIES (continued)		
Universal beams subject to compression	220	304
Universal columns subject to compression	224	308
Equal angles subject to compression	226	310
Unequal angles subject to compression	227	311
Equal angles back to back subject to compression	228	312
Unequal angles subject to compression	230	314
Universal beams subject to axial load and bending	232	316
Universal columns subject to axial load and bending	258	342
BOLT CAPACITIES		
Non-preloaded ordinary bolts	266	350
Non-preloaded countersunk bolts	268	352
Non-preloaded HSFG bolts	270	354
Preloaded HSFG bolts:		
Non-slip in service	271	355
Non-slip under factored loads	272	356
Non-slip in service - countersunk	273	357
Non-slip under factored loads - countersunk	274	358
WELDS		
Fillet welds	275	359

CHAPTER 1 GENERAL DESIGN CONSIDERATIONS

1.1 Design aims

The aim of any design process is the fulfilment of a purpose, and structural steelwork design is no exception. In building design, the purpose is most commonly the provision of space that is protected from the elements. Steelwork is also used to provide internal structures, particularly in industrial situations.

The designer must ensure that the structure is capable of resisting the anticipated loading with an adequate margin of safety and that it does not deform excessively during service. Due regard must be paid to economy which will involve consideration of ease of manufacture, including cutting, drilling and welding in the fabrication shop and transport to site. The provision and integration of services should be considered at an early stage and not merely added on when the structural design is complete. Under CDM requirements the designer has an obligation to consider how the structure will be erected, maintained and demolished. Sustainability issues such as recycling and reuse of materials should also be considered. Any likely extensions to the structure should be taken into account at this stage in the process.

1.2 Methods of design

Historically, engineers have been accustomed to assume that joints in structures behave as either pinned or rigid to render design calculations manageable. In 'simple design' the joints are idealised as perfect pins. 'Continuous design' assumes that joints are rigid and that no relative rotation of connected members occurs whatever the applied moment. The vast majority of designs carried out today make one of these two assumptions, but a more realistic alternative is now possible, which is known as semi-continuous design. As stated in BS 5950-1:2000 [1] Clause 2.1.2.1, the details of the joints used should fulfil the assumptions of the chosen design method.

1.2.1 Simple design

Simple design is the most traditional approach and is still commonly used. It is assumed that no moment is transferred from one connected member to another, except for the nominal moments which arise as a result of eccentricity at joints.

The resistance of the structure to lateral loads and sway is usually ensured by the provision of bracing or, in some multi-storey buildings, by concrete cores.

It is important that the designer recognises the assumptions regarding joint response and ensures that the detailing of the connections is such that no moments develop that can adversely affect the performance of the structure. Many years of experience have demonstrated the types of details that satisfy this criterion and the designer should refer to the standard connections given in the BCSA/SCI publication on joints in simple construction[2].

1.2.2 Continuous design

In continuous design, it is assumed that joints are rigid and transfer moment between members. The stability of the frame against sway is by frame action (i.e. by bending of beams and columns). Continuous design is more complex than simple design therefore software is commonly used to analyse the frame. Realistic combinations of pattern loading must be considered when designing continuous frames. The connections between members must have different characteristics depending on whether the design method for the frame is elastic or plastic.

In elastic design, the joints must possess sufficient rotational stiffness to ensure that the distribution of forces and moments around the frame are not significantly different to those calculated. The joint must be able to carry the moments, forces and shears arising from the frame analysis.

In plastic design, in determining the ultimate load capacity, the strength (not stiffness) of the joint is of prime importance. The strength of the joint will determine whether plastic hinges occur in the joints or in the members, and will have a significant effect on the collapse mechanism. If hinges are designed to occur in the joints, the joint must be detailed with sufficient ductility to accommodate the resulting rotations. The stiffness of the joints will be important when calculating beam deflections, sway deflections and sway stability.

1.2.3 Semi-continuous design

True semi-continuous design is more complex than either simple or continuous design as the real joint response is more realistically represented. Analytical routines to follow the true connection behaviour closely are highly involved and unsuitable for routine design, as they require the use of sophisticated computer programs. However, two simplified procedures do exist for both braced and unbraced frames; these are briefly referred to below. Braced frames are those where the resistance to lateral loads is provided by a bracing system or a core;

in unbraced frames this resistance is generated by bending moments in the columns and beams.

The simplified procedures are:

(i) The wind moment method, for unbraced frames.

In this procedure, the beam/column joints are assumed to be pinned when considering gravity loads. However, under wind loading they are assumed to be rigid, which means that lateral loads are carried by frame action. A fuller description of the method can be found in reference [3].

(ii) Semi-continuous design of braced frames.

In this procedure, account of the real joint behaviour is taken to reduce the bending moments applied to the beams and to reduce the deflections. Details of the method can be found in reference [4].

1.3 Loadings

The principal forms of loading associated with building design are:

(i) Dead loading

This is loading is of constant magnitude and location, and is mainly the self-weight of the structure itself.

(ii) Imposed loading

This is loading applied to the structure, other than wind, which is not of a permanent nature. Gravity loading due to occupants, equipment, furniture, material which might be stored within the building, demountable partitions and snow loads are the prime sources for imposed loads on building structures. BS 6399-1[5] should be consulted for imposed loadings. Note that in some cases clients may request that structures be designed for higher imposed loads than those specified in BS 6399-1.

(iii) Wind loading

Wind produces both lateral and (in some cases) vertical loads. Wind may blow in any direction, although usually only two orthogonal load-cases are considered.

Values to be adopted for each of these loads can be obtained from BS 6399 [5]. They are essentially the extreme loads that can be reasonably expected to occur on the structure, and are frequently described as the characteristic design loads.

1.4 Limit state design

1.4.1 Background

To cater for the inherent variability of loading and structural response, engineers apply factors to ensure the structure will carry the loads safely. Until about 20 years ago, design was largely based on an allowable stress approach. The maximum stress was calculated using the maximum anticipated loading on the structure and its value was limited to the yield stress of the material divided by a single global factor of safety. Serviceability deformations were calculated using these same maximum anticipated loadings. However, this approach gave inconsistent reserves of strength against collapse. The method is now superseded by a limit state approach in which the applied loads are multiplied by factors, capacities and resistances are determined using the design strength of the material. Limit states are the states beyond which the structure becomes unfit for its intended use. BS 5950-1 is a limit state design standard.

1.4.2 General

The values of the partial safety factors given in the Standard, which vary from load case to load case, reflect the probability of these values being exceeded for each specified situation. Reduced values of the partial safety factor are given when loadings are combined, as it is less likely that, for example, maximum wind will occur with maximum imposed load. This can be seen from Table 2 of BS 5950. The part of this table relevant to buildings not containing cranes is reproduced as Table 1.1.

1.4.3 Ultimate limit states

The ultimate limit state (ULS) concerns the safety of the whole or part of the structure. In buildings without cranes, the principal load combinations which should be considered are:

Load combination 1:   Dead load + imposed load

Load combination 2:   Dead load + wind load

Load combination 3:   Dead load + imposed load plus wind load.

Table 1.1 *Partial load factors γ_f for buildings without cranes*

Type of building and load combination	Factor γ_f
Dead load	1.4
Dead load with wind load and imposed load	1.2
Dead load when it counteracts the effects of other loads	1.0
Dead load when restraining sliding, overturning or uplift	1.0
Imposed load	1.6
Imposed load acting with wind load	1.2
Wind load	1.4
Wind load acting with imposed load	1.2
Storage tanks including contents	1.4
Storage tanks empty, when restraining sliding, overturning or uplift	1.0
Exceptional snow load (due to local drifting on roofs)	1.05

The limit states that need to be considered are described in turn.

(i) Limit state of strength

This limit state is reached when there is failure by yielding, buckling, rupture and any combination of these which limits the load carrying capacity of the structure. Each of the load combinations identified above should be taken into account.

(ii) Stability limit state

The Standard identifies two types of instability under this heading. The first involves overturning of the structure (or part of it) as a rigid body, lifting off its seating or sliding on its foundations. The second concerns the sway stiffness of the structure. If sway deflections due to horizontal forces become too large then excessive secondary effects can become significant. If the secondary effects are significant they must be taken into account in the design. This is discussed further in Section 1.5.

(iii) Fatigue

Generally this is rarely a problem in building structures as fatigue failure happens when a very large number (of the order of 2×10^6) of stress reversals of a significant magnitude occur. The only time that this is likely to cause concern is in buildings containing heavy vibrating plant or machinery, such as printing presses or indeed fatigue testing equipment.

(iv) Brittle fracture

This is a phenomenon in which steel loses its normal ductility and fails in a brittle manner. It is avoided by ensuring that the steel used (all components including welding materials) has adequate notch toughness. Brittle fracture is more likely with: low temperatures, large steel thickness, high tensile stresses, high strain rates and details that include stress raisers such as holes and welds. The higher the risk of brittle facture the tougher the specified steel must be. The requirement of BS 5950-1, Clause 2.4.4 is that the maximum thickness should be less than or equal to a factor K multiplied by t_1. The factor K (obtained from Table 3) is dependent upon the stress conditions, the detailing and the strain rate. The limiting thickness t_1 (obtained from Tables 4 or 5) is dependent upon the minimum service temperature and the steel specification. In practice, the required steel specification, including sub-grade, is identified for a particular design situation.

(v) Structural Integrity

Whilst this document covers the design of elements it must be remembered that structures are three dimensional and must act in a coherent fashion and be stable in all directions. In addition to having sufficient resistance to minimum horizontal loads, there are also requirements for minimum tying forces and checks against accidental damage which are covered in Clause 2.4.5 of BS 5950-1.

All buildings should be tied together at each floor and roof level. This is most effectively done using members approximately at right angles to each other (to provide three-dimensional robustness) or by steel reinforcement in concrete floor slabs, provided that they are properly secured to the columns. All ties should be able to resist a minimum force of 75 kN.

For frames of more than four storeys, there are additional requirements which can be found in Clause 2.4.5.3. They are designed to ensure that if a failure occurs at one location, then damage is limited to a small area and does not lead to a progressive collapse of the whole structure.

1.4.4 Serviceability limit state

Serviceability limit state (SLS) corresponds to the limit beyond which the specified service criteria are no longer met. Serviceability loads are generally taken as unfactored imposed loads, there are some exceptions. Further guidance is given in Clause 2.5.1 of BS 5950-1:2000. Serviceability criteria include deflection, vibration and durability which are considered in turn below.

(i) Deflections

Although a structure may have adequate strength, deflections at the specified serviceability design loading may still be unacceptable. Such distortion may result in doors or windows being inoperable, or plaster and other brittle finishes to cracking. Table 8 of the Standard gives limits for a variety of conditions – some of which are listed here as Table 1.2. Note that this table is titled "suggested limits for calculated deflections". This is because a general Standard cannot give definitive values to cater for all cases met within practice and it is essential for the engineer to exercise judgement in determining the requirements for each specific case considered.

Table 1.2 *Suggested limits for calculated deflection*

a) Vertical deflections of beams due to imposed load	
Cantilevers	Length / 180
Beams carrying plaster and other brittle finish	Span / 360
Other beams (except purlins and sheeting rails)	Span / 200
b) Horizontal deflection of columns due to imposed and wind load	
Tops of columns in single storey buildings, except portal frames	Height / 300
In each storey of a building with more than one storey	Storey height / 300

(ii) Vibrations and wind induced oscillations

Vibration and oscillation of structures should be limited to prevent damage to contents and discomfort to users. Traditionally, vibration has been deemed to be a problem only for masts and towers, when wind oscillations have needed attention, or in structures supporting vibrating machinery. Vibration is not usually a problem with normal buildings unless spans are large, say in excess of 9 m, or for the floors of dance halls or gymnasia, which are subject to rhythmic loading. The solution to any problem is not simply to over-design the members but rather to investigate the natural frequency of the structural system and to arrange that it differs significantly from the frequency of the disturbing forces, so that resonance does not occur. An SCI publication[6] gives guidance on this topic.

(iii) Durability

The durability of a structure should be considered for its intended use and intended life. Steel will corrode only if exposed to air and water together. The onus for ensuring suitable protection schemes lies with the design engineer and the use of BS 5493[7] is recommended. Consideration should be given to the environment and anticipated life of the structure and the degree of exposure for each component as well as the level and ease of maintenance after completion.

In particular, care should be taken to avoid detailing that produces pockets in which water and dirt can accumulate. Helpful information will be found in guides to corrosion[8], which show that in certain circumstances such as the interiors of multi-storey buildings, untreated steelwork may be acceptable.

1.5 Stability limit state

1.5.1 Resistance to horizontal forces

Structures should have an adequate resistance to horizontal forces to ensure a practical degree of robustness against incidental loading. For conventional structures, horizontal forces are frequently considered to be those arising from wind. Load combination 1 of Section 1.5.1 consists of pure gravity loading which does not contain any lateral force. However, the columns in buildings are never perfectly vertical. To generate an allowance for this effect without the necessity to explicitly include possible construction tolerances, a small horizontal force must also be applied at the head of the column. The value of this notional horizontal force is taken as 0.5% of the vertical force as described in Clause 2.4.2.4 of the Standard.

Thus all structures should be capable of resisting notional horizontal forces which should not be less than 0.5% of the factored dead plus imposed loads applied to the structure at that level. Because these forces are not externally applied forces they:

(i) do not contribute to the reactions required at the foundations

(ii) should not be applied when considering overturning

(iii) should not be combined with real horizontal loads

(iv) should not be combined with temperature effects

(v) should not be applied when considering pattern loading.

In load combinations 2 and 3 of Section 1.5.1 which contain real wind loads, to ensure robustness, there is a minimum value for the horizontal component of the wind load equal to 1% of the factored dead load.

These horizontal loads should be resisted by one (or more) of the following:

(i) triangulated bracing

(ii) moment resisting joints (frame action)

(iii) cantilever columns

(iv) shear walls

(v) specially designed staircase or lift-shaft enclosures or similar.

It must be remembered that consideration of load reversal must be included and, where horizontal loading is applied to roofs, cladding and other components, these, and their attachments to the structural frame, must be designed to resist such action. Where resistance to horizontal forces is provided by means other than the steel frame, e.g. by the concrete walls around the lift-shaft, this should be clearly stated in the design documents.

1.5.2 Sway stiffness

Horizontal forces will lead to a relative horizontal movement Δ between the upper and lower ends of vertical columns. In conjunction with the axial load P in the column, this will give rise to secondary moments. These are known as P-Δ moments. The new Standard draws special attention to such second order effects. The Standard therefore divides frames into non-sway and sway-sensitive frames. A frame is non-sway when the secondary effects are small enough to be ignored. Second order effects must be explicitly considered if the frame is classed as sway-sensitive. Sufficient stiffness should be provided also to limit twisting of the structure on plan, see Clause 2.4.2.5 of BS 5950-1.

Determination of sway sensitivity

Except for single storey frames, or other frames with sloping members and moment resisting joints, the process to evaluate sway sensitivity is as follows:

1. Define the maximum factored dead plus imposed vertical load at each floor and roof level.

2. Determine the notional horizontal forces (0.5% of the above) and apply these as horizontal point loads at each corresponding floor and roof level.

3. Carry out an elastic analysis of the frame under the notional horizontal forces alone to determine the horizontal deflection at each floor and roof level.

4. Evaluate the sway index λ of every storey as $h / 200\delta$

   δ is the relative horizontal deflection between the top and bottom of the column

   h is the storey height

5. The smallest value of λ for the entire frame is then taken as λ_{cr}.

6. If λ_{cr} is $\geq$ 10 then the frame is non-sway, and second order effects due to sway are small enough to be ignored. Otherwise, the frame is sway-sensitive and second order effects are not small enough to be ignored.

The above method for calculation of λ_{cr} may not always be practical i.e. where notional horizontal forces or floor levels are not readily identifiable. A second order elastic critical buckling load analysis is an alternative approach for obtaining λ_{cr}.

The Standard categorizes frames in to two types, clad frames where the stiffening effects of the cladding is ignored and bare steel frames or frames where the stiffening of the cladding was included in the calculation of λ_{cr}. This second category of frame is always classed sway-sensitive.

1.5.3 Non-sway frames

These frames are such that sway effects are so small as to be negligible. Forces and moments may be evaluated without allowances for sway effects and member design is straightforward. Effective length ratios for columns will be less than or equal to one.

1.5.4 Sway-sensitive frames

Provided that the frame is to be designed elastically there is a simple process to allow for sway effects. If the frame is designed plastically the process is more complex and is beyond the scope of this publication.

When λ_{cr} is less than 10 but not less than 4, the second order effects may be allowed for by a procedure which uses a magnification factor k_{amp}. For clad frames where the stiffening effects of the cladding is ignored, k_{amp} is evaluated very simply from the expression below:

$$k_{amp} = \lambda_{cr} / (1.15\lambda_{cr} - 1.5)$$

This magnification factor must be applied to the sway effects. The sway effects are the forces in the bracing system for a braced frame and they are the sway moments in a continuous frame. Two alternative procedures are set out in BS 5950-1 to implement this, which are set out below with additional comment.

(a) Deducting the non-sway effects

(i) Analyse the frame under the actual restraint conditions.

(ii) Add horizontal restraints at each floor or roof level to prevent sway and re-analyse (this will result in the non-sway moments being identified).

(iii) Obtain the sway effects by subtracting the set of moments and forces from ii) from those obtained in (i). These are the forces and moments to be amplified by k_{amp} and subsequently recombined with the forces and moments calculated in (ii).

(b) Direct calculation

(i) Analyse the frame with horizontal restraints at each floor and roof level to prevent sway.

(ii) Reverse the direction of the horizontal reactions produced at the added horizontal restraints.

(iii) Analyse the frame with these forces applied as loads to an otherwise unloaded frame under the actual restraint condition (as they are the forces causing sway to occur).

(iv) Adopt the forces and moments from iii) as the sway forces and moments, amplify them using k_{amp} and recombine with the non-sway forces and moments from (i).

Alternatively, if resistance to horizontal forces is provided by moment connections or cantilever columns, the second order effects can be allowed for by using the sway mode in-plane effective lengths (see Section 4.5, Table 4.2) for the columns and designing the beams to remain elastic under factored loads.

1.6 Design strengths

The minimum material design strength p_y is specified as being 1.0 Y_s but not greater than $U_s/1.2$ where Y_s and U_s are the minimum yield strength and the minimum tensile strength respectively. The value of the yield strength and thus the design strength decreases with thickness, and, for the most common grades of steel, the value may be determined from Table 9 of BS 5950-1, an extract from which is reproduced below as Table 1.3. For rolled sections, the design strength for the whole section is based on the thickest element (usually the flange).

The design resistances (capacities) of members are based on the material design strength without the application of any partial factor.

Table 1.3 *Design strength p_y for steel grades S275 and S355*

Thickness (in mm) Less than or equal to	S275	S355
16	275	355
40	265	345
63	255	335
80	245	325
100	235	315
150	225	295

CHAPTER 2 LOCAL RESISTANCE OF CROSS-SECTIONS

2.1 Local buckling

The cross-section of most structural members may be considered to be an assembly of flat plate elements. As these plate elements are relatively thin, they may buckle locally when subjected to compression. In turn, this may limit the axial load carrying capacity to a value below the squash load (cross-sectional area times yield strength) and the bending resistance to a value below the fully plastic moment of resistance (plastic section modulus times yield strength). This phenomenon is independent of the length of the member and hence is termed local buckling. It is dependant upon a number of parameters. The following are of particular importance:

(i) Width to thickness ratio of the element. This is often termed the aspect ratio. Wide, thin elements are more prone to buckling.

(ii) Support condition. This is dependent upon the edge restraint to the element. If the element is supported by other elements along both edges parallel to the direction of the member, then it is called an internal element, as both edges are prevented from distorting out of plane. If this condition only occurs along one edge, it is said to be an outstand element, as the free edge is able to distort out of plane. Each half of the flange of an I section is an outstand element, whilst the web is an internal element.

(iii) Yield strength of the material. The higher the yield strength of the material the greater is the likelihood of local buckling before yield is reached.

(iv) Stress distribution across the width of the plate element. The most severe form of stress distribution is uniform compression, which will occur throughout a cross-section under axial compressive loading or in the compression flange of an I section in bending. However, the web of an I section under flexure will be under a varying moment which is a less severe condition. This is because the maximum compressive stress will only occur at one location and the stress level will reduce across the width of the element possibly even changing to a tensile value.

(v) Residual stresses in rolled or welded sections. The presence of a weld within a cross-section can produce quite severe residual stresses that will adversely affect the behaviour with respect to local buckling.

All of these factors are included in the classification and design provisions of BS 5950-1.

2.2 Classification

2.2.1 Classes of cross-sections

BS 5950-1 sets out a practical conservative approach suitable for most design situations to ensure that local buckling does not occur. The Standard introduces four classes of cross-section which are defined below. These are initially described below in terms of the capacity of the cross-section under pure bending.

Class 1 plastic

Class 1 plastic cross-sections are sufficiently stocky that the material design strength may be attained throughout the cross-section. The moment of resistance is therefore equal to the fully plastic moment $p_y S$. This resistance can be maintained whilst rotation occurs at that cross-section. At the location of plastic hinges in plastic design Class 1 sections must be used.

Class 2 compact

Class 2 compact cross-sections can attain the fully plastic moment resistance but can not sustain significant rotations. Therefore, Class 2 compact sections can only be used for plastic design at locations where plastic hinges do not form and rotate.

Class 3 semi-compact

Class 3 semi-compact cross-sections are able to attain the material design strength at the extreme fibres of the cross-section and some way into the section but are unable to attain that stress throughout the entire cross-section. Such a cross-section can resist a moment equal to $p_y S_{eff}$, which is between the plastic moment capacity $p_y S$ and the elastic moment capacity $p_y Z$. S_{eff} is the effective plastic modulus and is calculated using the expressions given in Clause 3.5.6 of BS 5950. The conservative approach of using the elastic moment has been adopted in the worked examples.

Class 4 slender

Class 4 slender cross-sections contain elements that are so slender that local buckling is likely to occur before the attainment of the material design strength on the extreme fibres. Special procedures are needed to evaluate the capacity of the section; those procedures are beyond the scope of this document.

The differences in behaviour of the four classes may be seen in Figure 2.1, which illustrates the moment rotation behaviour of the cross-section.

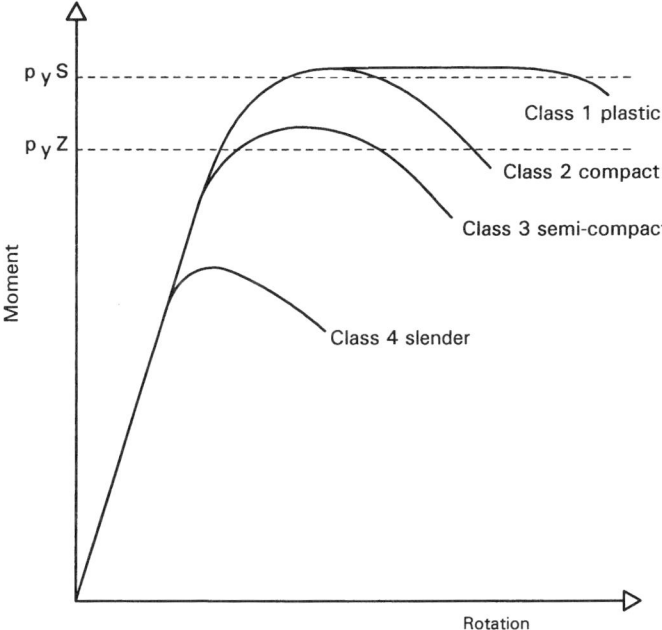

Figure 2.1 *Moment rotation behaviour of cross-sections of different classes*

If the section is under pure axial load instead of pure bending, then the criterion is simply whether the material design strength can be attained or whether local bucking occurs before the squash load is reached. Classes 1,2 and 3 are all able to develop the material strength in direct compression, so one set of limits is applicable for all three classes. If the section docs not meet the limit it is a Class 4 slender section and a more complex procedure is needed to evaluate the capacity; the procedure is beyond the scope of this document.

The situation when both axial load and bending are both present is a little more complex, but is covered by the clauses of BS 5950-1, as described below. In this situation, the classification will be dependent upon the values of axial load and moment, as will be illustrated in the example in Section 2.3.

When using hot rolled sections in steel grades S275 and S355, in the majority of cases in practice, the probability of the capacity being reduced by local buckling is quite small. If a more refined procedure is required then the reader is referred to BS 5950-5 [1], which deals specifically with cold formed sections that are more prone to local buckling because of their high aspect ratios and high yield stress.

2.2.2 Classification process

For the classification process, BS 5950-1 provides Figure 5, which is used in conjunction with Table 11 (for sections other than CHS and RHS). Figure 5 and Table 11 are reproduced here in part as Figure 2.2 and Table 2.1. Their use is illustrated in the examples forming part of this Chapter.

The cross-section classification process follows five basic steps, as listed below.

For each element in turn, carry out steps (i) to (iii)

(i) Evaluate the slenderness ratio (b/T or d/t) of all of the elements of the cross-section in which there is compressive stress. See Figure 2.2 for notation and relevant dimensions.

(ii) To allow for the influence of variation in the material design strength, evaluate the parameter ε as $(275/p_y)^{0.5}$, as indicated in note 2) at the foot of Table 2.1. For steel of grade S275 that is less than 16 mm thick, this parameter will be unity.

(iii) Where necessary (see below) evaluate the stress ratios r_1 and r_2.

(iv) In Table 2.1, identify the appropriate row of the table for the element under consideration and determine the class of that element, according to the limiting value of thickness ratio.

(v) Classify the complete cross-section according to the least favourable (highest) classification of the individual elements in the cross section.

The choice of the appropriate row of Table 2.1 depends on the boundary support conditions of the element and its stress condition (whether subject to uniform compressive stress or varying stress).

- For the compression flange of an I, H, channel or box section, the element is either an outstand element (supported along one edge only) or an internal element (supported along both edges). The stress is assumed to be uniform.

- For webs of I, H and box sections where the stress varies from tension to compression and the level of zero stress is at the mid-depth of the element, there is a simple set of three limits.

- For webs of I, H and box sections where the stress varies across the width of the element, other than for the simple case above, a stress ratio r_1 or r_2 must be determined. Expressions for the calculation of r_1 and r_2 are given in Clause 3.5.5 of BS 5950-1 and are repeated below for the case of I and H sections with equal flanges.
- For webs of channels, there is a simple set of three limits, irrespective of the stress condition.
- The elements of angles and Tees are all treated as outstand elements and there are simple sets of three limits for three cases.

Stress ratios r_1 and r_2

For I or H sections with equal flanges:

$$r_1 = \frac{F_c}{d\, t\, p_{yw}} \quad \text{but} \quad -1 < r_1 \leq 1$$

$$r_2 = \frac{F_c}{A_g\, p_{yw}}$$

where:

A_g is the gross cross-sectional area

d is the web depth

F_c is the axial compression (negative for tension)

p_{yw} is the design strength of the web

t is the web thickness.

Note: r_1 and r_2 are positive for compression and negative for tension.

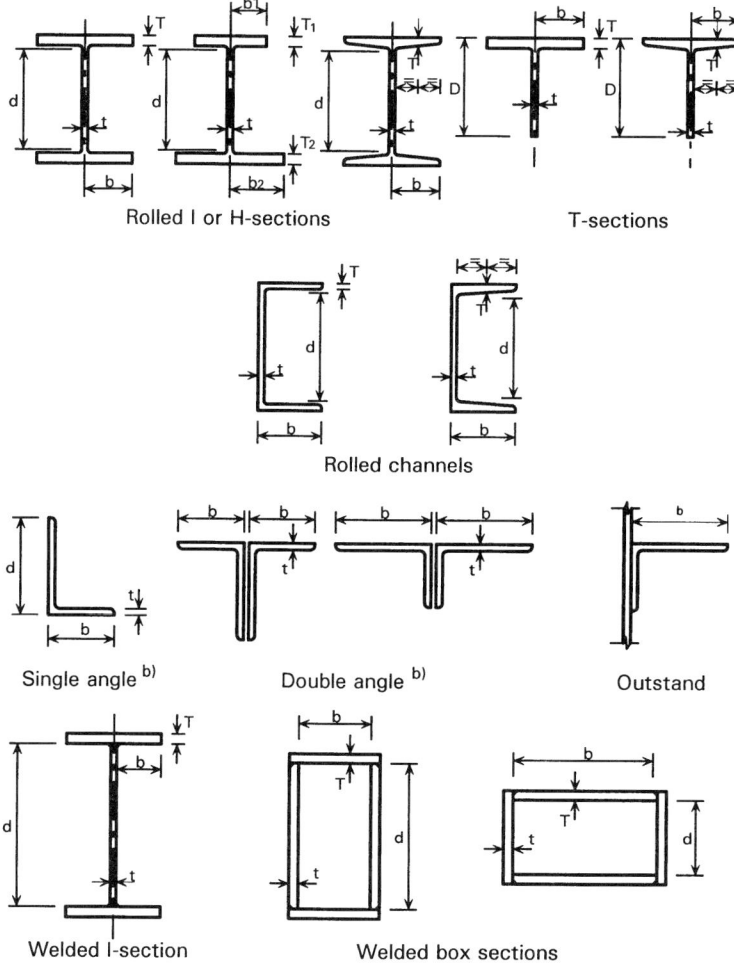

Notes:
a) For a box section, *B* and *b* are flange dimensions and *D* and *d* are web dimensions. The distinction between webs and flanges depends on whether the member is bent about its major axis or its minor axis.
b) For an angle, *b* is the width of the outstand leg and *d* is the width of the connected leg.

Figure 2.2 *Dimensions of compression elements*

Table 2.1 *Limiting width-to-thickness ratios for sections other than CHS and RHS*

Compression elements		Ratio [a]	Limiting value [b]		
			Class 1 plastic	Class 2 compact	Class 3 semi-compact
Outstand element of compression flange	Rolled section	b/T	9ε	10ε	15ε
	Welded section	b/T	8ε	9ε	13ε
Internal element of compression flange	Compression due to bending	b/T	28ε	32ε	40ε
	Axial compression	b/T	Not applicable		
Web of an I-, H- or box section [c]	Neutral axis at mid-depth		d/t	80ε	100ε	120ε
	Generally [d]	If r_1 is negative:	d/t	$\dfrac{80\varepsilon}{1+r_1}$ but $\geq 40\varepsilon$	$\dfrac{100\varepsilon}{1+r_1}$	$\dfrac{120\varepsilon}{1+2r_2}$ but $\geq 40\varepsilon$
		If r_1 is positive:	d/t		$\dfrac{100\varepsilon}{1+1.5r_1}$ but $\geq 40\varepsilon$	
	Axial compression [d]		d/t	Not applicable		
Web of a channel		d/t	40ε	40ε	40ε
Angle, compression due to bending (Both criteria should be satisfied)		b/t d/t	9ε 9ε	10ε 10ε	15ε 15ε
Single angle, or double angles with the components separated, axial compression (All three criteria should be satisfied)		b/t d/t $(b+d)/t$	Not applicable		15ε 15ε 24ε
Outstand leg of an angle in contact back-to-back in a double angle member		b/t	9ε	10ε	15ε
Outstand leg of an angle with its back in continuous contact with another component					
Stem of a T-section, rolled or cut from a rolled I- or H-section		D/t	8ε	9ε	18ε

a) Dimensions b, D, d, T and t are in Figure 2.2. For a box section b and T are flange dimensions and d and t are web dimensions, where the distinction between webs and flanges depends upon whether the box section is bent about its major axis or its minor axis.
b) The parameter $\varepsilon = (275/p_y)^{0.5}$
c) For the web of a hybrid section ε should be based on the design strength p_{yf} of the flanges.
d) The stress ratios r_1 and r_2 are defined in Section 2.2.2.

2.3 Example – Section classification

A 457 × 191 × 67 UC in steel grade S355 is to be used under three different conditions, as described below. Classify the section for each case and evaluate the local cross-sectional resistance.

Conditions:

(i)   under pure bending

(ii) under bending plus 700 kN axial compression

(iii) under pure axial compression of 700 kN.

The following section properties may be obtained from page 150.

$B = 189.9$ mm $T = 12.7$ mm

$d = 407.6$ mm $t = 8.5$ mm

$A_g = 85.5$ cm^2 $Z = 1300$ cm^3 $S = 1470$ cm^3

Slenderness ratios:

$b/T = 7.48$ and $d/t = 48.0$

Influence of material strength

Maximum material thickness = 12.7 mm, Table 1.3 gives p_y as 355 N/mm^2. Hence, $\varepsilon = (275/355)^{0.5} = 0.88$

Condition (i), Pure bending

Flanges

The limiting value of b/T for Class 1 is $9\varepsilon = 7.92$. The actual value is 7.48, therefore the flanges are Class 1 plastic.

Web

The limiting value of d/t for Class 1 is $80\varepsilon = 70.4$. The actual value is 48.0, therefore the web is Class 1 plastic.

The entire cross-section is classified as Class 1 plastic and thus the design strength of the material can be attained throughout the section. The moment capacity of the cross-section given by Clause 4.2.5.2 is thus,

$p_y S_x = 1470 \times 355 \times 10^{-3} = 522$ kNm.

Condition (ii), Bending plus 700kN axial compression

Flanges

The limiting value of b/T is as in condition i) above and the flanges are therefore Class 1 plastic.

Web

The level of zero stress will not be at mid depth of the web, so it is necessary to determine the stress ratios r_1 and r_2 from Table 2.1.

$r_1 = F_c / d\, t\, p_y = 700 \times 10^3 / 407.6 \times 8.5 \times 355 = 0.569$

$r_2 = F_c / A_g\, p_y = 700 \times 10^3 / 8550 \times 355 = 0.231$

The limiting value of d/t for Class 2 compact is

$100\varepsilon / (1+1.5r_1) = 88 / (1+1.5 \times 0.569) = 47.5$

The actual value is 48.0, therefore the web is not Class 2 compact.

The limiting value of d/t for Class 3 semi-compact is

$120\varepsilon / (1+2r_2) = 105.6 / (1+2 \times 0.231) = 72.2$

The actual value is 48.0, therefore the web is Class 3 semi-compact.

The entire cross-section is therefore Class 3 semi-compact and thus the design strength of the material can be attained at the extreme fibres. The moment capacity of the cross-section given by Clause 4.2.5.2 is thus,

$p_y Z_x = 1300 \times 355 \times 10^{-3} = 462$ kNm.

Condition (iii), Axial compression of 700kN

Flanges

The limiting value of b/T is as in condition i) above and the flanges are Class 1 plastic.

Web

When considering an I section in pure axial compression there is only one limit given in Table 2.1. The limit is the same as in condition (ii) above and web is therefore Class 3 semi-compact.

The entire cross-section therefore may be treated as Class 1, 2 or 3 under pure axial compression. The compression resistance (for a zero length strut) is therefore given by Clause 4.7.4 as:

$$A_g p_y = 8550 \times 355 \times 10^{-3} = 3035 \text{ kNm}.$$

This example has used one of the more slender UB sections. This has been done to illustrate the process and should not be taken as indicating that a large number of rolled sections will be unable to resist the full plastic moment.

It must also be remembered that these values are local capacities and overall buckling has yet to be considered.

2.4 General Guidance

All hot rolled I sections in grade S275, and most grade S355, are classified as Class 2 compact or better when in pure bending. They can therefore attain their full plastic moment capacity. The exceptions are shown below in Table 2.2. The majority of hot rolled I and H sections are classified as Class 1 plastic and are therefore suitable for plastic design.

Care should be exercised where a section is classified as Class 4 slender as special procedures to calculate member capacity, which are beyond the scope of this book, are required. No hot rolled I or H sections are Class 4 slender under pure bending.

The reader should examine the tables at the back of this book, which give the classification for both flanges and webs of most structural sections in grades S275 and S355 for a variety of conditions. These tables also enable the local cross-section capacities to be determined directly without the need to perform the calculations outlined above.

Table 2.2 *I and H sections that are Class 3 semi-compact under pure bending*

Section	Grade S275	Grade S355
Universal Beams	None	356 x 171 x 45
Universal Columns	356 x 368 x 129	356 x 368 x 153
	152 x 152 x 23	356 x 368 x 129
		305 x 305 x 97
		254 x 254 x 73
		203 x 203 x 46
		152 x 152 x 23
Joists	None	None

CHAPTER 3 BEAMS

3.1 Design considerations

3.1.1 General

A beam is a member that carries loading primarily in bending and which spans between supports or between connections to other members. This Chapter deals with the design of beams in steel-framed buildings, designed according to BS 5950-1. Guidance relates only to I, H and channel sections. The requirements at ultimate and serviceability limit states are discussed.

3.1.2 Span

In Clause 4.2.1.2 of BS 5950, the span of a beam is defined as the distance between effective points of support. In beam/column building frames, the difference between these support centres and the column centres is so small that it is customary to take the span as the distance between column centres when calculating moments, shears and deflections.

3.1.3 Loading

Loading may be classified as dead or imposed load, as described in Chapter 1. Dead loads are the permanent loads, typically including self weight of the steel, floors, roofs and walls. Imposed loads are variable, typically including crowd loading, storage, plant and machinery.

3.1.4 Lateral-torsional buckling

If an I section is subject to vertical loading that can move laterally with the beam, the imperfections of the beam mean it will tend to distort as indicated in Figure 3.1, which shows one half of a simply supported beam. Due to the bending action, the upper flange is in compression and acts like a strut. Being free to move, the compression flange will tend to buckle sideways dragging a reluctant tension flange behind it. The tension flange resists this sideways movement and therefore, as the beam buckles, the section also twists, with the web no longer vertical. This action is known as lateral-torsional buckling.

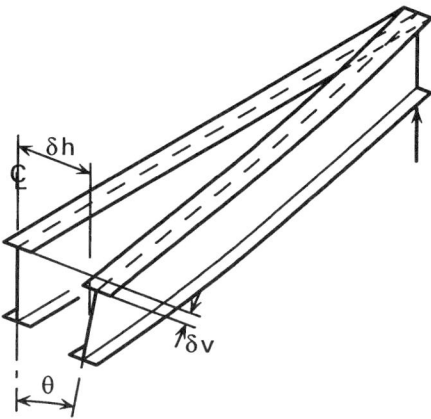

Figure 3.1 *Lateral torsional buckling – distorted shape of one half of a simply supported beam*

3.1.5 Fully restrained beams

Lateral-torsional buckling will be inhibited by the provision of lateral restraints to the compression flange. If the flange is restrained at intervals, lateral torsional buckling may occur between the restraints and this must be checked. If this restraint is continuous, the beam is fully restrained and lateral-torsional buckling will not occur.

Full (continuous) lateral restraint is provided by:

(i) in-situ and precast flooring or composite decking, provided that the flooring is supported directly on the top flange or is cast around it.

(ii) timber flooring, if the joists are fixed by cleats, bolts or other method providing a positive connection.

(iii) steel plate flooring, if it is bolted or welded at closely spaced intervals.

The continuous restraint should be designed to resist a force that is specified in the Standard as 2.5% of the maximum force in the compression flange. This restraining force may be assumed to be uniformly distributed along the compression flange. This force must be carried by the connection between the flooring and the beam.

Note that the restraint must be to the compression flange. Special care is required when considering regions where the bottom flange is in compression.

3.2 Moment and shear capacities

The calculation of shear capacity P_v is set out in BS 5950-1 in Clause 4.2.3. The shear capacity P_v of an I or H section is calculated as:

$$P_v = 0.6\, p_y A_v$$

where A_v is equal to the section depth times the web thickness.

The determination of the moment capacity of a beam M_c (effectively the moment capacity of the cross section, taking account of its classification) is given by Clause 4.2.5 of BS 5950-1. In the presence of low shear (applied shear $\leq 0.6\, P_v$), M_c is given by:

$M_c = p_y S_x$ for Class 1 plastic and Class 2 compact sections

$M_c = p_y S_{xeff}$ or $p_y Z_x$ (conservatively) for Class 3 semi-compact sections.

To avoid irreversible deformation at serviceability loads, M_c should be limited to $1.5 p_y Z$ generally and $1.2 p_y Z$ for simply supported beams.

If the shear force exceeds $0.6 P_v$ then the moment capacity M_c needs to be reduced, as set out in Clause 4.2.5.3 of BS 5950-1. It should be remembered that in most beams the maximum moment occurs at a position of low shear; the exception being cantilevers where maximum moment and maximum shear occur together at the support.

In beams with full restraint, the design bending moments in the beam are simply checked against the above moment capacity. In beams without full restraint, the design bending moments must also be checked against the buckling resistance moment, as discussed below.

3.3 Design of beams without full lateral restraint

When lateral-torsional buckling is possible, either over the full span of the beam or between intermediate restraints, the resistance of the beam to bending action will be reduced by its tendency to buckle. According to Clause 4.3.6.2, the beam is checked by calculating a buckling resistance moment M_b, and an equivalent uniform moment factor m_{LT}. The requirement is that, in addition to checking the moment capacity (as above), the following should be satisfied:

$M_x \leq M_{cx}$ and $M_x \leq M_b / m_{LT}$

The value of the buckling resistance depends on determination of a bending strength p_b (generally, less than the material design strength p_y). Values of p_b

may be obtained from Table 16 of BS 5950, reproduced in part here as Table 3.1 for selected material design strengths, depending on the value of the equivalent slenderness λ_{LT}.

Determination of these various parameters, and the conservative simplifications that BS 5950-1 allows to avoid excessive calculation, are described below.

Note that the bottom line of Table 3.1 lists λ_{L0}, the limiting slenderness. This is the value of slenderness below which p_b equals the material design p_y, which implies no reduction in capacity due to lateral-torsional buckling.

Table 3.1 *Bending strength p_b (N/mm^2) for rolled sections*

λ_{LT}	Steel grade and design strength p_y (N/mm^2)								
	S275					S355			
	235	245	255	265	275	315	325	335	345	355
30	235	245	255	265	275	315	325	335	345	355
40	229	238	246	254	262	294	302	309	317	325
50	210	217	224	231	238	265	272	279	285	292
60	189	195	201	207	213	236	241	246	251	257
70	169	174	179	184	188	206	210	214	218	222
80	150	154	158	161	165	178	181	184	187	190
90	132	135	138	141	144	153	156	158	160	162
100	116	118	121	123	125	132	134	136	137	139
110	102	104	106	107	109	115	116	117	119	120
120	90	91	93	94	96	100	101	102	103	104
130	80	81	82	83	84	88	89	90	90	91
140	71	72	73	74	75	78	78	79	80	80
150	64	64	65	66	67	69	70	70	71	71
160	57	58	59	59	60	62	62	62	63	63
170	52	52	53	53	54	56	56	56	57	57
180	47	47	48	48	49	50	51	51	51	51
190	43	43	44	44	44	46	46	46	46	47
200	39	39	40	40	40	42	42	42	42	42
210	36	36	37	37	37	38	38	38	39	39
230	31	31	31	31	31	32	32	33	33	33
250	26	27	27	27	27	28	28	28	28	28
λ_{L0}	37	36	36	35	34	32	32	31	31	30
Note: λ_{L0}, is the limiting slenderness. For slenderness below this value there is no reduction in capacity due to lateral-torsional buckling										

3.4 Equivalent slenderness

The value of the equivalent slenderness λ_{LT} is given by Clause 4.3.6.7, as follows:

$$\lambda_{LT} = uv\lambda(\beta_W)^{0.5}$$

The chief parameter in this expression is λ, which is the value of the effective length L_E divided by the radius of gyration r_y. See below for the determination of L_E.

For equal flange beams, the slenderness factor v may safely (and conservatively) be taken as 1.0.

Alternatively the slenderness factor v may be determined from Table 19 of BS 5950-1. This requires the designer to use the torsional index x, which may be found from section tables. The ratio λ/x is calculated. Table 19 of BS 5950-1 then gives a value of v which is less than unity for equal flanged sections.

The buckling parameter u may be found from section tables. For rolled I and H sections u may safely be taken as 0.9

The ratio β_W is defined in Clause 4.3.6.9 of the Standard as 1.0 if the section being used is Class 1 plastic or Class 2 compact. If the section is Class 3 semi-compact, β_W is the ratio Z_x / S_x if Z_x is used rather than S_{xeff} as the modulus for the section, otherwise β_W is S_{xeff} / S_x. Conservatively, β_W may always be taken as 1.0.

3.5 Effective length

The effective length L_E is determined from Table 3.2 for cantilevers and Table 3.3 for beams, where L_E is the effective length of the segment length under consideration. In Table 3.3 L_{LT} is the segment length which, for a simply supported beam without intermediate restraints, is its span. More generally L_{LT} is the length of segment over which lateral-torsional buckling can occur. It is therefore the distance between points of restraint. Two loading conditions are identified; normal and destabilising.

Destabilising refers to a situation where the loading is applied to the top flange of the beam or cantilever that is free to move laterally with the load. The normal condition thus refers to the situation where the load is applied to the web or the bottom flange. Longer effective lengths are associated with destabilising conditions, giving lower values of p_b and thus lower buckling resistance moments.

Table 3.2 Effective Length L_E for cantilevers

Restraint conditions		Loading Conditions	
At support	At tip	Normal	Destabilising
a) Continuous, with lateral restraint to top flange	1) Free	3.0L	7.5L
	2) Lateral restraint to top flange	2.7L	7.5L
	3) Torsional restraint	2.4L	4.5L
	4) Lateral and torsional restraint	2.1L	3.6L
b) Continuous, with partial torsional restraint	1) Free	2.0L	5.0L
	2) Lateral restraint to top flange	1.8L	5.0L
	3) Torsional restraint	1.6L	3.0L
	4) Lateral and torsional restraint	1.4L	2.4L
c) Continuous, with lateral and torsional restraint	1) Free	1.0L	2.5L
	2) Lateral restraint to top flange	0.9L	2.5L
	3) Torsional restraint	0.8L	1.5L
	4) Lateral and torsional restraint	0.7L	1.2L
d) Restrained laterally, torsionally and against rotation on plan	1) Free	0.8L	1.4L
	2) Lateral restraint to top flange	0.7L	1.4L
	3) Torsional restraint	0.6L	0.6L
	4) Lateral and torsional restraint	0.5L	0.5L
Tip restraint conditions			
1) Free (not braced on plan)	2) Lateral restraint to top flange (braced on plan in at least one bay)	3) Torsional restraint (not braced on plan)	4) Lateral and torsional restraint (braced on plan in at least one bay)

Table 3.3 *Effective length L_E for beams without intermediate restraint*

Conditions of restraint at supports		Loading conditions	
		Normal	Destabilising
Compression flange laterally restrained. Nominal torsional restraint against rotation about longitudinal axis	Both flanges fully restrained against rotation on plan	$0.7L_{LT}$	$0.85L_{LT}$
	Compression flange fully restrained against rotation on plan	$0.75L_{LT}$	$0.9L_{LT}$
	Both flanges partially restrained against rotation on plan	$0.8L_{LT}$	$0.95L_{LT}$
	Compression flange partially restrained against rotation on plan	$0.85L_{LT}$	$1.0L_{LT}$
	Both flanges free to rotate on plan	$1.0L_{LT}$	$1.2L_{LT}$
Compression flange laterally unrestrained. Both flanges free to rotate on plan	Partial torsional restraint against rotation about longitudinal axis provided by connection of bottom flange to supports	$1.0L_{LT} + 2D$	$1.2L_{LT} + 2D$
	Partial torsional restraint against rotation about longitudinal axis provided by pressure of bottom flange onto supports	$1.2L_{LT} + 2D$	$1.4L_{LT} + 2D$
D is the overall depth of the beam			

3.6 Equivalent uniform moment factor, m_{LT}

The values for p_b, based on λ_{LT}, have been derived assuming that the beam is under uniform moment throughout (as in Figure 3.2). In general, however, beams are subject to varying bending moment along their length, which is a less severe condition. It is possible to take advantage of this fact by using the equivalent uniform moment factor m_{LT}, which depends on the shape of the bending moment diagram.

The parameter m_{LT} is less than or equal to unity and is used to scale down the peak moment to an equivalent uniform moment. The value may conservatively be taken as unity but, to achieve more economy, m_{LT} may be determined from Table 18 of BS 5950-1, reproduced here as Table 3.4. For destabilizing loads m_{LT} must always be taken as 1.0, Clause 4.3.6.6 of BS 5950.

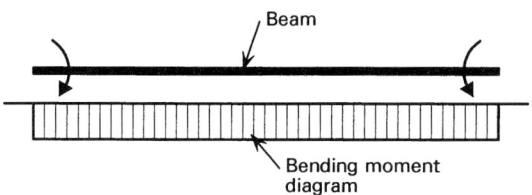

Figure 3.2 *Reference case – uniform moment throughout*

3.7 Calculation of bending resistance for beams without full restraint

The process to be adopted in a design is set out below in a step-by-step format.

1. Determine the bending moment diagram for the beam under factored loading and identify the maximum design moment M_x, and the maximum shear F_v.
2. Determine L_E from Table 3.2 for cantilevers or Table 3.3 for beams.
3. Look up r_y from section tables and evaluate λ as L_E / r_y.
4. Evaluate λ_{LT} as $uv\lambda(\beta_W)^{0.5}$ (see Section 3.4).
5. Determine p_b from Table 3.1.
6. Compute the buckling resistance moment M_b as:

   $M_b = p_b S_x$ for Class 1 plastic or Class 2 compact sections

   $M_b = p_b S_{xeff}$ or $p_b Z_x$ for Class 3 semi-compact sections.

7. Ensure that $M_x \leq M_b / m_{LT}$ (m_{LT} may be derived from Table 18 of BS 5950-1 which is reproduced in part as Table 3.4, or may conservatively be taken as 1.0)
8. Check that $M_x \leq M_{cx}$ (If $m_{LT} = 1.0$, this check is unnecessary). For calculation of M_{cx} see Section 3.2.

3.8 Calculation of bending resistance – a simpler approach

An alternative method that is even simpler, but which loses a little in economy, is available but is specifically restricted to rolled sections with equal flanges. In this method a value for p_b may be determined from Table 20 of BS 5950-1 with input parameters of $(\beta_W)^{0.5} L_E / r_y$ and D/T. The table is too extensive to repeat here. The buckling resistance moment can then be determined as in step 6 above.

Table 3.4 *Equivalent uniform moment factor m_{LT} for lateral torsional buckling*

Segments with end moments only (m_{LT} from formula for the general case)	β	m_{LT}
β positive [diagrams: M to βM with L_{LT}]	1.0	1.00
	0.9	0.96
	0.8	0.92
	0.7	0.88
	0.6	0.84
	0.5	0.80
	0.4	0.76
	0.3	0.72
	0.2	0.68
	0.1	0.64
	0.0	0.60
β negative [diagrams: M to βM with L_{LT}]	−0.1	0.56
	−0.2	0.52
	−0.3	0.48
	−0.4	0.46
	−0.5	0.44
	−0.6	0.44
	−0.7	0.44
	−0.8	0.44
	−0.9	0.44
	−1.0	0.44

Specific cases (no intermediate lateral restraints)

$m = 0.850$	$m = 0.925$	$m = 0.925$	$m = 0.744$

General case (segments between intermediate lateral restraints)

For beams: $m_{LT} = 0.2 + \dfrac{0.15M_2 + 0.5M_3 + 0.15M_4}{M_{max}}$ but $m_{LT} \geq 0.44$

All moments are taken as positive. The moments M_2 and M_4 are the values at the quarter points, the moments M_3 is the value at mid length and M_{max} is the maximum moment in the segment.

For cantilevers without intermediate lateral restraint: $m_{LT} = 1.00$.

3.9 Example – Beam with full lateral restraint

Design a simply supported beam carrying a concrete floor slab over a span of 5.0 m in grade S275 steel. The unfactored dead load, which includes an allowance for self weight, is 14 kN/m, and the ultimate unfactored imposed load is 19 kN/m. For ultimate load combination 1 the factored load is,

$1.4 \times 14 + 1.6 \times 19 = 50$ kN/m

Choice of section

Maximum moment = $wL^2/8 = 50 \times 5^2 / 8 = 156$ kNm

As the beam is fully restrained (due to the presence of the floor slab) the required moment capacity is $S_x p_y$ assuming that the section is at least Class 2 compact, given that most UB sections are at least Class 2.

Assuming that the maximum thickness is 16mm, $p_y = 275$ N/mm^2

Therefore $S_{required} = 156 \times 10^6 / 275 \times 10^{-3} = 568$ cm^3

The lightest rolled section to satisfy this criterion is a 356 × 127 × 39 UB. The plastic modulus $S_x = 659$ cm^3.

Determine section classification

Flange thickness $T = 10.7$ mm, which is less than 16mm, therefore p_y is 275 N/mm^2 and $\varepsilon = 1.00$.

Consider the flange. From section tables $b/T = 5.89$.
$5.89 < 9\varepsilon$, therefore classification is Class 1 plastic.

Consider web. From section tables $d/t = 47.2$.
$47.2 < 80\varepsilon$, therefore classification is Class 1 plastic.

Therefore, the section as a whole is Class 1 plastic.

Shear capacity check

Maximum shear force F_v is $wL/2 = 50 \times 5 / 2 = 125$ kN

From section tables, $D = 353.4$ mm $t = 6.6$ mm

Shear capacity, $P_v = 0.6\,p_y\,A_v = 0.6\,p_y\,t\,D$
$= 0.6 \times 275 \times 6.6 \times 353.4 / 10^3 = 385$ kN
$0.6\,P_v = 0.6 \times 385 = 231$ kN $> F_v\,(=125$ kN$)$

therefore the section is under 'low shear'.

Moment capacity

As the section is Class 1 plastic, and is under low shear, the moment capacity M_c is given by $M_c = S_x p_y = 659 \times 10^3 \times 275 / 10^6 = 181$ kNm.

Alternatively, the capacity may be obtained from the capacity tables (page 198), as $M_{cx} = 181$ kNm

Serviceability check

Serviceability load will be taken as unfactored imposed load = 19 kN/m

Deflection due to this load, w_i (Taking E as 205×10^3 N/mm² and I from section tables),

$= 5 w_i L^4 / 384\, E\, I$

$= (5 \times 19 \times 5000^4) / (384 \times 205 \times 10^3 \times 10170\, 10^4) = 7.4$ mm

As the recommended maximum deflection under imposed load only is span/360 (see Table 1.2) or 13.9 mm, the deflection is satisfactory. Deflections do have more significance for longer spans.

3.10 Example – Unrestrained beams

A beam is required to span 6.0 m and is to carry three point loads at the quarter points, 1.5 m apart. Each factored load is 80 kN. The three loads are applied to the top flange of the beam and they are free to move laterally. The compression flange is unrestrained over the entire span. At one end the compression flange has partial torsional restraint. At the other end both flanges are not restrained to any reliable degree against rotation on plan. Select a suitable UB section in grade S275 steel.

Determine design moment

Maximum moment due to point loads = 240 kNm (see pages 90 and 81)

Moment due to self weight. Guess self weight is 150 kg/m.

Maximum moment due to self weight = $1.4 \times 150 \times 9.81 \times 6^2 / 8 \times 10^3$

= 10 kNm

Maximum design moment = 240 + 10 = 250 kNm

Establish effective length L_E

Consult Table 3.3

The loading is to the top flange and is free to move and is therefore destabilising. Examination of the restraint conditions shows that the restraint at one end of the beam corresponds to the penultimate row of the table whilst the restraint at the other end corresponds to the final row. Take an average value and assume that the beam depth is about 600 mm. Therefore,

$L_E = 1.3\, L_{LT} + 2D = 1.3 \times 6.0 + 2 \times 0.6 = 9.0$ m

Select a trial section

Guess $p_b = 70$ N/mm². $S_{x\ required} = 247 \times 10^6 / 70 \times 10^3 = 3530$ cm³

Try a 610 x 229 x 140 UB.

From section tables, $S_x = 4140$ cm³, $T = 22.1$ mm, $r_y = 5.03$ cm

Check $M_x < M_{cx}$, $M_{cx} = p_y S_x = 1100$ kNm, therefore OK

Compute M_b for trial section

$T = 22.1$ mm therefore, from Table 2.1, $p_y = 265$ N/mm²

$\lambda = L_E / r_y = 9000 / 50.3 = 179$

$\lambda_{LT} = u\, v\, \lambda\, (\beta_W)^{0.5}$

Assume section is Class 1 plastic or Class 2 compact (this can be confirmed by calculation or by using Table 2.2). Therefore $\beta_W = 1.0$

Take $u = 0.9$ (its actual value from section tables is 0.875, which shows the simplification is safe).

Take $v = 1.0$ (using $x = 30.6$ from section tables, $\lambda / x = 5.85$, Table 19 of BS 5950-1, would give $v = 0.78$, which shows the simplification is safe).

Then $\lambda_{LT} = 0.9 \times 1.0 \times 179 \times 1.0 = 161$

From Table 3.1 $p_b = 58$ N/mm²

$M_b = p_b S_x = 58 \times 4140 \times 10^3 / 10^6 = 240$ kNm

The required buckling resistance is 250 kNm and the section fails this check.

The designer can now either choose a larger section or recognise the conservative assumptions made in the simple approach.

Allow for non-uniform moment

In this case the loads are destabilizing and the equivalent uniform moment factor m_{lt} must be taken as 1.0 (Clause 4.3.6.6).

Allowing for actual figures of u and v

Using actual values of, $u = 0.87$ and $v = 0.78$ gives,

$\lambda_{LT} = u\,v\,\lambda\,(\beta_W)^{0.5} = 0.87 \times 0.78 \times 179 \times 1.0 = 121$

This increases p_b to 93 N/mm², and, and hence $M_b = 385$ kNm. The check becomes,

$M_b / m_{LT} = 385 / 1.0 = 385$ kNm > 247 kNm, therefore OK.

3.11 Web bearing capacity and web buckling resistance

At locations where concentrated loads are applied to the flanges of beams, checks must be carried out to determine whether or not the web is sufficiently robust to disperse this loading. Two potential failure modes, indicated in Figure 3.3, are possible. Separate design checks are needed for web bearing and web buckling.

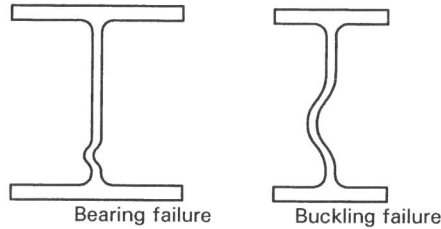

Bearing failure   Buckling failure

Figure 3.3 *Web Bearing and Web buckling failure*

A web bearing failure involves local crushing of the web by yielding at the most vulnerable location, which is in that part of the web closest to the applied load, adjacent to the root radius.

A web buckling failure involves a portion of the web buckling under or above the loaded point.

Web bearing and buckling checks are not required at locations where the load transfer is made directly to the beam web, for example by an end plate, angle cleats or fin plates.

3.11.2 Stiff bearing length

Both checks require the identification of a stiff bearing length b_1. This is the dimension, parallel to the longitudinal axis of the beam, through which the load is applied to the outer face of the flange. Where load is transferred through a solid cap plate or similar, it is the dimension of that plate. Where load is transferred through an I or H section, it is given by

$$b_1 = t + 1.6r + 2T$$

where:

- t is the thickness of the web
- r is the root radius
- T is the flange thickness
  (all relating to the beam applying the load.)

3.11.3 Web bearing capacity check

Figure 3.4 shows four examples of the length over which the concentrated load is assumed to disperse when it reaches the critical section at the tip of the root radius.

(i) At the member end where dispersion is only in one direction.

(ii) Where the beam is continuous over a support but where the overhang is insufficient to develop the full dispersion length.

(iii) Where the beam is continuous over a support and where the full dispersion length can be developed.

(iv) Under loading from a secondary beam.

The local capacity for web bearing P_{bw} is given in Clause 4.5.2.1 by

$$P_{bw} = (b_1 + n\,k)\,t\,p_{yw}$$

where:

- b_1 is the stiff bearing length (see above)
- n $= 5$ (except at the end of a member)
       $= 2 + (0.6\,b_e / k) \leq 5$ (at the end of a member)
- b_e is the distance to the end of the member from the nearest edge of the stiff bearing
- k $= T + r$ for a rolled I or H section
- T is the flange thickness
- r is the root radius
- t is the web thickness

(all relating to the beam being designed.)

If the value of the local compressive force exceeds P_{bw}, then web bearing stiffeners are required to carry the excess.

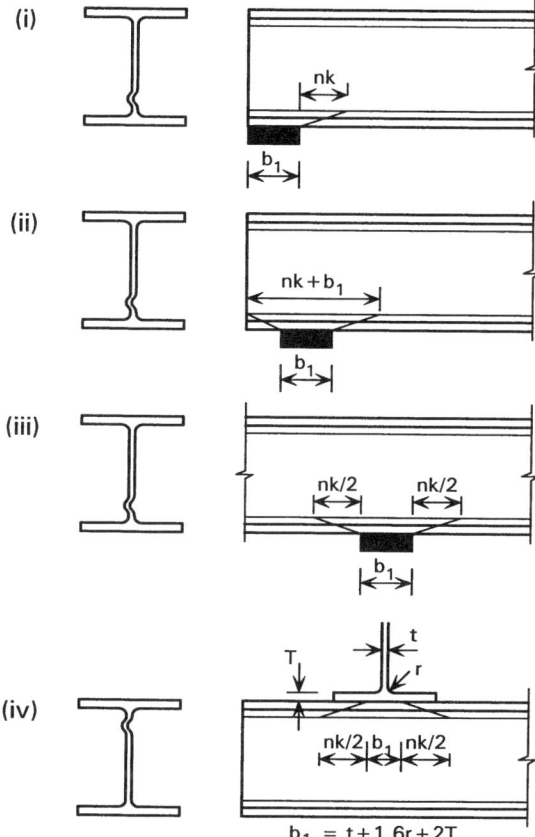

Figure 3.4 *Web bearing failure*

3.11.4 Web buckling check

At locations where concentrated loads are applied, the web of the beam is required to act as a strut. The check given in Clause 4.5.3.1 of the Standard for an unstiffened web assumes that the flange through which the load or reaction is applied is effectively restrained against both:

(a) rotation relative to the web (see Figure 3.5)

(b) lateral movement relative to the other flange (see Figure 3.5).

Flanges restrained against both conditions (a) and (b) is the situation most common in practice and implies that the portion of the web being checked as a strut is effectively held in position at both ends and effectively restrained in direction at both ends.

However, other cases could be envisaged. For example, a pair of beams may support other loaded beams on their upper flanges. If these beams can move laterally, then assumption (b) will be violated and additional computation will be required, as outlined at the end of this Section.

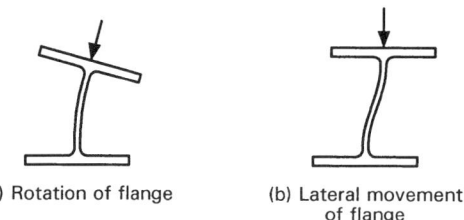

(a) Rotation of flange   (b) Lateral movement of flange

Figure 3.5 *Unrestrained flanges*

The width of the web that acts effectively as a strut is based on the stiff bearing length b_1 (as given in Section 3.11.3). The web buckling capacity is given in Clause 4.5.3.1 by:

$$P_x = \{25 \, \varepsilon \, t \, / \, [\, (b_1 + n k) \, d \,]^{0.5}\} \, P_{bw}$$

where:

$\varepsilon \quad = (275/p_y)^{0.5}$

$d \quad$ is the depth of the beam web

$P_{bw} \quad$ is the bearing capacity from 3.11.3

However, the dispersion can be limited by the proximity of the end of the beam. For values of a_e less than $0.7d$ a reduction factor should be applied, as given by the expression below,

$$(a_e + 0.7d) \, / \, 1.4 \, d$$

Where d is as defined above and a_e is as defined in Figure 3.6.

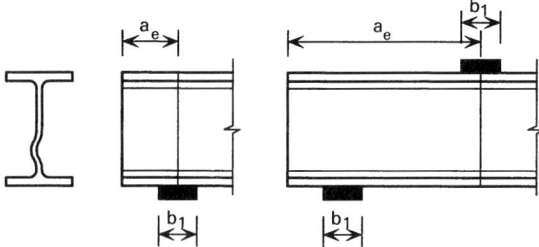

Figure 3.6 Web buckling failure

When the appropriate check has been carried out, if the capacity is not sufficient, then buckling stiffeners must be provided.

If the conditions of (a) or (b) set out above are not met (i.e. the flanges may move relative to each other), then another reduction in capacity is required. The reduced capacity P_{xr} is given by,

$$P_{xr} = [\,0.7d\,/\,L_E\,]\,P_x$$

L_E is the effective length of the web acting as a strut over its height and determined from its end conditions as set out in Table 22 of BS 5950-1 and Table 4.1 of the Chapter 4. Where the upper and lower flanges cannot move laterally relative to each other, L_E can safely be taken as d. The reduced capacity is then taken as 70% of P_x.

3.12 Web stiffeners

Web stiffeners generally take the form of flat plates welded to the web between the flanges of the beam. To accommodate the root radii or welds, they are cut or cropped at 45 degrees at their corners. The outstand for flat plate stiffeners is limited to a maximum value of 19 ε t_s but only the inner 13 ε t_s adjacent to the web is used in calculating resistances (Clause 4.5.1.2), to avoid problems with local buckling of the stiffener itself.

The capacity of a stiffener is determined by examining the capacity of a cruciform section constructed from the stiffeners plus a length of the web equal to a maximum value of 15 t_w on each side of the centreline of the stiffener. At an end support, the length on one side is limited by the end of the beam and is likely to be less than 15 t_w. Figure 3.7 shows such a cruciform section away from end effects.

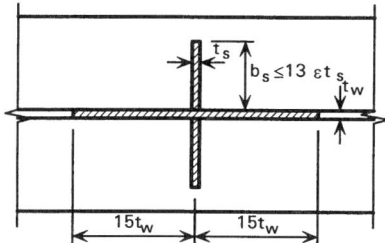

Figure 3.7 *Effective cruciform section for web buckling*

For web bearing, the capacity provided by the stiffener is given by the area of the stiffener in contact with the flange multiplied by the yield strength of the stiffener material. An allowance must be made for any cropping adjacent to the web. The capacity of the stiffener is added to that of the web to give the total capacity.

For web buckling, the resistance is determined by the cruciform section that buckles. As the cruciform section includes the web material, this gives the total capacity of the stiffened area. To calculate the capacity, it is necessary to determine the second moment of area I_s of the cruciform section. Using the dimensions shown in Figure 3.7,

$$I_s = [\, t_s\, (2\, b_s + t_w)^3 / 12\,] + [\, (30\, t_w - t_s)\, t_w^3 / 12\,]$$

The area of the section A_s (except at the extremities due to coping) is given by

$$A_s = 2\, b_s\, t_s + (30 t_w - t_s)\, t_w$$

(Note: Cope holes are not deducted in the determining A_s)

Thus the radius of gyration r may be calculated from

$$r = (I_s / A_s)^{0.5}$$

The slenderness λ is given by

$$\lambda = L_E / r$$

where:
    $L_E = 0.7L$ if the flange is restrained against rotation in the plane of the stiffener

    $L_E = 1.0L$ if it is not so restrained.

where:
    L is the clear length of stiffener between flanges.

The compressive strength of the section p_c is given by Table 24(c) of BS 5950-1. The buckling resistance may then be determined from $P_x = A_s p_c$.

3.13 Example – Web bearing and buckling

Check the beam in Section 3.10 at the loaded positions and at the supports

Under the central point load of 80 kN

Web bearing check

Assume that the load is transmitted to the beam via a 457 × 152 × 67 UB that sits on the top flange.

For this beam t = 9.0 mm; T = 15.0 mm; r = 10.2 mm

Stiff bearing length is given by,

$b_1 = t + 1.6r + 2T = 9.0 + 1.6 \times 10.2 + 2 \times 15.0 = 55.3$ mm

Local capacity for web bearing is given by,

P_{bw} = $(b_1 + n k) t p_{yw}$
n = 5 as this is not adjacent to a beam end
t = 13.1 mm; T = 22.1 mm; r = 12.7 mm
p_{yw} = 265 N/mm^2
k = $T + r$ = 22.1 + 12.7 = 34.8 mm (rolled section)
P_{bw} = $(55.3 + 5 \times 34.8) \times 13.1 \times 265 / 10^3$ = 796 kN.

Alternatively, the capacity may be obtained from the capacity tables (page 202), as follows:

For this section $C1$ = 604 and $C2$ = 3.47. Hence,

$P_{bw} = C1 + b_1 C2 = 604 + 55.3 \times 3.47 = 796$ kN

Applied load is 80 kN < 796 kN therefore satisfactory

Web buckling check

The load is not applied near the beam end, hence $a_e \geq 0.7d$ therefore,

P_x = $\{25 \, \varepsilon \, t / [\, (b_1 + n k) d \,]^{0.5}\} P_{bw}$
ε = 1.02 ; d = 547.6 mm
P_x = $\{25 \times 1.02 \times 13.1 / [(55.3 + 5 \times 34.8) \times 547.6]^{0.5}\}$ 796
 = 750 kN

Alternatively, the capacity may be obtained from the capacity tables (page 202), as follows:

For this section $C4 = 706$ and $K = 1.0$. Hence,

$$P_x = K(C4\ P_{bw})^{0.5} = 1.0\ (706 \times 796)^{0.5} = 749\ kN$$

Applied load is 80 kN < 749 kN therefore satisfactory.

At supports

The ULS reactions are each 120 kN and the dispersion length is limited by the end of the member. Assume that the support is on a 200 mm wide cap plate on a 203 UC, with a 100 mm overhang, as indicated below in Figure 3.8.

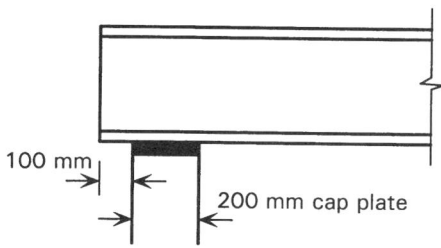

Figure 3.8 *Detail at end support*

Web bearing check

$P_{bw}\ \ = (b_1 + n\ k)\ t\ p_{yw}$
$n\ \ \ \ \ = 2 + (0.6\ b_e / k) = 2 + (0.6 \times 100 / 34.8) = 3.72$
$b_1\ \ \ = 200$ mm
$P_{bw}\ \ = (200 + 3.72 \times 34.8) \times 13.1 \times 265 / 10^3 = 1140$ kN

Alternatively, the capacity may be obtained from the capacity tables (page 202), as follows:

For this section $C1 = 242$ and $C2 = 3.47$. Hence,

$$P_{bw} = C1 + b_1\ C2 = 242 + 200 \times 3.47 = 936\ kN$$

The capacity tables give a conservative answer because b_e is assumed to be zero.

Applied load is 120 kN < 1140 kN therefore satisfactory.

Web buckling check

$a_e = 100$ mm

$a_e < 0.7d$ therefore the reduction factor for the close proximity of the beam end is given by,

$(a_e + 0.7d) / 1.4d = (100 + 0.7 \times 547.6)/1.4 \times 547.6 = 0.63$

Hence the buckling capacity,

$P_x = 0.63 \{25 \, \varepsilon \, t / [\, (\, b_1 + n \, k \,) \, d \,]^{0.5}\} \, P_{bw}$
$ = 0.63 \{25 \times 1.02 \times 13.1/[(200 + 3.72 \times 34.8) \, 547.6]^{0.5}\} 1140$
$ = 565$ kN

Alternatively, the capacity may be obtained from the capacity tables (page 202), as follows:

For this section $C4 = 706$ and $K = 0.63$. Hence,

$P_x = K \, (C4 \, P_{bw})^{0.5} = 0.63 \, (706 \times 1140)^{0.5} = 565$ kN

Applied load is 120 kN $<$ 565 kN therefore satisfactory.

3.14 Example – Web stiffeners

To demonstrate the design of stiffeners, consider a 610 x 229 x 101 UB carrying a point load of 800 kN via a 457 x 191 x 98 UB supported on its top flange remote from the end of the beam.

For a 457 x 191 x 98 UB: $t = 11.4$ mm; $T = 19.6$ mm; $r = 10.2$ mm

Stiff bearing length, $b_1 = 11.4 + 1.6 \times 10.2 + 2 \times 19.6 = 66.9$ mm

Web bearing check

For a 610 x 229 x 101 UB: $t = 10.5$ mm; $T = 14.8$; $r = 12.7$ mm; $p_{yw} = 265$ N/mm^2

$P_{bw} = (b_1 + n \, k \,) \, t \, p_{yw}$
$n = 5$ (*remote* from end effects)
$k = T + r = 14.8 + 12.7 = 27.5$ mm

$P_{bw} = (66.9 + 5 \times 27.5) \, 10.5 \times 275 / 10^3 = 590$ kN

This is less than the applied load of 800 kN. The extra bearing capacity required $= 800 - 590 = 210$ kN

Web buckling check

For a 610 x 229 x 101 UB : $d = 547.6$ mm

$P_x = \{25 \, \varepsilon \, t \, / \, [\,(b_1 + n\,k\,)\,d\,]^{0.5}\} \, P_{bw}$

$P_x = \{25 \times 1.0 \times 10.5 \, / \, [(66.9 + 5 \times 27.5) \, 547.6]^{0.5}\} 590 = 463$ kN

This is less than the applied load of 800 kN.

Try two plates, one on either side of the web as indicated in Figure 3.9

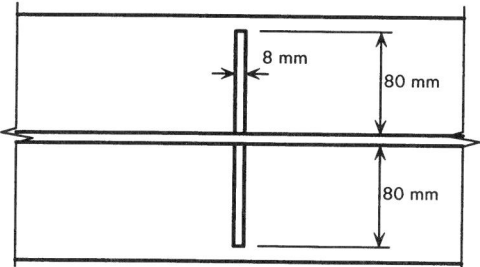

Figure 3.9 *Detail of trial stiffeners*

Width available for each plate for bearing is given by

$(B - t - 2r) \, / 2 = (227.6 - 10.5 - 2 \times 12\,7) \, / \, 2 = 95.8$ mm

Make each plate 80 mm wide with a 15 mm cope at the web corners. The width in bearing is thus 65 mm. The thickness of plate required, $t_p = 210 \times 10^3 \, / \, (2 \times 65 \times 275\,) = 5.8$ mm say 8 mm.

Check outstand dimension

At mid-height plate outstand is 80 mm and t= 8 mm. Therefore,

outstand $/ \, t_p \, \varepsilon = 80 \, / \, (8 \times 1.0\,) = 10.0 < 13$

Thus the full dimensions can be used without restriction due to local buckling of the stiffener plate.

Bearing capacity of pair of stiffeners is given by

contact area $\times$ material strength $= 2 \times 65 \times 8 \times 275 \, / \, 10^3 = 286$ kN

Total bearing capacity $= 590 + 286 = 876$ kN

Applied load is 800 kN $<$ 879 kN therefore satisfactory.

Buckling capacity of cruciform strut,

A_s = (2 × 80) 8 + (30 × 10.5 + 8) 10.5 = 4670 mm²

I_s = 8 (2 × 80 + 10.5)³ / 12 + (30 × 10.5 − 8) 10.5³ / 12
    = 3330 × 10³ mm⁴

Radius of gyration $r = (I_s / A_s)^{0.5}$ = (3330 × 10³ / 4670)$^{0.5}$ = 26.7 mm

Length L for buckling is the clear distance between the flanges.

$L = D - 2T$ = 602.6 − 2 × 14.8 = 573 mm

Assume flange is restrained against rotation

L_E = 0.7 L = 0.7 × 573 = 401 mm

$\lambda = L_E / r$ = 401 / 26.7 = 15.0

Use BS 5950-1 Table 24(c) to determine compressive strength p_c. At this slenderness there is no reduction below p_y, therefore p_c = 275 N/mm².

Then $P_x = A_s \, p_c$ = 4670 × 275 / 10³ = 1280 kN

Applied load is 800 kN < 1280 kN therefore satisfactory.

CHAPTER 4 MEMBERS IN TENSION AND COMPRESSION

4.1 Introduction

Members, which carry pure tension, generally referred to as ties, are relatively simple to design. In reality tension forces are frequently accompanied by moments and the member must be designed for the combined effects.

Compression members, which carry pure compression, are often referred to as struts. Building columns generally carry both axial compression and bending and must be designed for the combined effects.

4.2 Ties

The tension capacity P_t of a tie is given in Clause 4.6.1 as:

$$P_t = p_y A_e$$

where:

A_e is the effective net area of the member.

A_e is found by the addition of the effective net areas a_e of all the elements in the cross-section. For each element, a_e is given in Clause 3.4.3 as:

$$a_e = K_e a_n \text{ but } \leq a_g$$

where:

a_e is the net area of the element allowing for bolt holes

a_g is the gross area of the element

K_e is a factor depending on the grade of steel, being 1.2 for S275 and 1.1 for grade S355.

In members where the bolt holes are not staggered, the area to be deducted is the sum of the cross-sectional areas of the holes in a cross-section perpendicular to the direction of the axial force.

In members where the holes are staggered, the deduction should be the greater of the above and the sum of the sectional areas of a chain of holes on any diagonal or zigzag line (see Figure 4.1) less a further allowance of $0.25\, s_2\, t\, /g$ for each gauge spacing, where (as indicated in Fig 4.1):

g is the gauge spacing perpendicular to the direction of the tensile force

s is the staggered pitch

t is the thickness of the material.

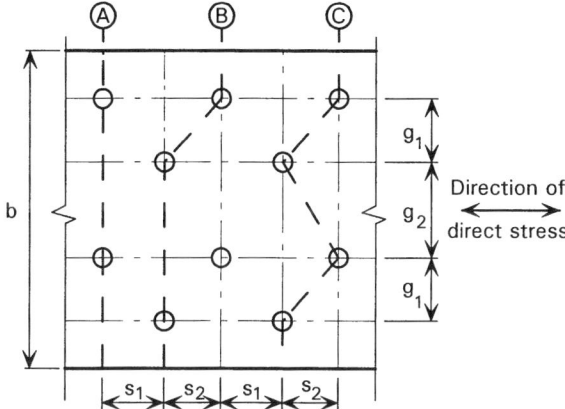

Figure 4.1 *Net area at staggered holes*

4.3 Simple tension members

Angles, channels or T-sections subject to tension with moments caused by eccentric end connections can be designed as 'simple tension members' using the provisions of Clause 4.6.3. Other members subject to tension and bending should be designed using Clause 4.8.2, see Section 4.4.

The design method of simple tension members used in BS 5950-1 is to account for the bending moment by using a reduced tension capacity as given below.

For single angle, channels or T-section members (Clause 4.6.3.1)

$P_t = p_y (A_e - 0.5\ a_2)$ for bolted connections

$P_t = p_y (A_g - 0.3\ a_2)$ for welded connections

For double angle, channels or T-section members (Clause 4.6.3.2)

$P_t = p_y (A_e - 0.25\ a_2)$ for bolted connections

$P_t = p_y (A_g - 0.15\ a_2)$ for welded connections

where:

p_y is the design strength
A_e is the effective net area of the member
A_g is the gross cross-sectional area
$a_2 = A_g - a_1$
a_1 is the overall gross area of the connected element.

4.4 Tension members also subjected to moments

Members that are under both tension and moment must be checked to ensure that the combined actions may be safely resisted. The necessary checks are in two parts. The first is a cross-section capacity check given in Clause 4.8.2.2 and reproduced below:

$$F_t / P_t + M_x / M_{cx} + M_y / M_{cy} \leq 1$$

where:

F_t is the axial tension at the section
P_t is the tension capacity (described above)
M_x is the moment about the major axis
M_y is the moment about the minor axis
M_{cx} is the moment capacity about the major axis
M_{cy} is the moment capacity about the minor axis.

However, this is not a sufficient check in all circumstances as, if the axial load is small, then it can readily be seen that the member is essentially a beam and lateral-torsional buckling can occur. Clause 4.8.2.2 therefore requires lateral-torsional buckling under the bending moments alone to be checked, as outlined in Chapter 3.

4.5 Struts

4.5.1 General

Members in compression have a limit on their load carrying capacity, known as the squash load, which is equal to the yield strength multiplied by the cross-sectional area. Long slender struts will fail at much lower loads by elastic buckling. However most practical compression members have a slenderness between these two extremes and will fail by a combination of yielding and buckling.

4.5.2 Effective Length

The end restraint conditions of a strut will effect the buckled shape of a strut (see Table 4.1) and also the buckling resistance. The effective length is best described as the length of a pin-ended member that would behave like the real member with its actual end restraints. Thus a vital step in the design of any compression member is the identification of the effective length.

Table 4.1 shows the buckled shapes and effective lengths (as given in BS 5950, Table 22) for some reference conditions. Firstly, they are separated into non-sway and sway conditions. Relative movement of the ends of the strut are restricted in a non-sway frame, this can be achieved by effective diagonal bracing or by the provision of shear walls - possibly as the concrete core around lift shafts and stair wells. However, if the building relies on frame action for its lateral stability, it is more likely to be a sway-sensitive frame. (See Chapter 1 for a fuller description of the distinction between sway and non-sway conditions).

Table 4.1 Strut effective lengths L_E

Restraint	Position	Position and Direction	Position and Direction	None	Direction
Shape (L is strut length)					
Restraint	Position	Position	Position and Direction	Position and Direction	Position and Direction
Practical L_E	1.0L	0.85L	0.7L	2.0L	1.2L

If the ends of a non-sway member have no rotational restraint, then the effective length of the strut is the actual length – by definition. If effective rotational restraint is present – for example from stiff beams that are effectively fastened to the column by stiff end plate connections - then the strut will be more resistant to buckling and the effective length will be reduced. In the extreme case of a non-sway strut which is fully restrained against rotation the effective length will be one half of the actual length. This is an idealised reference case, because full rotational restraint is not achievable in practice and therefore the effective length is taken as 0.7L. It is important to recognise that rotational restraint is provided by the members connected to the beam and is also reliant upon the stiffness of the connections to transmit this restraint.

The two cases at the right hand side of Table 4.1 show columns that can sway. Under these conditions the effective length ratios can never be less than unity; this lower limit implies complete rotational restraint at the column ends, which is not achievable in practice.

To design a column, it is necessary to determine the length over which it can buckle, termed the segment length. The length over which a strut can buckle is the length in any plane between restrained points in that plane. This is the distance between the intersections of the column and the restraining members and will usually be the storey height in a building frame. The restraining members will inhibit movement and/or rotation at the specific location. From the segment length, the effective length may be determined using Table 22 of BS 5950-1 which is reproduced here as Table 4.2.

If the beams are attached to the columns using flexible connections, such as fin plates, then it would be unwise to assume any rotational restraint, whatever the stiffness of the beam. With connections such as partial depth end plates or double angle cleats, provided that the beams are reasonably sized, partial restraint may be assumed. Stiff beams connected to the columns using substantial connections such as flush or extended end plates will provide effective rotational restraint. However the above is general advice based upon normal circumstances and the engineer must view each case on its merits.

Table 4.2 *Nominal effective length L_E for a compression member*

a) non-sway mode		
Restraint (in the plane under consideration) by other parts of the structure		Effective length L_E
Effectively held in Position at both ends	Effectively restrained in direction at both ends	$0.7L$
	Partially restrained in direction at both ends	$0.85L$
	Restrained in direction at one end	$0.85L$
	Not restrained in direction at either end	$1.0L$
b) sway mode		
One end	Other end	L_E
Effectively held in position and restrained in direction	Not held in position	Effectively restrained in direction	$1.2L$
		Partially restrained in direction	$1.5L$
		Not restrained in direction	$2.0L$

4.5.3 Strut curves

The interaction of yielding and instability effects is influenced by a number of parameters including the section shape, the axis of bending, the initial out of straightness and the residual stresses within the section. Considerable research has shown that the effect of these parameters may be efficiently incorporated by simply using the appropriate strut curve from a family of four (tabulated in BS 5950-1 in Table 24 as curves a, b, c and d). Table 23 enables the designer to determine the appropriate curve. Part of this table is reproduced here as Table 4.3.

Table 4.3 *Allocation of strut curve*

Type of section	Maximum thickness	Axis of bending	
		x - x	y - y
Rolled I - section	Not greater than 40 mm	(a)	(b)
	Greater than 40 mm	(b)	(c)
Rolled H - section	Not greater than 40 mm	(b)	(c)
	Greater than 40 mm	(c)	(d)

4.5.4 Strut slenderness and axial capacity

Once the appropriate strut curve has been selected, the value of the compressive strength p_c is obtained from Table 24 of BS 5950-1 (repeated here as Table 4.4 to Table 4.11) using input parameters of slenderness and the material design strength p_y. The slenderness λ of the strut is calculated as the effective length (as given by Section 4.5.2) divided by the radius of gyration of the column section about the appropriate axis, i.e. $\lambda = L_E / r$. For values of λ below 15, p_c may be taken as the yield strength p_y of the material. Provided that the strut section is Class 1, 2 or 3, the compression resistance P_c of the strut is given by Clause 4.7.4 as:

$$P_c = A_g p_c$$

where:

- A_g is the gross cross section area of the section
- p_c is the compression strength.

Table 4.4 Compressive strength p_c (N/mm²)

1) Values of p_c in N/mm² with $\lambda < 110$ for strut curve a

λ	Steel grade and design strength p_y (in N/mm²)								
	S 275				S 355				
	235	245	255	265	275	315	325	335	345	355
15	235	245	255	265	275	315	325	335	345	355
20	234	244	254	264	273	312	322	332	342	351
25	232	241	251	261	270	309	318	328	338	347
30	229	239	248	258	267	305	315	324	333	343
35	226	236	245	254	264	301	310	320	329	338
40	223	233	242	251	260	296	305	315	324	333
42	222	231	240	249	258	294	303	312	321	330
44	221	230	239	248	257	292	301	310	319	327
46	219	228	237	246	255	290	299	307	316	325
48	218	227	236	244	253	288	296	305	313	322
50	216	225	234	242	251	285	293	302	310	318
52	215	223	232	241	249	282	291	299	307	315
54	213	222	230	238	247	279	287	295	303	311
56	211	220	228	236	244	276	284	292	300	307
58	210	218	226	234	242	273	281	288	295	303
60	208	216	224	232	239	269	277	284	291	298
62	206	214	221	229	236	266	273	280	286	293
64	204	211	219	226	234	262	268	275	281	288
66	201	209	216	223	230	257	264	270	276	282
68	199	206	213	220	227	253	259	265	270	276
70	196	203	210	217	224	248	254	259	265	270
72	194	201	207	214	220	243	248	253	258	263
74	191	198	204	210	216	238	243	247	252	256
76	188	194	200	206	212	232	237	241	245	249
78	185	191	197	202	208	227	231	235	239	242
80	182	188	193	198	203	221	225	229	232	235
82	179	184	189	194	199	215	219	222	225	228
84	176	181	185	190	194	209	213	216	219	221
86	172	177	181	186	190	204	207	209	212	214
88	169	173	177	181	185	198	200	203	205	208
90	165	169	173	177	180	192	195	197	199	201
92	162	166	169	173	176	186	189	191	193	194
94	158	162	165	168	171	181	183	185	187	188
96	154	158	161	164	166	175	177	179	181	182
98	151	154	157	159	162	170	172	173	175	176
100	147	150	153	155	157	165	167	168	169	171
102	144	146	149	151	153	160	161	163	164	165
104	140	142	145	147	149	155	156	158	159	160
106	136	139	141	143	145	150	152	153	154	155
108	133	135	137	139	141	146	147	148	149	150

Table 4.5 Compressive strength p_c (N/mm^2)

	2) Values of p_c in N/mm^2 with $\lambda \geq 110$ for strut curve a								
	Steel grade and design strength p_y (in N/mm2)								
λ	S 275					S 355			
	235	245	255	265	275	315	325	335	345	355
110	130	132	133	135	137	142	143	144	144	145
112	126	128	130	131	133	137	138	139	140	141
114	123	125	126	128	129	133	134	135	136	136
116	120	121	123	124	125	129	130	131	132	132
118	117	118	120	121	122	126	126	127	128	128
120	114	115	116	118	119	122	123	123	124	125
122	111	112	113	114	115	119	119	120	120	121
124	108	109	110	111	112	115	116	116	117	117
126	105	106	107	108	109	112	113	113	114	114
128	103	104	105	105	106	109	109	110	110	111
130	100	101	102	103	103	106	106	107	107	108
135	94	95	95	96	97	99	99	100	100	101
140	88	89	90	90	91	93	93	93	94	94
145	83	84	84	85	85	87	87	87	88	88
150	78	79	79	80	80	82	82	82	82	83
155	74	74	75	75	75	77	77	77	77	78
160	70	70	70	71	71	72	72	73	73	73
165	66	66	67	67	67	68	68	69	69	69
170	62	63	63	63	64	64	65	65	65	65
175	59	59	60	60	60	61	61	61	61	62
180	56	56	57	57	57	58	58	58	58	58
185	53	54	54	54	54	55	55	55	55	55
190	51	51	51	51	52	52	52	52	53	53
195	48	49	49	49	49	50	50	50	50	50
200	46	46	46	47	47	47	47	47	48	48
210	42	42	42	43	43	43	43	43	43	43
220	39	39	39	39	39	39	39	40	40	40
230	35	36	36	36	36	36	36	36	36	36
240	33	33	33	33	33	33	33	33	33	33
250	30	30	30	30	30	31	31	31	31	31
260	28	28	28	28	28	28	29	29	29	29
270	26	26	26	26	26	26	27	27	27	27
280	24	24	24	24	24	25	25	25	25	25
290	23	23	23	23	23	23	23	23	23	23
300	21	21	21	21	21	22	22	22	22	22
310	20	20	20	20	20	20	20	20	20	20
320	19	19	19	19	19	19	19	19	19	19
330	18	18	18	18	18	18	18	18	18	18
340	17	17	17	17	17	17	17	17	17	17
350	16	16	16	16	16	16	16	16	16	16

Table 4.6 Compressive strength p_c (N/mm^2)

3) Values of p_c in N/mm^2 with $\lambda < 110$ for strut curve b									
	Steel grade and design strength p_y (in N/mm^2)								
λ	S 275					S 355			
	235	245	255	265	275	315	325	335	345	355
15	235	245	255	265	275	315	325	335	345	355
20	234	243	253	263	272	310	320	330	339	349
25	229	239	248	258	267	304	314	323	332	342
30	225	234	243	253	262	298	307	316	325	335
35	220	229	238	247	256	291	300	309	318	327
40	216	224	233	241	250	284	293	301	310	318
42	213	222	231	239	248	281	289	298	306	314
44	211	220	228	237	245	278	286	294	302	310
46	209	218	226	234	242	275	283	291	298	306
48	207	215	223	231	239	271	279	287	294	302
50	205	213	221	229	237	267	275	283	290	298
52	203	210	218	226	234	264	271	278	286	293
54	200	208	215	223	230	260	267	274	281	288
56	198	205	213	220	227	256	263	269	276	283
58	195	202	210	217	224	252	258	265	271	278
60	193	200	207	214	221	247	254	260	266	272
62	190	197	204	210	217	243	249	255	261	266
64	187	194	200	207	213	238	244	249	255	261
66	184	191	197	203	210	233	239	244	249	255
68	181	188	194	200	206	228	233	239	244	249
70	178	185	190	196	202	223	228	233	238	242
72	175	181	187	193	198	218	223	227	232	236
74	172	178	183	189	194	213	217	222	226	230
76	169	175	180	185	190	208	212	216	220	223
78	166	171	176	181	186	203	206	210	214	217
80	163	168	172	177	181	197	201	204	208	211
82	160	164	169	173	177	192	196	199	202	205
84	156	161	165	169	173	187	190	193	196	199
86	153	157	161	165	169	182	185	188	190	193
88	150	154	158	161	165	177	180	182	185	187
90	146	150	154	157	161	172	175	177	179	181
92	143	147	150	153	156	167	170	172	174	176
94	140	143	147	150	152	162	165	167	169	171
96	137	140	143	146	148	158	160	162	164	165
98	134	137	139	142	145	153	155	157	159	160
100	130	133	136	138	141	149	151	152	154	155
102	127	130	132	135	137	145	146	148	149	151
104	124	127	129	131	133	141	142	144	145	146
106	121	124	126	128	130	137	138	139	141	142
108	118	121	123	125	126	133	134	135	137	138

Table 4.7 *Compressive strength* p_c *(N/mm²)*

	4) Values of p_c in N/mm² with $\lambda \geq 110$ for strut curve b								
	Steel grade and design strength p_y (in N/mm²)								
λ	S 275				S 355				
	235	245	255	265	275	315	325	335	345	355
110	115	118	120	121	123	129	130	131	133	134
112	113	115	117	118	120	125	127	128	129	130
114	110	112	114	115	117	122	123	124	125	126
116	107	109	111	112	114	119	120	121	122	122
118	105	106	108	109	111	115	116	117	118	119
120	102	104	105	107	108	112	113	114	115	116
122	100	101	103	104	105	109	110	111	112	112
124	97	99	100	101	102	106	107	108	109	109
126	95	96	98	99	100	103	104	105	106	106
128	93	94	95	96	97	101	101	102	103	103
130	90	92	93	94	95	98	99	99	100	101
135	85	86	87	88	89	92	93	93	94	94
140	80	81	82	83	84	86	87	87	88	88
145	76	77	78	78	79	81	82	82	83	83
150	72	72	73	74	74	76	77	77	78	78
155	68	69	69	70	70	72	72	73	73	73
160	64	65	65	66	66	68	68	69	69	69
165	61	62	62	62	63	64	65	65	65	65
170	58	58	59	59	60	61	61	61	62	62
175	55	55	56	56	57	58	58	58	59	59
180	52	53	53	53	54	55	55	55	56	56
185	50	50	51	51	51	52	52	53	53	53
190	48	48	48	48	49	50	50	50	50	50
195	45	46	46	46	46	47	47	48	48	48
200	43	44	44	44	44	45	45	45	46	46
210	40	40	40	40	41	41	41	41	42	42
220	36	37	37	37	37	38	38	38	38	38
230	34	34	34	34	34	35	35	35	35	35
240	31	31	31	31	32	32	32	32	32	32
250	29	29	29	29	29	30	30	30	30	30
260	27	27	27	27	27	27	28	28	28	28
270	25	25	25	25	25	26	26	26	26	26
280	23	23	23	23	24	24	24	24	24	24
290	22	22	22	22	22	22	22	22	22	22
300	20	20	21	21	21	21	21	21	21	21
310	19	19	19	19	19	20	20	20	20	20
320	18	18	18	18	18	18	18	19	19	19
330	17	17	17	17	17	17	17	17	17	18
340	16	16	16	16	16	16	16	16	17	17
350	15	15	15	15	15	16	16	16	16	16

Table 4.8 Compressive strength p_c (N/mm²)

	5) Values of p_c in N/mm² with λ < 110 for strut curve c								
	Steel grade and design strength p_y (in N/mm²)								
λ	S 275					S 355			
	235	245	255	265	275	315	325	335	345	355
15	235	245	255	265	275	315	325	335	345	355
20	233	242	252	261	271	308	317	326	336	345
25	226	235	245	254	263	299	308	317	326	335
30	220	228	237	246	255	289	298	307	315	324
35	213	221	230	238	247	280	288	296	305	313
40	206	214	222	230	238	270	278	285	293	301
42	203	211	219	227	235	266	273	281	288	296
44	200	208	216	224	231	261	269	276	284	291
46	197	205	213	220	228	257	264	271	279	286
48	195	202	209	217	224	253	260	267	274	280
50	192	199	206	213	220	248	255	262	268	275
52	189	196	203	210	217	244	250	257	263	270
54	186	193	199	206	213	239	245	252	258	264
56	183	189	196	202	209	234	240	246	252	258
58	179	186	192	199	205	229	235	241	247	252
60	176	183	189	195	201	225	230	236	241	247
62	173	179	185	191	197	220	225	230	236	241
64	170	176	182	188	193	215	220	225	230	235
66	167	173	178	184	189	210	215	220	224	229
68	164	169	175	180	185	205	210	214	219	223
70	161	166	171	176	181	200	204	209	213	217
72	157	163	168	172	177	195	199	203	207	211
74	154	159	164	169	173	190	194	198	202	205
76	151	156	160	165	169	185	189	193	196	200
78	148	152	157	161	165	180	184	187	191	194
80	145	149	153	157	161	176	179	182	185	188
82	142	146	150	154	157	171	174	177	180	183
84	139	142	146	150	154	167	169	172	175	178
86	135	139	143	146	150	162	165	168	170	173
88	132	136	139	143	146	158	160	163	165	168
90	129	133	136	139	142	153	156	158	161	163
92	126	130	133	136	139	149	152	154	156	158
94	124	127	130	133	135	145	147	149	151	153
96	121	124	127	129	132	141	143	145	147	149
98	118	121	123	126	129	137	139	141	143	145
100	115	118	120	123	125	134	135	137	139	140
102	113	115	118	120	122	130	132	133	135	136
104	110	112	115	117	119	126	128	130	131	133
106	107	110	112	114	116	123	125	126	127	129
108	105	107	109	111	113	120	121	123	124	125

Table 4.9 Compressive strength p_c (N/mm^2)

6) Values of p_c in N/mm^2 with $\lambda \geq 110$ for strut curve c									
	Steel grade and design strength p_y (in N/mm^2)								
λ	S 275					S 355			
	235	245	255	265	275	315	325	335	345	355
110	102	104	106	108	110	116	118	119	120	122
112	100	102	104	106	107	113	115	116	117	118
114	98	100	101	103	105	110	112	113	114	115
116	95	97	99	101	102	108	109	110	111	112
118	93	95	97	98	100	105	106	107	108	109
120	91	93	94	96	97	102	103	104	105	106
122	89	90	92	93	95	99	100	101	102	103
124	87	88	90	91	92	97	98	99	100	100
126	85	86	88	89	90	94	95	96	97	98
128	83	84	86	87	88	92	93	94	95	95
130	81	82	84	85	86	90	91	91	92	93
135	77	78	79	80	81	84	85	86	87	87
140	72	74	75	76	76	79	80	81	81	82
145	69	70	71	71	72	75	76	76	77	77
150	65	66	67	68	68	71	71	72	72	73
155	62	63	63	64	65	67	67	68	68	69
160	59	59	60	61	61	63	64	64	65	65
165	56	56	57	58	58	60	60	61	61	61
170	53	54	54	55	55	57	57	58	58	58
175	51	51	52	52	53	54	54	55	55	55
180	48	49	49	50	50	51	52	52	52	53
185	46	46	47	47	48	49	49	50	50	50
190	44	44	45	45	45	47	47	47	47	48
195	42	42	43	43	43	45	45	45	45	45
200	40	41	41	41	42	43	43	43	43	43
210	37	37	38	38	38	39	39	39	40	40
220	34	34	35	35	35	36	36	36	36	36
230	31	32	32	32	32	33	33	33	33	34
240	29	29	30	30	30	30	31	31	31	31
250	27	27	27	28	28	28	28	28	29	29
260	25	25	26	26	26	26	26	26	27	27
270	23	24	24	24	24	24	25	25	25	25
280	22	22	22	22	22	23	23	23	23	23
290	21	21	21	21	21	21	21	22	22	22
300	19	19	20	20	20	20	20	20	20	20
310	18	18	18	19	19	19	19	19	19	19
320	17	17	17	17	18	18	18	18	18	18
330	16	16	16	16	17	16	17	17	17	17
340	15	15	15	16	16	16	16	16	16	16
350	15	15	15	15	15	15	15	15	15	15

Table 4.10 Compressive strength p_c (N/mm^2)

7) Values of p_c in N/mm^2 with $\lambda < 110$ for strut curve d

λ	Steel grade and design strength p_y (in N/mm^2)								
	S 275					S 355			
	235	245	255	265	275	315	325	335	345	355
15	235	245	255	265	275	315	325	335	345	355
20	232	241	250	259	269	305	314	323	332	341
25	223	231	240	249	257	292	301	309	318	326
30	213	222	230	238	247	279	287	296	304	312
35	204	212	220	228	236	267	274	282	290	297
40	195	203	210	218	225	254	261	268	275	283
42	192	199	206	214	221	249	256	263	270	277
44	188	195	202	209	216	244	251	257	264	271
46	185	192	199	205	212	239	245	252	258	265
48	181	188	195	201	208	234	240	246	252	259
50	178	184	191	197	204	228	235	241	247	253
52	174	181	187	193	199	223	229	235	241	246
54	171	177	183	189	195	218	224	229	235	240
56	167	173	179	185	191	213	219	224	229	234
58	164	170	175	181	187	208	213	218	224	229
60	161	166	172	177	182	203	208	213	218	223
62	157	163	168	173	178	198	203	208	212	217
64	154	159	164	169	174	193	198	202	207	211
66	150	156	160	165	170	188	193	197	201	205
68	147	152	157	162	166	184	188	192	196	200
70	144	149	153	158	162	179	183	187	190	194
72	141	145	150	154	158	174	178	182	185	189
74	138	142	146	150	154	170	173	177	180	183
76	135	139	143	147	151	165	169	172	175	178
78	132	136	139	143	147	161	164	167	170	173
80	129	132	136	140	143	156	160	163	165	168
82	126	129	133	136	140	152	155	158	161	163
84	123	126	130	133	136	148	151	154	156	159
86	120	123	127	130	133	144	147	149	152	154
88	117	120	123	127	129	140	143	145	148	150
90	114	118	121	123	126	137	139	141	144	146
92	112	115	118	120	123	133	135	137	139	142
94	109	112	115	117	120	129	132	134	136	138
96	107	109	112	115	117	126	128	130	132	134
98	104	107	109	112	114	123	125	126	128	130
100	102	104	107	109	111	119	121	123	125	126
102	99	102	104	106	108	116	118	120	121	123
104	97	99	102	104	106	113	115	116	118	120
106	95	97	99	101	103	110	112	113	115	116
108	93	95	97	99	101	107	109	110	112	113

Table 4.11 Compressive strength p_c (N/mm^2)

	8) Values of p_c in N/mm^2 with $\lambda \geq 110$ for strut curve d								
	Steel grade and design strength p_y (in N/mm^2)								
λ	S 275					S 355			
	235	245	255	265	275	315	325	335	345	355
110	91	93	95	96	98	105	106	108	109	110
112	88	90	92	94	96	102	103	105	106	107
114	86	88	90	92	94	99	101	102	103	104
116	85	86	88	90	91	97	98	99	101	102
118	83	84	86	88	89	95	96	97	98	99
120	81	82	84	86	87	92	93	94	95	96
122	79	81	82	84	85	90	91	92	93	94
124	77	79	80	82	83	88	89	90	91	92
126	76	77	78	80	81	86	87	88	89	89
128	74	75	77	78	79	84	85	85	86	87
130	72	74	75	76	77	82	83	83	84	85
135	68	70	71	72	73	77	78	79	79	80
140	65	66	67	68	69	73	73	74	75	75
145	62	63	64	65	65	69	69	70	71	71
150	59	60	60	61	62	65	66	66	67	67
155	56	57	57	58	59	62	62	63	63	64
160	53	54	55	55	56	58	59	59	60	60
165	50	51	52	53	53	55	56	56	57	57
170	48	49	49	50	51	53	53	54	54	54
175	46	47	47	48	48	50	51	51	51	52
180	44	45	45	46	46	48	48	49	49	49
185	42	43	43	44	44	46	46	46	47	47
190	40	41	41	42	42	44	44	44	44	45
195	38	39	39	40	40	42	42	42	42	43
200	37	37	38	38	39	40	40	40	41	41
210	34	34	35	35	35	37	37	37	37	37
220	31	32	32	32	33	34	34	34	34	34
230	29	29	30	30	30	31	31	31	32	32
240	27	27	28	28	28	29	29	29	29	29
250	25	25	26	26	26	27	27	27	27	27
260	24	24	24	24	24	25	25	25	25	25
270	22	22	22	23	23	23	23	23	24	24
280	21	21	21	21	21	22	22	22	22	22
290	19	20	20	20	20	20	21	21	21	21
300	18	18	19	19	19	19	19	19	19	20
310	17	17	17	18	18	18	18	18	18	18
320	16	16	16	17	17	17	17	17	17	17
330	15	15	16	16	16	16	16	16	16	16
340	15	15	15	15	15	15	15	15	15	15
350	14	14	14	14	14	14	14	15	15	15

4.6 Columns in simple construction

In non-sway frames using simple construction, joints are designed to be flexible. The distribution of forces and moments in the frame are determined assuming that the connections between beams and columns are pinned. The joint flexibility may include distortions, which arise as a consequence of plastic deformations in all components of the connections except the bolts. The beams are designed on the basis of being simply supported at their ends. A beam of span L measured between the column centrelines and subjected to uniformly distributed loading w, will be designed for a maximum moment of $w\,L^2/8$. In reality the beams are not supported on the column centre-lines and thus some eccentricity will occur, leading to moments in the columns. A nominal eccentricity is therefore assumed when designing columns in simple construction.

BS 5950-1, Clause 4.7.7 presents a well-established approach to the design of such columns which is outlined below.

1. All beams should be taken as fully loaded and pattern loading may be ignored.

2. The nominal moments are determined using the following.

    i) For a typical beam to column connection, the eccentricity should be taken as the distance from the centreline to the face of the column, plus 100 mm.

    ii) For a beam supported on a cap plate, the reaction should be taken as at the face of the member or the edge of any packing.

    iii) For a roof truss on a cap plate the eccentricity may be taken as zero provided that the simple connections do not develop adverse moments.

3. In multi-storey frames that are effectively continuous at their splices, the out of balance moments at every beam column joint may be divided equally between the column lengths above and below that point in proportion to their stiffness. However, if the value of I/L for these two lengths does not differ by a factor exceeding 1.5, then the out of balance moment may be divided equally. No moments should be carried over to adjacent levels (both above and below the beam level under consideration).

4. The nominal moments applied to the column, together with the applied axial load, should satisfy the following equation.

    $F_c / P_c \;+\; M_x / M_{bs} \;+\; M_y / p_y Z_y \;\leq\; 1$

where:

F_c is the compressive force

P_c is the compression resistance of the member as a strut (see Section 4.5.4 above)

M_x is the nominal moment about the major axis

M_{bs} is the buckling resistance moment for simple columns as described in section 3.6, using $\lambda_{LT} = 0.5\,L/r_y$

M_y is the nominal moment about the minor axis

p_y is the design strength

Z_y is the section modulus about the minor axis.

4.7 Compression members with moments

Compression members with moments are designed using a more comprehensive interaction formulation, as given in Clause 4.8.3.3. Two separate expressions are needed, the first to deal primarily with in-plane buckling, the second to deal with out-of-plane buckling.

These expressions, given in Clause 4.8.3.3.1, are reproduced below

$$F_c / P_c + m_x M_x / p_y Z_x + m_y M_y / p_y Z_y \leq 1$$

$$F_c / P_{cy} + m_{LT} M_{LT} / M_b + m_y M_y / p_y Z_y \leq 1$$

where:

M_b is the buckling resistance moment (see Section 3.3)

M_{LT} is the maximum major axis moment within the segment length

M_x, M_y are now the maximum moments in the segment length

All other terms are as defined in Section 4.6.

These two expressions permit the use of equivalent uniform moments via the reduction factors m_x, m_y, and m_{LT}. Descriptions of how to determine m_{LT} are to be found in Chapter 3 on beams. The other equivalent uniform moment parameters m_x and m_y are found in a similar way by using Table 26 of BS 5950-1, reproduced here as Table 4.12. It is used in a similar manner to the table for m_{LT}, the only difference being that the numeric values are somewhat different for a given bending moment diagram, as Table 4.12 relates to the influence of flexural buckling rather than lateral-torsional buckling. Note that values of m are likely to be different for buckling about the major and minor principal axes; each relates to the bending moment distribution in that particular plane.

Table 4.12 Equivalent uniform moment factor m_x and m_y

Segments with end moments only (m from formula for the general case)		β	m
β positive	β negative	1.0	1.00
		0.9	0.96
		0.8	0.92
		0.7	0.88
		0.6	0.84
		0.5	0.80
		0.4	0.76
		0.3	0.72
		0.2	0.68
		0.1	0.64
		0.0	0.60
		−0.1	0.58
		−0.2	0.56
		−0.3	0.54
		−0.4	0.52
		−0.5	0.50
		−0.6	0.48
		−0.7	0.46
		−0.8	0.44
		−0.9	0.42
		−1.0	0.40

Segments between intermediate lateral restraints

Specific cases	General case
$m = 0.90$	$M_1, M_2, M_3, M_4, M_5, M_{max}$
$m = 0.95$	
$m = 0.95$	$m = 0.2 + \dfrac{0.1M_2 + 0.6M_3 + 0.1M_4}{M_{max}}$ but $m \geq \dfrac{0.8M_{24}}{M_{max}}$
$m = 0.80$	

The moments M_2 and M_4 are the values at the quarter points and the moments M_3 is the value at mid length.

If M_2, M_3 and M_4 all lie on he same side of the axis, their values are all taken as positive. If they lie both side of the axis, the side leading to the larger value of m is taken as the positive side.

The values of M_{max} and M_{24} are always taken as positive. M_{max} is the maximum moment in the segment and M_{24} is the maximum moment in the central half of the segment.

It is also necessary to carry out cross-section capacity checks at the positions where the moments and the axial forces have their largest values. The capacity check for Class 1, 2 and 3 cross-sections is given by Clause 4.8.3.2 as:

$$F_c / A_g p_y + M_x / M_{cx} + M_y / M_{cy} \leq 1$$

All symbols are as described earlier in this chapter.

Alternatively, BS 5950-1 Clause 4.8.3.3.2 presents the *More exact method* that can be used for the design of compression members with moments, however, a greater amount of calculation is required with the *More exact method*.

4.8 Example – Angle section used as a tie

A 200 x 200 x 16 mm angle section in grade S275 is to be used as a tie. Firstly the connection will be made by a welded gusset plate and secondly by two M24 bolts in a line across the width of the member. Determine the tension capacity in each case.

From section tables: $A_g = 61.8$ cm^2

T_{max} is $\leq$ 16mm; therefore $p_y = 275$ N/mm^2.

Determine a_1 and a_2,

for connected leg $a_1 = 200 \times 16 = 3200$ mm^2

for unconnected leg $a_2 = 6180 - 3200 = 2980$ mm^2.

Connection using a welded gusset

Tension capacity from Clause 4.6.3.1

$P_t = p_y (A_g - 0.3 a_2) = 275 (6180 - 0.3 \times 2980) \, 10^{-3} = 1454$ kN

Alternatively, the capacity may be obtained from the capacity tables (page 208), as $P_t = 1450$ kN.

Connection using bolts.

For the unconnected leg, the net area a_{n1} is given by

a_{n1}	$= a_1 - 2 \times D_h \times t$	$= 3200 - 2 \times 26 \times 16$	$= 2368$ mm^2
a_{e1}	$= K_e \, a_{n1}$ but $\leq a_g$	$= 1.2 \times 2368 \leq 3200$	$= 2842$ mm^2
A_e	$= a_{e1} + a_2$	$= 2842 + 2980$	$= 5822$ mm^2

Tension capacity from Clause 4.6.3.1

$P_t = p_y (A_e - 0.5 a_2) = 275 (5822 - 0.5 \times 2980) \, 10^{-3} = 1191$ kN

Alternatively, the capacity may be obtained from the capacity tables (page 208), as $P_t = 1190$ kN.

4.9 Example – Axially loaded strut, 1

An 7.0 m long 152 x 152 x 30 UC in grade S275 steel is to be used as a strut with pinned ends and will carry axial load only. Determine its compression resistance.

From section tables:

$b/T = 8.13$; $\quad d/t = 19.0$; $\quad T_{max} = 9.4$ mm;
$A_g = 38.3$ cm^2; $\quad r_x = 6.76$ cm; $\quad r_y = 3.83$ cm
$T_{max} = 9.4$ mm; therefore $p_y = 275$ N/mm^2; thus $\varepsilon = 1.00$.

Section classification

From Table 2.1, the Class 3 semi-compact limit for b/T for the outstand of a compression flange = 15ε, therefore the flange is not slender.

From Table 2.1, the Class 3 semi-compact limit for d/t for the web of an H-section = 40ε, therefore the web is not slender.

Therefore the whole section is Class 3 semi-compact.

Effective length

For a section with pinned ends, from Table 4.2, $L_E = 1.0L = 7.0$ m for both axes.

Compression strength (y axis)

For y axis buckling, slenderness $\lambda = L_E / r_y = 7000/38.3 = 183$

Select strut curve from Table 4.3. For a UC buckling about the y axis, use curve (c).

Table 4.9 with $\lambda = 183$ and $p_y = 275$ N/mm^2 gives, $p_c = 49$ N/mm^2.

Compression resistance (y axis)

For a non-slender section $P_c = A_g \, p_c$

$P_{cy} = 38.3 \times 100 \times 49 / 1000 = 188$ kN

Compression strength (x axis)

For x axis buckling, slenderness $\lambda = L_E / r_x = 7000/67.6 = 104$

Select strut curve from Table 4.3. For a UC buckling about the x axis, use curve (b).

Table 4.6 with $\lambda = 104$ and $p_y = 275$ N/mm² gives, $p_c = 133$ N/mm².

Compression resistance (x axis)

For a non-slender section $P_c = A_g p_c$

$P_{cx} = 38.3 \times 100 \times 133 / 1000 = 509$ kN

Compression resistance is the lesser of P_{cx} and P_{cy}, which is 188 kN.

Alternatively, the resistance may be obtained from the capacity tables (page 225), as P_{cx} and P_{cy} equal to 514 kN and 187 kN respectively. (The discrepancies are due to rounding errors.)

In this example it is obvious that the minor axis buckling is the controlling factor but this is not always the case.

4.10 Example – Axially loaded strut, 2

A 254 x 254 x 89 UC in grade S275 steel is 6.0 m long and is pinned at its ends in both planes. It has a positional restraint located at its mid-height that prevents lateral movement parallel to the flanges. Determine its compression resistance.

From section tables:

$b/T = 7.41;$ $\quad d/t = 19.4;$ $\quad T_{max} = 17.3$ mm;

$A_g = 113$ cm²; $\quad r_x = 11.2$ cm; $\quad r_y = 6.55$ cm.

$T_{max} = 17.3$ mm; therefore $p_y = 265$ N/mm²; thus $\varepsilon = (275/265)^{0.5} = 1.02$.

Section classification

From Table 2.1, the Class 3 semi-compact limit for b/T for the outstand of a compression flange = 15ε, therefore the flange is not slender.

From Table 2.1, the Class 3 semi-compact limit for d/t for the web of an H-section = 40ε, therefore the web is not slender.

Therefore the whole section is Class 3 semi-compact.

Effective length

For a section with pinned ends, from Table 4.2, $L_E = 1.0L = 6.0$ m for the major axis and, due to restraint at mid-height $L = 3.0$ m for minor axis.

Compression strength (y axis)

For y axis buckling, slenderness $\lambda = L_E / r_y = 3000/65.5 = 45.8$

Select strut curve from Table 4.3. For a UC buckling about the y axis, use curve (c).

From Table 4.9 with $\lambda = 45.8$ and $p_y = 265$ N/mm^2 gives, $p_c = 220$ N/mm^2.

Compression resistance (y axis)

For a non-slender section $P_c = A_g \, p_c$

$P_{cy} = 113 \times 100 \times 220 / 1000 = 2490$ kN

Compression strength (x axis)

For x axis buckling, slenderness $\lambda = L_E / r_x = 6000/112 = 53.6$

Select strut curve from Table 4.3. For a UC buckling about the x axis, use curve (b).

From Table 4.7 with $\lambda = 53.6$ and $p_y = 265$ N/mm^2 gives, $p_c = 224$ N/mm^2.

Compression resistance

For a non-slender section $P_c = A_g \, p_c$

$P_{cx} = 113 \times 100 \times 224 / 1000 = 2530$ kN

Compression resistance is the lesser of P_{cx} and P_{cy} which is 2490 kN.

Alternatively, the resistances may be obtained from the capacity tables (page 225), as P_{cx} and P_{cy} equal to 2530 kN and 2490 kN respectively.

Note that in this example the lower buckling load does not correspond to the larger slenderness.

4.11 Example – Column in simple construction

A continuous column in a regular multi-storey frame supports beams at 4.0 m centres. The beams are connected to the column using partial depth end plates. At the upper end, two beams frame in to the web without eccentricity from the major axis. One transfers a load of 200 kN and the other a load of 300 kN. Only one beam frames into the major axis, transferring a load of 400 kN without eccentricity about the minor axis. All loads given are factored. The conditions at the lower end of the length to be checked are identical to those at the upper end.

Design moments and forces

Axial load in column = 450 + 200 + 300 + 400 = 1350 kN
Moments at both upper and lower ends of the column
Moment about major axis $M_x = 400 \, (D/2 + 100)$
Guess D = 320 mm, then $\quad M_x = 400 \, (0.16 + 0.1) = 104$ kNm
Moment about minor axis $M_y = (300 - 200)(t/2 + 100)$
Guess t = 20 mm, then $\quad M_y = 100 \, (0.01 + 0.1) \quad = 11$ kNm

As the column is continuous through the three stories under consideration the adjacent column lengths have I/L ratios differing by less than 1.5. Therefore, the moments may be divided equally between the upper and lower column lengths at each beam level and no carry-overs to adjacent joints are made. Thus the column needs to be designed for:

Axial load, $\quad F_c \quad = 1350$ kN
Moment, $M_x \quad = 104/2 \quad = 52$ kNm
Moment, $M_y \quad = 11/2 \quad = 5.5$ kNm

Select trial section

Try a 254 × 254 × 73 UC. From section tables:
$T \quad = 14.2$ mm; $\quad b/T \quad = 8.96$; $\quad d/t = 23.3$;
$r_y \quad = 6.48$ cm; $\quad S_x = 992$ cm^3; $\quad Z_y = 307$ cm^3;
$A \quad = 93.1$ cm^2

$T_{max} \le 16$ mm; therefore $p_y = 275$ N/mm^2; thus $\varepsilon = 1.00$.
From Table 2.1 neither web nor flanges are slender.

Effective length

The beams are connected to the column using partial depth end plates therefore little rotational restraint will be provided and L_E should be taken as 1.0 L i.e.,

$L_E = 4.0$ m.

Compression resistance

$\lambda = L_E/r_y = 4000 / 64.8 = 61.7$ and, from Table 4.8, $p_c = 197$ N/mm^2
$P_c = A_g \, p_c = 9310 \times 197 \times 10^{-3} = 1834$ kN

Buckling resistance moment

M_{bs} is calculated as M_b using a slenderness $\lambda_{LT} = 0.5L/r_y$.

$\lambda_{LT} = 0.5 \times 4000 / 64.8$

From Table 3.1, $p_{bs} = 275$ N/mm^2

$M_{bs} = p_{bs} S_x = 275 \times 992 \times 10^{-3} = 273$ kNm

$p_y Z_y = 275 \times 307 \times 10^{-3} = 84.4$ kNm

Axial load and bending interaction

Check the interaction according to Clause 4.7.7 (see Section 4.6)

$F_c / P_c + M_x / M_{bs} + M_y / p_y Z_y \leq 1$

$1350/1834 + 52/273 + 5.5/84.4 = 0.99 < 1.0$ ∴ satisfactory.

Check assumptions concerning depth of section and thickness of flange on eccentricities. Actual $D/2 = 254.1/2 = 127$ mm which is less than guessed 160 mm. Actual $t/2 = 8.6/2 = 4.3$ mm which is less than guessed 10 mm. The assumption is therefore satisfactory.

Adopt trial section 254 x 254 x 73 UC.

4.12 Example – Column under axial load and moment

A 356 x 368 x 153 UC in grade S275 steel is part of a braced multi-storey frame which has been shown to be non-sway. The storey height between beam centres is 6.0 m. The column is attached to the beams using flush end plate connections and the beams support concrete floor slabs, thus providing partial restraint against bending in both principal planes and full restraint against rotation in plan. The axial load in the column is 1500 kN. At the upper end of the column segment the applied moment is 300 kNm about the major axis and 60 kNm about the minor axis. The corresponding values at the lower end are 200 kNm and 80 kNm respectively (as shown in the bending moment diagrams in Figure 4.2). Check the adequacy of the column section for this storey.

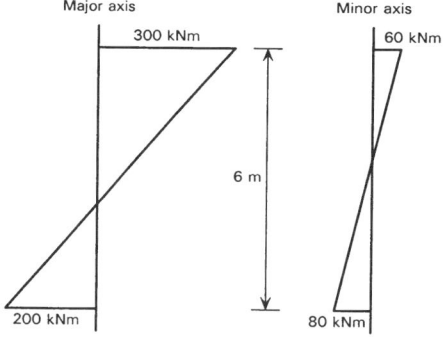

Figure 4.2 *Column bending moments*

Section properties, from section tables

d	= 290.2 mm	t = 12.3 mm		T	= 20.7 mm
b/T	= 8.95	d/t = 23.6		r_y	= 9.49 cm
A_g	= 195 cm^2	Z_x = 2680 cm^3		Z_y	= 948 cm^3
S_x	= 2970 cm^3	S_y = 1440 cm^3		D	= 362 mm

T_{max} = 20.7 mm, T_{max} > 16 mm therefore p_y = 265 N/mm^2

Section classification.

$\varepsilon = (275/p_y)^{0.5} = (275/265)^{0.5} = 1.02$

From Table 2.1, the Class 1 plastic limit for b/T for the outstand of a compression flange = 9ε = 9.18, therefore the flange is Class 1 plastic.

From Table 2.1, the Class 1 plastic limit for d/t for the web generally condition = $80\,\varepsilon/(1 + r_1)$.

$r_1 = F_c/(d\,t\,p_{yw})$ but $-1 \leq r_1 \leq 1$

$r_1 = 1500 \times 10^3 / (290.2 \times 12.3 \times 265) = 1.59$ therefore $r_1 = 1.0$

Thus the web d/t limit is $80 \times 1.02 / (1 + 1) = 40.8$, so the web is also Class 1 plastic.

Shear Capacity

Maximum shear force arising from the bending moments is given by:

$F_v = 300 / (6 \times (200/300)) = 75$ kN

The shear capacity is given by Clause 4.2.3 (see Section 3.2) as:

$P_v = 0.6\,p_y\,D\,t = 0.6 \times 265 \times 10^{-3} \times 362 \times 12.3 = 708$ kN

$F_v < 0.6\,P_v$ therefore, the section is subject to low shear.

Cross-section capacity

It is necessary to carry out cross-section capacity checks at each end of the member. This is because the maximum major axis moment occurs at the upper end and the maximum minor axis moment occurs at the lower end. Except for slender sections, the local capacity check is

$F_c / A_g\,p_y + M_x / M_{cx} + M_y / M_{cy} \leq 1.00$

$M_{cx} = p_y\,S_x = 265 \times 2970 \times 10^3 / 10^6 = 787$ kNm

$M_{cy} = p_y\,S_y = 265 \times 1440 \times 10^3 / 10^6 = 382$ kNm

Note: To avoid irreversible deformation at serviceability loads M_c should be limited to $1.5 p_y Z$ generally. Therefore, M_{cy} is limited to 377 kNm.

At the upper end, the expression becomes:

$$\frac{1500}{5170} + \frac{300}{787} + \frac{60}{377} = 0.29 + 0.38 + 0.16 = 0.83 \ < \ 1.0$$

The upper end satisfies the cross-section capacity check.

At the lower end, the expression becomes:

$$\frac{1500}{5170} + \frac{200}{787} + \frac{80}{377} = 0.29 + 0.25 + 0.21 = 0.75 \ < \ 1.0$$

The lower end satisfies the cross-section capacity check.

Axial load and in-plane buckling interaction

The interaction requirement is

$$F_c / P_c \ + \ m_x M_x / p_y Z_x \ + \ m_y M_y / p_y Z_y \ \leq \ 1$$

The member length is 6.00 m. From Table 4.2, consideration of the end constraints for in-plane buckling gives

$L_E = 0.85L = 0.85 \times 6.00 = 5.10$ m

Slenderness $\lambda = L_E / r_y = 5.10 / 94.9 = 53.7$

Select strut curve from Table 4.3. For a UC buckling about the y axis, use curve (c).

Table 4.8 with $\lambda = 53.7$ and $p_y = 265$ N/mm^2 gives, $p_c = 206$ N/mm^2.

$P_{cy} = A_g p_c = 195 \times 100 \times 206 / 1000 = 4017$ kN

Consider moment distribution for major axis bending. Maximum moment is 300 kNm and the moment at the other end is –200 kNm.
Therefore $\beta = -200/300 = -0.667$, from Table 4.12 $m_x = 0.47$.

Consider moment distribution for minor axis bending. Maximum moment is 80 kNm and the moment at the other end is –60 kNm.
Therefore $\beta = -60/80 = -0.75$, from Table 4.12 $m_y = 0.45$.

Substituting in the interaction expression above

$$\frac{1500}{4017} + \frac{0.47 \times 300 \times 10^6}{265 \times 2680 \times 10^3} + \frac{0.45 \times 80 \times 10^6}{265 \times 948 \times 10^3} = 0.37 + 0.20 + 0.14 = 0.71 \ < \ 1.0$$

Therefore the section satisfies this check.

Axial load and lateral-torsional buckling interaction

The interaction requirement is

$$F_c / P_{cy} + m_{LT}M_{LT} / M_b + m_y M_y / p_y Z_y \leq 1$$

Using Table 4.2, consideration of the end conditions of the column gives $L_E = 0.7L$. L is the storey height of 6.0 m. Therefore $L_E = 0.7 \times 6.0 = 4.2$ m.

$$\lambda_{LT} = u\, v\, \lambda\, \beta_w^{0.5}$$

u may be safely taken as 0.9 (or taken from section tables)

v may be safely taken as 1.0

As the section is at least Class 2 compact, $\beta_w = 1.0$

$\lambda \quad = L_E / r_y = 4200 / 94.9 = 44.3$

$\lambda_{LT} = 0.9 \times 1.0 \times 44.3 \times 1.0 = 39.8$

Table 3.1 with $p_y = 265$ and $\lambda = 39.8$ gives, $p_b = 254$ N/mm^2.

For a Class 1 plastic section, buckling resistance moment $M_b = p_b\, S_x$

$M_b = 254 \times 2970 \times 10^3 / 10^6 = 754$ kNm

From Table 3.4 and the pattern of moments corresponding to the major axis, when $\beta = 0.667$ $m_{LT} = 0.44$.

Substituting in the interaction expression above,

$$\frac{1500}{4017} + \frac{0.44 \times 300}{754} + \frac{0.45 \times 80 \times 10^6}{265 \times 948 \times 10^3} = 0.37 + 0.18 + 0.14 = 0.69 < 1.0$$

Therefore section satisfies this check.

CHAPTER 5   TRUSSES

5.1   Introduction

A truss is a triangulated framework of members in which loads are primarily resisted by axial forces in the individual members. The most commonly used truss is single span, simply supported and statically determinate with joints assumed to act as pins. Trusses can be pitched with sloping rafters as shown in Figure 5.1 or can have parallel top and bottom chords. Trusses with parallel chords are often referred to as lattice girders.

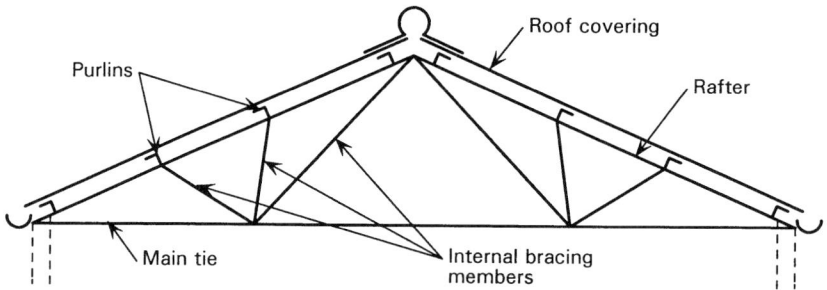

Figure 5.1 *Typical roof structure*

5.2   Typical uses

A common application of pitched trusses is for roofs. Lattice girders have a wider variety of uses including support of roofs and floors particularly with longer spans or heavier loads.

The support of long span flat roofs is generally accomplished by using trusses with parallel chords. Pitched roofs are normally supported by pitched trusses, even for modest spans, the exception being the specialised area of pitched roof portal frames. Portal frames are beyond the scope of this publication and will not be considered further.

One advantage of trusses is that they can be delivered to site as one complete unit, as several smaller units or even as individual elements. The choice will depend upon the size of the truss, the ease of transport between the fabrication shop and the site and the availability of space on site.

5.2.1 Spans

The most efficient form of truss to be employed in any given situation is usually controlled by the span to be covered. Figure 5.2 shows a variety of pitched roof trusses together with the spans over which they are customarily used. For spans in excess of these values, lattice girders may be more practical. However, lattice girders are used for a whole range of spans (greater than approximately 7 m).

Figure 5.3 shows two types of lattice girder – the N-girder or Pratt truss and the Warren girder. These trusses have depth to span ratios typically in the range of 1:10 to 1:14.

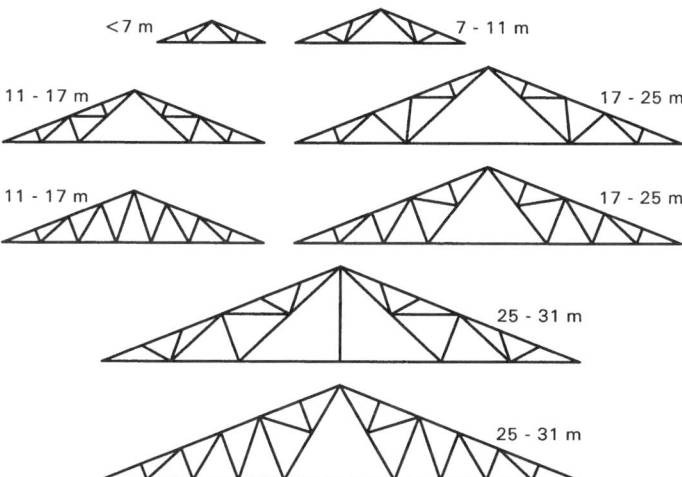

Figure 5.2 *Typical roof trusses and associated spans*

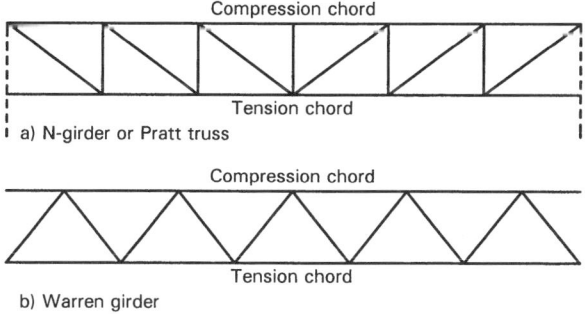

Figure 5.3 *Lattice girders*

5.3 Design concept

Typical roof trusses are plane frames consisting of sloping rafters which meet at the apex or ridge of the frame (see Figure 5.1). The lower ends of the rafters are prevented from spreading by a horizontal main tie, whilst internal bracing members triangulate the truss and carry primarily axial forces. The internal members also reduce the segment lengths of the chords which enables lighter weight and therefore more efficient chords to be used.

5.3.1 Roof arrangement

The roof coverings may be made from a variety of materials ranging from traditional slates or tiles, profiled steel sheeting or more exotic materials. These coverings are supported on purlins (members running between the trusses), which are supported by the rafters and therefore apply loads to the rafters. The purlins also provide out of plane stability-to-the truss. Stability to the truss must be provided at all times, including during erection, when temporary bracing may be used.

The spacing of the purlins (which can range from as little as 900 mm to over 3.5 m) is normally dictated by the roofing material. If the purlins are only located at points where internal members meet, (the panel points) then the truss members will be subjected primarily to axial forces. However, if the spacing is such that the purlins are supported between the panel points, then rafters will need to be designed for combined axial load and bending. Figure 5.4 shows the two possible options.

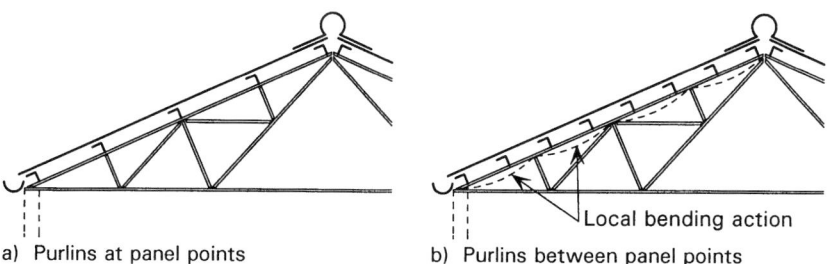

a) Purlins at panel points   b) Purlins between panel points

Figure 5.4 *Purlins at or between panel points*

5.3.2 Pre-cambering

Deflections of nominally flat trusses (Pratt trusses or Warren trusses) must be considered if ponding and therefore overloading are to be avoided. Two possible solutions are to either pre-camber the truss or to have a shallow slope in the top chord. The concept of pre-cambering is often extended to longer span pitched roof trusses where the nominally horizontal bottom chord may in fact slope upwards slightly from the supports. This is carried out so that under loading, the bottom chord does not deflect below the horizontal.

5.3.3 Typical sections

The sections used for the members of a typical roof truss may be single angles, double angles (single angles fastened back to back), single channels, double channels or single T sections. For members with more than one component (double angles or double channels), the elements may be connected directly to each other. Alternatively a gusset plate may be inserted between them which enables a connection to be made to other members so that eccentricities at the connections are minimised. For single component members this is not possible and a lapped joint with its consequent eccentricity is unavoidable.

If, as is normally the case, the members consist of angles, channels and T sections then the axial loads should be determined assuming that the joints are pinned. The moments caused by eccentricities at the ends need not be considered explicitly and the individual members may be checked using Clauses 4.7.10 to 4.7.13 of BS 5950-1. These clauses give values for the effective lengths to be taken for buckling about the various axes. Care must be taken to ensure that all possible axes of buckling are recognised and this will often involve consideration of buckling about the a-a, b-b, u-u and v-v axes. The assumption implied in this approach is that the members may be represented by lines meeting at a point located at the nodes. If the frame is welded, it is customary to detail the frame so that the centroidal axes of the members lie on these lines. If the frame is bolted, then it is usual to ensure that the lines of the bolt holes meet at the nodes. Any moments arising from minor eccentricities are allowed for in the choice of effective lengths. Figure 5.5 shows some typical details from an example of bolted roof truss, using back to back angles for the members, with gusset plates at the connections.

Figure 5.6 shows a welded truss using T sections, detail 2 show that the members node without any eccentricity. Figures 5.5 and 5.6 are only examples of a number of typical details from a wide variety of solutions which may be adopted.

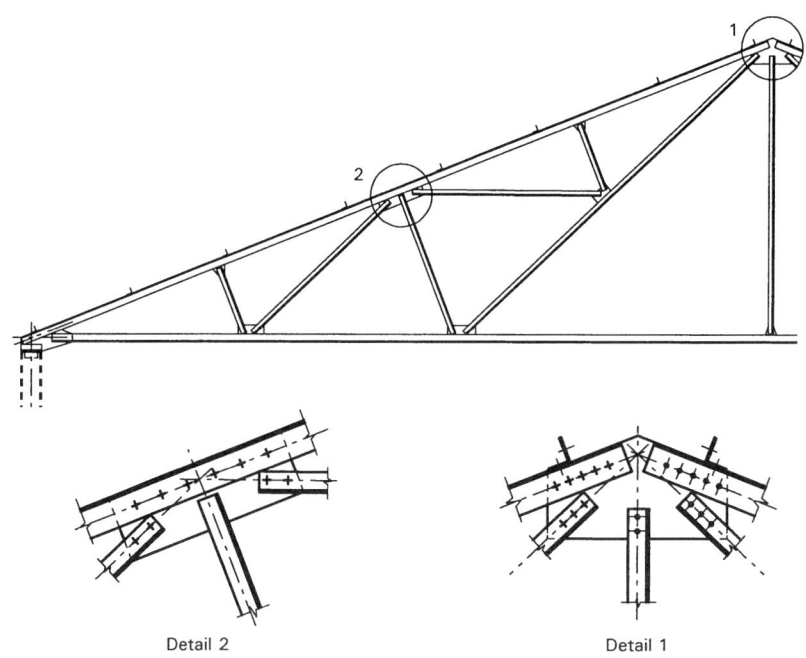

Figure 5.5 *Bolted roof truss and typical details*

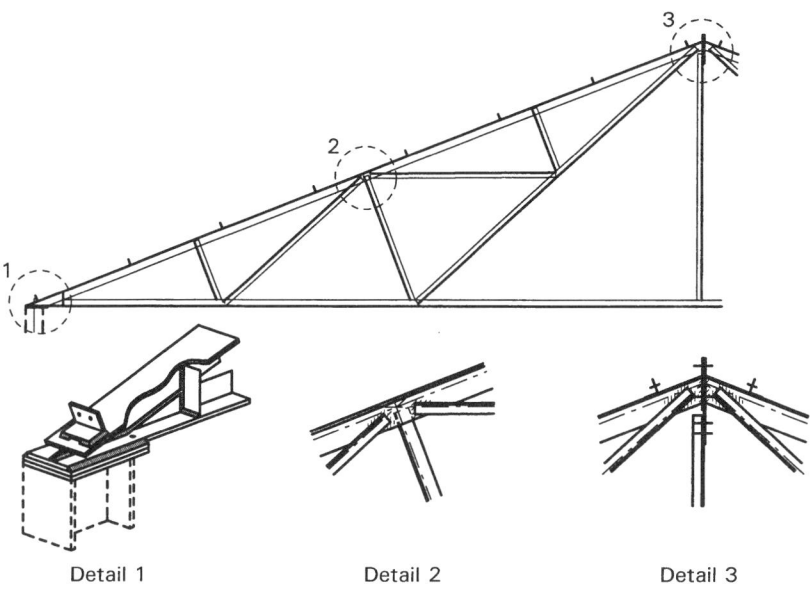

Figure 5.6 *Welded roof truss and typical details*

5.3.4 Joint Capacities

The detailing of the joints is a vital part of truss design. The capacity of the truss may be controlled by the capacity of the joints as much as by the capacity of the members If members are selected so that their capacity is almost fully utilised, the resulting joint details required to transmit the applied forces, can be very impractical. The joints should therefore be considered at an early stage in the design, in conjunction with the selection of the members. As mentioned above, the joint eccentricities will effect design of the truss and its members. The joints adopted in practice must not invalidate the assumptions made at the design stage.

GENERAL DESIGN DATA

BENDING MOMENT AND DEFLECTION FORMULAE FOR BEAMS

NOTATION

L	= Length of span in millimetres
W	= Total distributed or point load in Newtons
W_1 or W_2	= Point load in Newtons
Σ	= Resultant of point loads in Newtons
R_A, R_B, R_C, etc.	= Reaction at A, B or C, etc. in Newtons
F	= Shearing force in Newtons
m	= Applied moment in Newton millimetres
M_X	= Bending moment in Newton millimetres ⎫
δ_X	= Deflection in millimetres ⎬ At distance X from the left hand support A
i_X	= Slope in radians ⎭
M_A or M_B	= End fixing moments in Newton millimetres
M_{max}	= Maximum bending moment in Newton millimetres
$M_{max\,max}$	= Absolute maximum bending moment in Newton millimetres
M_{load}	= Bending moment under the load in Newton millimetres
δ_{max}	= Maximum deflection in millimetres
$\delta_{max\,max}$	= Absolute maximum deflection in millimetres
$\delta_{negative}$	= Negative, i.e. upward, deflection in millimetres
i_A or i_B	= Slope at A or at B in radians
E	= Modulus of elasticity, 2.05×10^5 N/mm^2
I	= Constant moment of inertia of uniform section beam in mm^4

SIGN CONVENTION

Loads	+	Positive when acting downward
Support Reaction	+	Positive when acting upward
Shearing force	+	Positive on a section where the upward left hand support reaction is greater than the algebraic sum of external loads located left of that section
Bending Moment	+	Positive (shown above base line on diagrams) when causing convexity downward
Deflection	+	Positive when downward
Slope		Appropriate values in radians are given, but the signs depend upon which support or which section is being considered, and can be readily ascertained by inspection

Where space permits, general equations for M_X and i_X at any point of the beam, and also the equation to the elastic line (δ_X) have been included.

Values for Slope. These may be used in evaluating the angle of rotation for rubber bearings and similar constructional elements.

FORMULAE FOR BEAMS

SIMPLY SUPPORTED BEAM — UNIFORM LOAD ON FULL SPAN

Span $= L$
Total Uniform Load $= W$

$$R_A = R_B = \frac{W}{2}$$

At mid-span: $M_{max} = \dfrac{WL}{8}$

$$\delta_{max} = \frac{5}{384} \cdot \frac{WL^3}{EI}$$

$$i_A = i_B = \frac{WL^2}{24EI}$$

At X from A: $M_X = \dfrac{WX}{2L}(L - X)$

$$\delta_X = \frac{WX}{24EIL}\left(X^3 - 2X^2L + L^3\right)$$

$$i_X = \frac{W}{24EIL}\left(4X^3 - 6X^2L + L^3\right)$$

SIMPLY SUPPORTED BEAM — UNIFORM LOAD ON PART OF SPAN

Span $= L$
Total Uniform Load $= W$

Let $r = \dfrac{0.5b + c}{L}$

$R_A = Wr$ $\qquad R_B = W(1-r)$

at $X = a + rb$ $\qquad M_{max} = Wr(a + 0.5rb)$

$$I_A = \frac{Wr}{6EI}\left(L^2 - c^2 - Lbr\right)$$

$$I_B = \frac{W(1-r)}{6EI}\left[L^2 - a^2 - Lb(1-r)\right]$$

Equation to elastic line between C and D, i.e. $a \leq X \leq a + b$

$$\delta_X = -\frac{W}{24EIb}\left[X^4 - 4(a+rb)X^3 + 6a^2X^2 + 4\left\{rb\left(L^2 - c^2 - cb - \frac{b^2}{2}\right) - a^3\right\}X + a^4\right]$$

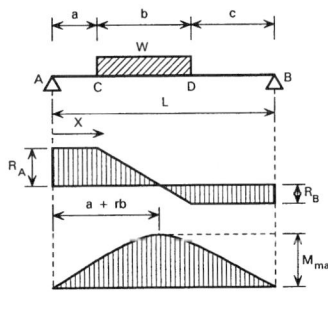

FORMULAE FOR BEAMS

SIMPLY SUPPORTED BEAM — TRIANGULAR LOAD ON FULL SPAN

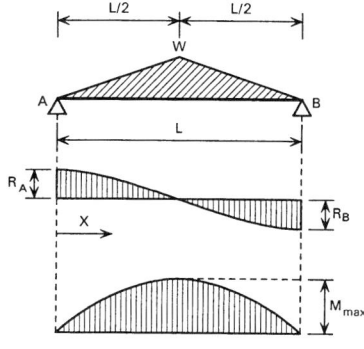

Span $= L$
Total Load $= W$

$R_A = R_B = \dfrac{W}{2}$

At mid-span: $M_{max} = \dfrac{WL}{6}$

$\delta_{max} = \dfrac{WL^3}{60EI}$

$i_A = i_B = \dfrac{5WL^2}{96EI}$

At X from A between A & centre:

$M_X = \dfrac{WX}{6L^2}\left(3L^2 - 4X^2\right)$

$\delta_X = \dfrac{WX}{480EIL^2}\left(16X^4 - 40X^2L^2 + 25L^4\right)$

$i_X = \dfrac{W}{96EIL^2}\left(16X^4 - 24X^2L^2 + 5L^4\right)$

SIMPLY SUPPORTED BEAM — POINT LOAD AT MID-SPAN

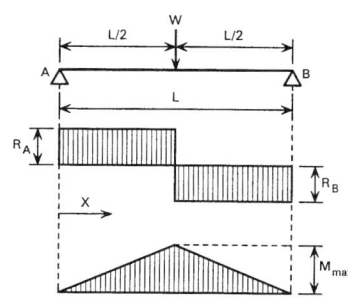

Span $= L$
Point Load $= W$

$R_A = R_B = \dfrac{W}{2}$

at mid-span: $M_{max} = \dfrac{WL}{4}$

$\delta_{max} = \dfrac{1}{48} \cdot \dfrac{WL^3}{EI}$

$i_A = i_B = \dfrac{WL^2}{16EI}$

at X from A between A & centre:

$M_X = \dfrac{WX}{2}$

$\delta_X = \dfrac{WX}{48EI}\left(3L^2 - 4X^2\right)$

$i_X = \dfrac{W}{16EI}\left(L^2 - 4X^2\right)$

FORMULAE FOR BEAMS

SIMPLY SUPPORTED BEAM — POINT LOAD AT ANY POSITION

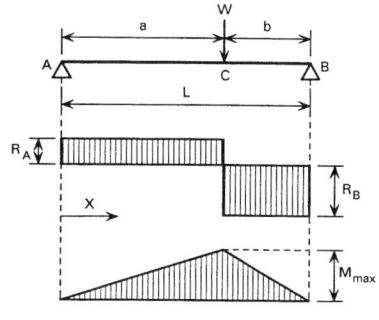

Span $= L$
Point Load $= W$

$$R_A = \frac{Wb}{L}$$

$$R_B = \frac{Wa}{L}$$

at C under load: $M_{max} = \dfrac{Wab}{L}$

$$\delta_C = \frac{Wa^2b^2}{3EIL}$$

$$i_A = \frac{Wab}{6EIL}(L+b) \qquad i_B = \frac{Wab}{6EIL}(L+a)$$

When $a > b$, δ_{max} is at X from A:

$$\delta_{max} = \frac{Wab(L+b)}{27EIL}\sqrt{3a(L+b)}$$

$$X = \sqrt{\frac{a(L+b)}{3}}$$

SIMPLY SUPPORTED BEAM — TWO EQUAL SYMMETRICAL POINT LOADS

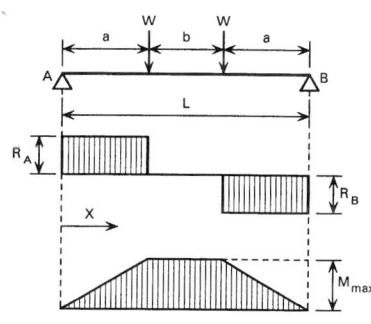

Span $= L$
Two Point Loads each $= W$
$R_A = R_B = W$
M_{max} over length b $= Wa$

$$\delta_{max} \text{ at mid-span} = \frac{Wa}{24EI}\left(3L^2 - 4a^2\right)$$

$$\delta \text{ under either load} = \frac{Wa^2}{6EI}(3L - 4a) ;$$

$$i_A = i_B = \frac{Wa}{2EI}(L - a)$$

If $a = b = \dfrac{L}{3}$, $\delta_{max} = \dfrac{23}{648} \cdot \dfrac{WL^3}{EI}$

FORMULAE FOR BEAMS

BEAM FIXED AT BOTH ENDS — UNIFORM LOAD ON FULL SPAN

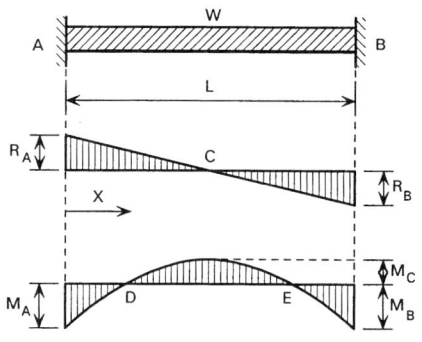

Span $= L$

Total uniform load $= W$

$R_A = R_B = \dfrac{W}{2}$

$M_A = M_B = \dfrac{WL}{12}$

at mid-span: $M_C = \dfrac{WL}{24}$

$\delta_{max} = \dfrac{WL^3}{384EI}$

at X from A: $M_x = \dfrac{W}{12L}(L^2 - 6LX + 6X^2)$

$\delta_x = \dfrac{WX^2}{24EIL}(L - X)^2$

$i_x = \dfrac{WX}{12EIL}(L^2 - 3LX + 2X^2)$

at 0.211 L from either end $M_D = M_E = 0$

BEAM FIXED AT BOTH ENDS — POINT LOAD AT MID-SPAN

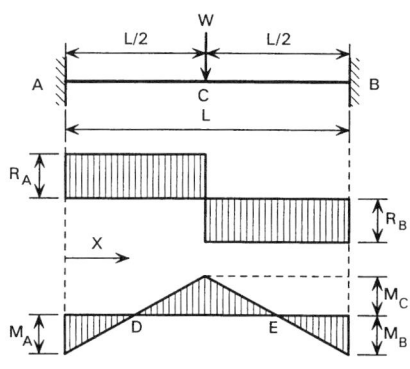

Span $= L$

Point Load $= W$

$R_A = R_B = \dfrac{W}{2}$

$M_A = M_B = -\dfrac{WL}{8}$

at mid-span: $M_C = \dfrac{WL}{8}$

$\delta_{max} = \dfrac{WL^3}{192EI}$

at X from A between A & C:

$M_x = \dfrac{W}{8}(4X - L)$

$\delta_x = \dfrac{WX^2}{48EI}(3L - 4X)$

$i_x = \dfrac{WX}{8EI}(L - 2X)$

at 0.25L from either end $M_D = M_E = 0$

FORMULAE FOR BEAMS

BEAM FIXED AT BOTH ENDS

POINT LOAD AT ANY POSITION

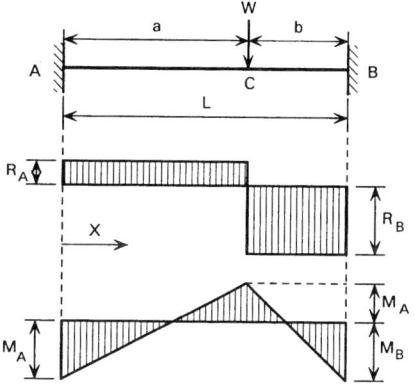

Span $= L$
Point Load $= W$

$$R_A = \frac{Wb^2(L+2a)}{L^3} \quad R_B = \frac{Wa^2(L+2b)}{L^3}$$

$$M_A = \frac{Wab^2}{L^2} \quad M_B = -\frac{Wa^2b}{L^2}$$

at C under load $\quad M_C = \frac{2Wa^2b^2}{L^3}$

at X from A between A & C:

$$M_x = -\frac{Wab^2}{L^2} + \frac{Wb^2(L+2a)X}{L^3}$$

$$\delta_x = \frac{Wb^2X^2(3La - (L+2a)X)}{6EIL^3}$$

$$i_x = \frac{Wb^2X(2La - (L+2a)X)}{2EIL^3}$$

When a > b, the maximum deflection is at

$$X = \frac{2La}{L+2a}$$

$$\delta_{max} = \frac{2Wa^3b^2}{3EI(L+2a)^2}$$

CANTILEVER

UNIFORM LOAD ON PART OF SPAN

SPECIAL CASE: UNIFORM LOAD ON FULL SPAN

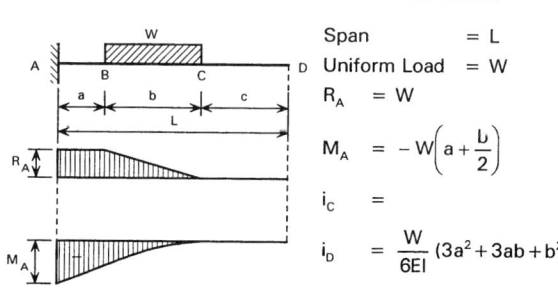

Span $= L$
Uniform Load $= W$

$R_A = W$

$M_A = -W\left(a + \dfrac{b}{2}\right)$

$i_C =$

$i_D = \dfrac{W}{6EI}(3a^2 + 3ab + b^2)$

$\delta_D = \dfrac{W}{24EI}[8a^3 + 18a^2b + 12ab^2 + 3b^3 + 4c(3a^2 + 3ab + b^2)]$

Span $= L = b$
$a = c = 0$
Uniform Load $= W$

$R_A = W$

$M_A = -\dfrac{WL}{2}$

$i_D = \dfrac{WL^2}{6EI}$

$\delta_D = \dfrac{WL^3}{8EI}$

FORMULAE FOR BEAMS

CANTILEVER	TRIANGULAR LOAD ON PART OF SPAN	SPECIAL CASE: TRIANGULAR LOAD ON FULL SPAN
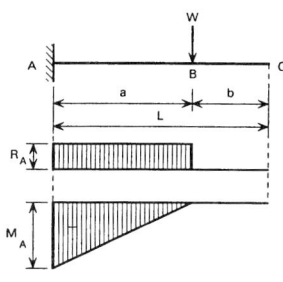	Span $= L$ Triangular Load $= W$ $R_A = W$ $M_A = -\dfrac{Wa}{3}$ $\delta_C = \dfrac{Wa^2}{15EI}\left(L + \dfrac{b}{4}\right)$ $i_B = i_C = \dfrac{Wa^2}{12EI}$	Span $= L = a$ $b = 0$ Triangular Load $= W$ $R_A = W$ $M_A = -\dfrac{WL}{3}$ $\delta_C = \dfrac{WL^3}{15EI}$ $i_C = \dfrac{WL^2}{12EI}$
CANTILEVER	POINT LOAD AT ANY POSITION	SPECIAL CASE: POINT LOAD AT FREE END
	Span $= L$ Point Load $= W$ $R_A = W$ $M_A = -Wa$ $\delta_C = \dfrac{Wa^2}{3EI}\left(L + \dfrac{b}{2}\right)$ $i_B = i_C = \dfrac{Wa^2}{2EI}$	Span $= L = a$ $b = 0$ Point Load $= W$ $R_A = W$ $M_A = -WL$ $\delta_C = \dfrac{WL^3}{3EI}$ $i_C = \dfrac{WL^2}{2EI}$

FORMULAE FOR BEAMS

PROPPED CANTILEVER

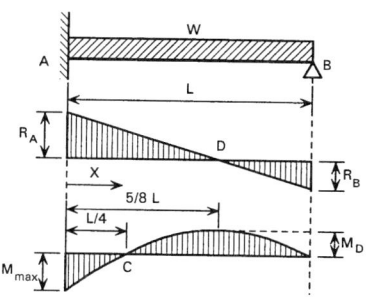

UNIFORM LOAD ON FULL SPAN

Span $= L$
Total Uniform Load $= W$

$R_A = \dfrac{5}{8} W$ $\qquad R_B = \dfrac{3}{8} W$

at A $\qquad M_{max} = -\dfrac{WL}{8}$

at $\dfrac{5}{8} L$ from A $\qquad M_D = \dfrac{9}{128} WL$

at 0.5785L from A $\qquad \delta_{max} = \dfrac{WL^3}{185EI}$

at B $\qquad i_B = \dfrac{WL^2}{48EI}$

at X from A: $M_X = -\dfrac{W}{8L}\left(L^2 - 5LX + 4X^2\right)$

$\delta_X = \dfrac{WX^2}{48EIL}\left(3L^2 - 5LX + 2X^2\right)$

$i_X = \dfrac{WX}{48EIL}\left(6L^2 - 15LX + 8X^2\right)$

PROPPED CANTILEVER

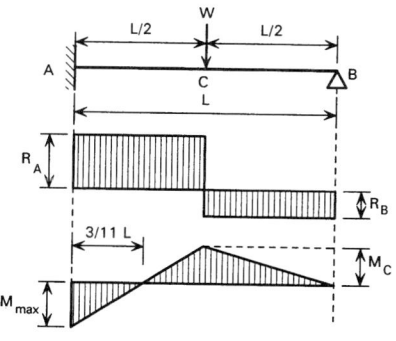

at $\dfrac{3}{11} L$ from A, $M = 0$

POINT LOAD AT MID-SPAN

Span $= L$
Point Load $= W$

$R_A = \dfrac{11}{16} W$ $\qquad R_B = \dfrac{5}{16} W$

at A, $M_{max} = -\dfrac{3}{16} WL$

at mid-span under load $\qquad M_C = \dfrac{5}{32} WL$

$\qquad \delta_C = \dfrac{7WL^3}{768EI}$

at 0.5528L from A, $\qquad \delta_{max} = \dfrac{WL^3}{107EI}$

at B $\qquad i_B = \dfrac{WL^2}{32EI}$

FORMULAE FOR BEAMS

PROPPED CANTILEVER

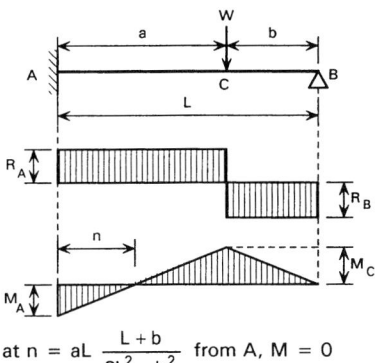

at $n = aL\dfrac{L+b}{3L^2-b^2}$ from A, M = 0

POINT LOAD AT ANY POSITION

Span $= L$
Point Load $= W$

$$R_A = \frac{Wb(3L^2-b^2)}{2L^3} \quad R_B = \frac{Wa^2(2L+b)}{2L^3}$$

$$M_A = \frac{Wab(L+b)}{2L^2} \quad M_C = \frac{Wa^2b(2L+b)}{2L^3}$$

$$i_B = \frac{Wa^2b}{4EIL}$$

Absolute max deflection is under the load when
$a = b\sqrt{2} = 0.5858L \quad \delta_{max\,max} = \dfrac{WL^3}{102EI}$

When $a > b\sqrt{2}$ max deflection is between A & C

$$\delta_{max} = \frac{Wa^3b}{3EI} \cdot \frac{(L+b)^3}{(3L^2-b^2)^2}$$

When $a < b\sqrt{2}$ max deflection is between C & B

$$\delta_{max} = \frac{Wa^2b}{6EI}\sqrt{\frac{b}{2L+b}}$$

PROPPED CANTILEVER

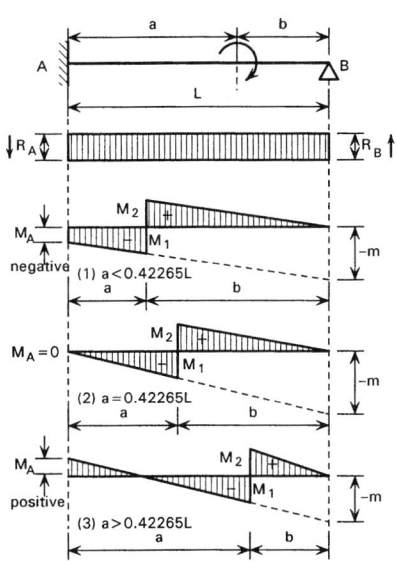

(1) $a < 0.42265L$ negative
(2) $a = 0.42265L$ $M_A = 0$
(3) $a > 0.42265L$ positive

MOMENT APPLIED AT ANY POINT

Span $= L \qquad$ Applied Moment $= m$

$$M_A = \frac{L^2-3b^2}{2L^2}m \quad i_B = \frac{ma}{4EIL}(2b-a)$$

$$R_A = -R_B = -\frac{3(L^2-b^2)}{2L^3}m = -\frac{m+M_A}{L}$$

For case (1)

$$M_1 = -\frac{m}{L^3}\left(a^3 + \frac{3}{2}a^2b + b^3\right)$$

$$M_2 = -\frac{3\,mab}{L^3}\left(b+\frac{a}{2}\right) = m + M_1$$

For case (2)
$M_1 = -0.42265\,m$
$M_2 = 0.57735\,m$

For case (3)
$M_1 = $ as for (1)
$M_2 = $ as for (1)

FORMULAE FOR BEAMS

PROPPED CANTILEVER   UNIFORM LOAD ON LENGTH BEYOND PROP

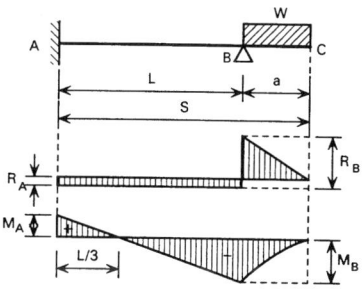

at $X = \dfrac{L}{3}$ from A, $M = 0$

Span = L   Full Length = S
Uniform Load = W

$$R_A = -\frac{3Wa}{4L} \qquad R_B = \frac{W}{L}\left(S - \frac{a}{4}\right)$$

$$M_A = \frac{Wa}{4} \qquad M_B = -\frac{Wa}{2}$$

Deflection at C $= \delta_{max} = -\dfrac{Wa^2 S}{8EI}$

Max. Negative Deflection at $X = \dfrac{2}{3}L$

$$\delta_{neg} = -\frac{WL^2 a}{54EI}$$

Slope at C $= I_c = \dfrac{Wa}{8EI}\left(S + \dfrac{a}{3}\right)$

PROPPED CANTILEVER   POINT LOAD AT FREE END

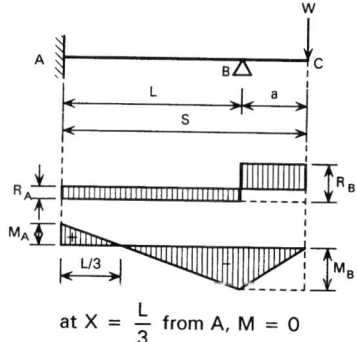

at $X = \dfrac{L}{3}$ from A, $M = 0$

Span = L   Full Length = S
Point Load = W

$$R_A = -\frac{3Wa}{2L} \qquad R_B = \frac{W}{L}\left(S + \frac{a}{2}\right)$$

$$M_A = \frac{Wa}{2} \qquad M_B = -Wa$$

Deflection at C $= \delta_{max} = \dfrac{Wa^2}{4EI}\left(S + \dfrac{a}{3}\right)$

Max. Negative Deflection at $X = \dfrac{2}{3}L$

$$\delta_{neg} = -\frac{WL^2 a}{27EI}$$

Slope at C $= I_c = \dfrac{Wa}{4EI}(S + a)$

FORMULAE FOR BEAMS

Concentrated Load Conditions

The load tables on pages 196 to 201 and 280 to 285 are also applicable to laterally supported simple span beams with equal concentrated loads spaced as shown in the accompanying table of equivalent uniform loads. Except for short spans where shear controls the design, the beam load tables may be entered with this equivalent uniform load.

Note:

The equivalent uniform loads produce the same maximum bending moment as the concentrated load arrangement, but will not produce equivalent shear forces or deflections.

TABLE OF EQUIVALENT UNIFORM LOADS	
Type of Loading: Equal Loads, Equal Spaces	Equivalent Uniform Load
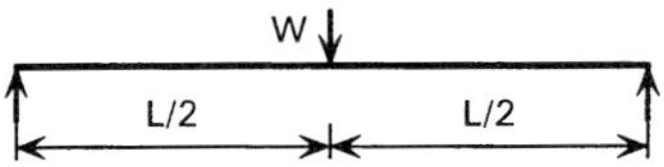 W at L/2, L/2	2.00 W
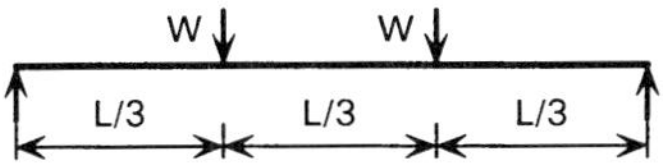 W at L/3, L/3, L/3 (2 loads)	2.67 W
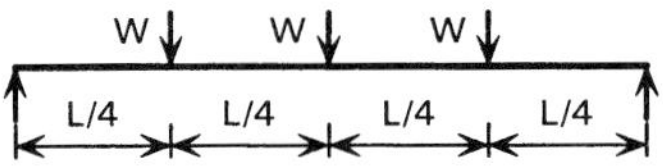 W at L/4, L/4, L/4, L/4 (3 loads)	4.00 W
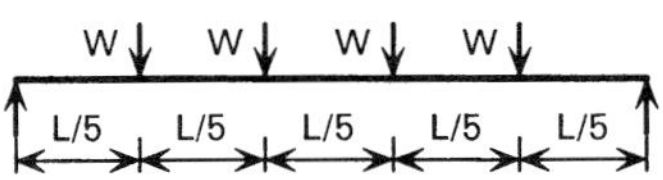 W at L/5, L/5, L/5, L/5, L/5 (4 loads)	4.80 W

FORMULAE FOR BEAMS

SIMPLY SUPPORTED BEAM

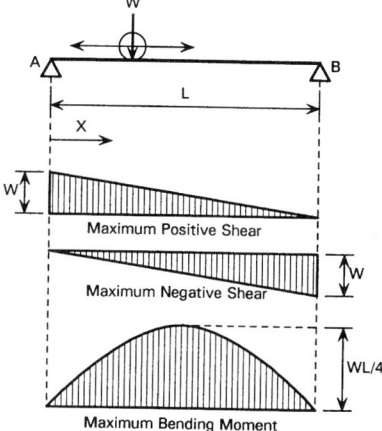

SINGLE CONCENTRATED MOVING LOAD

Maximum Positive Shear at any section occurs when the load is immediately to the right of the section. Similarly, Maximum Negative Shear occurs when the load is to the left. For a section distance X from A:

Positive $F_{Xmax} = W \dfrac{L-X}{L}$

Negative $F_{Xmax} = -W \dfrac{X}{L}$

Maximum Bending Moment at any section occurs when the load is over the section. For a section distance X from A:

$M_{Xmax} = W \dfrac{X(L-X)}{L}$

The Absolute Maximum Bending Moment and Deflection occur under the load at mid span:

$M_{max.max} = \dfrac{WL}{4}$

$\delta_{max\,max} = \dfrac{WL^3}{48EI}$

Maximum end slope at A occurs with the load at X = 0.42265L from A

$i_{Amax} = 0.06415 \dfrac{WL^2}{EI}$

FORMULAE FOR BEAMS

TWO CONCENTRATED MOVING LOADS

Assume:
$W_1 > W_2 \qquad W_1 + W_2 = \Sigma \qquad W_2 = n\Sigma$

Fixed Distance $b = mL$

$b_1 = \dfrac{W_2}{W_1 + W_2} b = nmL \qquad b_2 = (m - nm)L$

Maximum Reaction at A and Absolute Maximum Positive Shear occur when W_1 is immediately to the right of A:

$R_{A\,max} = \text{Positive } F_{max\,max} = W_1 + W_2 \dfrac{L-b}{L}$

For a section distance X from A:

$X \leq L - b$

Positive $F_{max} = \dfrac{L-X}{L}\Sigma - mW_2$

$L - b \leq x$

Positive $F_{max} = \dfrac{L-X}{L}W_1$

Note:

For R_{Bmax}, interchange values of W_1 and W_2 in the formula for R_{Amax}.

For Negative Shear, interchange W_1 and W_2 in formulae for Positive Shear, measuring X from B towards A

If $m > \dfrac{n}{1-n}$, calculate R_{Bmax} and Negative Shear values for W_1 only as single load.

Absolute Maximum Bending Moment occurs under W_1 when that load and the resultant of both loads are equidistant from mid-span (see loading diagram):

$M_{max\,max} = \dfrac{(L-b_1)^2}{4L}\Sigma$

If $m < n$, the Maximum Bending Moment at any section occurs under one of the loads. For a section distance X from A:

$X \leq L(l-n)\, M_{max}$ under $W_1 = \dfrac{(L-b_1-X)X}{L}\Sigma$

$L(l-n) \leq X\, M_{max}$ under $W_2 = \dfrac{(X-b_2)(L-X)}{L}\Sigma$

SIMPLY SUPPORTED BEAM

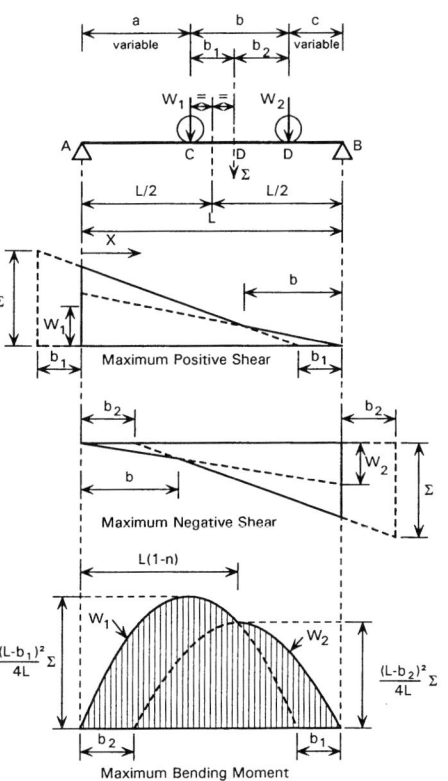

FORMULAE FOR BEAMS

SIMPLY SUPPORTED BEAM

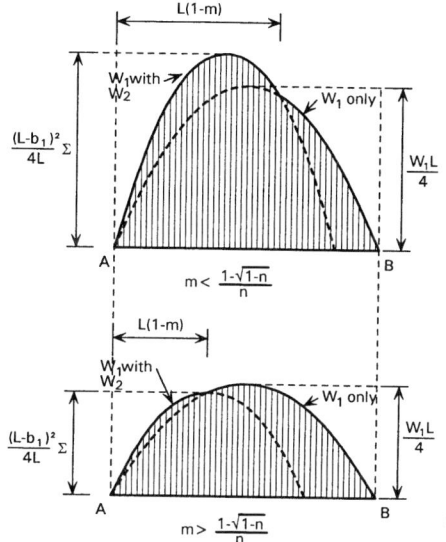

TWO CONCENTRATED MOVING LOADS
(continued)

If m > n, the Maximum Bending Moment at any section always occurs under W_1 (the heavier load), whether W_2 is on or off the span.

For a section distance X from A:

$X \leq L(l-m)$

$$M_{max} = \frac{(L - b_1 - X)X}{L} \Sigma$$

$L(l-m) \leq X$

$$M_{max} = \frac{(L - X)X}{L} W_1$$

If $n < m < \dfrac{1-\sqrt{l-n}}{n}$ the Absolute Maximum Bending Moment occurs under W_1 with W_2 on the span.

If $n < m > \dfrac{1-\sqrt{l-n}}{n}$ the Absolute Maximum Bending Moment occurs under W_1 at mid-span with W_2 off the span.

Note: When the two loads are equal ($W_1 = W_2$ and $n = \frac{1}{2}$) the critical value of $\dfrac{1-\sqrt{l-n}}{n} = 0.5858$.

SIMPLY SUPPORTED BEAMS
CARRYING SEVERAL MOVING CONCENTRATED LOADS

The Maximum Reaction and the Maximum Shear due to several moving concentrated loads occur at one support with one of the loads at that support. The location producing the Absolute Maximum must be found by trial.

The Maximum Bending Moment due to several moving concentrated loads occurs under one of the loads when that load and the gravity centre of all loads are equidistant from mid-span. The Absolute Maximum must be determined by trial.

FIXED END MOMENTS
For use in analysis by 'Moment Distribution'

Fixing Moment at LH End		LOADING	Fixing Moment at RH End	
Both Ends Fixed	This End only Fixed	Span = L (in all cases)	Both Ends Fixed	This End only Fixed
$\dfrac{WL}{8}$	$\dfrac{3WL}{16}$	L/2, W, L/2 (point load at centre)	$\dfrac{WL}{8}$	$\dfrac{3WL}{16}$
$\dfrac{Wab^2}{L^2}$	$\dfrac{Wab(a+2b)}{2L^2}$	a, W, b (point load)	$\dfrac{Wa^2b}{L^2}$	$\dfrac{Wab(2a+b)}{2L^2}$
$\dfrac{Wa(a+c)}{2L}$	$\dfrac{3Wa(a+c)}{4L}$	a, W/2, c, W/2, a	$\dfrac{Wa(a+c)}{2L}$	$\dfrac{3Wa(a+c)}{4L}$
$\dfrac{WL}{9}$	$\dfrac{WL}{6}$	L/3, W/2, L/3, W/2, L/3	$\dfrac{WL}{9}$	$\dfrac{WL}{6}$
$\dfrac{5WL}{48}$	$\dfrac{5WL}{32}$	L/4, W/3, L/4, W/3, L/4, W/3, L/4	$\dfrac{5WL}{48}$	$\dfrac{5WL}{32}$
$\dfrac{WL}{12}\left(\dfrac{n+1}{n}\right)$	$\dfrac{WL}{8}\left(\dfrac{n+1}{n}\right)$	Total load W in (n-1) parts. Span divided into n equal parts	$\dfrac{WL}{12}\left(\dfrac{n+1}{n}\right)$	$\dfrac{WL}{8}\left(\dfrac{n+1}{n}\right)$
$\dfrac{WL}{12}$	$\dfrac{WL}{8}$	Total load W (UDL full span)	$\dfrac{WL}{12}$	$\dfrac{WL}{8}$
$\dfrac{11WL}{96}$	$\dfrac{9WL}{64}$	L/2, L/2, Total = W (UDL on left half)	$\dfrac{5WL}{96}$	$\dfrac{7WL}{64}$
$\dfrac{Wa}{12L^2}(6L^2 - 8aL + 3a^2)$	$\dfrac{Wa}{8L^2}(2L-a)^2$	a, L-a, Total = W (UDL over length a from LH)	$\dfrac{Wa}{12L^2}(4L - 3a)$	$\dfrac{Wa}{8L^2}(2L-a)^2$
$\dfrac{5WL}{48}$	$\dfrac{5WL}{32}$	L/2, L/2, Total = W (triangular load, peak at centre)	$\dfrac{5WL}{48}$	$\dfrac{5WL}{32}$
$\dfrac{WL}{15}$	$\dfrac{7WL}{60}$	Total = W (triangular load, peak at RH)	$\dfrac{WL}{10}$	$\dfrac{2WL}{15}$
$\dfrac{Mb}{L^2}(3a-L)$	$\dfrac{M}{2}(2 - 6n + 3n^2)$	a, b, a/L = n, b/L = m, M (applied moment)	$\dfrac{Ma}{L^2}(3b-L)$	$\dfrac{M}{2}(2 - 6m + 3m^2)$

TRIGONOMETRICAL FORMULAE

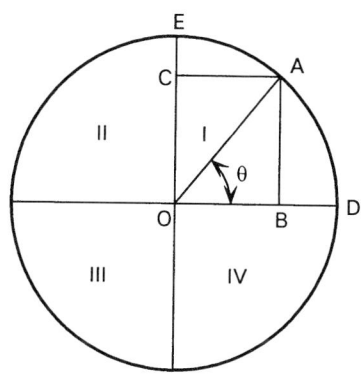

$\sin \theta = \dfrac{AB}{OA} = \dfrac{1}{\operatorname{cosec} \theta} = \cos(90° - \theta)$

$\cos \theta = \dfrac{OB}{OA} = \dfrac{1}{\sec \theta} = \sin(90° - \theta)$

$\tan \theta = \dfrac{AB}{OB} = \dfrac{\sin \theta}{\cos \theta} = \cot(90° - \theta)$

$\operatorname{cosec} \theta = \dfrac{OA}{AB} = \dfrac{1}{\sin \theta} = \sec(90° - \theta)$

$\sec \theta = \dfrac{OA}{AB} = \dfrac{1}{\cos \theta} = \operatorname{cosec}(90° - \theta)$

$\cot \theta = \dfrac{OB}{AB} = \dfrac{1}{\tan \theta} = \tan(90° - \theta)$

In quadrant
- I   All ratios are positive
- II Sin, Cosec are positive   Cos, Tan, Sec, Cotan are negative
- III Tan, Cotan are positive   Sin, Cosec, Cos, Sec are negative
- IV Cos, Sec are positive   Sin, Tan, Cosec, Cotan are negative

$\sin^2 \theta + \cos^2 \theta = 1$

$\sin(\theta \pm \varphi) = \sin \theta \cos \varphi \pm \cos \theta \sin \varphi$

$\cos(\theta \pm \varphi) = \cos \theta \cos \varphi \mp \sin \theta \sin \varphi$

$\tan(\theta + \varphi) = \dfrac{\tan \theta + \tan \varphi}{1 - \tan \theta \tan \varphi}$

$\tan(\theta - \varphi) = \dfrac{\tan \theta - \tan \varphi}{1 + \tan \theta \tan \varphi}$

$\sin 2\theta = 2 \sin \theta \cos \theta \qquad \sin \tfrac{1}{2}\theta = \sqrt{\dfrac{1 - \cos \theta}{2}}$

$\cos 2\theta = \cos^2 \theta - \sin^2 \theta \qquad \cos \tfrac{1}{2}\theta = \sqrt{\dfrac{1 + \cos \theta}{2}}$

$\tan 2\theta = \dfrac{2 \tan \theta}{1 - \tan^2 \theta} \qquad \tan \tfrac{1}{2}\theta = \dfrac{1 - \cos \theta}{\sin \theta}$

$\sin \theta + \sin \varphi = 2 \sin\left(\dfrac{\theta + \varphi}{2}\right) \cos\left(\dfrac{\theta - \varphi}{2}\right)$

$\sin \theta - \sin \varphi = 2 \cos\left(\dfrac{\theta + \varphi}{2}\right) \sin\left(\dfrac{\theta - \varphi}{2}\right)$

$\cos \theta + \cos \varphi = 2 \cos\left(\dfrac{\theta + \varphi}{2}\right) \cos\left(\dfrac{\theta - \varphi}{2}\right)$

$\cos \theta - \cos \varphi = -2 \sin\left(\dfrac{\theta + \varphi}{2}\right) \sin\left(\dfrac{\theta - \varphi}{2}\right)$

$\tan \theta + \tan \varphi = \sin(\theta + \varphi) \div \cos \theta \cos \varphi$

$\tan \theta - \tan \varphi = \sin(\theta - \varphi) \div \cos \theta \cos \varphi$

SOLUTION OF TRIANGLES

General

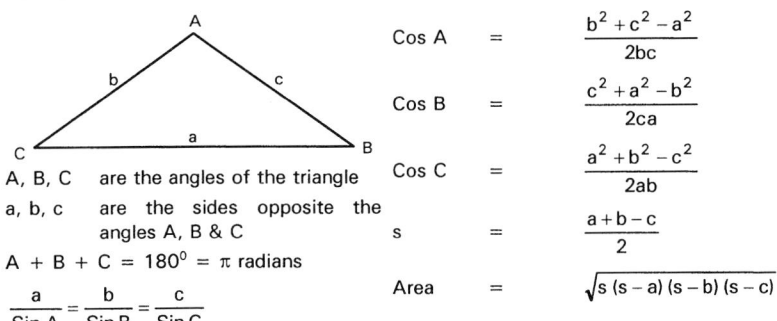

A, B, C   are the angles of the triangle
a, b, c   are the sides opposite the angles A, B & C

$A + B + C = 180° = \pi$ radians

$$\frac{a}{\sin A} = \frac{b}{\sin B} = \frac{c}{\sin C}$$

$$\cos A = \frac{b^2 + c^2 - a^2}{2bc}$$

$$\cos B = \frac{c^2 + a^2 - b^2}{2ca}$$

$$\cos C = \frac{a^2 + b^2 - c^2}{2ab}$$

$$s = \frac{a+b-c}{2}$$

$$\text{Area} = \sqrt{s(s-a)(s-b)(s-c)}$$

Oblique Angled Triangles

Known	Required					
	A	B	C	b	c	Area
a, b, c	$\cos A = \frac{b^2+c^2-a^2}{2bc}$	$\cos B = \frac{c^2+a^2-b^2}{2ca}$	$\cos C = \frac{a^2+b^2-c^2}{2ab}$	–	–	$\sqrt{s(s-a)(s-b)(s-c)}$
a, A, B	–	–	$180° - (A+B)$	$\frac{a \sin B}{\sin A}$	$\frac{a \sin C}{\sin A}$	–
a, b, A	–	$\sin B = \frac{b \sin A}{a}$	–	–	$\frac{b \sin C}{\sin B}$	–
a, b, C	$\tan A = \frac{a \sin C}{b - a \cos C}$	–	–	–	$\sqrt{a^2 + b^2 - 2ab \cos C}$	½ ab sin C

SOLUTION OF TRIANGLES

Right Angled Triangles

$a^2 = c^2 - b^2$
$b^2 = c^2 - a^2$
$c^2 = a^2 + b^2$

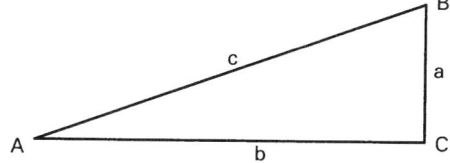

Known	Required					
	A	B	a	b	c	Area
a, b	$\tan A = \dfrac{a}{b}$	$\tan B = \dfrac{b}{a}$	–	–	$\sqrt{a^2 + b^2}$	$\dfrac{ab}{2}$
a, c	$\sin A = \dfrac{a}{c}$	$\cos B = \dfrac{a}{c}$	–	$\sqrt{c^2 - a^2}$	–	$\dfrac{a\sqrt{c^2 - a^2}}{2}$
A, a	–	$90° - A$	–	$a \cot A$	$\dfrac{a}{\sin A}$	$\dfrac{a^2 \cot A}{2}$
A, b	–	$90° - A$	$b \tan A$	–	$\dfrac{b}{\cos A}$	$\dfrac{b^2 \tan A}{2}$
A, c	–	$90° - A$	$c \sin A$	$c \cos A$	–	$\dfrac{c^2 \sin 2A}{4}$

PROPERTIES OF GEOMETRICAL FIGURES

Square
Axis of moments through centre

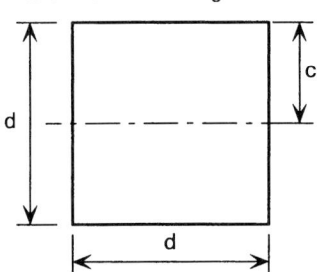

$A = d^2$

$c = \dfrac{d}{2}$

$I = \dfrac{d^4}{12}$

$z = \dfrac{d^3}{6}$

$r = \dfrac{d}{\sqrt{12}} = 0.288675\,d$

Square
Axis of moments on base

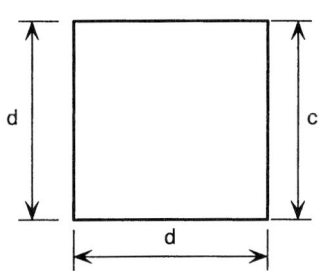

$A = d^2$

$c = d$

$I = \dfrac{d^4}{3}$

$z = \dfrac{d^3}{3}$

$r = \dfrac{d}{\sqrt{3}} = 0.577350\,d$

Square
Axis of moments on diagonal

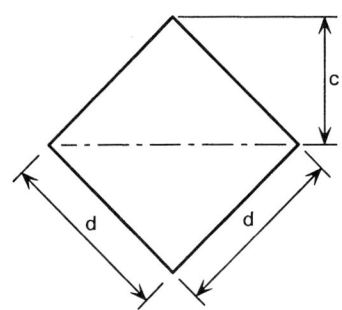

$A = d^2$

$c = \dfrac{d}{\sqrt{2}} = 0.707107\,d$

$I = \dfrac{d^4}{12}$

$z = \dfrac{d^3}{6\sqrt{2}} = 0.117851\,d^3$

$r = \dfrac{d}{\sqrt{12}} = 0.288675\,d$

PROPERTIES OF GEOMETRICAL FIGURES

Rectangle
Axis of moments through centre

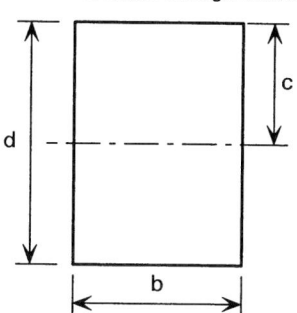

$A = bd$

$c = \dfrac{d}{2}$

$I = \dfrac{bd^3}{12}$

$z = \dfrac{bd^2}{6}$

$r = \dfrac{d}{\sqrt{12}} = 0.288675\ d$

Rectangle
Axis of moments on base

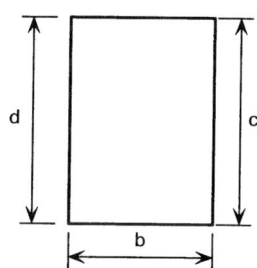

$A = bd$

$c = d$

$I = \dfrac{bd^3}{3}$

$z = \dfrac{bd^2}{3}$

$r = \dfrac{d}{\sqrt{3}} = 0.577350\ d$

Rectangle
Axis of moments on diagonal

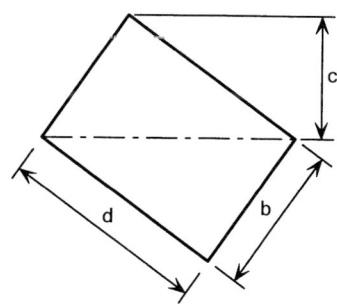

$A = bd$

$c = \dfrac{bd}{\sqrt{b^2 + d^2}}$

$I = \dfrac{b^3 d^3}{6(b^2 + d^2)}$

$z = \dfrac{b^2 d^2}{6\sqrt{b^2 + d^2}}$

$r = \dfrac{bd}{\sqrt{6(b^2 + d^2)}}$

PROPERTIES OF GEOMETRICAL FIGURES

Rectangle
Axis of moments any line through centre of gravity

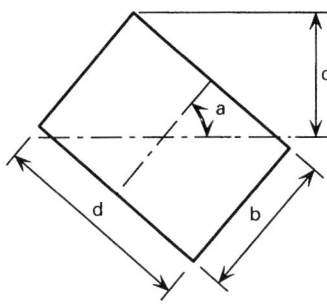

$A = bd$

$c = \dfrac{b \sin a + d \cos a}{2}$

$I = \dfrac{bd(b^2 \sin^2 a + d^2 \cos^2 a)}{12}$

$z = \dfrac{bd(b^2 \sin^2 a + d^2 \cos^2 a)}{6(b \sin a + d \cos a)}$

$r = \sqrt{\dfrac{b^2 \sin^2 a + d^2 \cos^2 a}{12}}$

Hollow Rectangle
Axis of moments through centre

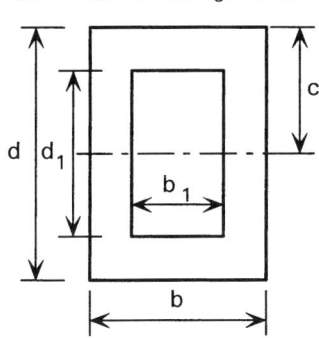

$A = bd - b_1 d_1$

$c = \dfrac{d}{2}$

$I = \dfrac{bd^3 - b_1 d_1^3}{12}$

$z = \dfrac{bd^3 - b_1 d_1^3}{6d}$

$r = \sqrt{\dfrac{bd^3 - b_1 d_1^3}{12A}}$

Equal Rectangles
Axis of moments through centre of gravity

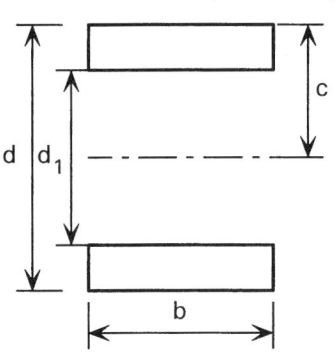

$A = b(d - d_1)$

$c = \dfrac{d}{2}$

$I = \dfrac{b(d^3 - d_1^3)}{12}$

$z = \dfrac{b(d^3 - d_1^3)}{6d}$

$r = \sqrt{\dfrac{d^3 - d_1^3}{12(d - d_1)}}$

PROPERTIES OF GEOMETRICAL FIGURES

Unequal Rectangles
Axis of moments through centre of gravity

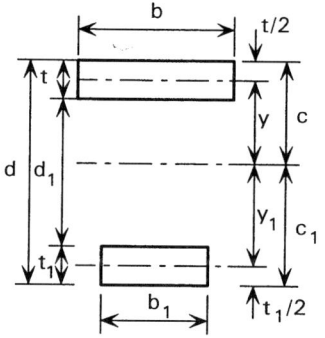

$A = bt + b_1 t_1$

$c = \dfrac{\dfrac{1}{2}bt^2 + b_1 t_1 \left(d - \dfrac{1}{2}t_1\right)}{A}$

$I = \dfrac{bt^3}{12} + bty^2 + \dfrac{b_1 t_1^3}{12} + b_1 t_1 y_1^2$

$z = \dfrac{I}{c} \quad z_1 = \dfrac{I}{c_1}$

$r = \sqrt{\dfrac{I}{A}}$

Triangle
Axis of moments through centre of gravity

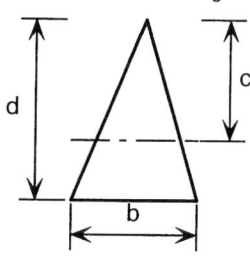

$A = \dfrac{bd}{2}$

$c = \dfrac{2d}{3}$

$I = \dfrac{bd^3}{36}$

$z = \dfrac{bd^2}{24}$

$r = \dfrac{d}{\sqrt{18}} = 0.235702\,d$

Triangle
Axis of moments on base

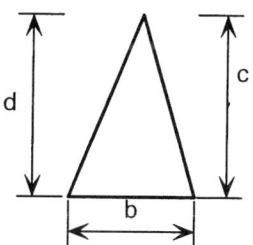

$A = \dfrac{bd}{2}$

$c = d$

$I = \dfrac{bd^3}{12}$

$z = \dfrac{bd^2}{12}$

$r = \dfrac{d}{\sqrt{6}} = 0.408248\,d$

PROPERTIES OF GEOMETRICAL FIGURES

Trapezoid
Axis of moments through centre of gravity

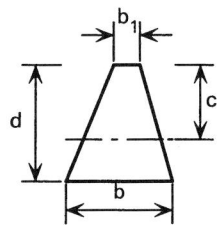

$$A = \frac{d(b + b_1)}{2}$$

$$c = \frac{d(2b + b_1)}{3(b + b_1)}$$

$$I = \frac{d^3(b^2 + 4bb_1 + b_1^2)}{36(b + b_1)}$$

$$z = \frac{d^2(b^2 + 4bb_1 + b_1^2)}{12(2b + b_1)}$$

$$r = \frac{d}{6(b + b_1)}\sqrt{2(b^2 + 4bb_1 + b_1^2)}$$

Circle
Axis of moments through centre

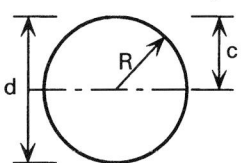

$$A = \frac{\pi d^2}{4} = \pi R^2 = 0.785398\ d^2 = 3.141593\ R^2$$

$$c = \frac{d}{2} = R$$

$$I = \frac{\pi d^4}{64} = \frac{\pi R^4}{4} = 0.049087\ d^4 = 0.785398\ R^4$$

$$z = \frac{\pi d^3}{32} = \frac{\pi R^3}{4} = 0.098175\ d^3 = 0.785398\ R^3$$

$$r = \frac{d}{4} = \frac{R}{2}$$

Hollow Circle
Axis of moments through centre

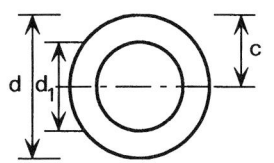

$$A = \frac{\pi(d^2 - d_1^2)}{4} = 0.785398\ (d^2 - d_1^2)$$

$$c = \frac{d}{2}$$

$$I = \frac{\pi(d^4 - d_1^4)}{64} = 0.049087\ (d^4 - d_1^4)$$

$$z = \frac{\pi(d^4 - d_1^4)}{32d} = 0.098175\ \frac{d^4 - d_1^4}{d}$$

$$r = \frac{\sqrt{d^2 + d_1^2}}{4}$$

PROPERTIES OF GEOMETRICAL FIGURES

Half Circle
Axis of moments through centre of gravity

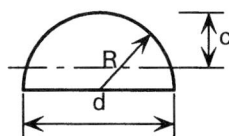

$$A = \frac{\pi R^2}{2} = 1.570796\ R^2$$

$$c = R\left(1 - \frac{4}{3\pi}\right) = 0.575587\ R$$

$$I = R^4\left(\frac{\pi}{8} - \frac{8}{9\pi}\right) = 0.109757\ R^4$$

$$z = \frac{R^3(9\pi^2 - 64)}{24(3\pi - 4)} = 0.190687\ R^3$$

$$r = R\frac{\sqrt{9\pi^2 - 64}}{6\pi} = 0.264336\ R$$

Parabola

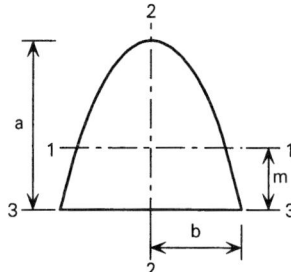

$$A = \frac{4}{3}ab$$

$$m = \frac{2}{5}a$$

$$I_1 = \frac{16}{175}a^3b$$

$$I_2 = \frac{4}{15}ab^3$$

$$I_3 = \frac{32}{105}a^3b$$

PROPERTIES OF GEOMETRICAL FIGURES

Half Parabola

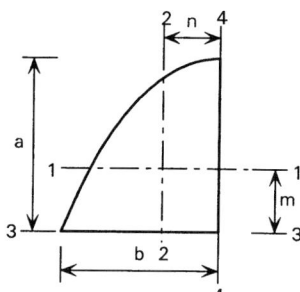

$A = \frac{2}{3}ab$

$m = \frac{2}{5}a$

$n = \frac{3}{8}b$

$I_1 = \frac{8}{175}a^3b$

$I_2 = \frac{19}{480}ab^3$

$I_3 = \frac{16}{105}a^3b$

$I_4 = \frac{2}{15}ab^3$

Complement of half parabola

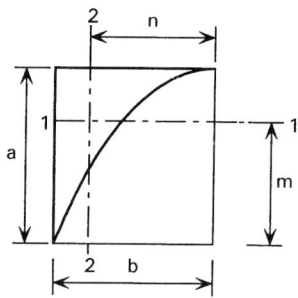

$A = \frac{1}{3}ab$

$m = \frac{7}{10}a$

$n = \frac{3}{4}b$

$I_1 = \frac{37}{2100}a^3b$

$I_2 = \frac{1}{80}ab^3$

PROPERTIES OF GEOMETRICAL FIGURES

Half Ellipse or Circle   **Ellipse**   **Circle**

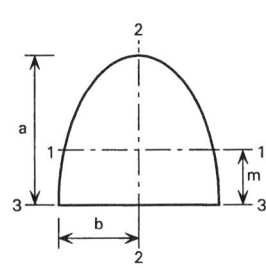

$A = \dfrac{1}{2}\pi ab \qquad\qquad = \dfrac{\pi}{2} R^2 \text{ or } 1.5708 R^2$

$m = \dfrac{4a}{3\pi} \qquad\qquad\qquad = \dfrac{4}{3\pi} R \text{ or } 0.5756 R$

$I_1 = a^3 b \left(\dfrac{\pi}{8} - \dfrac{8}{9\pi} \right) \qquad = \left(\dfrac{\pi}{8} - \dfrac{8}{9\pi} \right) R^4 \text{ or } 0.1096 R^4$

$I_2 = \dfrac{1}{8}\pi ab^3$

$\qquad\qquad\qquad\qquad\qquad = \dfrac{\pi}{8} R^4 \text{ or } 0.3927 R^4$

$I_3 = \dfrac{1}{8}\pi a^3 b$

Quarter Ellipse or Circle

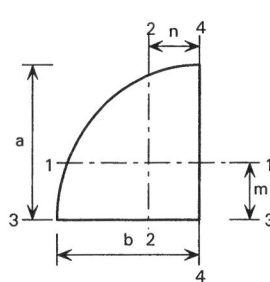

$A = \dfrac{1}{4}\pi ab \} \qquad\qquad = \dfrac{\pi}{4} R^2 \text{ or } 0.7854 R^2$

$m = \dfrac{4a}{3\pi} \} \qquad\qquad\quad = \dfrac{4}{3\pi} R \text{ or } 0.5756 R^2$

$n = \dfrac{4b}{3\pi} \}$

$I_1 = a^2 b \left(\dfrac{\pi}{16} - \dfrac{4}{9\pi} \right)$

$\qquad\qquad\qquad\qquad\qquad = \left(\dfrac{\pi}{16} - \dfrac{4}{9\pi} \right) R^4 \text{ or } 0.0548 R^4$

$I_2 = ab^3 \left(\dfrac{\pi}{16} - \dfrac{4}{9\pi} \right)$

$I_3 = \dfrac{1}{16}\pi a^3 b$

$\qquad\qquad\qquad\qquad\qquad = \dfrac{\pi}{16} R^4 \text{ or } 0.0625 R^4$

$I_4 = \dfrac{1}{16}\pi ab^3$

Elliptic Complement or ¼ circle $A = ab\left(1 - \dfrac{\pi}{4}\right)$

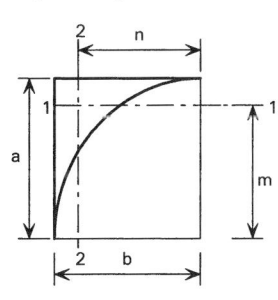

$\qquad\qquad\qquad\qquad\qquad = \left(1 - \dfrac{\pi}{4}\right) R^2 \text{ or } 0.2146 R^2$

$m = \dfrac{a}{6\left(1 - \dfrac{\pi}{4}\right)}, \; n = \dfrac{b}{6\left(1 - \dfrac{\pi}{4}\right)} \quad = \dfrac{R}{6(1 - \pi/4)} \text{ or } 0.7767 R$

$I_1 = a^3 b \left(\dfrac{1}{3} - \dfrac{\pi}{16} - \dfrac{1}{36\left(1 - \dfrac{\pi}{4}\right)} \right)$

$\qquad\qquad\qquad\qquad\qquad = \left(\dfrac{1}{3} - \dfrac{\pi}{16} - \dfrac{1}{36\left(1 - \dfrac{\pi}{4}\right)} \right) R^4$

$I_2 = ab^3 \left(\dfrac{1}{3} - \dfrac{\pi}{16} - \dfrac{1}{36\left(1 - \dfrac{\pi}{4}\right)} \right) \qquad\qquad \text{or } 0.0075 R^4$

METRIC CONVERSIONS

Basic conversion factors

The following equivalents of SI units are given in imperial and, where applicable, metric technical units.

1mm = 0.03937in	1in = 25.4mm	1 hect. = 2.471 acres	1 acre = 0.405 hect
1m = 3.281ft	1ft = 0.3048m	$1mm^3$ = 0.00006102 in^3	$1in^3$ = 16.390mm^3
= 1.094yd	1yd = 0.9144m	$1m^3$ = 35.31ft^2	$1ft^3$ = 0.02832m^3
1km = 0.6214 mile	1 mile = 1.609km	= 1.308yd^3	$1yd^3$ = 0.7646m^3
$1mm^2$ = 0.00155 in^2	$1in^2$ = 645.2mm^2	$1mm^4$ = 0.000002403in^4	$1in^4$ = 416,200mm^4
$1m^2$ = 10.76ft^2	$1ft^2$ = 0.0929m^2		
$1m^2$ = 1.196yd^2	$1yd^2$ = 0.8361m^2		

Force

1 N	= 0.2248 lbf	= 0.1020 kgf	1 kN	= 0.1004 tonf	= 102.0 kgf	= 0.1020 tonf
4.448 N	= 1 lbf	= 0.4536 kgf	9.964 kN	= 1 tonf	= 1,16 kgf	= 1016 tonf
9.807 N	= 2.205 lbf	= 1 kgf	9.807 kN	= 0.9842 tonf	= 1,000 kgf	= 1 tonf

Force per unit length

1 N/m	= 0.06852 lbf/ft	= 0.1020 kgf/m	1 kN/m	= 0.0306 tonf/ft	= 0.1020 tonf/m
14.59 N/m	= 1 lbf/ft	= 1.488 kgf/m	32.69 kN/m	= 1 ton/ft	= 3.33 tonne f/m
9.807 N/m	= 0.672 lbf/ft	= 1 kgf/m	9.807 kN/m	= 0.3000 tonf/ft	= 1 tonf/m

Force per unit area

1 N/mm^2	= 145 lbf/in^2	= 10.20 kgf/cm^2	1 N/mm^2	= 0.0648 $tonf/in^2$	= 10.2 kgf/cm^2
0.0069 N/mm^2 = 1 lbf/in^2		= 0.070kgf/cm^2	15.44 N/mm^2	= 1 $tonf/in^2$	= 157.5 kgf/cm^2
0.0981 N/mm^2 = 14.22lbf/in^2		= 1kgf/cm^2	0.0981 N/mm^2 = 0.00635 $tonf/in^2$		= 1 kgf/cm^2
1 N/m^2	= 0.0210 lbf/ft^2	= 0.102 kgf/m^2	1 N/mm^2	= 9.32 $tonf/ft^2$	= 10.2 kgf/cm^2
47.88 N/m^2 = 1 lbf/ft^2		= 4.882 kgf/m^2	0.1073 N/mm^2	= 1 $tonf/ft^2$	= 1.094 kgf/cm^2
9.807 N/m^2	= 0.2048 lbf/ft^2	= 1kgf/m^2	0.0981 N/mm^2	= 0.914 $tonf/ft^2$	= 1kgf/cm^2

Force per unit volume

1 N/m^3	= 0.00637 lbf/ft^2	= 0.102 kgf/m^3	1 kN/m^3	= 0.00284 $tonf/ft^3$	= 0.102 $tonf/m^3$
157.1 N/m^3	= 1lbf/ft^3	= 16.02 kgf/m^3	351.9 kN/m^3	= 1 $tonf/ft^3$	= 35.88 $tonf/m^3$
9.807 N/m^3	= 0.0624 lbf/ft^3	= 1 kgf/m^3	9.807 kN/m^3	= 0.0279 $tonf/ft^3$	= 1 $tonf/m^3$
1 kN/m^3	= 0.0037 lbf/in^3	= 0.102 $tonf/m^3$			
271 kN/m^3	= 1 lbf/in^3	= 27.68 $tonf/m^3$			
9.81 kN/m^3	= 0.0362 lbf/in^3 = 1 $tonf/m^3$				

Moment

1 N-m	= 8.851 lbf-in	= 0.7376 lbf-ft	= 0.1020 kgf-m
0.1130 N-m	= 1 lbf-in	= 0.08333 lbf-ft	= 0.01152 kgf-m
1.356 N-m	= 12 lbf-in	1 lbf-ft	= 0.1383 kgf-m
9.807 N-m	= 86.80 lbf-in	= 7.233 lbf-ft	= 1 kgf-m

Fluid capacity

1 litre	= 0.22 imperial gallons	= 0.2642 USA gallons
4.564 litres	= 1 imperial gallon	= 1.201 USA gallons
3.785 litres	= 0.8327 imperial gallons	= 1 USA gallon

EXPLANATORY NOTES

EXPLANATORY NOTES

1 GENERAL

This publication is a design guide to BS 5950-1:2000[1]. The symbols used are generally the same as those in that Standard.

1.1 Material, section dimensions and tolerances

The structural sections referred to in this design guide are of weldable structural steels conforming to the relevant British Standards given in the table below:

Structural steel products

Product	Technical delivery requirements		Dimensions	Tolerances
	Non alloy steels	Fine grain steels		
Universal beams, Universal columns, and Universal bearing piles	BS EN 10025[9]	BS EN 10113-1[10]	BS 4-1[11]	BS EN 10034[12]
Joists			BS 4-1[11]	BS 4-1[11] BS EN 10024[13]
Parallel Flange Channels			BS 4-1[11]	BS EN 10279[14]
Angles			BS EN 10056-1[15]	BS EN 10056-2[15]
Structural tees cut from universal beams and universal columns			BS 4-1[11]	—
Castellated universal beams Castellated universal columns			—	—
ASB (asymmetric beams) *Slimdek*® beam	Generally BS EN 10025[9], but also see note b)	See note a)		Generally BS EN 10034[12], but also see note b)
Hot Finished Hollow Sections	BS EN 10210-1[16]		BS EN 10210-2[16]	BS EN 10210-2[16]
Cold Formed Hollow Sections	BS EN 10219-1[17]		BS EN 10219-2[17]	BS EN 10219-2[17]
Notes: For full details of the British Standards, see the reference list at the end of the Explanatory Notes. a) See Corus publication[18]. b) For further details consult Corus.				

1.2 Dimensional units
The dimensions of sections are given in millimetres (mm).

1.3 Property units
Generally, the centimetre (cm) is used for the calculated properties but for surface areas and for the warping constant (H), the metre (m) and the decimetre (dm) respectively are used.

Note: 1 dm   =   0.1 m   =   100 mm

1 dm^6 = $1 \times 10^{-6} \text{ m}^6$ = $1 \times 10^{12} \text{ mm}^6$

1.4 Mass and force units
The units used are the kilogram (kg), the Newton (N) and the metre per second per second (m/s^2) so that $1 \text{ N} = 1 \text{ kg} \times 1 \text{ m/s}^2$. For convenience, a standard value of the acceleration due to gravity has been generally accepted as 9.80665 m/s^2. Thus, the force exerted by 1 kg under the action of gravity is 9.80665 N and the force exerted by 1 tonne (1000 kg) is 9.80665 kilonewtons (kN).

2 DIMENSIONS OF SECTIONS

2.1 Masses
The masses per metre have been calculated assuming that the density of steel is 7850 kg/m^3.

In all cases, including compound sections, the tabulated masses are for the steel section alone and no allowance has been made for connecting material or fittings.

2.2 Ratios for local buckling
The ratios of the flange outstand to thickness (b/T) and the web depth to thickness (d/t) are given for I, H and channel sections. The ratios of the outside diameter to thickness (D/t) are given for circular hollow sections. The ratios d/t and b/t are also given for square and rectangular hollow sections. All the ratios for local buckling have been calculated using the dimensional notation given in Figure 5 of BS 5950-1:2000 and are for use when element

and section class are being checked to the limits given in Tables 11 and 12 of BS 5950-1:2000.

2.3 Dimensions for detailing

The dimensions C, N and n have the meanings given in the figures at the heads of the tables and have been calculated according to the formulae below. The formulae for N and C make allowance for rolling tolerances, whereas the formulae for n make no such allowance.

2.3.1 Universal beams and universal columns

$N = (B - t)/2 + 10$ mm (rounded to the nearest 2 mm above)
$n = (D - d)/2$ (rounded to the nearest 2 mm above)
$C = t/2 + 2$ mm (rounded to the nearest mm)

2.3.2 Joists

$N = (B - t)/2 + 6$ mm (rounded to the nearest 2 mm above)
$n = (D - d)/2$ (rounded to the nearest 2 mm above)
$C = t/2 + 2$ mm (rounded to the nearest mm)

Note: Flanges of BS 4-1 joists have an 8° taper.

2.3.3 Parallel flange channels

$N = (B - t) + 6$ mm (rounded up to the nearest 2 mm above)
$n = (D - d)/2$ (to the next higher multiple of 2 mm)
$C = t + 2$ mm (rounded up to the nearest mm)

3 SECTION PROPERTIES

3.1 General

All section properties have been accurately calculated and rounded to three significant figures. They have been calculated from the metric dimensions given in the appropriate standards (see Section 1.2). For angles, BS EN 10056-1 assumes that the toe radius equals half the root radius.

3.2 Sections other than hollow sections

3.2.1 Second moment of area (I)

The second moment of area of the section, often referred to as moment of inertia, has been calculated taking into account all tapers, radii and fillets of the sections.

3.2.2 Radius of gyration (r)

The radius of gyration is a parameter used in buckling calculation and is derived as follows:

$$r = (I/A)^{1/2}$$

where:

A is the cross-sectional area.

For castellated sections, the radius of gyration given is calculated at the net section as required in design to BS 5950-1.

3.2.3 Elastic modulus (Z)

The elastic modulus is used to calculate the elastic moment capacity based on the design strength of the section or the stress at the extreme fibre of the section from a known moment. It is derived as follows:

$$Z = I/y$$

where:

y is the distance to the extreme fibre of the section from the elastic neutral axis.

For castellated sections, the elastic moduli given are those at the net section. The elastic moduli of the tee are calculated at the outer face of the flange and toe of the tee formed at the net section.

For parallel flange channels, the elastic modulus about the minor (y-y) axis is given at the toe of the section, i.e.:

$$y = B - c_y$$

where:

B is the width of the section

c_y is the distance from the back of the web to the centroidal axis.

For angles, the elastic moduli about both axes are given at the toes of the section, i.e:

$y_x = A - c_x$

$y_y = B - c_y$

where:

- A is the leg length perpendicular to x-x axis
- B is the leg length perpendicular to y-y axis
- c_x is the distance from the back of the angle to the centre of gravity, referred to as the x-x axis
- c_y is the distance from the back of the angle to the centre of gravity, referred to as the y-y axis.

3.2.4 Buckling parameter (u) and torsional index (x)

The buckling parameter and torsional index used in buckling calculations are derived as follows:

(i) For bi-symmetric flanged sections and flanged sections symmetrical about the minor axis only:

$u = [(4 S_x^2 \gamma) / (A^2 h^2)]^{1/4}$

$x = 0.566 h [A/J]^{1/2}$

(ii) For flanged sections symmetric about the major axis only:

$u = [(I_y S_x^2 \gamma) / (A^2 H)]^{1/4}$

$x = 1.132 [(A H) / (I_y J)]^{1/2}$

where:

- S_x is the plastic modulus about the major axis
- $\gamma = [1 - I_y/I_x]$
- I_x is the second moment of area about the major axis
- I_y is the second moment of area about the minor axis
- A is the cross-sectional area
- h is the distance between shear centres of flanges (for T sections, h is the distance between the shear centre of the flange and the toe of the web)
- H is the warping constant
- J is the torsion constant.

3.2.5 Warping constant (*H*) and torsion constant (*J*)

(a) I and H sections

The warping constant and torsion constant for I and H sections are calculated using the formulae given in the SCI publication P057 *Design of members subject to combined bending and torsion*[19].

(b) Tee sections

For Tee sections cut from UB and UC sections, the warping constant (*H*) and torsion constant (*J*) have been derived as given below.

$$H = \frac{1}{144} T^3 B^3 + \frac{1}{36}\left(d - \frac{T}{2}\right)^3 t^3$$

$$J = \frac{1}{3} BT^3 + \frac{1}{3}(d - T) t^3 + \alpha_1 D_1^4 - 0.21 T^4 - 0.105 t^4$$

where:

$$\alpha_1 = -0.042 + 0.2204 \frac{t}{T} + 0.1355 \frac{r}{T} - 0.0865 \frac{t\,r}{T^2} - 0.0725 \frac{t^2}{T^2}$$

$$D_1 = \frac{(T+r)^2 + (r + 0.25\,t)\,t}{2r + T}$$

Note: These formulae do not apply to tee sections cut from joists which have tapered flanges. For such sections, details are given in SCI Publication P057[19].

(c) Parallel flange channels

For parallel flange channels, the warping constant (*H*) and torsion constant (*J*) are calculated as follows:

$$H = \frac{h^2}{4}\left[I_y - A\left(c_y - \frac{t}{2}\right)^2 \left(\frac{h^2 A}{4 I_x} - 1\right)\right]$$

$$J = \frac{2}{3} BT^3 + \frac{1}{3}(D - 2T) t^3 + 2\alpha_3 D_3^4 - 0.42 T^4$$

where:

c_y is the distance from the back of the web to the Centroidal axis

$$\alpha_3 = -0.0908 + 0.2621 \frac{t}{T} + 0.1231 \frac{r}{T} - 0.0752 \frac{t\,r}{T^2} - 0.0945 \left(\frac{t}{T}\right)^2$$

$$D_3 = 2\left[(3r+t+T)-\sqrt{2(2r+t)(2r+T)}\right]$$

Note: The formula for the torsion constant (J) is applicable to parallel flange channels only and does not apply to tapered flange channels.

(d) Angles

For angles, the torsion constant (J) is calculated as follows:

$$J = \frac{1}{3}bt^3 + \frac{1}{3}(d-t)t^3 + \alpha_3 D_3^4 - 0.21t^4$$

where:

$$\alpha_3 = 0.0768 + 0.0479\frac{r}{t}$$

$$D_3 = 2\left[(3r+2t)-\sqrt{2(2r+t)^2}\right]$$

(e) ASB sections

For ASB (asymmetric beams) *Slimdek*® beam, the warping constant (H) and torsion constant (J) are as given in Corus brochure, *Structural sections* [18].

3.2.6 Plastic modulus (*S*)

The full plastic moduli about both principal axes are tabulated for all sections except angle sections. For angle sections, BS 5950-1:2000 requires design using the elastic modulus.

The reduced plastic moduli under axial load are tabulated for both principal axes for all sections except asymmetric beams and angle sections. For angle sections, BS 5950-1:2000 requires design using the elastic modulus.

When a section is loaded to full plasticity by a combination of bending and axial compression about the major axis, the plastic neutral axis shifts and may be located either in the web or the tension flange (or in the taper part of the flange for a joist) depending on the relative values of bending and axial compression. Formulae giving the reduced plastic modulus under combined loading have to be used, which use a parameter n as follows:

$$n = \frac{F}{A\,p_y} \quad \text{(This is shown in the member capacity tables as } F/P_z\text{)}$$

where:

F is the factored axial load

A is the cross-sectional area

p_y is the design strength of the steel.

For each section, there is a "change" value of n. Formulae for reduced plastic modulus and the "change" value are given below.

(a) Universal beams and universal columns

If the value of n calculated is less than the change value, the plastic neutral axis is in the web and the formula for lower values of "n" must be used. If n is greater than the change value, the plastic neutral axis lies in the tension flange and the formula for higher values of n must be used. The same principles apply when the sections are loaded axially and bent about the minor axis, lower and higher values of n indicating that the plastic neutral axis lies inside or outside the web respectively.

Major axis bending:

Reduced plastic modulus: Change value:

$S_{rx} = K_1 - K_2 n^2$ for $n < \dfrac{(D-2T)t}{A}$

$S_{rx} = K_3 (1-n)(K_4 + n)$ for $n \geq \dfrac{(D-2T)t}{A}$

where:

$K_1 = S_x$ $K_2 = \dfrac{A^2}{4t}$

$K_3 = \dfrac{A^2}{4B}$ $K_4 = \dfrac{2DB}{A} - 1$

Minor axis bending:

Reduced plastic modulus: Change value:

$S_{ry} = K_1 - K_2 n^2$ for $n < \dfrac{tD}{A}$

$S_{ry} = K_3 (1-n)(K_4 + n)$ for $n \geq \dfrac{tD}{A}$

where:

$K_1 = S_y$ $K_2 = \dfrac{A^2}{4D}$

$$K_3 = \frac{A^2}{8T} \qquad K_4 = \frac{4BT}{A} - 1$$

(b) Joists

Major axis bending:

If the value of n calculated is less than the lower change value (n_1), the plastic neutral axis is in the web and the formula for lower values of n must be used. If n is greater than the higher change value (n_2), the plastic neutral axis lies in the part of the tension flange that is not tapered and the formula for higher values of n must be used. If the value of n calculated lies between the lower change value (n_1) and the higher change value (n_2), the plastic neutral axis lies in the tapered part of the flange and then a linear interpolation between the two formulae is used to calculate the reduced plastic modulus.

Reduced plastic modulus   Change value

$$S_{rx} = S_{rx1} = K_1 - K_2 n^2 \quad \text{for } n \leq n_1 = \left\{ \frac{D}{A} - \frac{2}{A}\left(T + \frac{B-t}{4} tan(\theta)\right) \right\} t$$

$$S_{rx} = S_{rx2} = K_3 (1-n)(K_4 + n)$$

$$\text{for } n \geq n_2 = 1 - \frac{2B}{A}\left(T - \frac{B-t}{4} tan(\theta)\right)$$

$$S_{rx} = S_{rx1} + (S_{rx2} - S_{rx1})\left(\frac{n-n_1}{n_2 - n_1}\right) \quad \text{for } n_1 < n < n_2$$

where:

$$K_1 = S_x \qquad K_2 = \frac{A^2}{4t}$$

$$K_3 = \frac{A^2}{4B} \qquad K_4 = \frac{2DB}{A} - 1$$

$$\theta = 8° \text{ (flange taper)}$$

Minor axis bending:

The same principles apply when the sections are loaded axially and bent about the minor axis, lower and higher values of n indicating that the plastic neutral axis lies inside or outside the web respectively.

Reduced plastic modulus   Change value

$S_{ry} = K_1 - K_2 n^2$ for $n < \dfrac{tD}{A}$

$S_{ry} = K_3 (1 - n)(K_4 + n)$ for $n \geq \dfrac{tD}{A}$

where:

$K_1 = S_y$ $K_2 = \dfrac{A^2}{4D}$

$K_3 = 0.87 \dfrac{A^2}{8T}$ $K_4 = \dfrac{4BT}{A} - 1$

(c) Parallel flange channels

Major axis bending:

If the value of n calculated is less than the change value, the plastic neutral axis is in the web and the formula for lower values of n must be used. If n is greater than the change value, the plastic neutral axis lies in the flange and the formula for higher values of n must be used.

Reduced plastic modulus   Change value

$S_{rx} = K_1 - K_2 n^2$ for $n < \dfrac{(D - 2T)t}{A}$

$S_{rx} = K_3 (1 - n)(K_4 + n)$ for $n \geq \dfrac{(D - 2T)t}{A}$

where:

$K_1 = S_x$ $K_2 = \dfrac{A^2}{4t}$

$K_3 = \dfrac{A^2}{4B}$ $K_4 = \dfrac{2DB}{A} - 1$

Minor axis bending:

In calculating the reduced plastic modulus of a channel for axial force combined with bending about the minor axis, the axial force is considered as acting at the centroidal axis of the cross-section whereas it is considered to be resisted at the plastic neutral axis. The value of the reduced plastic modulus takes account of the resulting moment due to eccentricity relative to the net centroidal axis.

The reduced plastic modulus of a parallel flange channel bending about the minor axis depends on whether the stresses induced by the axial force and applied moment are the same or of opposite kind towards the back of the channel. Where the stresses are of the same kind, an initial increase in axial force may cause a small initial rise of the "reduced" plastic modulus, due to the eccentricity of the axial force.

For each section there is again a change value of n. For minor axis bending, the position of the plastic neutral axis when there is no axial load may be either in the web or the flanges. When the value of n is less than the change value, the formula for lower values of n must be used. If n is greater than the change value, the formula for higher values of n must be used.

The formulae concerned are complex and are therefore not quoted here.

3.2.7 Equivalent slenderness coefficient (ϕ_a) and monosymmetry index (ψ_a)

The equivalent slenderness coefficient (ϕ_a) is tabulated for both equal and unequal angles. Two values of the equivalent slenderness coefficient are given for each unequal angle. The larger value is based on the major axis elastic modulus (Z_u) to the toe of the short leg and the lower value is based on the major axis elastic modulus to the toe of the long leg.

The equivalent slenderness coefficient (ϕ_a) is calculated as follows:

$$\phi_a = \left[\frac{Z_u^2 \gamma_a}{AJ}\right]^{0.5}$$

Definitions of all the individual terms are given in BS 5950-1[1], Clause B.2.9.

The monosymmetry index (ψ_a) is only applicable for unequal angles and is calculated as follows:

$$\psi_a = \left[2v_0 - \frac{\int v_i\left(u_i^2 + v_i^2\right)dA}{I_u}\right]\frac{1}{t}$$

Definitions of all the individual terms are given in BS 5950-1[1], Clause B.2.9.

3.3 Hollow sections

Section properties are given for both hot-finished and cold-formed hollow sections. The ranges of hot-finished and cold-formed sections covered are different. The section ranges listed are in line with sections that are readily available from the major section manufacturers. For the same overall dimensions and wall thickness, the section properties for hot-finished and cold-formed sections are different because the corner radii are different.

3.3.1 Common properties

For comment on second moment of area, radius of gyration and elastic modulus, see Sections 3.2.1, 3.2.2 and 3.2.3.

For hot-finished square and rectangular hollow sections, the sectional properties have been calculated, using corner radii of $1.5t$ externally and $1.0t$ internally, as specified by BS EN 10210-2[16].

For cold-formed square and rectangular hollow sections, the sectional properties have been calculated, using the external corner radii of $2t$ if $t \leq 6$ mm, 2.5t if 6 mm $<$ t $\leq$ 10 mm and $3t$ if $t > 10$ mm as specified by BS EN 10219-2[17]. The internal corner radii used is $1.0t$ if $t \leq 6$ mm, $1.5t$ if 6 mm $<$ t $\leq$ 10 mm and $2t$ if $t > 10$ mm, as specified by BS EN 10219-2[17].

3.3.2 Torsion constant (J)

For circular hollow sections:

$$J = 2I$$

For square and rectangular hollow sections:

$$J = \frac{4A_h^2 t}{h} + \frac{t^3 h}{3}$$

where:

I is the second moment of area

t is the thickness of section

h is the mean perimeter $= 2[(B - t) + (D - t)] - 2R_c(4 - \pi)$

A_h is the area enclosed by mean perimeter $= (B - t)(D - t) - R_c^2(4 - \pi)$

B is the breadth of section

D is the depth of section

R_c is the average of internal and external corner radii.

3.3.3 Torsion modulus constant (C)

For circular hollow sections

$$C = 2Z$$

For square and rectangular hollow sections

$$C = J / \left(t + \frac{2A_h}{h}\right)$$

where:

Z is the elastic modulus and J, t, A_h and h are as defined in Section 3.3.2.

3.3.4 Plastic modulus of hollow sections (S)

The full plastic modulus (S) is given in the tables. When a member is subject to a combination of bending and axial load the plastic neutral axis shifts. Formulae giving the reduced plastic modulus under combined loading have to be used, which use the parameter n as defined below.

$$n = \frac{F}{A\,p_y} \quad \text{(This is shown in the member capacity tables as } F/P_z\text{)}$$

where:

F is the factored axial load

A is the cross-sectional area

p_y is the design strength of the steel.

For square and rectangular hollow sections there is a "change" value of n. Formulae for reduced plastic modulus and "change" value are given below.

(a) Circular hollow sections

$$S_r = S \cos\left(\frac{n\pi}{2}\right)$$

(b) Square and rectangular hollow sections

If the value of n calculated is less than the change value, the plastic neutral axis is in the webs and the formula for lower values of n must be used. If n is greater than the change value, the plastic neutral axis lies in the flange and the formula for higher values of n must be used.

Major axis bending:

Reduced plastic modulus   Change value

$$S_{rx} = S_x - \frac{A^2 n^2}{8t} \quad \text{for} \quad n \leq \frac{2t(D-2t)}{A}$$

$$S_{rx} = \frac{A^2(1-n)}{4(B-t)}\left[\frac{2D(B-t)}{A} + n - 1\right] \quad \text{for} \quad n > \frac{2t(D-2t)}{A}$$

Minor axis bending:

Reduced plastic modulus   Change value

$$S_{ry} = S_y - \frac{A^2 n^2}{8t} \quad \text{for} \quad n \leq \frac{2t(B-2t)}{A}$$

$$S_{ry} = \frac{A^2(1-n)}{4(D-t)}\left[\frac{2B(D-t)}{A} + n - 1\right] \quad \text{for} \quad n > \frac{2t(B-2t)}{A}$$

where:

S, S_x, S_y are the full plastic moduli about the relevant axes
A is the gross cross-sectional area
D, B and t are as defined in Section 3.3.2.

4　CAPACITY AND RESISTANCE TABLES

Code Ref.

4.1　General

The values displayed in the member capacity and resistance tables have been rounded to three significant figures.

Capacity and resistance tables are given for strength grades S275 and S355. The scope of the load capacity tables in this publication follows that originally adopted for the BCSA publication *Structural Steelwork Handbook*. Where further information is required, including load capacities of both hot finished and cold formed structural hollow sections, reference should be made to the comprehensive SCI/BCSA publication P202 *Steelwork Design Guide to BS 5950-1:2000. Volume 1, 6th Edition* [20] (the Blue Book). Further design guidance concerning structural hollow sections, including composite concrete filled sections is availabe from Corus Tubes Technical helpline 0500 123 133.

Code Re

4.2 Design strength

The member capacity and resistance tables have been based on the following values of design strength p_y.

3.1.

Steel Grade	Flange Thickness less than or equal to (mm)	Design strength p_y (N/mm^2)
S275	16	275
	40	265
	63	255
	80	245
	100	235
	150	225
S355	16	355
	40	345
	63	335
	80	325
	100	315
	150	295

Table

5 BENDING TABLES

5.1 Bending: UB sections, UC sections, joists and parallel flange channels

(i) Moment capacity, assuming shear load is low (<60% of the shear capacity):

$M_{cx} = p_y S_x$ but $\leq 1.2 p_y Z_x$ for class 1 (plastic) and class 2 (compact sections)   4.2.5.2

$M_{cx} = p_y S_{xeff}$ but $\leq 1.2 p_y Z_x$ for class 3 (semi-compact sections)

If the moment capacity is governed by $1.2 p_y Z_x$ the values in the tables have been printed in italic type because higher values may be used in some circumstances. This limit is only appropriate for   4.2.5.1

Code Ref.

simply supported beams and cantilevers. For other cases, a general limit of $1.5\, p_y\, Z_x$ should be applied.

(ii) Where the shear load is high (>60% of the shear capacity), the values should be checked and reduced, if necessary.

(iii) The moment capacity M_{cx} has been based on the section classification given but it should be noted that this classification applies to members subject to bending about the x-x axis only

Note: None of the universal beams, universal columns, joists or parallel flange channels in grade S275 or S355 are slender under bending only.

(iv) The ultimate UDL for restrained beams is given for a range of beam lengths. The values tabulated take account of high and low shear and also the beam shear capacity.

Ultimate UDL = Minimum $\{\, 2P_v\,,\, W_{mom}\, \}$

where:

P_v is the beam shear capacity (see Section 6.1(c))

W_{mom} is the UDL capacity of the beam $= 8\, M_{cx}\, /\, L$

M_{cx} is the moment capacity for the beam for high shear or low shear as appropriate

L is the length of the beam

If the resulting shear force is low ($\leq 0.6 P_v$) the moment capacity is calculated as in Section 5.1(a). Otherwise, the modulus of the section, used to calculate the moment capacity, is reduced by $\rho\, S_v$. S_v is the plastic modulus of the shear area and ρ is a factor to take account of the level of shear force (F_v) which is given by,

4.2.5.3

$\rho = [\, 2\, (F_v\, /\, P_v) - 1\,]^{0.5}$

4.2.5.3

Note: See table of equivalent factors on page 90 for use of tables with point loads

Code Ref

(v) Deflections for serviceability have been calculated assuming an unfactored imposed load (W_{imp}) of 40% of the ultimate load (W_{ult}).

In general,

$W_{ult} = 1.4\ W_{dead} + 1.6\ W_{imp}$

Therefore a value of $W_{imp} = 0.4\ W_{ult}$ is equivalent to,

$W_{imp} / W_{dead} = 1.556$

Values of Ultimate UDL to the right of the solid zigzag line produce serviceability deflections > span/360 and shaded values produce serviceability deflections > span/200.

6 WEB BEARING AND BUCKLING TABLES

6.1 UB sections, UC sections and joists: bearing, buckling and shear capacities for unstiffened webs

Values have been calculated as follows:

(a) Bearing

The bearing capacity P_w, of the unstiffened web is given by:

$P_w = (b_1 + n\ k)\ t\ p_{yw}$ (BS 5950-1 notation)   4.5.2.1

$\quad = b_1 C2 + C1$ (Capacity table notation)

where:

b_1 is the stiff bearing length

$n\ \ = 5$, for continuous over bearing   4.5.2.1

$\quad = 2 + 0.6\ b_e / k$, for end bearing

$\quad = 2$, for end bearing (b_e taken as zero in the tables)

Code Ref.

b_e is the distance to the nearer end of the member from the end of the stiff bearing   4.5.2.1

k = $(T + r)$ for rolled sections

t is the thickness of the web

T is the thickness of the flange

r is the root radius

p_{yw} is the design strength of the web.

(i) Bearing factor $C1$ is due to the beam alone

Generally, $C1 = n\,k\,t\,p_{yw}$.

$\therefore \quad C1 = 5\,(T + r)\,t\,p_{yw}$ for continuous over bearing

$\therefore \quad C1 = 2\,(T + r)\,t\,p_{yw}$ for end bearing.

(ii) Bearing factor $C2$ is equal to $t\,p_{yw}$ and must be multiplied by b_1 to give the stiff bearing contribution.

(b) Buckling

Generally the buckling resistance, P_x, of the unstiffened web is given by:

$$P_x = K \frac{25\,\varepsilon\,t}{\sqrt{(b_1 + n\,k)d}} P_w \qquad \text{Combining with (a), this can be re-written as:} \qquad 4.5.3.1$$

$$P_x = K \left[\left(\frac{25\sqrt{275}\,t}{\sqrt{d/t}} \right)^2 P_w \right]^{0.5} \qquad \text{(BS 5950-1 notation)}$$

$$= K\,(C4\,P_w)^{0.5} \qquad \text{(Capacity table notation)} \qquad 4.5.3.1$$

if $a_e \geq 0.7d$ then $K = 1$

if $a_e < 0.7d$ then $K = \dfrac{a_e}{1.4d} + 0.5$

125

where:

> d is the depth between fillets
> t is the web thickness
> P_w is the web bearing capacity from (a) above
> a_e is the distance from the centre of the load or reaction to the nearer end of the member.

(i) Buckling factor $C4$ is the same for end bearing and continuous over bearing

$$C4 = \left(\frac{25 \sqrt{275}\, t}{\sqrt{d/t}} \right)^2$$

(c) Shear

The shear capacity of the section is given by:

$$P_v = 0.6\, p_y\, t\, D \qquad 4.2.3$$

where:

D = Total depth of section.

Note: Since none of the rolled sections have $d/t > 70\varepsilon$, there is no need to check for shear buckling of the web.

6.2 Parallel flange channels: bearing, buckling and shear capacities for unstiffened webs

The nominal cross-section dimensions of parallel flange channels give a square heel, but the tolerances include a small heel radius (see Corus brochure *Structural sections*[18]), based on BS EN 10279:2000[14]. The heel radius can be between zero and 0.3 times the flange thickness.

In most cases the loads and reactions are applied directly to the web (by angle cleats, end plates or fin plates), so this is usually of no significance. However, if a force from a load or reaction is applied to the channel through the flange, the presence of a heel radius may produce eccentricity of this force relative to the web, so reduction

Code Ref.

factors, K_b for bearing and K_w for buckling to allow for this have been included in the tables.

The use of these reduction factors will provide acceptable design, but if necessary the eccentricity can be eliminated. A simple method is to use a continuously welded flange plate local to the stiff bearing, extending at least to the back of the channel. Depending on details, other methods may be appropriate.

Values have been calculated as follows:

(a) Bearing capacity, P_w

$P_w = K_b (b_1 + nk) t p_{yw}$ (BS 5950-1 notation)

$ = K_b (b_1 C2 + C1)$ (Capacity table notation)

where:

K_b is a reduction factor to allow for heel radius.

All other notation as in Section 9.1.

(b) Buckling resistance, P_x

$$P_x = K_w K \left[\left(\frac{25\sqrt{275t}}{\sqrt{d/t}} \right)^2 P_w \right]^{0.5}$$ (BS 5950-1 notation)

$ = K_w K (C4\ P_w)^{0.5}$ (Capacity table notation)

where:

K_w is a reduction factor to allow for heel radius.

All other notation as in Section 6.1.

(c) Shear capacity   4.2.3

$P_v = 0.6 p_y t D$

7 TENSION TABLES

7.1 Tension members: Single angles

The value of tension capacity P_t is given generally by equivalent tension area $\times p_y$:

    (ii)   For bolted sections, $P_t = (A_e - 0.5a_2)p_y$ 4.6.3.

    (iii)   For welded sections, $P_t = (A_g - 0.3a_2)p_y$

where:

    A_e is the effective net area of the angle   4.6.1/3.4.:

    A_g is the gross area of the angle

    p_y is the design strength of the angle

    a_2 is defined below.

Note: A block shear check (BS 5959-1:2000, clause 6.2.4 and Figure 22) is also required for tension members. However, 'block shear' capacities have not been tabulated, as there are too many variables in the possible bolt arrangements.

The effective net area of the section A_e is given by:

    For bolted sections, $A_e = a_{e1} + a_{e2}$ but $\leq 1.2(a_{n1} + a_{n2})$ 4.6.1/3.4.3

where:

a_{e1}	$= K_e\, a_{n1}$ but	$\leq a_1$	3.4.3
a_{e2}	$= K_e\, a_{n2}$ but	$\leq a_2$	
a_{n1}	$= a_1 -$ area of bolt holes in connected leg		
a_{n2}	$= a_2$		
A_g	$=$ Gross area of single angle		
K_e	$= 1.2$ for grade S275		3.4.3
	$= 1.1$ for grade S355		
a_1	$=$ Gross area of connected leg		
	$= A \times t$ if long leg connected		

$$= B \times t \quad \text{if short leg connected}$$

$$a_2 = A_g - a_1$$ 4.6.3.1

7.2 Compound tension members: Two angles

The values of tension capacity are based on the effective net area of the section, calculated as for single angles in Section 7.1.

The value of tension capacity P_t is given generally by equivalent tension area $\times p_y$:

(i) For bolted sections,

$P_t = 2(A_e - 0.25a_2)p_y$ For a gusset between the angles   4.6.3.2 (a)

$P_t = 2(A_e - 0.5a_2)p_y$ For a gusset on the back of the angles   4.6.3.2 (b)

(ii) For welded sections,

$P_t = 2(A_g - 0.15a_2)p_y$ For a gusset between the angles   4.6.3.2 (a)

$P_t = 2(A_g - 0.3a_2)p_y$ For a gusset on the back of the angles   4.6.3.2 (b)

Symbols as defined in Section 7.1.

Note: A block shear check (BS 5950-1:2000, Clause 6.2.4 and Figure 22) is also required for tension members. However, 'block shear' capacities have not been tabulated as there are too many variables in the possible bolt arrangements.

8 COMPRESSION TABLES

8.1 Compression members: UB and UC sections

(a) Compression Resistance, P_c   4.7.4 (a)

(i) For non-slender (Class 1, 2 or 3) cross-sections:

$$P_c = A_g p_c$$

where:

A_g is the gross cross-sectional area

p_c is the compressive strength.

The compressive strength p_c is obtained using the following values of Robertson constant, a.

4.7.
Annex. C.

Table 2

Type of Section	Robertson constant a	
	Axis of buckling	
	x-x	y-y
UB sections (flange thickness up to 40 mm)	2.0	3.5
UB sections (flange thickness over 40 mm)	3.5	5.5
UC sections (flange thickness up to 40 mm)	3.5	5.5
UC sections (flange thickness over 40 mm)	5.5	8.0

For I and H sections with a flange thickness between 40 mm and 50 mm, the value of p_c is taken as the average of the values obtained for thickness up to 40 mm and over 40 mm, as noted in Table 23.

Table 2. Note

(ii) For class 4 slender cross-sections:

$P_c = A_{eff} p_{cs}$ for class 4 slender cross-sections

4.7.4 (b)

where:

A_{eff} is the effective cross-sectional area

p_{cs} is the value of p_c for a reduced slenderness
$= \lambda (A_{eff} / A_g)^{0.5}$

The section classification of a section is partly dependent on the level of axial load applied. None of the universal columns, joists or parallel flange channels can be slender under axial compression only, but some universal beams and hollow sections can be slender. Sections that can be slender under axial compression are marked thus *.

UB sections can be slender when,

$d/t > 40\varepsilon$.

Table 11

where:

d is the depth of the web

t is the thickness of the web or wall

$\varepsilon = (275/p_y)^{0.5}$

p_y is the design strength.

If a cross-section can be slender under axial load, the tabulated compression resistance is only based on the slender cross-section equation (given above) if the value from this equation is greater than the axial load required to make the cross-section slender. Otherwise, the compression resistance of a potentially slender section is given as the smaller of the non-slender compression resistance and the axial load required to make the section slender. Tabulated values based on the equation for slender cross-sections are printed in italic type.

An example is given below:

686 × 254 × 170 UB S275

For this section, $d/t = 42.4 > 40\varepsilon$

Hence, the section can become slender if axial load is sufficiently high. The axial load at which the section becomes slender is 5410 kN.

This value is calculated by setting d/t for the section equal to the class 3 limit from BS 5950-1[1], Table 11 and then solving for the value of axial load (F_c).

For $L_E = 4$ m,

$P_{cx} = p_{cs} A_{eff} = 5660$ kN (slender cross-section)

Hence table shows *5660* kN in italic type because, $p_{cs} A_{eff} >$ value at which cross-section becomes slender, 5410 kN.

For $L_E = 12$ m,

$P_{cx} = p_c A = 5400$ kN (non-slender cross-section)

Hence the table shows 5400 kN in normal type because, $p_c A <$ value at which cross-section becomes slender, 5410 kN.

(b) **Compression resistances P_{cx} and P_{cy}**

The values of compression resistance P_{cx} and P_{cy} for buckling about the two principal axes are based on:

- The effective lengths (L_E) given at the head of the table. 4.7.

- The slenderness (λ), calculated as follows:

    For UB, UC, joist and hollow sections,

    $\lambda = L_E/r_x$ for x axis buckling   4.7.
    $\lambda = L_E/r_y$ for y axis buckling

8.2 Compression members: single angles

(a) **Compression resistance, $P_c = A_g\, p_c$**   4.7.

where:

A_g is the gross cross-section area

p_c is the compressive strength and has been obtained using a Robertson constant, a, of 5.5.   Table 2. Annex C.

In the case of a single bolt at each end, the compression resistance should be taken as 80% of that for an axially loaded member with the same slenderness. (Note: no values are given for this case).

Sections which are slender are marked * and their resistances have been calculated using a reduced design strength.   3.6.!

An angle cross-sections is slender if (using code notation):

d/t or $b/t > 15\varepsilon$ or $(d + b)/t > 24\varepsilon$ Table 1.

Or using the notation in these tables, the requirements become:

A/t or $B/t > 15\varepsilon$ or $(A + B)/t > 24\varepsilon$

In these circumstances, the design strength is reduced by the least of these factors:

$$\left(\frac{15\varepsilon}{A/t}\right)^2 : \left(\frac{15\varepsilon}{B/t}\right)^2 : \left(\frac{24\varepsilon}{(A+B)/t}\right)^2 \qquad 3.6.5$$

where:

$\varepsilon = (275/p_y)^{0.5}$

(b) The values of compression resistance are based on:

- The length (L) between intersections of centroidal axes or setting out line of the bolts given at the head of the tables.

- The slenderness (λ), calculated as follows:

(i) For two or more bolts in standard clearance holes in line along the angle at each end or an equivalent welded connection, the slenderness should be taken as the greater of: 4.7.10.2 (a)

Table 25

$0.85\, L_v/r_v$ but $\geq 0.7\, L_v/r_v + 15$;   and

$1.0\, L_a/r_a$ but $\geq 0.7\, L_a/r_a + 30$;   and

$0.85\, L_b/r_b$ but $\geq 0.7\, L_b/r_b + 30$.

(ii) For a single bolt at each end, the λ should be taken as the greater of: 4.7.10.2 (c)

Table 25

$1.0\, L_v/r_v$ but $\geq 0.7\, L_v/r_v + 15$;   and

$1.0\, L_a/r_a$ but $\geq 0.7\, L_a/r_a + 30$;   and

$1.0\, L_b/r_b$ but $\geq 0.7\, L_b/r_b + 30$.

where:

r_v is the minimum radius of gyration

r_a is the radius of gyration about the axis parallel to the connected leg

r_b is the radius of gyration about the axis perpendicular to the connected leg.

8.3 Compound compression members: two angles

The tables assume that the angles are interconnected back to back, as recommended in Clause 4.7.13 of the code.   4.7.1

(a) Compression resistance, $P_c = A_g\, p_c$

where:

A_g is the gross cross-section area of the two angles

p_c is the compressive strength and has been obtained using a Robertson constant of 5.5.   Table 2 Annex C.

Sections which are slender are marked * and their resistances have been calculated using a reduced design strength.   3.6.

An angle cross-sections is slender if (using code notation):

d/t or $b/t > 15\varepsilon$ or $(d + b)/t > 24\varepsilon$ Table 1

Or using the notation in these tables, the requirements become:

A/t or $B/t > 15\varepsilon$ or $(A + B)/t > 24\varepsilon$

In these circumstances, the design strength is reduced by the least of these factors:

$$\left(\frac{15\varepsilon}{A/t}\right)^2 \quad : \quad \left(\frac{15\varepsilon}{B/t}\right)^2 \quad : \quad \left(\frac{24\varepsilon}{(A+B)/t}\right)^2 \qquad 3.6.$$

where:

$\varepsilon = (275/p_y)^{0.5}$

(b) The values of compression resistance are based on:

- The length (L) between intersections of centroidal axes or setting out line of the bolts given at the head of the tables.

- The slenderness (λ), calculated as follows:

    (i) For double angles connected to one side of a gusset or member by two or more bolts in line along each   4.7.10.3 (a angle, or by an equivalent weld at each end, the slenderness λ should be taken as the greater of: Table 25

$1.0\ L_x/r_{xm}$ but $\geq 0.7\ L_x/r_{xm} + 30$; and

$[(0.85\ L_y/r_{ym})^2 + \lambda_c^2]^{0.5}$ but $\geq 1.4\ \lambda_c$

(ii) For double angles connected to one side of a gusset or member by one bolt in each angle, the slenderness λ should be taken as the greater of: 4.7.10.3 (b)
Table 25

$1.0\ L_x/r_{xm}$ but $\geq 0.7\ L_x/r_{xm} + 30$;   and

$[(1.0\ L_y/r_{ym})^2 + \lambda_c^2]^{0.5}$ but $\geq 1.4\ \lambda_c$

(iii) For double angles connected to both sides of a gusset or member by two or more bolts in line along each angle, the slenderness λ should be taken as the greater of: 4.7.10.3 (c)
Table 25

$0.85\ L_x/r_{xm}$ but $\geq 0.7\ L_x/r_x + 30$;   and

$[(1.0\ L_y/r_{ym})^2 + \lambda_c^2]^{0.5}$ but $\geq 1.4\ \lambda_c$

(iv) For double angles connected to both sides of a gusset or member by a single bolt in each angle, the slenderness λ should be taken as the greater of: 4.7.10.3 (e)
Table 25

$1.0\ L_x/r_{xm}$ but $\geq 0.7\ L_x/r_{xm} + 30$;   and

$[(1.0\ L_y/r_{ym})^2 + \lambda_c^2]^{0.5}$ but $\geq 1.4\ \lambda_c$

For double angles connected to both sides of a gusset or member by a single bolt in each angle, the compression resistance should be taken as 80% of that for an axially loaded member with the same slenderness. (Note: no values are given for this case).

where:

λ_c = L_e/r_v but not greater than 50

r_v is the minimum radius of gyration of a single angle

r_{xm} and r_{ym} are the radii of gyration of the double angles about the x and y axes.

L_c is the length L divided by the number of bays. There are a sufficient number of bays so that $\lambda_c \leq 50$. The number of bays is at least three and if there are more than three, the compression resistance is printed in bold type.

4.7.9 (c

9 AXIAL AND BENDING TABLES

9.1 Axial load and bending: UB and UC sections

Generally, members subject to axial compression and bending should be checked for cross-section capacity (Clause 4.8.3.2) and member buckling (Clause 4.8.3.3).

Columns in simple construction should be checked in accordance with Clause 4.7.7.

All the relevant parameters required to evaluate the interaction equations given in the above clauses have been presented in tabular form, as follows:

(a) Cross-section capacity check

4.8.3.

The tables are applicable to members subject to combined tension and bending and also to members subject to combined compression and bending. However, the values in the tables are conservative for tension, as the more onerous compression section classification limits have been used.

Values are given in the tables for:

(i) $P_z = A_g p_y$ (p_y is the design strength)

(ii) F/P_z limits

The compact and semi-compact limits are the maximum values of F/P_z up to which the section is either compact or semi-compact, respectively. The compact limit is given in bold type.

(iii) M_{cx} and M_{cy}

These are the moment capacities (with low shear load) about the major and minor axes respectively. They have been calculated as in Section 8.1 using S_x, S_{xeff}, and S_y, S_{yeff}, as appropriate.

Note: S_{xeff} and S_{yeff} can change with F/P_Z values.

When F/P_z exceeds the semi-compact limit, the section is slender, due to the web and the moment capacities tabulated are based on a reduced design strength and the gross section properties, instead of $p_y Z_{eff}$

The symbol $ indicates that the section would be overloaded due to axial load alone i.e. the section is slender and $F > A_{eff} p_y$.

(iv) M_{rx} and M_{ry}

These have been determined using the reduced plastic moduli given in the section property tables for the values of F/P_z at the head of the table. Values of M_{rx} and M_{ry} are not valid for semi-compact and slender sections hence, no values are shown when F/P_z exceeds the limit for a compact section (shown as " - " in the tables).

$M_{rx} = p_y S_{rx} \leq M_{cx}$		4.8.2.3
$M_{ry} = p_y S_{ry} \leq M_{cy}$		Annex I.2.1

(b)   Member buckling check   4.8.3.3

The symbol * denotes that the section is slender when fully stressed under axial compression only (due to the web becoming slender). None of the sections listed are slender due to the flanges being slender. Under combined axial compression and bending, the section would be compact or semi-compact up to the given F/P_z limits.

Values are given in the table for:

(i) $P_z = A_g p_y$

(ii) $p_y Z_x$

This is used in the simplified method for member buckling.   4.8.3.3.1

(iii) $p_y Z_y$

This is used for columns in simple construction and in the   4.7.7

simplified method for member buckling. 4.8.3.3.

(iv) F/P_z limit

The limits in normal and bold type are the maximum values up to which the section is either semi-compact or compact, respectively. The tabulated resistances are only valid up to the given F/P_z limit.

(v) P_{cx} and P_{cy}

These are the compression resistances for buckling about the major and minor axes respectively. The adjacent F/P_z limit, ensures that the section is not slender and have been calculated as in Section 6.1.

(vi) M_b is the buckling resistance moment, used in both the simplified and the more exact method. Values of M_b are given for two F/P_z limits the higher limit ensures the section is semi-compact and the lower limit (in bold) ensures the section is compact.

M_b = $p_b S_x$ ≤ M_{cx} for class 1 and class 2   4.3.6.4 (c

M_b = $p_b S_{xeff}$ ≤ M_{cx} for class 3

p_b has been obtained for particular values of $\lambda_{LT} = u\,v\,\lambda\,\beta_w^{0.5}$ 4.3.6.

where:

β_w = 1.0   for class 1 and class 2   4.3.6.9

     = S_{xeff} / S_x for class 3

λ is the slenderness = L_E/r_y 4.3.6.7

u is a Buckling parameter (as defined in the code)   4.3.6.8

v is a Slenderness factor (as defined in the code)   4.3.6.7

10 BOLTS AND WELDS

10.1 Bolt capacities

The types of bolts covered are:

- Grades 4.6, 8.8 and 10.9, as specified in BS 4190[21]: ISO metric black hexagon bolts, screws and nuts.

- Non-preloaded and preloaded HSFG bolts as specified in BS 4395[22]: High strength friction grip bolts and associated nuts and washers for structural engineering. Part 1: General grade and Part 2: Higher grade. Preloaded HSFG bolts should be tightened to minimum shank tension (P_o) as specified in BS 4604[23]

- Countersunk bolts as specified in BS 4933[24]: ISO metric black cup and countersunk bolts and screws with hexagon nuts.

Information on assemblies of matching bolts, nuts and washers is given in BS 5950-2[1]

(a) Non-preloaded bolts, Ordinary (Grades 4.6, 8.8 and 10.9) and HSFG (General and Higher Grade):

    (i)   The tensile stress area (A_t) is obtained from the above standards:

    (ii)   The tension capacity of the bolt is given by:

P_{nom}	$= 0.8 p_t A_t$	Nominal	6.3.4.2
P_t	$= p_t A_t$	Exact	6.3.4.3

where:

        p_t is the tension strength of the bolt.   Table 34

    (iii)   The shear capacity of the bolt is given by:

$$P_s = p_s A_s \quad\quad 6.3.2.1$$

where:

p_s is the shear strength of the bolt — Table 3

A_s is the shear area of the bolt.

In the tables, A_s has been taken as equal to A_t.

The shear capacity given in the tables must be reduced for large packings, large grip lengths, kidney shaped slots or long joints when applicable. — 6.3.2. 6.3.2. 6.3.2. 6.3.2.

(iv) The effective bearing capacity given is the lesser of the bearing capacity of the bolt given by:

$$P_{bb} = d\, t_p\, p_{bb}$$ 6.3.3.

and the bearing capacity of the connected ply given by:

$$P_{bs} = k_{bs}\, d\, t_p\, p_{bs}$$ 6.3.3.

assuming that the end distance is greater than or equal to twice the bolt diameter to meet the requirement that $P_{bs} \leq 0.5\, k_{bs}\, e\, t_p\, p_{bs}$

where:

d is the nominal diameter of the bolt

t_p is the thickness of the ply.

For countersunk bolts, t_p is taken as the ply thickness minus half the depth of countersinking. Depth of countersinking is taken as half the bolt diameter based on a 90° countersink. — 6.3.3.

p_{bb} is the bearing strength of the bolt — Table 31

p_{bs} is the bearing strength of the ply — Table 32

e is the end distance

k_{bs} is a coefficient to allow for hole type. — 6.3.3.3

Tables assume standard clearance holes, therefore k_{bs} is taken as 1.0. For oversize holes and short slots, $k_{bs} = 0.7$. For long slots and kidney shaped slots, $k_{bs} = 0.5$.

(b) Preloaded HSFG bolts (general grade and higher grade):

(i) The proof load of the bolt (P_o) is obtained from BS 4604[23]. The same proof load is used for countersunk bolts as for non-countersunk bolts. For this to be acceptable, the head dimensions must be as specified in BS 4933[24].

(ii) The tension capacity (P_t) of the bolt is taken as: 6.4.5

$1.1 \, P_o$ for non-slip in service

$0.9 \, P_o$ for non-slip under factored load.

(iii) The slip resistance of the bolt is given by: 6.4.2

$P_{SL} = 1.1 \, K_s \mu P_o$ for non-slip in service

$P_{SL} = 0.9 \, K_s \mu P_o$ for non-slip under factored load

where:

K_s is taken as 1.0 for fasteners in standard clearance holes   6.4.2

μ is the slip factor.   Table 35

(iv) The bearing resistance is only applicable for non-slip in service and is taken as:

$P_{bg} = 1.5 \, d \, t_p \, p_{bs}$ 6.4.4

assuming that the end distance is greater than or equal to three times the bolt diameter, to meet the requirement that $P_{bg} \leq 0.5 \, e \, t_p \, p_{bs}$.

where:

d is the nominal diameter of the bolt

t_p is the thickness of the ply

p_{bs}	is the bearing strength of the ply.	Table

(v) The shear capacity of the bolt is given by: 6.4.1

$$P_S = p_s A_s$$ 6.3.2

where:

p_s	is the shear strength of the bolt	Table
A_s	is the shear area of the bolt	

In the tables, A_s has been taken as equal to A_t.

10.2 Welds

Capacities of longitudinal and transverse fillet welds per unit length are tabulated. The weld capacities are given by,

Longitudinal shear capacity	$P_L = p_w a$	
Transverse capacity	$P_T = K p_w a$	6.8.7

where:

p_w	is the weld design strength	Table 3
a	is the throat thickness, taken as 0.7 x the leg length	
K	is the enhancement factor for transverse welds.	6.8.7

The plates are assumed to be at 90° and therefore $K = 1.25$. Electrode classifications of E35 and E42 are assumed for steel grade S275 and S355 respectively. Welding consumables are in accordance with BS EN 440[25], BS EN 449[26], BS EN 756[27], BS EN 758[28], or BS EN 1668[29] as appropriate.

Table 3

REFERENCES

1. BRITISH STANDARDS INSTITUTION
   BS 5950 Structural use of steelwork in building
   BS 5950-1:2000 Code of Practice for design. Rolled and welded sections
   BS 5950-2:2000 Specification for materials, fabrication and erection. Rolled and welded sections
   BS 5950-5:1988 Code of practice for design of cold formed thin gauge sections

2. THE STEEL CONSTRUCTION INSTITUTE and THE BRITISH CONSTRUCTIONAL STEEL WORK ASSOCIACTION
   Joints in Simple Construction Volume 1: Design Methods (P205)
   The Steel Construction Institute, 1993

3. SALTER, P.R., COUCHMAN, G.H. and ANDERSON, D.
   Wind moment design of low rise frames (P263)
   The Steel Construction Institute, 1999

4. COUCHMAN, G.H.
   Design of semi-continuous braced frames (P183)
   The Steel Construction Institute, 1997

5. BRITISH STANDARDS INSTITUTION
   BS 6399: Loadings for buildings
   BS 6399-1:1996 Code of practice for dead and imposed loads
   BS 6399-2:1997 Code of practice for wind loads
   BS 6399-3:1998 Code of practice for imposed roof loads

6. WYATT, T.A.
   Design guide on the vibration of floors (P076)
   The Steel Construction Institute, 1989

7. BRITISH STANDARDS INSTITUTION
   BS 5493:1977 Code of practice for protective coating of iron and steel structures against corrosion

8. CORUS
   Corrosion protection of structural steelwork
   Corus, 2000

9   BRITISH STANDARDS INSTITUTION
    BS EN 10025:1993 Hot rolled products of non-alloy structural steels. Technical delivery conditions (including amendment 1995)

10 BRITISH STANDARDS INSTITUTION
    BS EN 10113 Hot rolled products in weldable fine grain structural steels
    BS EN 10113-1:1993 General delivery conditions. (Replaces BS 4360:1990)

11 BRITISH STANDARDS INSTITUTION
    BS4 Structural steel sections
    BS4-1:1993 Specification for hot rolled sections
    (Including amendment 2001)

12 BRITISH STANDARDS INSTITUTION
    BS EN 10034:1993 Structural steel I and H sections. Tolerances on shape and dimensions
    (Replaces BS 4-1: 1980)

13 BRITISH STANDARDS INSTITUTION
    BS EN 10024:1995 Hot rolled taper flange I sections. Tolerances on shape and dimensions

14 BRITISH STANDARDS INSTITUTION
    BS EN 10279:2000 Hot rolled steel channels. Tolerances on shape, dimension and mass
    (Including amendment 1, 2000)

15 BRITISH STANDARDS INSTITUTION
    BS EN 10056 Specification for structural steel equal and unequal angles
    BS EN 10056-1:1999 Dimensions (Replaces BS 4848-4: 1972)
    BS EN 10056-2:1999 Tolerances on shape and dimensions (Replaces BS 4848-4: 1972)

16 BRITISH STANDARDS INSTITUTION
    BS EN 10210 Hot finished structural hollow sections of non-alloy and fine grain structural steels
    BS EN 10210-1:1994 Technical delivery requirements (Replaces BS 4360: 1990)
    BS EN 10210-2:1997 Tolerances, dimensions and sectional properties. (Replaces BS 4848-2: 1991)

17 BRITISH STANDARDS INSTITUTION
BS EN 10219 Cold formed welded structural sections of non-alloy and fine grain steels
BS EN 10219-1:1997 Technical delivery requirements
BS EN 10219-2:1997 Tolerances and sectional properties. (Replaces BS 6363: 1983)

18 Structural sections to BS 4: Part 1: 1963 and BS EN 10056: 1999
Corus Construction and Industrial Sections, 03/2001

19 NETHERCOT, D.A., SALTER, P.R. and MALIK, A.S.
Design of members subject to combined bending and torsion (P057)
The Steel Construction Institute, 1989

20 THE STEEL CONSTRUCTION INSTITUTE and THE BRITISH CONSTRUCTIONAL STEELWORK ASSOCIATION LTD.
Steelwork design guide to BS 5950-1:2000. Volume 1 Section Properties, Member Capacities
SCI/BCSA, 2001

21 BRITISH STANDARDS INSTITUTION
BS 4190:2001 ISO metric black hexagon bolts, screws and nuts. Specification

22 BRITISH STANDARDS INSTITUTION
BS 4395 Specification for high strength friction grip bolts and associated nuts and washers for structural engineering
BS 4395-1:1969 General grade (including amendments 1, amendments 2: 1997)
BS 4395-2:1969 Higher grade bolts and nuts and general grade washers (including amendment 1, amendment 2: 1976)

23 BRITISH STANDARDS INSTITUTION
BS 4604 Specification for the use of high strength friction grip bolts in structural steelwork. Metric series
BS 4604-1:1970 General grade (including amendment 1, amendment 2, and amendment 3: 1982)
BS 4604-2:1970 Higher grade (parallel shank) (including amendment 1, amendment 2: 1972)

24 BRITISH STANDARDS INSTITUTION
BS 4933:1973 Specification for ISO metric black cup and countersunk head bolts and screws with hexagon nuts

25 BRITISH STANDARDS INSTITUTION
BS EN 440:1995 Welding consumables. Wire electrodes and deposits for gas shielded metal arc welding of non alloy and fine grain steels. Classification

26 BRITISH STANDARDS INSTITUTION
BS EN 499:1995 Welding consumables. Covered electrodes for manual metal arc welding of non alloy and fine grain steels. Classification

27 BRITISH STANDARDS INSTITUTION
BS EN 756:1996 Welding consumables. Wire electrodes and wire-flux combinations for submerged arc welding of non alloy and fine grain steels. Classification

28 BRITISH STANDARDS INSTITUTION
BS EN 758:1997 Welding consumables. Tubular cored electrodes for metal arc welding with and without a gas shield of non-alloy and fine grain steels. Classification

29 BRITISH STANDARDS INSTITUTION
BS EN 1668:1997 Welding consumables. Rods, wires and deposits for tungsten inert gas welding of non alloy and fine grain steels. Classification

TABLES OF DIMENSIONS AND GROSS SECTION PROPERTIES

BS 5950-1: 2000
BS 4-1: 1993

UNIVERSAL BEAMS

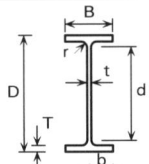
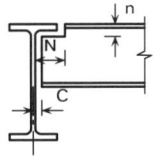

DIMENSIONS

Section Designation	Mass per Metre	Depth of Section	Width of Section	Thickness		Root Radius	Depth between Fillets	Ratios for Local Buckling		Dimensions for Detailing			Surface Area	
				Web	Flange			Flange	Web	End Clearance	Notch		Per Metre	Per Tonne
		D	B	t	T	r	d	b/T	d/t	C	N	n		
	kg/m	mm	mm	mm	mm	mm	mm			mm	mm	mm	m^2	m^2
1016x305x487 # +	486.6	1036.1	308.5	30.0	54.1	30.0	867.9	2.85	28.9	17	150	86	3.19	6.57
1016x305x437 # +	436.9	1025.9	305.4	26.9	49.0	30.0	867.9	3.12	32.3	16	150	80	3.17	7.25
1016x305x393 # +	392.7	1016.0	303.0	24.4	43.9	30.0	868.2	3.45	35.6	14	150	74	3.14	8.01
1016x305x349 # +	349.4	1008.1	302.0	21.1	40.0	30.0	868.1	3.77	41.1	13	150	70	3.13	8.96
1016x305x314 # +	314.3	1000.0	300.0	19.1	35.9	30.0	868.2	4.18	45.5	12	150	66	3.11	9.90
1016x305x272 # +	272.3	990.1	300.0	16.5	31.0	30.0	868.1	4.84	52.6	10	152	63	3.10	11.4
1016x305x249 # +	248.7	980.2	300.0	16.5	26.0	30.0	868.2	5.77	52.6	10	152	56	3.08	12.4
1016x305x222 # +	222.0	970.3	300.0	16.0	21.1	30.0	868.1	7.11	54.3	10	152	52	3.06	13.8
914 x 419 x 388 #	388.0	921.0	420.5	21.4	36.6	24.1	799.6	5.74	37.4	13	210	62	3.44	8.87
914 x 419 x 343 #	343.3	911.8	418.5	19.4	32.0	24.1	799.6	6.54	41.2	12	210	58	3.42	9.95
914 x 305 x 289 #	289.1	926.6	307.7	19.5	32.0	19.1	824.4	4.81	42.3	12	156	52	3.01	10.4
914 x 305 x 253 #	253.4	918.4	305.5	17.3	27.9	19.1	824.4	5.47	47.7	11	156	48	2.99	11.8
914 x 305 x 224 #	224.2	910.4	304.1	15.9	23.9	19.1	824.4	6.36	51.8	10	156	44	2.97	13.3
914 x 305 x 201 #	200.9	903.0	303.3	15.1	20.2	19.1	824.4	7.51	54.6	10	156	40	2.96	14.7
838 x 292 x 226 #	226.5	850.9	293.8	16.1	26.8	17.8	761.7	5.48	47.3	10	150	46	2.81	12.4
838 x 292 x 194 #	193.8	840.7	292.4	14.7	21.7	17.8	761.7	6.74	51.8	9	150	40	2.79	14.4
838 x 292 x 176 #	175.9	834.9	291.7	14.0	18.8	17.8	761.7	7.76	54.4	9	150	38	2.78	15.8
762 x 267 x 197	196.8	769.8	268.0	15.6	25.4	16.5	686.0	5.28	44.0	10	138	42	2.55	13.0
762 x 267 x 173	173.0	762.2	266.7	14.3	21.6	16.5	686.0	6.17	48.0	9	138	40	2.53	14.6
762 x 267 x 147	146.9	754.0	265.2	12.8	17.5	16.5	686.0	7.58	53.6	8	138	34	2.51	17.1
762 x 267 x 134	133.9	750.0	264.4	12.0	15.5	16.5	686.0	8.53	57.2	8	138	32	2.51	18.7
686 x 254 x 170	170.2	692.9	255.8	14.5	23.7	15.2	615.1	5.40	42.4	9	132	40	2.35	13.8
686 x 254 x 152	152.4	687.5	254.5	13.2	21.0	15.2	615.1	6.06	46.6	9	132	38	2.34	15.4
686 x 254 x 140	140.1	683.5	253.7	12.4	19.0	15.2	615.1	6.68	49.6	8	132	36	2.33	16.6
686 x 254 x 125	125.2	677.9	253.0	11.7	16.2	15.2	615.1	7.81	52.6	8	132	32	2.32	18.5
610 x 305 x 238	238.1	635.8	311.4	18.4	31.4	16.5	540.0	4.96	29.3	11	158	48	2.45	10.3
610 x 305 x 179	179.0	620.2	307.1	14.1	23.6	16.5	540.0	6.51	38.3	9	158	42	2.41	13.5
610 x 305 x 149	149.2	612.4	304.8	11.8	19.7	16.5	540.0	7.74	45.8	8	158	38	2.39	16.0
610 x 229 x 140	139.9	617.2	230.2	13.1	22.1	12.7	547.6	5.21	41.8	9	120	36	2.11	15.1
610 x 229 x 125	125.1	612.2	229.0	11.9	19.6	12.7	547.6	5.84	46.0	8	120	34	2.09	16.7
610 x 229 x 113	113.0	607.6	228.2	11.1	17.3	12.7	547.6	6.60	49.3	8	120	30	2.08	18.4
610 x 229 x 101	101.2	602.6	227.6	10.5	14.8	12.7	547.6	7.69	52.2	7	120	28	2.07	20.5
533 x 210 x 122	122.0	544.5	211.9	12.7	21.3	12.7	476.5	4.97	37.5	8	110	34	1.89	15.5
533 x 210 x 109	109.0	539.5	210.8	11.6	18.8	12.7	476.5	5.61	41.1	8	110	32	1.88	17.2
533 x 210 x 101	101.0	536.7	210.0	10.8	17.4	12.7	476.5	6.03	44.1	7	110	32	1.87	18.5
533 x 210 x 92	92.1	533.1	209.3	10.1	15.6	12.7	476.5	6.71	47.2	7	110	30	1.86	20.2
533 x 210 x 82	82.2	528.3	208.8	9.6	13.2	12.7	476.5	7.91	49.6	7	110	26	1.85	22.5

+ Section is not given in BS 4-1: 1993.
# Check availability.
FOR EXPLANATION OF TABLES SEE NOTE 2

BS 5950-1: 2000
BS 4-1: 1993

UNIVERSAL BEAMS

PROPERTIES

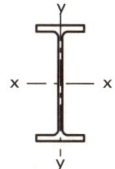

Section Designation	Second Moment of Area		Radius of Gyration		Elastic Modulus		Plastic Modulus		Buckling Parameter	Torsional Index	Warping Constant	Torsional Constant	Area of Section
	Axis x-x	Axis y-y	Axis x-x	Axis y-y	Axis x-x	Axis y-y	Axis x-x	Axis y-y	u	x	H	J	A
	cm^4	cm^4	cm	cm	cm^3	cm^3	cm^3	cm^3			dm^6	cm^4	cm^2
1016x305x487 # +	1020000	26700	40.6	6.57	19700	1730	23200	2800	0.867	21.1	64.4	4300	620
1016x305x437 # +	910000	23500	40.4	6.49	17700	1540	20800	2470	0.868	23.1	55.9	3190	557
1016x305x393 # +	808000	20500	40.2	6.40	15900	1350	18500	2170	0.868	25.5	48.4	2330	500
1016x305x349 # +	723000	18500	40.3	6.44	14400	1220	16600	1940	0.872	27.9	43.3	1720	445
1016x305x314 # +	644000	16200	40.1	6.37	12900	1080	14900	1710	0.872	30.7	37.7	1260	400
1016x305x272 # +	554000	14000	40.0	6.35	11200	934	12800	1470	0.872	35.0	32.2	835	347
1016x305x249 # +	481000	11800	39.0	6.09	9820	784	11400	1250	0.861	39.9	26.8	582	317
1016x305x222 # +	408000	9550	38.0	5.81	8410	636	9810	1020	0.849	45.8	21.5	390	283
914 x 419 x 388 #	720000	45400	38.2	9.59	15600	2160	17700	3340	0.885	26.7	88.9	1730	494
914 x 419 x 343 #	626000	39200	37.8	9.46	13700	1870	15500	2890	0.883	30.1	75.8	1190	437
914 x 305 x 289 #	504000	15600	37.0	6.51	10900	1010	12600	1600	0.867	31.9	31.2	926	368
914 x 305 x 253 #	436000	13300	36.8	6.42	9500	871	10900	1370	0.865	36.2	26.4	626	323
914 x 305 x 224 #	376000	11200	36.3	6.27	8270	739	9540	1160	0.861	41.3	22.1	422	286
914 x 305 x 201 #	325000	9420	35.7	6.07	7200	621	8350	982	0.853	46.9	18.4	291	256
838 x 292 x 226 #	340000	11400	34.3	6.27	7990	773	9160	1210	0.869	35.0	19.3	514	289
838 x 292 x 194 #	279000	9070	33.6	6.06	6640	620	7640	974	0.862	41.6	15.2	306	247
838 x 292 x 176 #	246000	7800	33.1	5.90	5890	535	6810	842	0.856	46.5	13.0	221	224
762 x 267 x 197	240000	8180	30.9	5.71	6230	610	7170	959	0.868	33.2	11.3	404	251
762 x 267 x 173	205000	6850	30.5	5.58	5390	514	6200	807	0.865	38.1	9.39	267	220
762 x 267 x 147	169000	5460	30.0	5.40	4470	411	5160	647	0.858	45.2	7.40	159	187
762 x 267 x 134	151000	4790	29.7	5.30	4020	362	4640	570	0.853	49.8	6.46	119	171
686 x 254 x 170	170000	6630	28.0	5.53	4920	518	5630	811	0.872	31.8	7.42	308	217
686 x 254 x 152	150000	5780	27.8	5.46	4370	455	5000	710	0.871	35.4	6.42	220	194
686 x 254 x 140	136000	5180	27.6	5.39	3990	409	4560	638	0.869	38.6	5.72	169	178
686 x 254 x 125	118000	4380	27.2	5.24	3480	346	3990	542	0.863	43.8	4.80	116	159
610 x 305 x 238	210000	15800	26.3	7.23	6590	1020	7490	1570	0.887	21.3	14.5	785	303
610 x 305 x 179	153000	11400	25.9	7.07	4940	743	5550	1140	0.886	27.7	10.2	340	228
610 x 305 x 149	126000	9310	25.7	7.00	4110	611	4590	937	0.886	32.7	8.17	200	190
610 x 229 x 140	112000	4510	25.0	5.03	3620	391	4140	611	0.875	30.6	3.99	216	178
610 x 229 x 125	98600	3930	24.9	4.97	3220	343	3680	535	0.874	34.1	3.45	154	159
610 x 229 x 113	87300	3430	24.6	4.88	2870	301	3280	469	0.870	38.1	2.99	111	144
610 x 229 x 101	75800	2920	24.2	4.75	2520	256	2880	400	0.863	43.1	2.52	77.0	129
533 x 210 x 122	76000	3390	22.1	4.67	2790	320	3200	500	0.878	27.6	2.32	178	155
533 x 210 x 109	66800	2940	21.9	4.60	2480	279	2830	436	0.874	31.0	1.99	126	139
533 x 210 x 101	61500	2690	21.9	4.57	2290	256	2610	399	0.873	33.2	1.81	101	129
533 x 210 x 92	55200	2390	21.7	4.51	2070	228	2360	356	0.873	36.4	1.60	75.7	117
533 x 210 x 82	47500	2010	21.3	4.38	1800	192	2060	300	0.863	41.6	1.33	51.5	105

+ Section is not given in BS 4-1: 1993.
\# Check availability.
FOR EXPLANATION OF TABLES SEE NOTE 3

BS 5950-1: 2000
BS 4-1: 1993

UNIVERSAL BEAMS

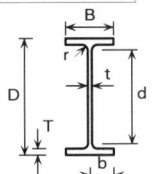
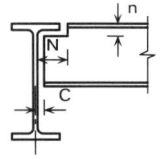

DIMENSIONS

Section Designation	Mass per Metre	Depth of Section	Width of Section	Thickness		Root Radius	Depth between Fillets	Ratios for Local Buckling		Dimensions for Detailing			Surface Area	
				Web	Flange			Flange	Web	End Clearance	Notch		Per Metre	Per Tonne
		D	B	t	T	r	d	b/T	d/t	C	N	n		
	kg/m	mm	mm	mm	mm	mm	mm			mm	mm	mm	m²	m²
457 x 191 x 98	98.3	467.2	192.8	11.4	19.6	10.2	407.6	4.92	35.8	8	102	30	1.67	16.9
457 x 191 x 89	89.3	463.4	191.9	10.5	17.7	10.2	407.6	5.42	38.8	7	102	28	1.66	18.5
457 x 191 x 82	82.0	460.0	191.3	9.9	16.0	10.2	407.6	5.98	41.2	7	102	28	1.65	20.1
457 x 191 x 74	74.3	457.0	190.4	9.0	14.5	10.2	407.6	6.57	45.3	7	102	26	1.64	22.1
457 x 191 x 67	67.1	453.4	189.9	8.5	12.7	10.2	407.6	7.48	48.0	6	102	24	1.63	24.3
457 x 152 x 82	82.1	465.8	155.3	10.5	18.9	10.2	407.6	4.11	38.8	7	84	30	1.51	18.4
457 x 152 x 74	74.2	462.0	154.4	9.6	17.0	10.2	407.6	4.54	42.5	7	84	28	1.50	20.3
457 x 152 x 67	67.2	458.0	153.8	9.0	15.0	10.2	407.6	5.13	45.3	7	84	26	1.50	22.3
457 x 152 x 60	59.8	454.6	152.9	8.1	13.3	10.2	407.6	5.75	50.3	6	84	24	1.49	24.9
457 x 152 x 52	52.3	449.8	152.4	7.6	10.9	10.2	407.6	6.99	53.6	6	84	22	1.48	28.2
406 x 178 x 74	74.2	412.8	179.5	9.5	16.0	10.2	360.4	5.61	37.9	7	96	28	1.51	20.3
406 x 178 x 67	67.1	409.4	178.8	8.8	14.3	10.2	360.4	6.25	41.0	6	96	26	1.50	22.3
406 x 178 x 60	60.1	406.4	177.9	7.9	12.8	10.2	360.4	6.95	45.6	6	96	24	1.49	24.8
406 x 178 x 54	54.1	402.6	177.7	7.7	10.9	10.2	360.4	8.15	46.8	6	96	22	1.48	27.4
406 x 140 x 46	46.0	403.2	142.2	6.8	11.2	10.2	360.4	6.35	53.0	5	78	22	1.34	29.2
406 x 140 x 39	39.0	398.0	141.8	6.4	8.6	10.2	360.4	8.24	56.3	5	78	20	1.33	34.2
356 x 171 x 67	67.1	363.4	173.2	9.1	15.7	10.2	311.6	5.52	34.2	7	94	26	1.38	20.6
356 x 171 x 57	57.0	358.0	172.2	8.1	13.0	10.2	311.6	6.62	38.5	6	94	24	1.37	24.1
356 x 171 x 51	51.0	355.0	171.5	7.4	11.5	10.2	311.6	7.46	42.1	6	94	22	1.36	26.7
356 x 171 x 45	45.0	351.4	171.1	7.0	9.7	10.2	311.6	8.82	44.5	6	94	20	1.36	30.1
356 x 127 x 39	39.1	353.4	126.0	6.6	10.7	10.2	311.6	5.89	47.2	5	70	22	1.18	30.2
356 x 127 x 33	33.1	349.0	125.4	6.0	8.5	10.2	311.6	7.38	51.9	5	70	20	1.17	35.4
305 x 165 x 54	54.0	310.4	166.9	7.9	13.7	8.9	265.2	6.09	33.6	6	90	24	1.26	23.3
305 x 165 x 46	46.1	306.6	165.7	6.7	11.8	8.9	265.2	7.02	39.6	5	90	22	1.25	27.1
305 x 165 x 40	40.3	303.4	165.0	6.0	10.2	8.9	265.2	8.09	44.2	5	90	20	1.24	30.8
305 x 127 x 48	48.1	311.0	125.3	9.0	14.0	8.9	265.2	4.47	29.5	7	70	24	1.09	22.7
305 x 127 x 42	41.9	307.2	124.3	8.0	12.1	8.9	265.2	5.14	33.1	6	70	22	1.08	25.8
305 x 127 x 37	37.0	304.4	123.4	7.1	10.7	8.9	265.2	5.77	37.4	6	70	20	1.07	29.0
305 x 102 x 33	32.8	312.7	102.4	6.6	10.8	7.6	275.9	4.74	41.8	5	58	20	1.01	30.8
305 x 102 x 28	28.2	308.7	101.8	6.0	8.8	7.6	275.9	5.78	46.0	5	58	18	1.00	35.4
305 x 102 x 25	24.8	305.1	101.6	5.8	7.0	7.6	275.9	7.26	47.6	5	58	16	0.992	40.0
254 x 146 x 43	43.0	259.6	147.3	7.2	12.7	7.6	219.0	5.80	30.4	6	82	22	1.08	25.1
254 x 146 x 37	37.0	256.0	146.4	6.3	10.9	7.6	219.0	6.72	34.8	5	82	20	1.07	29.0
254 x 146 x 31	31.1	251.4	146.1	6.0	8.6	7.6	219.0	8.49	36.5	5	82	18	1.06	34.2
254 x 102 x 28	28.3	260.4	102.2	6.3	10.0	7.6	225.2	5.11	35.7	5	58	18	0.904	31.9
254 x 102 x 25	25.2	257.2	101.9	6.0	8.4	7.6	225.2	6.07	37.5	5	58	16	0.897	35.6
254 x 102 x 22	22.0	254.0	101.6	5.7	6.8	7.6	225.2	7.47	39.5	5	58	16	0.890	40.5
203 x 133 x 30	30.0	206.8	133.9	6.4	9.6	7.6	172.4	6.97	26.9	5	74	18	0.923	30.8
203 x 133 x 25	25.1	203.2	133.2	5.7	7.8	7.6	172.4	8.54	30.2	5	74	16	0.915	36.4
203 x 102 x 23	23.1	203.2	101.8	5.4	9.3	7.6	169.4	5.47	31.4	5	60	18	0.790	34.2
178 x 102 x 19	19.0	177.8	101.2	4.8	7.9	7.6	146.8	6.41	30.6	4	60	16	0.738	38.8
152 x 89 x 16	16.0	152.4	88.7	4.5	7.7	7.6	121.8	5.76	27.1	4	54	16	0.638	39.8
127 x 76 x 13	13.0	127.0	76.0	4.0	7.6	7.6	96.6	5.00	24.1	4	46	16	0.537	41.3

FOR EXPLANATION OF TABLES SEE NOTE 2

BS 5950-1: 2000
BS 4-1: 1993

UNIVERSAL BEAMS

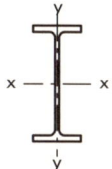

PROPERTIES

Section Designation	Second Moment of Area		Radius of Gyration		Elastic Modulus		Plastic Modulus		Buckling Parameter	Torsional Index	Warping Constant	Torsional Constant	Area of Section
	Axis x-x	Axis y-y	Axis x-x	Axis y-y	Axis x-x	Axis y-y	Axis x-x	Axis y-y	u	x	H	J	A
	cm^4	cm^4	cm	cm	cm^3	cm^3	cm^3	cm^3			dm^6	cm^4	cm^2
457 x 191 x 98	45700	2350	19.1	4.33	1960	243	2230	379	0.882	25.7	1.18	121	125
457 x 191 x 89	41000	2090	19.0	4.29	1770	218	2010	338	0.879	28.3	1.04	90.7	114
457 x 191 x 82	37100	1870	18.8	4.23	1610	196	1830	304	0.879	30.8	0.922	69.2	104
457 x 191 x 74	33300	1670	18.8	4.20	1460	176	1650	272	0.877	33.8	0.818	51.8	94.6
457 x 191 x 67	29400	1450	18.5	4.12	1300	153	1470	237	0.872	37.9	0.705	37.1	85.5
457 x 152 x 82	36600	1190	18.7	3.37	1570	153	1810	240	0.871	27.4	0.591	89.2	105
457 x 152 x 74	32700	1050	18.6	3.33	1410	136	1630	213	0.873	30.2	0.518	65.9	94.5
457 x 152 x 67	28900	913	18.4	3.27	1260	119	1450	187	0.868	33.6	0.448	47.7	85.6
457 x 152 x 60	25500	795	18.3	3.23	1120	104	1290	163	0.868	37.5	0.387	33.8	76.2
457 x 152 x 52	21400	645	17.9	3.11	950	84.6	1100	133	0.859	43.8	0.311	21.4	66.6
406 x 178 x 74	27300	1550	17.0	4.04	1320	172	1500	267	0.882	27.6	0.608	62.8	94.5
406 x 178 x 67	24300	1370	16.9	3.99	1190	153	1350	237	0.880	30.5	0.533	46.1	85.5
406 x 178 x 60	21600	1200	16.8	3.97	1060	135	1200	209	0.880	33.8	0.466	33.3	76.5
406 x 178 x 54	18700	1020	16.5	3.85	930	115	1060	178	0.871	38.3	0.392	23.1	69.0
406 x 140 x 46	15700	538	16.4	3.03	778	75.7	888	118	0.872	39.0	0.207	19.0	58.6
406 x 140 x 39	12500	410	15.9	2.87	629	57.8	724	90.8	0.858	47.5	0.155	10.7	49.7
356 x 171 x 67	19500	1360	15.1	3.99	1070	157	1210	243	0.886	24.4	0.412	55.7	85.5
356 x 171 x 57	16000	1110	14.9	3.91	896	129	1010	199	0.882	28.8	0.330	33.4	72.6
356 x 171 x 51	14100	968	14.8	3.86	796	113	896	174	0.881	32.1	0.286	23.8	64.9
356 x 171 x 45	12100	811	14.5	3.76	687	94.8	775	147	0.874	36.8	0.237	15.8	57.3
356 x 127 x 39	10200	358	14.3	2.68	576	56.8	659	89.1	0.871	35.2	0.105	15.1	49.8
356 x 127 x 33	8250	280	14.0	2.58	473	44.7	543	70.3	0.863	42.2	0.081	8.79	42.1
305 x 165 x 54	11700	1060	13.0	3.93	754	127	846	196	0.889	23.6	0.234	34.8	68.8
305 x 165 x 46	9900	896	13.0	3.90	646	108	720	166	0.891	27.1	0.195	22.2	58.7
305 x 165 x 40	8500	764	12.9	3.86	560	92.6	623	142	0.889	31.0	0.164	14.7	51.3
305 x 127 x 48	9580	461	12.5	2.74	616	73.6	711	116	0.874	23.3	0.102	31.8	61.2
305 x 127 x 42	8200	389	12.4	2.70	534	62.6	614	98.4	0.872	26.6	0.0846	21.1	53.4
305 x 127 x 37	7170	336	12.3	2.67	471	54.5	539	85.4	0.871	29.7	0.0725	14.8	47.2
305 x 102 x 33	6500	194	12.5	2.15	416	37.9	481	60.0	0.867	31.6	0.0442	12.2	41.8
305 x 102 x 28	5370	155	12.2	2.08	348	30.5	403	48.5	0.859	37.4	0.0349	7.40	35.9
305 x 102 x 25	4460	123	11.9	1.97	292	24.2	342	38.8	0.846	43.4	0.0273	4.77	31.6
254 x 146 x 43	6540	677	10.9	3.52	504	92.0	566	141	0.890	21.2	0.103	23.9	54.8
254 x 146 x 37	5540	571	10.8	3.48	433	78.0	483	119	0.889	24.4	0.0857	15.3	47.2
254 x 146 x 31	4410	448	10.5	3.36	351	61.3	393	94.1	0.879	29.6	0.0660	8.55	39.7
254 x 102 x 28	4010	179	10.5	2.22	308	34.9	353	54.8	0.874	27.5	0.0280	9.57	36.1
254 x 102 x 25	3420	149	10.3	2.15	266	29.2	306	46.0	0.867	31.4	0.0230	6.42	32.0
254 x 102 x 22	2840	119	10.1	2.06	224	23.5	259	37.3	0.856	36.3	0.0182	4.15	28.0
203 x 133 x 30	2900	385	8.71	3.17	280	57.5	314	88.2	0.881	21.5	0.0374	10.3	38.2
203 x 133 x 25	2340	308	8.56	3.10	230	46.2	258	70.9	0.877	25.6	0.0294	5.96	32.0
203 x 102 x 23	2110	164	8.46	2.36	207	32.2	234	49.8	0.888	22.5	0.0154	7.02	29.4
178 x 102 x 19	1360	137	7.48	2.37	153	27.0	171	41.6	0.886	22.6	0.00987	4.41	24.3
152 x 89 x 16	834	89.8	6.41	2.10	109	20.2	123	31.2	0.889	19.6	0.00470	3.56	20.3
127 x 76 x 13	473	55.7	5.35	1.84	74.6	14.7	84.2	22.6	0.896	16.3	0.00199	2.85	16.5

FOR EXPLANATION OF TABLES SEE NOTE 3

BS 5950-1: 2000
BS 4-1: 1993

UNIVERSAL BEAMS

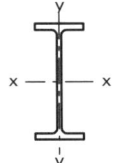

REDUCED PLASTIC MODULUS UNDER AXIAL LOAD

Section Designation	Plastic Modulus Axis x-x	Major Axis Reduced Modulus				Plastic Modulus Axis y-y	Minor Axis Reduced Modulus			
		Lower Values of n		Change Formula At n =	Higher Values of n		Lower Values of n		Change Formula At n =	Higher Values of n
	cm³	K1	K2	K3	K4	cm³	K1	K2	K3	K4
1016x305x487 # +	23200	23200	32000	0.449	3120	9.31	2800	2800	928	0.501	8880	0.0768
1016x305x437 # +	20800	20800	28800	0.448	2540	10.2	2470	2470	756	0.495	7910	0.0747
1016x305x393 # +	18500	18500	25600	0.453	2060	11.3	2170	2170	615	0.496	7120	0.0641
1016x305x349 # +	16600	16600	23500	0.440	1640	12.7	1940	1940	491	0.478	6190	0.0858
1016x305x314 # +	14900	14900	20900	0.443	1330	14.0	1710	1710	400	0.478	5570	0.0770
1016x305x272 # +	12800	12800	18200	0.441	1000	16.1	1470	1470	304	0.471	4860	0.0720
1016x305x249 # +	11400	11400	15200	0.483	837	17.6	1250	1250	256	0.510	4830	-0.0158
1016x305x222 # +	9810	9810	12500	0.525	667	19.6	1020	1020	206	0.549	4740	-0.105
914 x 419 x 388 #	17700	17700	28500	0.367	1450	14.7	3340	3340	663	0.399	8340	0.246
914 x 419 x 343 #	15500	15500	24600	0.376	1140	16.5	2890	2890	524	0.405	7470	0.225
914 x 305 x 289 #	12600	12600	17400	0.457	1100	14.5	1600	1600	366	0.491	5300	0.0695
914 x 305 x 253 #	10900	10900	15100	0.462	853	16.4	1370	1370	284	0.492	4670	0.0561
914 x 305 x 224 #	9540	9540	12800	0.480	671	18.4	1160	1160	224	0.507	4270	0.0178
914 x 305 x 201 #	8350	8350	10800	0.509	540	20.4	982	982	181	0.533	4050	-0.0424
838 x 292 x 226 #	9160	9160	12900	0.445	709	16.3	1210	1210	245	0.475	3880	0.0915
838 x 292 x 194 #	7640	7640	10400	0.475	521	18.9	974	974	181	0.501	3510	0.0283
838 x 292 x 176 #	6810	6810	8960	0.498	430	20.7	842	842	150	0.522	3340	-0.0208
762 x 267 x 197	7170	7170	10100	0.448	586	15.5	959	959	204	0.479	3090	0.0863
762 x 267 x 173	6200	6200	8490	0.467	455	17.4	807	807	159	0.495	2810	0.0457
762 x 267 x 147	5160	5160	6840	0.492	330	20.4	647	647	116	0.516	2500	-0.00820
762 x 267 x 134	4640	4640	6060	0.506	275	22.2	570	570	97.0	0.528	2350	-0.0390
686 x 254 x 170	5630	5630	8110	0.432	459	15.3	811	811	170	0.463	2480	0.118
686 x 254 x 152	5000	5000	7130	0.439	370	17.0	710	710	137	0.468	2240	0.102
686 x 254 x 140	4560	4560	6420	0.449	314	18.4	638	638	116	0.475	2100	0.0806
686 x 254 x 125	3990	3990	5440	0.474	251	20.5	542	542	93.8	0.497	1960	0.0280
610 x 305 x 238	7490	7490	12500	0.348	739	12.1	1570	1570	362	0.386	3660	0.289
610 x 305 x 179	5550	5550	9220	0.354	423	15.7	1140	1140	210	0.383	2760	0.271
610 x 305 x 149	4590	4590	7650	0.356	296	18.6	937	937	147	0.380	2290	0.264
610 x 229 x 140	4140	4140	6060	0.421	345	14.9	611	611	129	0.454	1800	0.142
610 x 229 x 125	3680	3680	5330	0.428	277	16.6	535	535	104	0.457	1620	0.127
610 x 229 x 113	3280	3280	4670	0.442	227	18.3	469	469	85.3	0.469	1500	0.0970
610 x 229 x 101	2880	2880	3960	0.467	183	20.3	400	400	69.0	0.491	1400	0.0451
533 x 210 x 122	3200	3200	4750	0.410	285	13.8	500	500	111	0.445	1420	0.162
533 x 210 x 109	2830	2830	4160	0.419	229	15.4	436	436	89.4	0.451	1280	0.142
533 x 210 x 101	2610	2610	3830	0.421	197	16.5	399	399	77.1	0.450	1190	0.136
533 x 210 x 92	2360	2360	3410	0.432	165	18.0	356	356	64.6	0.459	1100	0.113
533 x 210 x 82	2060	2060	2850	0.460	131	20.1	300	300	51.9	0.484	1040	0.0531

+ Section is not given in BS 4-1: 1993.
# Check availability.
$n = F/(A\,p_y)$, where F is the Factored axial load, A is the gross cross sectional area and p_y is the design strength of the section.
For lower values of n, the reduced plastic modulus, $S_r = K1 - K2.n^2$, for both major and minor axis bending.
For higher values of n, the reduced plastic modulus, $S_r = K3(1-n)(K4+n)$, for both major and minor axis bending.
FOR EXPLANATION OF TABLES SEE NOTE 3

BS 5950-1: 2000
BS 4-1: 1993

UNIVERSAL BEAMS

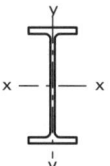

REDUCED PLASTIC MODULUS UNDER AXIAL LOAD

Section Designation	Plastic Modulus Axis x-x	Major Axis Reduced Modulus				Plastic Modulus Axis y-y	Minor Axis Reduced Modulus			
		Lower Values of n		Change Formula At n =	Higher Values of n		Lower Values of n		Change Formula At n =	Higher Values of n
	cm³	K1	K2		K3	K4	cm³	K1	K2	K3	K4
457 x 191 x 98	2230	2230	3440	0.390	203	13.4	379	379	84.0	0.425	1000	0.207
457 x 191 x 89	2010	2010	3080	0.395	169	14.6	338	338	69.8	0.428	914	0.194
457 x 191 x 82	1830	1830	2760	0.406	143	15.8	304	304	59.3	0.436	853	0.172
457 x 191 x 74	1650	1650	2490	0.407	118	17.4	272	272	49.0	0.435	772	0.167
457 x 191 x 67	1470	1470	2150	0.425	96.3	19.1	237	237	40.3	0.451	720	0.128
457 x 152 x 82	1810	1810	2600	0.430	176	12.8	240	240	58.7	0.468	723	0.123
457 x 152 x 74	1630	1630	2320	0.435	145	14.1	213	213	48.3	0.469	656	0.111
457 x 152 x 67	1450	1450	2030	0.450	119	15.5	187	187	40.0	0.482	610	0.0786
457 x 152 x 60	1290	1290	1790	0.455	95.0	17.2	163	163	32.0	0.483	546	0.0670
457 x 152 x 52	1100	1100	1460	0.488	72.9	19.6	133	133	24.7	0.513	509	-0.00290
406 x 178 x 74	1500	1500	2350	0.383	124	14.7	267	267	54.1	0.415	698	0.216
406 x 178 x 67	1350	1350	2080	0.392	102	16.1	237	237	44.7	0.421	640	0.196
406 x 178 x 60	1200	1200	1850	0.393	82.3	17.9	209	209	36.0	0.420	572	0.190
406 x 178 x 54	1060	1060	1540	0.425	66.9	19.8	178	178	29.5	0.450	545	0.124
406 x 140 x 46	888	888	1260	0.442	60.5	18.6	118	118	21.3	0.468	384	0.0864
406 x 140 x 39	724	724	963	0.491	43.5	21.7	90.8	90.8	15.5	0.513	358	-0.0176
356 x 171 x 67	1210	1210	2010	0.353	105	13.7	243	243	50.3	0.387	582	0.272
356 x 171 x 57	1010	1010	1630	0.371	76.4	16.0	199	199	36.8	0.400	506	0.234
356 x 171 x 51	896	896	1420	0.379	61.4	17.8	174	174	29.7	0.405	458	0.215
356 x 171 x 45	775	775	1170	0.405	48.0	20.0	147	147	23.4	0.429	423	0.158
356 x 127 x 39	659	659	938	0.440	49.1	16.9	89.1	89.1	17.5	0.469	289	0.0836
356 x 127 x 33	543	543	740	0.473	35.4	19.8	70.3	70.3	12.7	0.497	261	0.0120
305 x 165 x 54	846	846	1500	0.325	70.8	14.1	196	196	38.1	0.357	431	0.330
305 x 165 x 46	720	720	1290	0.323	52.1	16.3	166	166	28.1	0.350	366	0.331
305 x 165 x 40	623	623	1100	0.331	39.9	18.5	142	142	21.7	0.355	323	0.312
305 x 127 x 48	711	711	1040	0.416	74.8	11.7	116	116	30.1	0.457	335	0.146
305 x 127 x 42	614	614	891	0.424	57.4	13.3	98.4	98.4	23.2	0.460	295	0.127
305 x 127 x 37	539	539	784	0.426	45.1	14.9	85.4	85.4	18.3	0.458	260	0.119
305 x 102 x 33	481	481	663	0.459	42.7	14.3	60.0	60.0	14.0	0.493	202	0.0576
305 x 102 x 28	403	403	536	0.487	31.6	16.5	48.5	48.5	10.4	0.516	183	-0.00120
305 x 102 x 25	342	342	431	0.534	24.6	18.6	38.8	38.8	8.18	0.560	178	-0.100
254 x 146 x 43	566	566	1040	0.308	50.9	13.0	141	141	28.9	0.341	295	0.366
254 x 146 x 37	483	483	883	0.313	38.0	14.9	119	119	21.7	0.342	255	0.353
254 x 146 x 31	393	393	656	0.354	26.9	17.5	94.1	94.1	15.7	0.380	229	0.267
254 x 102 x 28	353	353	517	0.420	31.8	13.8	54.8	54.8	12.5	0.455	163	0.133
254 x 102 x 25	306	306	428	0.450	25.2	15.4	46.0	46.0	9.98	0.482	153	0.0686
254 x 102 x 22	259	259	344	0.489	19.3	17.4	37.3	37.3	7.73	0.517	144	-0.0136
203 x 133 x 30	314	314	570	0.314	27.3	13.5	88.2	88.2	17.7	0.346	190	0.346
203 x 133 x 25	258	258	448	0.334	19.2	15.9	70.9	70.9	12.6	0.362	164	0.300
203 x 102 x 23	234	234	400	0.339	21.2	13.1	49.8	49.8	10.6	0.373	116	0.288
178 x 102 x 19	171	171	307	0.321	14.5	13.8	41.6	41.6	8.28	0.352	93.1	0.318
152 x 89 x 16	123	123	229	0.303	11.6	12.3	31.2	31.2	6.77	0.337	67.0	0.344
127 x 76 x 13	84.2	84.2	171	0.271	8.98	10.7	22.6	22.6	5.37	0.308	44.9	0.399

$n = F/(A\,p_y)$, where F is the Factored axial load, A is the gross cross sectional area and p_y is the design strength of the section.
For lower values of n, the reduced plastic modulus, $S_r = K1 - K2.n^2$, for both major and minor axis bending.
For higher values of n, the reduced plastic modulus, $S_r = K3(1-n)(K4+n)$, for both major and minor axis bending.
FOR EXPLANATION OF TABLES SEE NOTE 3

BS 5950-1: 2000
BS 4-1: 1993

UNIVERSAL COLUMNS

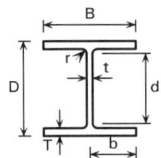
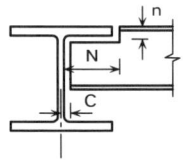

DIMENSIONS

Section Designation	Mass per Metre	Depth of Section	Width of Section	Thickness		Root Radius	Depth between Fillets	Ratios for Local Buckling		Dimensions for Detailing			Surface Area	
				Web	Flange			Flange	Web	End Clearance	Notch		Per Metre	Per Tonne
		D	B	t	T	r	d	b/T	d/t	C	N	n		
	kg/m	mm	mm	mm	mm	mm	mm			mm	mm	mm	m²	m²
356 x 406 x 634 #	633.9	474.6	424.0	47.6	77.0	15.2	290.2	2.75	6.10	26	200	94	2.52	3.98
356 x 406 x 551 #	551.0	455.6	418.5	42.1	67.5	15.2	290.2	3.10	6.89	23	200	84	2.47	4.49
356 x 406 x 467 #	467.0	436.6	412.2	35.8	58.0	15.2	290.2	3.55	8.11	20	200	74	2.42	5.19
356 x 406 x 393 #	393.0	419.0	407.0	30.6	49.2	15.2	290.2	4.14	9.48	17	200	66	2.38	6.05
356 x 406 x 340 #	339.9	406.4	403.0	26.6	42.9	15.2	290.2	4.70	10.9	15	200	60	2.35	6.90
356 x 406 x 287 #	287.1	393.6	399.0	22.6	36.5	15.2	290.2	5.47	12.8	13	200	52	2.31	8.05
356 x 406 x 235 #	235.1	381.0	394.8	18.4	30.2	15.2	290.2	6.54	15.8	11	200	46	2.28	9.69
356 x 368 x 202 #	201.9	374.6	374.7	16.5	27.0	15.2	290.2	6.94	17.6	10	190	44	2.19	10.8
356 x 368 x 177 #	177.0	368.2	372.6	14.4	23.8	15.2	290.2	7.83	20.2	9	190	40	2.17	12.3
356 x 368 x 153 #	152.9	362.0	370.5	12.3	20.7	15.2	290.2	8.95	23.6	8	190	36	2.16	14.1
356 x 368 x 129 #	129.0	355.6	368.6	10.4	17.5	15.2	290.2	10.50	27.9	7	190	34	2.14	16.6
305 x 305 x 283	282.9	365.3	322.2	26.8	44.1	15.2	246.7	3.65	9.21	15	158	60	1.94	6.86
305 x 305 x 240	240.0	352.5	318.4	23.0	37.7	15.2	246.7	4.22	10.7	14	158	54	1.91	7.94
305 x 305 x 198	198.1	339.9	314.5	19.1	31.4	15.2	246.7	5.01	12.9	12	158	48	1.87	9.46
305 x 305 x 158	158.1	327.1	311.2	15.8	25.0	15.2	246.7	6.22	15.6	10	158	42	1.84	11.6
305 x 305 x 137	136.9	320.5	309.2	13.8	21.7	15.2	246.7	7.12	17.9	9	158	38	1.82	13.3
305 x 305 x 118	117.9	314.5	307.4	12.0	18.7	15.2	246.7	8.22	20.6	8	158	34	1.81	15.3
305 x 305 x 97	96.9	307.9	305.3	9.9	15.4	15.2	246.7	9.91	24.9	7	158	32	1.79	18.5
254 x 254 x 167	167.1	289.1	265.2	19.2	31.7	12.7	200.3	4.18	10.4	12	134	46	1.58	9.45
254 x 254 x 132	132.0	276.3	261.3	15.3	25.3	12.7	200.3	5.16	13.1	10	134	38	1.55	11.7
254 x 254 x 107	107.1	266.7	258.8	12.8	20.5	12.7	200.3	6.31	15.6	8	134	30	1.52	14.2
254 x 254 x 89	88.9	260.3	256.3	10.3	17.3	12.7	200.3	7.41	19.4	7	134	30	1.50	16.9
254 x 254 x 73	73.1	254.1	254.6	8.6	14.2	12.7	200.3	8.96	23.3	6	134	28	1.49	20.4
203 x 203 x 86	86.1	222.2	209.1	12.7	20.5	10.2	160.8	5.10	12.7	8	110	32	1.24	14.4
203 x 203 x 71	71.0	215.8	206.4	10.0	17.3	10.2	160.8	5.97	16.1	7	110	28	1.22	17.2
203 x 203 x 60	60.0	209.6	205.8	9.4	14.2	10.2	160.8	7.25	17.1	7	110	26	1.21	20.1
203 x 203 x 52	52.0	206.2	204.3	7.9	12.5	10.2	160.8	8.17	20.4	6	110	24	1.20	23.0
203 x 203 x 46	46.1	203.2	203.6	7.2	11.0	10.2	160.8	9.25	22.3	6	110	22	1.19	25.8
152 x 152 x 37	37.0	161.8	154.4	8.0	11.5	7.6	123.6	6.71	15.5	6	84	20	0.912	24.7
152 x 152 x 30	30.0	157.6	152.9	6.5	9.4	7.6	123.6	8.13	19.0	5	84	18	0.901	30.0
152 x 152 x 23	23.0	152.4	152.2	5.8	6.8	7.6	123.6	11.2	21.3	5	84	16	0.889	38.7

Check availability.
FOR EXPLANATION OF TABLES SEE NOTE 2

BS 5950-1: 2000
BS 4-1: 1993

UNIVERSAL COLUMNS

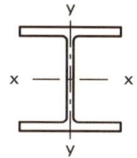

PROPERTIES

Section Designation	Second Moment of Area		Radius of Gyration		Elastic Modulus		Plastic Modulus		Buckling Parameter	Torsional Index	Warping Constant	Torsional Constant	Area of Section
	Axis x-x	Axis y-y	Axis x-x	Axis y-y	Axis x-x	Axis y-y	Axis x-x	Axis y-y	u	x	H	J	A
	cm^4	cm^4	cm	cm	cm^3	cm^3	cm^3	cm^3			dm^6	cm^4	cm^2
356 x 406 x 634 #	275000	98100	18.4	11.0	11600	4630	14200	7110	0.843	5.46	38.8	13700	808
356 x 406 x 551 #	227000	82700	18.0	10.9	9960	3950	12100	6060	0.841	6.05	31.1	9240	702
356 x 406 x 467 #	183000	67800	17.5	10.7	8380	3290	10000	5030	0.839	6.86	24.3	5810	595
356 x 406 x 393 #	147000	55400	17.1	10.5	7000	2720	8220	4150	0.837	7.87	18.9	3550	501
356 x 406 x 340 #	123000	46900	16.8	10.4	6030	2330	7000	3540	0.836	8.84	15.5	2340	433
356 x 406 x 287 #	99900	38700	16.5	10.3	5080	1940	5810	2950	0.834	10.2	12.3	1440	366
356 x 406 x 235 #	79100	31000	16.3	10.2	4150	1570	4690	2380	0.835	12.0	9.54	812	299
356 x 368 x 202 #	66300	23700	16.1	9.60	3540	1260	3970	1920	0.844	13.4	7.16	558	257
356 x 368 x 177 #	57100	20500	15.9	9.54	3100	1100	3460	1670	0.843	15.0	6.09	381	226
356 x 368 x 153 #	48600	17600	15.8	9.49	2680	948	2970	1440	0.844	17.0	5.11	251	195
356 x 368 x 129 #	40300	14600	15.6	9.43	2260	793	2480	1200	0.845	19.8	4.18	153	164
305 x 305 x 283	78900	24600	14.8	8.27	4320	1530	5110	2340	0.856	7.65	6.35	2030	360
305 x 305 x 240	64200	20300	14.5	8.15	3640	1280	4250	1950	0.854	8.74	5.03	1270	306
305 x 305 x 198	50900	16300	14.2	8.04	3000	1040	3440	1580	0.854	10.2	3.88	734	252
305 x 305 x 158	38800	12600	13.9	7.90	2370	808	2680	1230	0.852	12.5	2.87	378	201
305 x 305 x 137	32800	10700	13.7	7.83	2050	692	2300	1050	0.852	14.1	2.39	249	174
305 x 305 x 118	27700	9060	13.6	7.77	1760	589	1960	895	0.851	16.2	1.98	161	150
305 x 305 x 97	22300	7310	13.4	7.69	1450	479	1590	726	0.852	19.2	1.56	91.2	123
254 x 254 x 167	30000	9870	11.9	6.81	2080	744	2420	1140	0.851	8.50	1.63	626	213
254 x 254 x 132	22500	7530	11.6	6.69	1630	576	1870	878	0.850	10.3	1.19	319	168
254 x 254 x 107	17500	5930	11.3	6.59	1310	458	1480	697	0.849	12.4	0.898	172	136
254 x 254 x 89	14300	4860	11.2	6.55	1100	379	1220	575	0.851	14.5	0.717	102	113
254 x 254 x 73	11400	3910	11.1	6.48	898	307	992	465	0.849	17.3	0.562	57.6	93.1
203 x 203 x 86	9450	3130	9.28	5.34	850	299	977	456	0.849	10.2	0.318	137	110
203 x 203 x 71	7620	2540	9.18	5.30	706	246	799	374	0.853	11.9	0.250	80.2	90.4
203 x 203 x 60	6130	2070	8.96	5.20	584	201	656	305	0.846	14.1	0.197	47.2	76.4
203 x 203 x 52	5260	1780	8.91	5.18	510	174	567	264	0.848	15.8	0.167	31.8	66.3
203 x 203 x 46	4570	1550	8.82	5.13	450	152	497	231	0.846	17.7	0.143	22.2	58.7
152 x 152 x 37	2210	706	6.85	3.87	273	91.5	309	140	0.849	13.3	0.0399	19.2	47.1
152 x 152 x 30	1750	560	6.76	3.83	222	73.3	248	112	0.849	16.0	0.0308	10.5	38.3
152 x 152 x 23	1250	400	6.54	3.70	164	52.6	182	80.2	0.840	20.7	0.0212	4.63	29.2

Check availability.
FOR EXPLANATION OF TABLES SEE NOTE 3

BS 5950-1: 2000
BS 4-1: 1993

UNIVERSAL COLUMNS

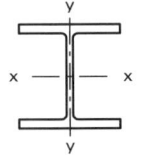

REDUCED PLASTIC MODULUS UNDER AXIAL LOAD

Section Designation	Plastic Modulus Axis x-x	Major Axis Reduced Modulus				Plastic Modulus Axis y-y	Minor Axis Reduced Modulus				
		Lower Values Of n		Change Formula At n =	Higher Values Of n			Lower Values Of n	Change Formula At n =	Higher Values Of n	
	cm³	K1	K2		K3	K4	cm³	K1	K2	K3	K4
356 x 406 x 634 #	14200	14200	34300	0.189	3850	3.98	7110	7110	3440	0.280	10600	0.617
356 x 406 x 551 #	12100	12100	29300	0.192	2940	4.43	6060	6060	2700	0.273	9120	0.610
356 x 406 x 467 #	10000	10000	24700	0.193	2150	5.05	5030	5030	2030	0.263	7630	0.607
356 x 406 x 393 #	8220	8220	20500	0.196	1540	5.81	4150	4150	1500	0.256	6370	0.600
356 x 406 x 340 #	7000	7000	17600	0.197	1160	6.56	3540	3540	1150	0.250	5460	0.597
356 x 406 x 287 #	5810	5810	14800	0.198	838	7.59	2950	2950	849	0.243	4580	0.593
356 x 406 x 235 #	4690	4690	12200	0.197	568	9.05	2380	2380	588	0.234	3710	0.593
356 x 368 x 202 #	3970	3970	10000	0.206	441	9.91	1920	1920	442	0.240	3060	0.573
356 x 368 x 177 #	3460	3460	8830	0.205	341	11.2	1670	1670	345	0.235	2670	0.573
356 x 368 x 153 #	2970	2970	7710	0.202	256	12.8	1440	1440	262	0.229	2290	0.575
356 x 368 x 129 #	2480	2480	6490	0.203	183	15.0	1200	1200	190	0.225	1930	0.570
305 x 305 x 283	5110	5110	12100	0.206	1010	5.53	2340	2340	889	0.272	3680	0.577
305 x 305 x 240	4250	4250	10200	0.208	734	6.34	1950	1950	663	0.265	3100	0.570
305 x 305 x 198	3440	3440	8340	0.210	506	7.47	1580	1580	469	0.257	2540	0.565
305 x 305 x 158	2680	2680	6420	0.217	326	9.11	1230	1230	310	0.257	2030	0.545
305 x 305 x 137	2300	2300	5510	0.219	246	10.4	1050	1050	237	0.254	1750	0.539
305 x 305 x 118	1960	1960	4700	0.221	183	11.9	895	895	179	0.251	1510	0.531
305 x 305 x 97	1590	1590	3850	0.222	125	14.2	726	726	124	0.247	1240	0.523
254 x 254 x 167	2420	2420	5900	0.204	427	6.20	1140	1140	392	0.261	1790	0.580
254 x 254 x 132	1870	1870	4620	0.205	270	7.59	878	878	256	0.251	1400	0.573
254 x 254 x 107	1480	1480	3630	0.212	180	9.12	697	697	174	0.250	1130	0.556
254 x 254 x 89	1220	1220	3120	0.205	125	10.8	575	575	123	0.237	928	0.565
254 x 254 x 73	992	992	2520	0.208	85.1	12.9	465	465	85.3	0.235	763	0.553
203 x 203 x 86	977	977	2370	0.210	144	7.48	456	456	135	0.257	733	0.564
203 x 203 x 71	799	799	2040	0.200	99.0	8.85	374	374	94.7	0.239	591	0.579
203 x 203 x 60	656	656	1550	0.223	70.9	10.3	305	305	69.6	0.258	513	0.531
203 x 203 x 52	567	567	1390	0.216	53.8	11.7	264	264	53.3	0.246	439	0.541
203 x 203 x 46	497	497	1200	0.222	42.4	13.1	231	231	42.4	0.249	392	0.525
152 x 152 x 37	309	309	694	0.236	35.9	9.61	140	140	34.3	0.275	241	0.508
152 x 152 x 30	248	248	563	0.236	23.9	11.6	112	112	23.2	0.268	195	0.503
152 x 152 x 23	182	182	369	0.275	14.0	14.9	80.2	80.2	14.0	0.302	157	0.416

\# Check availability.
$n = F/(A\, p_y)$, where F is the Factored axial load, A is the gross cross sectional area and p_y is the design strength of the section.
For lower values of n, the reduced plastic modulus, $S_r = K1 - K2.n^2$, for both major and minor axis bending.
For higher values of n, the reduced plastic modulus, $S_r = K3(1-n)(K4+n)$, for both major and minor axis bending.
FOR EXPLANATION OF TABLES SEE NOTE 3

[BLANK PAGE]

BS 5950-1: 2000
BS 4-1: 1993

JOISTS

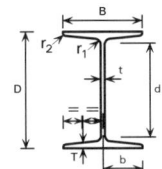

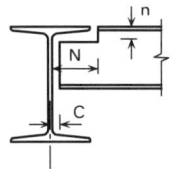

DIMENSIONS

Section Designation	Mass per Metre	Depth of Section	Width of Section	Thickness		Radii		Depth between Fillets	Ratios for Local Buckling		Dimensions for Detailing			Surface Area	
				Web	Flange	Root	Toe		Flange	Web	End Clearance	Notch		Per Metre	Per Tonne
		D	B	t	T	r_1	r_2	d	b/T	d/t	C	N	n		
	kg/m	mm	mm	mm	mm	mm	mm	mm			mm	mm	mm	m²	m²
254 x 203 x 82 #	82.0	254.0	203.2	10.2	19.9	19.6	9.7	166.6	5.11	16.3	7	104	44	1.21	14.8
254 x 114 x 37 ‡	37.2	254.0	114.3	7.6	12.8	12.4	6.1	199.3	4.46	26.2	6	60	28	0.899	24.2
203 x 152 x 52 #	52.3	203.2	152.4	8.9	16.5	15.5	7.6	133.2	4.62	15.0	6	78	36	0.932	17.8
152 x 127 x 37 #	37.3	152.4	127.0	10.4	13.2	13.5	6.6	94.3	4.81	9.07	7	66	30	0.737	19.8
127 x 114 x 29 #	29.3	127.0	114.3	10.2	11.5	9.9	4.8	79.5	4.97	7.79	7	60	24	0.646	22.0
127 x 114 x 27 #	26.9	127.0	114.3	7.4	11.4	9.9	5.0	79.5	5.01	10.7	6	60	24	0.650	24.2
127 x 76 x 16 ‡	16.5	127.0	76.2	5.6	9.6	9.4	4.6	86.5	3.97	15.4	5	42	22	0.512	31.0
114 x 114 x 27 ‡	27.1	114.3	114.3	9.5	10.7	14.2	3.2	60.8	5.34	6.40	7	60	28	0.618	22.8
102 x 102 x 23 #	23.0	101.6	101.6	9.5	10.3	11.1	3.2	55.2	4.93	5.81	6	54	24	0.549	23.9
102 x 44 x 7 #	7.5	101.6	44.5	4.3	6.1	6.9	3.3	74.6	3.65	17.3	4	28	14	0.350	46.6
89 x 89 x 19 #	19.5	88.9	88.9	9.5	9.9	11.1	3.2	44.2	4.49	4.65	7	46	24	0.476	24.4
76 x 76 x 15 ‡	15.0	76.2	80.0	8.9	8.4	9.4	4.6	38.1	4.76	4.28	6	42	20	0.419	27.9
76 x 76 x 13 #	12.8	76.2	76.2	5.1	8.4	9.4	4.6	38.1	4.54	7.47	5	42	20	0.411	32.1

‡ Not available from some leading producers. Check availability.
\# Check availability.
FOR EXPLANATION OF TABLES SEE NOTE 2

BS 5950-1: 2000
BS 4-1: 1993

JOISTS

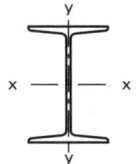

PROPERTIES

Section Designation	Second Moment of Area		Radius of Gyration		Elastic Modulus		Plastic Modulus		Buckling Parameter	Torsional Index	Warping Constant	Torsional Constant	Area of Section
	Axis x-x	Axis y-y	Axis x-x	Axis y-y	Axis x-x	Axis y-y	Axis x-x	Axis y-y	u	x	H	J	A
	cm⁴	cm⁴	cm	cm	cm³	cm³	cm³	cm³			dm⁶	cm⁴	cm²
254 x 203 x 82 #	12000	2280	10.7	4.67	947	224	1080	371	0.888	11.0	0.312	152	105
254 x 114 x 37 ‡	5080	269	10.4	2.39	400	47.1	459	79.1	0.885	18.7	0.0392	25.2	47.3
203 x 152 x 52 #	4800	816	8.49	3.50	472	107	541	176	0.890	10.7	0.0711	64.8	66.6
152 x 127 x 37 #	1820	378	6.19	2.82	239	59.6	279	99.8	0.867	9.33	0.0183	33.9	47.5
127 x 114 x 29 #	979	242	5.12	2.54	154	42.3	181	70.8	0.853	8.77	0.00807	20.8	37.4
127 x 114 x 27 #	946	236	5.26	2.63	149	41.3	172	68.2	0.868	9.31	0.00788	16.9	34.2
127 x 76 x 16 ‡	571	60.8	5.21	1.70	90.0	16.0	104	26.4	0.891	11.8	0.00210	6.72	21.1
114 x 114 x 27 ‡	736	224	4.62	2.55	129	39.2	151	65.8	0.839	7.92	0.00601	18.9	34.5
102 x 102 x 23 #	486	154	4.07	2.29	95.6	30.3	113	50.6	0.836	7.42	0.00321	14.2	29.3
102 x 44 x 7 #	153	7.82	4.01	0.907	30.1	3.51	35.4	6.03	0.872	14.9	0.000178	1.25	9.50
89 x 89 x 19 #	307	101	3.51	2.02	69.0	22.8	82.7	38.0	0.830	6.58	0.00158	11.5	24.9
76 x 76 x 15 ‡	172	60.9	3.00	1.78	45.2	15.2	54.2	25.8	0.820	6.42	0.000700	6.83	19.1
76 x 76 x 13 #	158	51.8	3.12	1.79	41.5	13.6	48.7	22.4	0.853	7.21	0.000595	4.59	16.2

‡ Not available from some leading producers. Check availability.
# Check availability.
FOR EXPLANATION OF TABLES SEE NOTE 3

| BS 5950-1: 2000 |
| BS 4-1: 1993 |

JOISTS

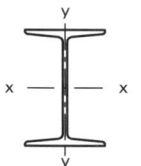

REDUCED PLASTIC MODULUS UNDER AXIAL LOAD

Section Designation	Plastic Modulus Axis x-x	Major Axis Reduced Modulus					Plastic Modulus Axis y-y	Minor Axis Reduced Modulus				
		Lower Values Of n		Change Formula	Higher Values Of n			Lower Values Of n		Change Formula	Higher Values Of n	
				At n_1 =	At n_2 =					At n =		
	cm³	K1	K2			K3	K4	cm³	K1	K2		K3	K4
254 x 203 x 82 #	1080	1080	2680	0.195	0.504	134	8.88	371	371	107	0.248	597	0.548
254 x 114 x 37 ‡	459	459	737	0.354	0.576	49.0	11.3	79.1	79.1	22.1	0.408	190	0.236
203 x 152 x 52 #	541	541	1250	0.213	0.490	72.7	8.30	176	176	54.5	0.272	292	0.511
152 x 127 x 37 #	279	279	542	0.256	0.533	44.4	7.15	99.8	99.8	37.0	0.334	186	0.412
127 x 114 x 29 #	181	181	342	0.262	0.542	30.5	6.77	70.8	70.8	27.5	0.347	132	0.407
127 x 114 x 27 #	172	172	396	0.208	0.507	25.6	7.48	68.2	68.2	23.1	0.275	112	0.523
127 x 76 x 16 ‡	104	104	198	0.272	0.500	14.6	8.18	26.4	26.4	8.75	0.337	50.3	0.388
114 x 114 x 27 ‡	151	151	313	0.234	0.557	26.0	6.58	65.8	65.8	26.0	0.315	121	0.420
102 x 102 x 23 #	113	113	226	0.239	0.534	21.2	6.04	50.6	50.6	21.2	0.329	90.8	0.428
102 x 44 x 7 #	35.4	35.4	52.5	0.391	0.575	5.07	8.52	6.03	6.03	2.22	0.460	16.1	0.143
89 x 89 x 19 #	82.7	82.7	163	0.240	0.515	17.4	5.36	38.0	38.0	17.4	0.340	67.9	0.416
76 x 76 x 15 ‡	54.2	54.2	103	0.250	0.533	11.4	5.37	25.8	25.8	12.0	0.354	47.4	0.404
76 x 76 x 13 #	48.7	48.7	129	0.170	0.463	8.66	6.15	22.4	22.4	8.66	0.239	34.2	0.576

‡ Not available from some leading producers. Check availability.
# Check availability.
$n = F/(A\, p_y)$, where F is the Factored axial load, A is the gross cross sectional area and p_y is the design strength of the section.
For values of n lower than n_1, the reduced plastic modulus, $S_{rx} = S_{rx1}$ = K1-K2.n^2, for major axis bending.
For values of n higher than n_2, the reduced plastic modulus, $S_{rx} = S_{rx2}$ = K3(1-n)(K4+n), for major axis bending.
For values of n between n_1 and n_2, the reduced plastic modulus, $S_{rx} = S_{rx1} + (S_{rx2} - S_{rx1})(n-n_1)/(n_2-n_1)$, for major axis bending.
For lower values of n, the reduced plastic modulus, S_{ry} = K1 - K2.n^2, for minor axis bending.
For higher values of n, the reduced plastic modulus, S_{ry} = K3(1-n)(K4+n), for minor axis bending.
FOR EXPLANATION OF TABLES SEE NOTE 3

[BLANK PAGE]

BS 5950-1: 2000
BS 4-1: 1993

PARALLEL FLANGE CHANNELS

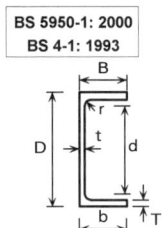

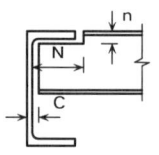

DIMENSIONS

Section Designation	Mass per Metre	Depth of Section	Width of Section	Thickness		Root Radius	Depth between Fillets	Ratios for Local Buckling		Dimensions for Detailing			Surface Area	
				Web	Flange			Flange	Web	End Clearance	Notch		Per Metre	Per Tonne
		D	B	t	T	r	d	b/T	d/t	C	N	n		
	kg/m	mm	mm	mm	mm	mm	mm			mm	mm	mm	m²	m²
430 x 100 x 64	64.4	430	100	11.0	19.0	15	362	5.26	32.9	13	96	36	1.23	19.0
380 x 100 x 54	54.0	380	100	9.5	17.5	15	315	5.71	33.2	12	98	34	1.13	20.9
300 x 100 x 46	45.5	300	100	9.0	16.5	15	237	6.06	26.3	11	98	32	0.969	21.3
300 x 90 x 41	41.4	300	90	9.0	15.5	12	245	5.81	27.2	11	88	28	0.932	22.5
260 x 90 x 35	34.8	260	90	8.0	14.0	12	208	6.43	26.0	10	88	28	0.854	24.5
260 x 75 x 28	27.6	260	75	7.0	12.0	12	212	6.25	30.3	9	74	26	0.796	28.8
230 x 90 x 32	32.2	230	90	7.5	14.0	12	178	6.43	23.7	10	90	28	0.795	24.7
230 x 75 x 26	25.7	230	75	6.5	12.5	12	181	6.00	27.8	9	76	26	0.737	28.7
200 x 90 x 30	29.7	200	90	7.0	14.0	12	148	6.43	21.1	9	90	28	0.736	24.8
200 x 75 x 23	23.4	200	75	6.0	12.5	12	151	6.00	25.2	9	76	26	0.678	28.9
180 x 90 x 26	26.1	180	90	6.5	12.5	12	131	7.20	20.2	9	90	26	0.697	26.7
180 x 75 x 20	20.3	180	75	6.0	10.5	12	135	7.14	22.5	8	76	24	0.638	31.4
150 x 90 x 24	23.9	150	90	6.5	12.0	12	102	7.50	15.7	9	90	26	0.637	26.7
150 x 75 x 18	17.9	150	75	5.5	10.0	12	106	7.50	19.3	8	76	24	0.579	32.4
125 x 65 x 15 #	14.8	125	65	5.5	9.5	12	82.0	6.84	14.9	8	66	22	0.489	33.1
100 x 50 x 10 #	10.2	100	50	5.0	8.5	9	65.0	5.88	13.0	7	52	18	0.382	37.5

Check availability.
FOR EXPLANATION OF TABLES SEE NOTE 2

BS 5950-1: 2000
BS 4-1: 1993

PARALLEL FLANGE CHANNELS

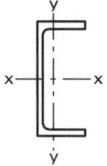

PROPERTIES

Section Designation	Second Moment of Area		Radius of Gyration		Elastic Modulus		Plastic Modulus		Buckling Parameter	Torsional Index	Warping Constant	Torsional Constant	Area of Section
	Axis x-x	Axis y-y	Axis x-x	Axis y-y	Axis x-x	Axis y-y	Axis x-x	Axis y-y	u	x	H	J	A
	cm^4	cm^4	cm	cm	cm^3	cm^3	cm^3	cm^3			dm^6	cm^4	cm^2
430 x 100 x 64	21900	722	16.3	2.97	1020	97.9	1220	176	0.917	22.5	0.219	63.0	82.1
380 x 100 x 54	15000	643	14.8	3.06	791	89.2	933	161	0.933	21.2	0.150	45.7	68.7
300 x 100 x 46	8230	568	11.9	3.13	549	81.7	641	148	0.944	17.0	0.0813	36.8	58.0
300 x 90 x 41	7220	404	11.7	2.77	481	63.1	568	114	0.934	18.4	0.0581	28.8	52.7
260 x 90 x 35	4730	353	10.3	2.82	364	56.3	425	102	0.943	17.2	0.0379	20.6	44.4
260 x 75 x 28	3620	185	10.1	2.30	278	34.4	328	62.0	0.932	20.5	0.0203	11.7	35.1
230 x 90 x 32	3520	334	9.27	2.86	306	55.0	355	98.9	0.949	15.1	0.0279	19.3	41.0
230 x 75 x 26	2750	181	9.17	2.35	239	34.8	278	63.2	0.945	17.3	0.0153	11.8	32.7
200 x 90 x 30	2520	314	8.16	2.88	252	53.4	291	94.5	0.952	12.9	0.0197	18.3	37.9
200 x 75 x 23	1960	170	8.11	2.39	196	33.8	227	60.6	0.956	14.7	0.0107	11.1	29.9
180 x 90 x 26	1820	277	7.40	2.89	202	47.4	232	83.5	0.950	12.8	0.0141	13.3	33.2
180 x 75 x 20	1370	146	7.27	2.38	152	28.8	176	51.8	0.945	15.3	0.00754	7.34	25.9
150 x 90 x 24	1160	253	6.18	2.89	155	44.4	179	76.9	0.937	10.8	0.00890	11.8	30.4
150 x 75 x 18	861	131	6.15	2.40	115	26.6	132	47.2	0.945	13.1	0.00467	6.10	22.8
125 x 65 x 15 #	483	80.0	5.07	2.06	77.3	18.8	89.9	33.2	0.942	11.1	0.00194	4.72	18.8
100 x 50 x 10 #	208	32.3	4.00	1.58	41.5	9.89	48.9	17.5	0.942	10.0	0.000491	2.53	13.0

Check availability.
FOR EXPLANATION OF TABLES SEE NOTE 3

BS 5950-1: 2000
BS 4-1: 1993

PARALLEL FLANGE CHANNELS

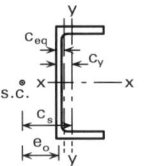

MAJOR AXIS REDUCED PLASTIC MODULUS UNDER AXIAL LOAD

Section Designation	Area of Section	Dimension				Plastic Modulus Axis x-x	Major Axis Reduced Modulus				
							Lower Values of n		Change Formula at n =	Higher Values of n	
	A cm^2	e_o cm	C_s cm	C_y cm	C_{eq} cm	cm^3	K1	K2		K3	K4
430 x 100 x 64	82.1	3.27	5.34	2.62	0.954	1220	1220	1530	0.525	168	9.48
380 x 100 x 54	68.7	3.48	5.79	2.79	0.904	933	933	1240	0.477	118	10.1
300 x 100 x 46	58.0	3.68	6.29	3.05	1.31	641	641	934	0.414	84.1	9.35
300 x 90 x 41	52.7	3.18	5.33	2.60	0.879	568	568	772	0.459	77.2	9.24
260 x 90 x 35	44.4	3.32	5.66	2.74	1.14	425	425	615	0.418	54.7	9.55
260 x 75 x 28	35.1	2.62	4.37	2.10	0.676	328	328	441	0.470	41.2	10.1
230 x 90 x 32	41.0	3.46	6.01	2.92	1.69	355	355	559	0.370	46.6	9.11
230 x 75 x 26	32.7	2.78	4.75	2.30	1.03	278	278	411	0.408	35.6	9.55
200 x 90 x 30	37.9	3.60	6.37	3.12	2.24	291	291	512	0.318	39.8	8.51
200 x 75 x 23	29.9	2.91	5.09	2.48	1.53	227	227	372	0.352	29.7	9.04
180 x 90 x 26	33.2	3.64	6.48	3.17	2.36	232	232	424	0.304	30.6	8.76
180 x 75 x 20	25.9	2.87	4.98	2.41	1.34	176	176	280	0.368	22.4	9.42
150 x 90 x 24	30.4	3.71	6.69	3.30	2.66	179	179	356	0.269	25.7	7.88
150 x 75 x 18	22.8	2.99	5.29	2.58	1.81	132	132	236	0.314	17.3	8.88
125 x 65 x 15 #	18.8	2.56	4.53	2.25	1.55	89.9	89.9	161	0.310	13.6	7.64
100 x 50 x 10 #	13.0	1.94	3.43	1.73	1.18	48.9	48.9	84.5	0.319	8.45	6.69

\# Check availability.
e_o is the distance from the centre of the web to the shear centre.
C_s is the distance from the centroidal axis to the shear centre.
C_y is the distance from the back of the web to the centroidal axis.
C_{eq} is the distance from the back of the web to the equal area axis.
$n = F/(A_g.p_y)$, where F is the Factored axial load, A_g is the gross cross sectional area and p_y is the design strength of the section.
For lower values of n, the reduced plastic modulus, $S_r = K1 - K2.n^2$
For higher values of n, the reduced plastic modulus, $S_r = K3(1-n)(K4+n)$
FOR EXPLANATION OF TABLES SEE NOTE 3

PARALLEL FLANGE CHANNELS

BS 5950-1: 2000
BS 4-1: 1993

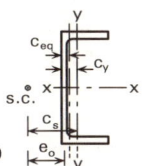

MINOR AXIS REDUCED PLASTIC MODULUS UNDER AXIAL LOAD

Section Designation	Dimension C_v cm	Plastic Modulus Axis y-y cm^3	Minor Axis reduced Modulus under axial load about centriodal axis									
			Axis load and moment inducing stresses of the same kind towards back of web						Change Formula at n =	Axial load and moment inducing stresses of the opposite kind towards back of web			
			Lower Values of n			Higher Values of n				Lower Values of n			Higher Values of n		
			K1	K2	K3	K1	K2	K3		K1	K2	K3	K1	K2	K3
430 x 100 x 64	2.62	176	176	39.2	3.49	162	443	0.634	0.152	176	39.2	3.49	176	39.2	3.49
380 x 100 x 54	2.79	161	161	31.1	4.17	158	338	0.532	0.0503	161	31.1	4.17	161	31.1	4.17
300 x 100 x 46	3.05	148	148	255	0.419	148	255	0.419	0.0689	148	255	0.419	149	28.0	4.32
300 x 90 x 41	2.60	114	114	23.2	3.92	113	224	0.495	0.0241	114	23.2	3.92	114	23.2	3.92
260 x 90 x 35	2.74	102	102	176	0.419	102	176	0.419	0.0626	102	176	0.419	103	18.9	4.42
260 x 75 x 28	2.10	62.0	62.0	11.9	4.22	61.0	129	0.525	0.0359	62.0	11.9	4.22	62.0	11.9	4.22
230 x 90 x 32	2.92	98.9	98.9	150	0.338	99.1	150	0.338	0.158	98.9	150	0.338	101	18.2	4.56
230 x 75 x 26	2.30	63.2	63.2	107	0.410	63.1	107	0.410	0.0854	63.2	107	0.410	63.6	11.6	4.47
200 x 90 x 30	3.12	94.5	94.5	128	0.261	94.5	128	0.261	0.260	94.5	128	0.261	100	17.9	4.60
200 x 75 x 23	2.48	60.6	60.6	89.2	0.318	60.8	89.2	0.318	0.196	60.6	89.2	0.318	62.8	11.2	4.64
180 x 90 x 26	3.17	83.5	83.5	110	0.242	83.5	110	0.242	0.295	83.5	110	0.242	89.8	15.3	4.87
180 x 75 x 20	2.41	51.8	51.8	79.9	0.350	51.9	79.9	0.350	0.166	51.8	79.9	0.350	53.1	9.32	4.70
150 x 90 x 24	3.30	76.9	76.9	96.3	0.201	76.9	96.3	0.201	0.359	76.9	96.3	0.201	85.0	15.4	4.72
150 x 75 x 18	2.58	47.2	47.2	64.8	0.271	47.2	64.8	0.271	0.275	47.2	64.8	0.271	50.1	8.64	4.80
125 x 65 x 15 #	2.25	33.2	33.2	46.5	0.281	33.4	46.5	0.281	0.269	33.2	46.5	0.281	35.2	7.07	3.98
100 x 50 x 10 #	1.73	17.5	17.5	24.8	0.291	17.6	24.8	0.291	0.231	17.5	24.8	0.291	18.3	4.22	3.33

Check availability.
C_v is the distance from the back of the web to the centroidal axis.
$n = F/(A_g.p_y)$, where F is the Factored axial load, A_g is the gross cross sectional area and p_y is the design strength of the section.
For axial load and moment inducing stresses of the same kind towards back of web, the reduced plastic modulus, $S_r = K1 + K2.n.(K3-n)$
For axial load and moment inducing stresses of the opposite kind towards back of web, the reduced plastic modulus, $S_r = K1 - K2.n.(K3+n$
FOR EXPLANATION OF TABLES SEE NOTE 3

BS 5950-1: 2000
Corus ASB: 2001

ASB (ASYMMETRIC BEAMS)

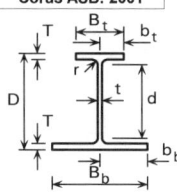
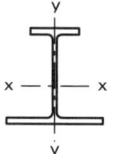

DIMENSIONS AND PROPERTIES

Section Designation	Mass per Metre	Depth of Section	Width of Flange		Thickness		Root Radius	Depth between Fillets	Ratios for Local Buckling			Second Moment of Area		Surface Area	
			Top	Bottom	Web	Flange			Flanges		Web	Axis x-x	Axis y-y	Per Metre	Per Tonne
		D	B_t	B_b	t	T	r	d	b_t/T	b_b/T	d/t				
	kg/m	mm	mm	mm	mm	mm	mm	mm				cm^4	cm^4	m^2	m^2
300 ASB 249 +	249	342	203	313	40.0	40.0	27.0	208	2.54	3.91	5.20	52900	13200	1.59	6.38
300 ASB 196	196	342	183	293	20.0	40.0	27.0	208	2.29	3.66	10.4	45900	10500	1.55	7.93
300 ASB 185 +	185	320	195	305	32.0	29.0	27.0	208	3.36	5.26	6.50	35700	8750	1.53	8.29
300 ASB 155	155	326	179	289	16.0	32.0	27.0	208	2.80	4.52	13.0	34500	7990	1.51	9.71
300 ASB 153 +	153	310	190	300	27.0	24.0	27.0	208	3.96	6.25	7.70	28400	6840	1.50	9.81
280 ASB 136 +	136	288	190	300	25.0	22.0	24.0	196	4.32	6.82	7.84	22200	6260	1.46	10.7
280 ASB 124	124	296	178	288	13.0	26.0	24.0	196	3.42	5.54	15.1	23500	6410	1.46	11.8
280 ASB 105	105	288	176	286	11.0	22.0	24.0	196	4.00	6.50	17.8	19200	5300	1.44	13.7
280 ASB 100 +	100	276	184	294	19.0	16.0	24.0	196	5.75	9.19	10.3	15500	4250	1.43	14.2
280 ASB 74	73.6	272	175	285	10.0	14.0	24.0	196	6.25	10.2	19.6	12800	3330	1.40	19.1

+ Sections are fire engineered with thick webs.
ASB sections are only available in S355.
FOR EXPLANATION OF TABLES SEE NOTES 2 AND 3

BS 5950-1: 2000
Corus ASB: 2001

ASB (ASYMMETRIC BEAMS)

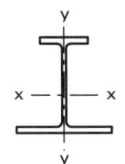
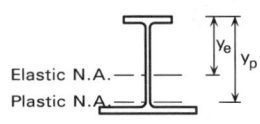

PROPERTIES (CONTINUED)

Section Designation	Radius of Gyration		Elastic Modulus			Neutral Axis Position		Plastic Modulus		Buckling Parameter	Torsional Index	Warping Constant	Torsional Constant	Area of Section
	Axis x-x	Axis y-y	Axis x-x Top	Axis x-x Bottom	Axis y-y	Elastic	Plastic	Axis x-x	Axis y-y	u	x	H	J	A
	cm	cm	cm^3	cm^3	cm^3	y_e cm	y_p cm	cm^3	cm^3			dm^6	cm^4	cm^2
300 ASB 249 +	12.9	6.40	2760	3530	843	19.2	22.6	3760	1510	0.820	6.80	2.00	2000	318
300 ASB 196	13.6	6.48	2320	3180	714	19.8	28.1	3060	1230	0.840	7.86	1.50	1180	249
300 ASB 185 +	12.3	6.10	1980	2540	574	18.0	21.0	2660	1030	0.820	8.56	1.20	871	235
300 ASB 155	13.2	6.35	1830	2520	553	18.9	27.3	2360	950	0.840	9.40	1.07	620	198
300 ASB 153 +	12.1	5.93	1630	2090	456	17.4	20.4	2160	817	0.820	9.97	0.895	513	195
280 ASB 136 +	11.3	6.00	1370	1770	417	16.3	19.2	1810	741	0.810	10.2	0.710	379	174
280 ASB 124	12.2	6.37	1360	1900	445	17.3	25.7	1730	761	0.830	10.5	0.721	332	158
280 ASB 105	12.0	6.30	1150	1610	370	16.8	25.3	1440	633	0.830	12.1	0.574	207	133
280 ASB 100 +	11.0	5.76	995	1290	289	15.6	18.4	1290	511	0.810	13.2	0.451	160	128
280 ASB 74	11.4	5.96	776	1060	234	15.7	21.3	978	403	0.830	16.7	0.338	72.0	93.7

+ Sections are fire engineered with thick webs.
ASB sections are only available in S355.
FOR EXPLANATION OF TABLES SEE NOTES 2 AND 3

[BLANK PAGE]

BS 5950-1: 2000
BS EN 10056-1: 1999

EQUAL ANGLES

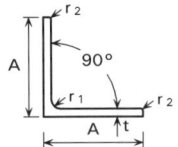

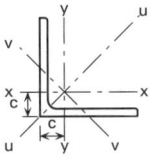

DIMENSIONS AND PROPERTIES

Section Designation Size	Mass per Metre	Radius Root	Radius Toe	Area of Section	Dimension c	Second Moment of Area Axis x-x, y-y	Second Moment of Area Axis u-u	Second Moment of Area Axis v-v	Radius of Gyration Axis x-x, y-y	Radius of Gyration Axis u-u	Radius of Gyration Axis v-v	Elastic Modulus Axis x-x, y-y	Torsional Constant	Equivalent Slenderness Coefficient
A x A mm	t mm	r_1 mm	r_2 mm	cm²	c cm	cm⁴	cm⁴	cm⁴	cm	cm	cm	cm³	J cm⁴	ϕ_a
200 x 200	24 #	71.1	18.0	9.00	90.6	5.84	3330	5280	1380	6.06	7.64	3.90	235	182	2.50
	20	59.9	18.0	9.00	76.3	5.68	2850	4530	1170	6.11	7.70	3.92	199	107	3.05
	18	54.3	18.0	9.00	69.1	5.60	2600	4150	1050	6.13	7.75	3.90	181	78.9	3.43
	16	48.5	18.0	9.00	61.8	5.52	2340	3720	960	6.16	7.76	3.94	162	56.1	3.85
150 x 150	18 #	40.1	16.0	8.00	51.2	4.38	1060	1680	440	4.55	5.73	2.93	99.8	58.6	2.48
	15	33.8	16.0	8.00	43.0	4.25	898	1430	370	4.57	5.76	2.93	83.5	34.6	3.01
	12	27.3	16.0	8.00	34.8	4.12	737	1170	303	4.60	5.80	2.95	67.7	18.2	3.77
	10	23.0	16.0	8.00	29.3	4.03	624	990	258	4.62	5.82	2.97	56.9	10.80	4.51
120 x 120	15 #	26.6	13.0	6.50	34.0	3.52	448	710	186	3.63	4.57	2.34	52.8	27.0	2.37
	12	21.6	13.0	6.50	27.5	3.40	368	584	152	3.65	4.60	2.35	42.7	14.2	2.99
	10	18.2	13.0	6.50	23.2	3.31	313	497	129	3.67	4.63	2.36	36.0	8.41	3.61
	8 #	14.7	13.0	6.50	18.8	3.24	259	411	107	3.71	4.67	2.38	29.5	4.44	4.56
100 x 100	15 #	21.9	12.0	6.00	28.0	3.02	250	395	105	2.99	3.76	1.94	35.8	22.3	1.92
	12	17.8	12.0	6.00	22.7	2.90	207	328	85.7	3.02	3.80	1.94	29.1	11.8	2.44
	10	15.0	12.0	6.00	19.2	2.82	177	280	73.0	3.04	3.83	1.95	24.6	6.97	2.94
	8	12.2	12.0	6.00	15.5	2.74	145	230	59.9	3.06	3.85	1.96	19.9	3.68	3.70
90 x 90	12 #	15.9	11.0	5.50	20.3	2.66	149	235	62.0	2.71	3.40	1.75	23.5	10.46	2.17
	10	13.4	11.0	5.50	17.1	2.58	127	201	52.6	2.72	3.42	1.75	19.8	6.20	2.64
	8	10.9	11.0	5.50	13.9	2.50	104	166	43.1	2.74	3.45	1.76	16.1	3.28	3.33
	7 #	9.61	11.0	5.50	12.2	2.45	92.6	147	38.3	2.75	3.46	1.77	14.1	2.24	3.80
80 x 80	10 ‡	11.9	10.0	5.00	15.1	2.34	87.5	139	36.4	2.41	3.03	1.55	15.4	5.45	2.33
	8 ‡	9.63	10.0	5.00	12.3	2.26	72.2	115	29.9	2.43	3.06	1.56	12.6	2.88	2.94
75 x 75	8 ‡	8.99	9.00	4.50	11.4	2.14	59.1	93.8	24.5	2.27	2.86	1.46	11.0	2.65	2.76
	6 ‡	6.85	9.00	4.50	8.73	2.05	45.8	72.7	18.9	2.29	2.89	1.47	8.41	1.17	3.70
70 x 70	7 ‡	7.38	9.00	4.50	9.40	1.97	42.3	67.1	17.5	2.12	2.67	1.36	8.41	1.69	2.92
	6 ‡	6.38	9.00	4.50	8.13	1.93	36.9	58.5	15.3	2.13	2.68	1.37	7.27	1.093	3.41
65 x 65	7 ‡	6.83	9.00	4.50	8.73	2.05	33.4	53.0	13.8	1.96	2.47	1.26	7.18	1.58	2.67
60 x 60	8 ‡	7.09	8.00	4.00	9.03	1.77	29.2	46.1	12.2	1.80	2.26	1.16	6.89	2.09	2.14
	6 †	5.42	8.00	4.00	6.91	1.69	22.8	36.1	9.44	1.82	2.29	1.17	5.29	0.922	2.90
	5 ‡	4.57	8.00	4.00	5.82	1.64	19.4	30.7	8.03	1.82	2.30	1.17	4.45	0.550	3.48
50 x 50	6 ‡	4.47	7.00	3.50	5.69	1.45	12.8	20.3	5.34	1.50	1.89	0.968	3.61	0.755	2.38
	5 ‡	3.77	7.00	3.50	4.80	1.40	11.0	17.4	4.55	1.51	1.90	0.973	3.05	0.450	2.88
	4 ‡	3.06	7.00	3.50	3.89	1.36	8.97	14.2	3.73	1.52	1.91	0.979	2.46	0.240	3.57
45 x 45	4.5 ‡	3.06	7.00	3.50	3.90	1.25	7.14	11.4	2.94	1.35	1.71	0.870	2.20	0.304	2.84
40 x 40	5 ‡	2.97	6.00	3.00	3.79	1.16	5.43	8.60	2.26	1.20	1.51	0.773	1.91	0.352	2.26
	4 ‡	2.42	6.00	3.00	3.08	1.12	4.47	7.09	1.86	1.21	1.52	0.777	1.55	0.188	2.83
35 x 35	4 ‡	2.09	5.00	2.50	2.67	1.00	2.95	4.68	1.23	1.05	1.32	0.678	1.18	0.158	2.50
30 x 30	4 ‡	1.78	5.00	2.50	2.27	0.878	1.80	2.85	0.754	0.892	1.12	0.577	0.850	0.137	2.07
	3 ‡	1.36	5.00	2.50	1.74	0.835	1.40	2.22	0.585	0.899	1.13	0.581	0.649	0.0613	2.75
25 x 25	4 ‡	1.45	3.50	1.75	1.85	0.762	1.02	1.61	0.430	0.741	0.931	0.482	0.586	0.1070	1.75
	3 ‡	1.12	3.50	1.75	1.42	0.723	0.803	1.27	0.334	0.751	0.945	0.484	0.452	0.0472	2.38
20 x 20	3 ‡	0.882	3.50	1.75	1.12	0.598	0.392	0.618	0.165	0.590	0.742	0.383	0.279	0.0382	1.81

‡ Not available from some leading producers. Check availability.
\# Check availability.
c is the distance from the back of the leg to the centre of gravity.
FOR EXPLANATION OF TABLES SEE NOTES 2 AND 3

BS 5950-1: 2000
BS EN 10056-1: 1999

UNEQUAL ANGLES

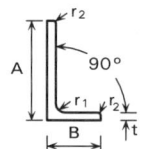

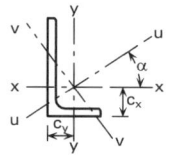

DIMENSIONS AND PROPERTIES

Section Designation Size A x B	Mass per Metre	Radius		Dimension		Second Moment of Area				Radius of Gyration			
	Thickness	Root	Toe			Axis x-x	Axis y-y	Axis u-u	Axis v-v	Axis x-x	Axis y-y	Axis u-u	Axis v-v
mm	t mm	r_1 mm	r_2 mm	c_x cm	c_y cm	cm^4	cm^4	cm^4	cm^4	cm	cm	cm	cm
200 x 150	18 #	47.1	15.0	7.50	6.34	3.86	2390	1160	2920	623	6.30	4.38	6.97	3.22
	15	39.6	15.0	7.50	6.21	3.73	2020	979	2480	526	6.33	4.40	7.00	3.23
	12	32.0	15.0	7.50	6.08	3.61	1650	803	2030	430	6.36	4.44	7.04	3.25
200 x 100	15	33.8	15.0	7.50	7.16	2.22	1760	299	1860	193	6.40	2.64	6.59	2.12
	12	27.3	15.0	7.50	7.03	2.10	1440	247	1530	159	6.43	2.67	6.63	2.14
	10	23.0	15.0	7.50	6.93	2.01	1220	210	1290	135	6.46	2.68	6.65	2.15
150 x 90	15	33.9	12.0	6.00	5.21	2.23	761	205	841	126	4.74	2.46	4.98	1.93
	12	21.6	12.0	6.00	5.08	2.12	627	171	694	104	4.77	2.49	5.02	1.94
	10	18.2	12.0	6.00	5.00	2.04	533	146	591	88.3	4.80	2.51	5.05	1.95
150 x 75	15	24.8	12.0	6.00	5.52	1.81	713	119	753	78.6	4.75	1.94	4.88	1.58
	12	20.2	12.0	6.00	5.40	1.69	588	99.6	623	64.7	4.78	1.97	4.92	1.59
	10	17.0	12.0	6.00	5.31	1.61	501	85.6	531	55.1	4.81	1.99	4.95	1.60
125 x 75	12	17.8	11.0	5.50	4.31	1.84	354	95.5	391	58.5	3.95	2.05	4.15	1.61
	10	15.0	11.0	5.50	4.23	1.76	302	82.1	334	49.9	3.97	2.07	4.18	1.61
	8	12.2	11.0	5.50	4.14	1.68	247	67.6	274	40.9	4.00	2.09	4.21	1.63
100 x 75	12	15.4	10.0	5.00	3.27	2.03	189	90.2	230	49.5	3.10	2.14	3.42	1.59
	10	13.0	10.0	5.00	3.19	1.95	162	77.6	197	42.2	3.12	2.16	3.45	1.59
	8	10.6	10.0	5.00	3.10	1.87	133	64.1	162	34.6	3.14	2.18	3.47	1.60
100 x 65	10 #	12.3	10.0	5.00	3.36	1.63	154	51.0	175	30.1	3.14	1.81	3.35	1.39
	8 #	9.94	10.0	5.00	3.27	1.55	127	42.2	144	24.8	3.16	1.83	3.37	1.40
	7 #	8.77	10.0	5.00	3.23	1.51	113	37.6	128	22.0	3.17	1.83	3.39	1.40
100 x 50	8 ‡	8.97	8.00	4.00	3.60	1.13	116	19.7	123	12.8	3.19	1.31	3.28	1.06
	6 ‡	6.84	8.00	4.00	3.51	1.05	89.9	15.4	95.4	9.92	3.21	1.33	3.31	1.07
80 x 60	7 ‡	7.36	8.00	4.00	2.51	1.52	59.0	28.4	72.0	15.4	2.51	1.74	2.77	1.28
80 x 40	8 ‡	7.07	7.00	3.50	2.94	0.963	57.6	9.61	60.9	6.34	2.53	1.03	2.60	0.838
	6 ‡	5.41	7.00	3.50	2.85	0.884	44.9	7.59	47.6	4.93	2.55	1.05	2.63	0.845
75 x 50	8 ‡	7.39	7.00	3.50	2.52	1.29	52.0	18.4	59.6	10.8	2.35	1.40	2.52	1.07
	6 ‡	5.65	7.00	3.50	2.44	1.21	40.5	14.4	46.6	8.36	2.37	1.42	2.55	1.08
70 x 50	6 ‡	5.41	7.00	3.50	2.23	1.25	33.4	14.2	39.7	7.92	2.20	1.43	2.40	1.07
65 x 50	5 ‡	4.35	6.00	3.00	1.99	1.25	23.2	11.9	28.8	6.32	2.05	1.47	2.28	1.07
60 x 40	6 ‡	4.46	6.00	3.00	2.00	1.01	20.1	7.12	23.1	4.16	1.88	1.12	2.02	0.855
	5 ‡	3.76	6.00	3.00	1.96	0.972	17.2	6.11	19.7	3.54	1.89	1.13	2.03	0.860
60 x 30	5 ‡	3.36	5.00	2.50	2.17	0.684	15.6	2.63	16.5	1.71	1.91	0.784	1.97	0.633
50 x 30	5 ‡	2.96	5.00	2.50	1.73	0.741	9.36	2.51	10.3	1.54	1.57	0.816	1.65	0.639
45 x 30	4 ‡	2.25	4.50	2.25	1.48	0.740	5.78	2.05	6.65	1.18	1.42	0.850	1.52	0.640
40 x 25	4 ‡	1.93	4.00	2.00	1.36	0.623	3.89	1.16	4.35	0.700	1.26	0.687	1.33	0.534
40 x 20	4 ‡	1.77	4.00	2.00	1.47	0.480	3.59	0.600	3.80	0.393	1.26	0.514	1.30	0.417
30 x 20	4 ‡	1.46	4.00	2.00	1.03	0.541	1.59	0.553	1.81	0.330	0.925	0.546	0.988	0.421
	3 ‡	1.12	4.00	2.00	0.990	0.502	1.25	0.437	1.43	0.256	0.935	0.553	1.00	0.424

‡ Not available from some leading producers. Check availability.
# Check availability.
c_x is the distance from the back of the short leg to the centre of gravity.
c_y is the distance from the back of the long leg to the centre of gravity.
FOR EXPLANATION OF TABLES SEE NOTES 2 AND 3

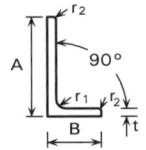
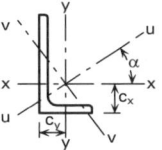

BS 5950-1: 2000
BS EN 10056-1: 1999

UNEQUAL ANGLES

DIMENSIONS AND PROPERTIES (CONTINUED)

Section Designation		Elastic Modulus		Angle Axis x-x to Axis u-u	Torsional Constant	Equivalent Slenderness Coefficient		Mono-symmetry Index	Area of Section
Size	Thickness	Axis x-x	Axis y-y			Min	Max		
A x B	t			Tan α	J	ϕ_a	ϕ_a	ψ_a	
mm	mm	cm^3	cm^3		cm^4				cm^2
200 x 150	18 #	175	104	0.549	67.9	2.93	3.72	4.60	60.1
	15	147	86.9	0.551	39.9	3.53	4.50	5.55	50.5
	12	119	70.5	0.552	20.9	4.43	5.70	6.97	40.8
200x100	15	137	38.5	0.260	34.3	3.54	5.17	9.19	43.0
	12	111	31.3	0.262	18.0	4.42	6.57	11.5	34.8
	10	93.2	26.3	0.263	10.66	5.26	7.92	13.9	29.2
150x90	15	77.7	30.4	0.354	26.8	2.58	3.59	5.96	33.9
	12	63.3	24.8	0.358	14.1	3.24	4.58	7.50	27.5
	10	53.3	21.0	0.360	8.30	3.89	5.56	9.03	23.2
150x75	15	75.2	21.0	0.253	25.1	2.62	3.74	6.84	31.7
	12	61.3	17.1	0.258	13.2	3.30	4.79	8.60	25.7
	10	51.6	14.5	0.261	7.80	3.95	5.83	10.4	21.7
125x75	12	43.2	16.9	0.354	11.6	2.66	3.73	6.23	22.7
	10	36.5	14.3	0.357	6.87	3.21	4.55	7.50	19.1
	8	29.6	11.6	0.360	3.62	4.00	5.75	9.43	15.5
100x75	12	28.0	16.5	0.540	10.05	2.10	2.64	3.46	19.7
	10	23.8	14.0	0.544	5.95	2.54	3.22	4.17	16.6
	8	19.3	11.4	0.547	3.13	3.18	4.08	5.24	13.5
100x65	10 #	23.2	10.5	0.410	5.61	2.52	3.43	5.45	15.6
	8 #	18.9	8.54	0.413	2.96	3.14	4.35	6.86	12.7
	7 #	16.6	7.53	0.415	2.02	3.58	5.00	7.85	11.2
100x50	8 ‡	18.2	5.08	0.258	2.61	3.30	4.80	8.61	11.4
	6 ‡	13.8	3.89	0.262	1.14	4.38	6.52	11.6	8.71
80x60	7 ‡	10.7	6.34	0.546	1.66	2.92	3.72	4.78	9.38
80x40	8 ‡	11.4	3.16	0.253	2.05	2.61	3.73	6.85	9.01
	6 ‡	8.73	2.44	0.258	0.899	3.48	5.12	9.22	6.89
75x50	8 ‡	10.4	4.95	0.430	2.14	2.36	3.18	4.92	9.41
	6 ‡	8.01	3.81	0.435	0.935	3.18	4.34	6.60	7.19
70x50	6 ‡	7.01	3.78	0.500	0.899	2.96	3.89	5.44	6.89
65x50	5 ‡	5.14	3.19	0.577	0.498	3.38	4.26	5.08	5.54
60x40	6 ‡	5.03	2.38	0.431	0.735	2.51	3.39	5.26	5.68
	5 ‡	4.25	2.02	0.434	0.435	3.02	4.11	6.34	4.79
60x30	5 ‡	4.07	1.14	0.257	0.382	3.15	4.56	8.26	4.28
50x30	5 ‡	2.86	1.11	0.352	0.340	2.51	3.52	5.99	3.78
45x30	4 ‡	1.91	0.910	0.436	0.166	2.85	3.87	5.92	2.87
40x25	4 ‡	1.47	0.619	0.380	0.142	2.51	3.48	5.75	2.46
40x20	4 ‡	1.42	0.393	0.252	0.131	2.57	3.68	6.86	2.26
30x20	4 ‡	0.807	0.379	0.421	0.1096	1.79	2.39	3.95	1.86
	3 ‡	0.621	0.292	0.427	0.0486	2.40	3.28	5.31	1.43

‡ Not available from some leading producers. Check availability.
\# Check availability.
FOR EXPLANATION OF TABLES SEE NOTES 2 AND 3

| BS 5950-1: 2000 |
| BS EN 10056-1: 1999 |

EQUAL ANGLES
BACK TO BACK

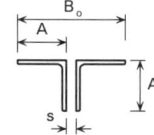
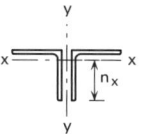

DIMENSIONS AND PROPERTIES

Overall Size		Composed of Two Angles		Total Mass per Metre	Space between Angles	Distance	Total Area	Second Moment of Area		Radius of Gyration		Elastic Modulus	
A	B_o	A x A	t		s	n_x		I_x	I_y	r_x	r_y	Z_x	Z_y
mm	mm	mm	mm	kg/m	mm	cm	cm^2	cm^4	cm^4	cm	cm	cm^3	cm^3
200	415	200 x 200	24 #	142	15	14.2	181	6660	14500	6.06	8.95	470	699
			20	120	15	14.3	153	5700	12000	6.11	8.87	398	578
			18	109	15	14.4	138	5200	10800	6.13	8.83	362	520
			16	97.0	15	14.5	124	4680	9540	6.16	8.79	324	460
150	312	150 x 150	18 #	80.2	12	10.6	102	2120	4660	4.55	6.75	200	299
			15	67.6	12	10.8	86.0	1800	3820	4.57	6.66	167	245
			12	54.6	12	10.9	69.6	1470	3020	4.60	6.59	135	194
			10	46.0	12	11.0	58.6	1250	2500	4.62	6.54	114	160
120	252	120 x 120	15 #	53.2	12	8.48	68.0	896	2050	3.63	5.49	106	163
			12	43.2	12	8.60	55.0	736	1620	3.65	5.42	85.4	129
			10	36.4	12	8.69	46.4	626	1340	3.67	5.36	72.0	106
			8 #	29.4	12	8.76	37.6	518	1070	3.71	5.34	59.0	84.9
100	210	100 x 100	15 #	43.8	10	6.98	56.0	500	1190	2.99	4.62	71.6	113
			12	35.6	10	7.10	45.4	414	939	3.02	4.55	58.2	89.4
			10	30.0	10	7.18	38.4	354	777	3.04	4.50	49.2	74.0
			8	24.4	10	7.26	31.0	290	615	3.06	4.46	39.8	58.6
90	190	90 x 90	12 #	31.8	10	6.34	40.6	298	703	2.71	4.16	47.0	74.0
			10	26.8	10	6.42	34.2	254	578	2.72	4.11	39.6	60.9
			8	21.8	10	6.50	27.8	208	458	2.74	4.06	32.2	48.2
			7 #	19.2	10	6.55	24.4	185	398	2.75	4.04	28.2	41.8
80	170	80 x 80	10 ‡	23.8	8	5.66	30.2	175	402	2.41	3.65	30.8	47.3
			8 ‡	19.3	8	5.74	24.6	144	318	2.43	3.60	25.2	37.5
75	160	75 x 75	8 ‡	18.0	8	5.36	22.8	118	265	2.27	3.41	22.0	33.2
			6 ‡	13.7	8	5.45	17.5	91.6	196	2.29	3.35	16.8	24.6
70	150	70 x 70	7 ‡	14.8	8	5.03	18.8	84.6	190	2.12	3.18	16.8	25.4
			6 ‡	12.8	8	5.07	16.3	73.8	162	2.13	3.16	14.5	21.6
65	140	65 x 65	7 ‡	13.7	8	4.45	17.5	66.8	172	1.96	3.14	14.4	24.5
60	130	60 x 60	8 ‡	14.2	8	4.23	18.1	58.4	143	1.80	2.82	13.8	22.1
			6 ‡	10.8	8	4.31	13.8	45.6	106	1.82	2.77	10.6	16.3
			5 ‡	9.14	8	4.36	11.6	38.8	87.2	1.82	2.74	8.90	13.4
50	110	50 x 50	6 ‡	8.94	8	3.55	11.4	25.6	64.5	1.50	2.38	7.22	11.7
			5 ‡	7.54	8	3.60	9.60	22.0	53.1	1.51	2.35	6.10	9.66
			4 ‡	6.12	8	3.64	7.78	17.9	42.0	1.52	2.32	4.92	7.64

‡ Not available from some leading producers. Check availability.
# Check availability.
Properties about y-y axis:
$I_y = (Total\ Area).(r_y)^2$
$Z_y = I_y / (0.5B_o)$
FOR EXPLANATION OF TABLES SEE NOTES 2 AND 3

BS 5950-1: 2000
BS EN 10056-1: 1999

UNEQUAL ANGLES
LONG LEGS BACK TO BACK

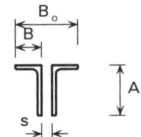
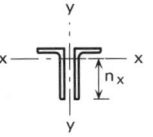

DIMENSIONS AND PROPERTIES

Overall Size		Composed of Two Angles		Total Mass per Metre	Space between Angles	Distance	Total Area	Second Moment of Area		Radius of Gyration		Elastic Modulus	
A	B_o	A x B	t		s	n_x		I_x	I_y	r_x	r_y	Z_x	Z_y
mm	mm	mm	mm	kg/m	mm	cm	cm^2	cm^4	cm^4	cm	cm	cm^3	cm^3
200	315	200 x 150	18 #	94.2	15	13.7	120	4780	4860	6.30	6.36	350	309
			15	79.2	15	13.8	101	4040	3990	6.33	6.28	294	253
			12	64.0	15	13.9	81.6	3300	3160	6.36	6.22	238	201
200	215	200 x 100	15	67.5	15	12.8	86.0	3520	1360	6.40	3.97	274	127
			12	54.6	15	13.0	69.6	2880	1060	6.43	3.90	222	98.6
			10	46.0	15	13.1	58.4	2440	865	6.46	3.85	186	80.5
150	192	150 x 90	15	53.2	12	9.79	67.8	1520	953	4.74	3.75	155	99.3
			12	43.2	12	9.92	55.0	1250	749	4.77	3.69	127	78.0
			10	36.4	12	10.0	46.4	1070	615	4.80	3.64	107	64.1
150	162	150 x 75	15	49.6	12	9.48	63.4	1430	606	4.75	3.09	150	74.8
			12	40.4	12	9.60	51.4	1180	469	4.78	3.02	123	57.9
			10	34.0	12	9.69	43.4	1000	383	4.81	2.97	103	47.3
125	162	125 x 75	12	35.6	12	8.19	45.4	708	461	3.95	3.19	86.4	56.9
			10	30.0	12	8.27	38.2	604	377	3.97	3.14	73.0	46.5
			8	24.4	12	8.36	31.0	494	296	4.00	3.09	59.2	36.6
100	160	100 x 75	12	30.8	10	6.73	39.4	378	433	3.10	3.31	56.0	54.1
			10	26.0	10	6.81	33.2	324	354	3.12	3.27	47.6	44.3
			8	21.2	10	6.90	27.0	266	280	3.14	3.22	38.6	35.0
100	140	100 x 65	10 #	24.6	10	6.64	31.2	308	244	3.14	2.79	46.4	34.8
			8 #	19.9	10	6.73	25.4	254	191	3.16	2.74	37.8	27.3
			7 #	17.5	10	6.77	22.4	226	166	3.17	2.72	33.2	23.7
100	110	100 x 50	8 ‡	17.9	10	6.40	22.8	232	100.0	3.19	2.09	36.4	18.2
			6 ‡	13.7	10	6.49	17.4	180	72.7	3.21	2.04	27.6	13.2
80	128	80 x 60	7 ‡	14.7	8	5.49	18.8	118	126	2.51	2.59	21.4	19.7
80	88	80 x 40	8 ‡	14.1	8	5.06	18.0	115	52.7	2.53	1.71	22.8	12.0
			6 ‡	10.8	8	5.15	13.8	89.8	37.9	2.55	1.66	17.5	8.61
75	108	75 x 50	8 ‡	14.8	8	4.98	18.8	104	90.6	2.35	2.19	20.8	16.8
			6 ‡	11.3	8	5.06	14.4	81.0	66.1	2.37	2.14	16.0	12.2
70	100	70 x 50	6 †	10.8	8	4.77	13.8	66.8	65.9	2.20	2.19	14.0	12.2
65	108	65 x 50	5 ‡	8.70	8	4.51	11.1	46.4	54.0	2.05	2.21	10.3	9.99
60	88	60 x 40	6 ‡	8.92	8	4.00	11.4	40.2	36.8	1.88	1.80	10.1	8.37
			5 ‡	7.52	8	4.04	9.58	34.4	30.3	1.89	1.78	8.50	6.88

‡ Not available from some leading producers. Check availability.
# Check availability.
Properties about y-y axis:
I_y = (Total Area).$(r_y)^2$
Z_y = I_y / $(0.5 B_o)$
FOR EXPLANATION OF TABLES SEE NOTES 2 AND 3

BS 5950-1: 2000
BS 4-1: 1993

STRUCTURAL TEES
CUT FROM UNIVERSAL BEAMS

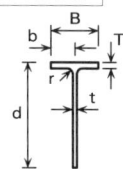
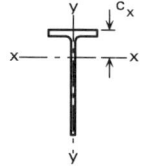

DIMENSIONS AND PROPERTIES

Section Designation	Cut from Universal Beam Section Designation	Mass per Metre	Width of Section	Depth of Section	Thickness		Root Radius	Ratios for Local Buckling		Dimension	Second Moment of Area	
					Web	Flange		Flange	Web		Axis x-x	Axis y-y
			B	d	t	T	r	b/T	d/t	c_x		
		kg/m	mm	mm	mm	mm	mm			cm	cm^4	cm^4
305 x 457 x 127	914 x 305 x 253	126.7	305.5	459.1	17.3	27.9	19.1	5.47	26.5	12.0	32700	6650
305 x 457 x 112	914 x 305 x 224	112.1	304.1	455.1	15.9	23.9	19.1	6.36	28.6	12.1	29100	5620
305 x 457 x 101	914 x 305 x 201	100.4	303.3	451.4	15.1	20.2	19.1	7.51	29.9	12.5	26400	4710
292 x 419 x 113	838 x 292 x 226	113.3	293.8	425.4	16.1	26.8	17.8	5.48	26.4	10.8	24600	5680
292 x 419 x 97	838 x 292 x 194	96.9	292.4	420.3	14.7	21.7	17.8	6.74	28.6	11.1	21300	4530
292 x 419 x 88	838 x 292 x 176	87.9	291.7	417.4	14.0	18.8	17.8	7.76	29.8	11.4	19600	3900
267 x 381 x 99	762 x 267 x 197	98.4	268.0	384.8	15.6	25.4	16.5	5.28	24.7	9.89	17500	4090
267 x 381 x 87	762 x 267 x 173	86.5	266.7	381.0	14.3	21.6	16.5	6.17	26.6	9.98	15500	3430
267 x 381 x 74	762 x 267 x 147	73.5	265.2	376.9	12.8	17.5	16.5	7.58	29.4	10.2	13200	2730
267 x 381 x 67	762 x 267 x 134	66.9	264.4	374.9	12.0	15.5	16.5	8.53	31.2	10.3	12100	2390
254 x 343 x 85	686 x 254 x 170	85.1	255.8	346.4	14.5	23.7	15.2	5.40	23.9	8.67	12100	3320
254 x 343 x 76	686 x 254 x 152	76.2	254.5	343.7	13.2	21.0	15.2	6.06	26.0	8.61	10800	2890
254 x 343 x 70	686 x 254 x 140	70.0	253.7	341.7	12.4	19.0	15.2	6.68	27.6	8.63	9910	2590
254 x 343 x 63	686 x 254 x 125	62.6	253.0	338.9	11.7	16.2	15.2	7.81	29.0	8.85	8980	2190
305 x 305 x 119	610 x 305 x 238	119.0	311.4	317.8	18.4	31.4	16.5	4.96	17.3	7.11	12300	7920
305 x 305 x 90	610 x 305 x 179	89.5	307.1	310.0	14.1	23.6	16.5	6.51	22.0	6.69	9040	5700
305 x 305 x 75	610 x 305 x 149	74.6	304.8	306.1	11.8	19.7	16.5	7.74	25.9	6.45	7420	4650
229 x 305 x 70	610 x 229 x 140	69.9	230.2	308.5	13.1	22.1	12.7	5.21	23.5	7.61	7740	2250
229 x 305 x 63	610 x 229 x 125	62.5	229.0	306.0	11.9	19.6	12.7	5.84	25.7	7.54	6900	1970
229 x 305 x 57	610 x 229 x 113	56.5	228.2	303.7	11.1	17.3	12.7	6.60	27.4	7.58	6270	1720
229 x 305 x 51	610 x 229 x 101	50.6	227.6	301.2	10.5	14.8	12.7	7.69	28.7	7.78	5690	1460
210 x 267 x 61	533 x 210 x 122	61.0	211.9	272.2	12.7	21.3	12.7	4.97	21.4	6.66	5160	1690
210 x 267 x 55	533 x 210 x 109	54.5	210.8	269.7	11.6	18.8	12.7	5.61	23.3	6.61	4600	1470
210 x 267 x 51	533 x 210 x 101	50.5	210.0	268.3	10.8	17.4	12.7	6.03	24.8	6.53	4250	1350
210 x 267 x 46	533 x 210 x 92	46.1	209.3	266.5	10.1	15.6	12.7	6.71	26.4	6.55	3890	1200
210 x 267 x 41	533 x 210 x 82	41.1	208.8	264.1	9.6	13.2	12.7	7.91	27.5	6.75	3530	1000
191 x 229 x 49	457 x 191 x 98	49.2	192.8	233.5	11.4	19.6	10.2	4.92	20.5	5.53	2970	1170
191 x 229 x 45	457 x 191 x 89	44.6	191.9	231.6	10.5	17.7	10.2	5.42	22.1	5.47	2680	1050
191 x 229 x 41	457 x 191 x 82	41.0	191.3	229.9	9.9	16.0	10.2	5.98	23.2	5.47	2470	935
191 x 229 x 37	457 x 191 x 74	37.1	190.4	228.4	9.0	14.5	10.2	6.57	25.4	5.38	2220	836
191 x 229 x 34	457 x 191 x 67	33.6	189.9	226.6	8.5	12.7	10.2	7.48	26.7	5.46	2030	726
152 x 229 x 41	457 x 152 x 82	41.0	155.3	232.8	10.5	18.9	10.2	4.11	22.2	5.96	2600	592
152 x 229 x 37	457 x 152 x 74	37.1	154.4	230.9	9.6	17.0	10.2	4.54	24.1	5.88	2330	523
152 x 229 x 34	457 x 152 x 67	33.6	153.8	228.9	9.0	15.0	10.2	5.13	25.4	5.91	2120	456
152 x 229 x 30	457 x 152 x 60	29.9	152.9	227.2	8.1	13.3	10.2	5.75	28.0	5.84	1880	397
152 x 229 x 26	457 x 152 x 52	26.2	152.4	224.8	7.6	10.9	10.2	6.99	29.6	6.04	1670	322

FOR EXPLANATION OF TABLES SEE NOTES 2 AND 3

STRUCTURAL TEES
CUT FROM UNIVERSAL BEAMS

PROPERTIES (CONTINUED)

Section Designation	Radius of Gyration		Elastic Modulus				Plastic Modulus		Buckling Parameter	Torsional Index	Mono-symmetry Index	Warping Constant (*)	Torsional Constant	Area of Section
	Axis x-x	Axis y-y	Axis x-x			Axis y-y	Axis x-x	Axis y-y						
			Flange	Toe					u	x	ψ	H	J	A
	cm	cm	cm³	cm³		cm³	cm³	cm³				cm⁶	cm⁴	cm²
305 x 457 x 127	14.2	6.42	2720	965		435	1730	685	0.656	18.1	0.749	17000	313	161
305 x 457 x 112	14.3	6.27	2400	871		369	1570	582	0.666	20.6	0.753	12400	211	143
305 x 457 x 101	14.4	6.07	2110	808		311	1460	491	0.685	23.4	0.759	9820	146	128
292 x 419 x 113	13.1	6.27	2280	776		387	1380	606	0.640	17.5	0.742	11500	257	144
292 x 419 x 97	13.1	6.06	1930	689		310	1240	487	0.660	20.8	0.747	7830	153	123
292 x 419 x 88	13.2	5.90	1720	644		267	1160	421	0.675	23.2	0.751	6320	111	112
267 x 381 x 99	11.8	5.71	1770	613		305	1090	479	0.641	16.6	0.741	7620	202	125
267 x 381 x 87	11.9	5.58	1550	550		257	986	404	0.654	19.0	0.745	5450	134	110
267 x 381 x 74	11.9	5.40	1300	481		206	867	324	0.670	22.6	0.749	3600	79.5	93.6
267 x 381 x 67	11.9	5.30	1180	445		181	806	285	0.679	24.9	0.753	2850	59.2	85.3
254 x 343 x 85	10.5	5.53	1390	464		259	826	406	0.624	15.9	0.731	4720	154	108
254 x 343 x 76	10.5	5.46	1250	417		227	743	355	0.627	17.7	0.732	3420	110	97.0
254 x 343 x 70	10.5	5.39	1150	388		204	691	319	0.633	19.3	0.734	2720	84.3	89.2
254 x 343 x 63	10.6	5.24	1010	358		173	643	271	0.651	21.9	0.740	2090	58.1	79.7
305 x 305 x 119	9.02	7.23	1740	500		509	894	787	0.483	10.6	0.661	11300	393	152
305 x 305 x 90	8.91	7.07	1350	372		371	657	572	0.485	13.8	0.664	4710	170	114
305 x 305 x 75	8.83	7.00	1150	307		305	539	469	0.483	16.3	0.666	2690	100	95.0
229 x 305 x 70	9.32	5.03	1020	333		196	592	306	0.613	15.3	0.727	2560	108	89.1
229 x 305 x 63	9.31	4.97	915	299		172	531	268	0.617	17.0	0.728	1840	77.1	79.7
229 x 305 x 57	9.33	4.88	826	275		150	489	235	0.626	19.0	0.731	1400	55.7	72.0
229 x 305 x 51	9.40	4.76	732	255		128	456	200	0.645	21.5	0.736	1080	38.5	64.4
210 x 267 x 61	8.15	4.67	775	251		160	446	250	0.600	13.8	0.719	1660	89.2	77.7
210 x 267 x 55	8.14	4.60	697	226		140	401	218	0.605	15.4	0.721	1200	63.2	69.4
210 x 267 x 51	8.12	4.57	650	209		128	371	200	0.606	16.6	0.722	951	50.5	64.3
210 x 267 x 46	8.14	4.51	593	193		114	343	178	0.613	18.2	0.724	737	37.8	58.7
210 x 267 x 41	8.21	4.38	523	179		96.1	320	150	0.634	20.8	0.730	565	25.8	52.3
191 x 229 x 49	6.88	4.33	536	167		122	296	189	0.573	12.9	0.705	835	60.6	62.6
191 x 229 x 45	6.87	4.29	491	152		109	269	169	0.576	14.1	0.706	628	45.3	56.9
191 x 229 x 41	6.88	4.23	452	141		97.8	250	152	0.583	15.4	0.708	494	34.6	52.2
191 x 229 x 37	6.86	4.20	413	127		87.8	225	130	0.583	16.9	0.709	365	25.9	47.3
191 x 229 x 34	6.90	4.12	372	118		76.5	209	119	0.597	18.9	0.713	280	18.6	42.7
152 x 229 x 41	7.05	3.37	436	150		76.3	267	120	0.634	13.7	0.740	534	44.6	52.3
152 x 229 x 37	7.03	3.33	397	135		67.8	242	107	0.637	15.1	0.742	396	33.0	47.2
152 x 229 x 34	7.04	3.27	359	125		59.3	223	93.3	0.646	16.8	0.745	305	23.8	42.8
152 x 229 x 30	7.02	3.23	322	111		52.0	199	81.5	0.649	18.7	0.746	217	16.9	38.1
152 x 229 x 26	7.08	3.11	276	102		42.3	183	66.7	0.671	21.9	0.753	161	10.7	33.3

(*) Note units are cm⁶ and not dm⁶.
FOR EXPLANATION OF TABLES SEE NOTES 2 AND 3

BS 5950-1: 2000
BS 4-1: 1993

BS 5950-1: 2000
BS 4-1: 1993

STRUCTURAL TEES
CUT FROM UNIVERSAL BEAMS

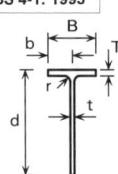
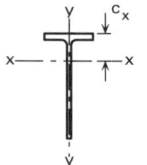

DIMENSIONS AND PROPERTIES

Section Designation	Cut from Universal Beam Section Designation	Mass per Metre	Width of Section	Depth of Section	Thickness		Root Radius	Ratios for Local Buckling		Dimension	Second Moment of Area	
					Web	Flange		Flange	Web		Axis x-x	Axis y-y
			B	d	t	T	r	b/T	d/t	c_x		
		kg/m	mm	mm	mm	mm	mm			cm	cm^4	cm^4
178 x 203 x 37	406 x 178 x 74	37.1	179.5	206.3	9.5	16.0	10.2	5.61	21.7	4.76	1740	773
178 x 203 x 34	406 x 178 x 67	33.6	178.8	204.6	8.8	14.3	10.2	6.25	23.2	4.73	1570	682
178 x 203 x 30	406 x 178 x 60	30.0	177.9	203.1	7.9	12.8	10.2	6.95	25.7	4.64	1400	602
178 x 203 x 27	406 x 178 x 54	27.1	177.7	201.2	7.7	10.9	10.2	8.15	26.1	4.83	1290	511
140 x 203 x 23	406 x 140 x 46	23.0	142.2	201.5	6.8	11.2	10.2	6.35	29.6	5.02	1120	269
140 x 203 x 20	406 x 140 x 39	19.5	141.8	198.9	6.4	8.6	10.2	8.24	31.1	5.32	979	205
171 x 178 x 34	356 x 171 x 67	33.5	173.2	181.6	9.1	15.7	10.2	5.52	20.0	4.00	1150	681
171 x 178 x 29	356 x 171 x 57	28.5	172.2	178.9	8.1	13.0	10.2	6.62	22.1	3.97	986	554
171 x 178 x 26	356 x 171 x 51	25.5	171.5	177.4	7.4	11.5	10.2	7.46	24.0	3.94	882	484
171 x 178 x 23	356 x 171 x 45	22.5	171.1	175.6	7.0	9.7	10.2	8.82	25.1	4.05	798	406
127 x 178 x 20	356 x 127 x 39	19.5	126.0	176.6	6.6	10.7	10.2	5.89	26.8	4.43	728	179
127 x 178 x 17	356 x 127 x 33	16.5	125.4	174.4	6.0	8.5	10.2	7.38	29.1	4.56	626	140
165 x 152 x 27	305 x 165 x 54	27.0	166.9	155.1	7.9	13.7	8.9	6.09	19.6	3.21	642	531
165 x 152 x 23	305 x 165 x 46	23.1	165.7	153.2	6.7	11.8	8.9	7.02	22.9	3.07	536	448
165 x 152 x 20	305 x 165 x 40	20.1	165.0	151.6	6.0	10.2	8.9	8.09	25.3	3.03	468	382
127 x 152 x 24	305 x 127 x 48	24.0	125.3	155.4	9.0	14.0	8.9	4.47	17.3	3.94	662	231
127 x 152 x 21	305 x 127 x 42	21.0	124.3	153.5	8.0	12.1	8.9	5.14	19.2	3.87	573	194
127 x 152 x 19	305 x 127 x 37	18.5	123.4	152.1	7.1	10.7	8.9	5.77	21.4	3.78	501	168
102 x 152 x 17	305 x 102 x 33	16.4	102.4	156.3	6.6	10.8	7.6	4.74	23.7	4.14	487	97.0
102 x 152 x 14	305 x 102 x 28	14.1	101.8	154.3	6.0	8.8	7.6	5.78	25.7	4.20	420	77.7
102 x 152 x 13	305 x 102 x 25	12.4	101.6	152.5	5.8	7.0	7.6	7.26	26.3	4.43	377	61.5
146 x 127 x 22	254 x 146 x 43	21.5	147.3	129.7	7.2	12.7	7.6	5.80	18.0	2.64	343	339
146 x 127 x 19	254 x 146 x 37	18.5	146.4	127.9	6.3	10.9	7.6	6.72	20.3	2.55	292	285
146 x 127 x 16	254 x 146 x 31	15.6	146.1	125.6	6.0	8.6	7.6	8.49	20.9	2.66	259	224
102 x 127 x 14	254 x 102 x 28	14.2	102.2	130.1	6.3	10.0	7.6	5.11	20.7	3.24	277	89.3
102 x 127 x 13	254 x 102 x 25	12.6	101.9	128.5	6.0	8.4	7.6	6.07	21.4	3.32	250	74.3
102 x 127 x 11	254 x 102 x 22	11.0	101.6	126.9	5.7	6.8	7.6	7.47	22.3	3.45	223	59.7
133 x 102 x 15	203 x 133 x 30	15.0	133.9	103.3	6.4	9.6	7.6	6.97	16.1	2.11	154	192
133 x 102 x 13	203 x 133 x 25	12.5	133.2	101.5	5.7	7.8	7.6	8.54	17.8	2.10	131	154

FOR EXPLANATION OF TABLES SEE NOTES 2 AND 3

| BS 5950-1: 2000 |
| BS 4-1: 1993 |

STRUCTURAL TEES
CUT FROM UNIVERSAL BEAMS

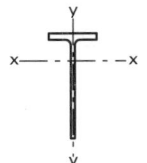

PROPERTIES (CONTINUED)

Section Designation	Radius of Gyration		Elastic Modulus				Plastic Modulus		Buckling Parameter	Torsional Index	Mono-symmetry Index	Warping Constant (*)	Torsional Constant	Area of Section
	Axis x-x	Axis y-y	Axis x-x		Axis y-y		Axis x-x	Axis y-y						
			Flange	Toe					u	x	ψ	H	J	A
	cm	cm	cm^3	cm^3	cm^3		cm^3	cm^3				cm^6	cm^4	cm^2
178 x 203 x 37	6.06	4.04	365	109	86.1		194	133	0.556	13.8	0.696	350	31.4	47.2
178 x 203 x 34	6.07	3.99	332	100	76.3		177	118	0.561	15.2	0.698	262	23.1	42.8
178 x 203 x 30	6.04	3.97	301	89.0	67.6		157	105	0.561	16.9	0.699	186	16.7	38.3
178 x 203 x 27	6.13	3.85	268	84.6	57.5		150	89.1	0.588	19.1	0.705	146	11.6	34.5
140 x 203 x 23	6.19	3.03	224	74.2	37.8		132	59.1	0.633	19.5	0.740	93.7	9.51	29.3
140 x 203 x 20	6.28	2.87	184	67.2	28.9		121	45.4	0.668	23.7	0.750	66.3	5.35	24.8
171 x 178 x 34	5.20	3.99	288	81.5	78.6		145	121	0.500	12.2	0.672	249	27.8	42.7
171 x 178 x 29	5.21	3.91	248	70.9	64.4		125	99.4	0.514	14.4	0.676	154	16.7	36.3
171 x 178 x 26	5.21	3.86	224	63.9	56.5		113	87.1	0.522	16.0	0.677	110	11.9	32.4
171 x 178 x 23	5.28	3.76	197	59.1	47.4		104	73.3	0.545	18.4	0.683	79.2	7.92	28.7
127 x 178 x 20	5.41	2.68	164	55.0	28.4		98.1	44.5	0.632	17.6	0.739	57.1	7.55	24.9
127 x 178 x 17	5.45	2.58	137	48.6	22.3		87.2	35.1	0.654	21.1	0.746	38.0	4.40	21.1
165 x 152 x 27	4.32	3.93	200	52.2	63.7		92.9	97.8	0.389	11.8	0.636	128	17.4	34.4
165 x 152 x 23	4.27	3.91	174	43.7	54.1		77.2	82.8	0.380	13.6	0.636	78.6	11.1	29.4
165 x 152 x 20	4.27	3.86	155	38.6	46.3		67.7	70.9	0.393	15.5	0.638	52.0	7.37	25.7
127 x 152 x 24	4.65	2.74	168	57.1	36.8		102	58.0	0.602	11.7	0.714	104	15.9	30.6
127 x 152 x 21	4.63	2.70	148	49.9	31.3		88.9	49.2	0.606	13.2	0.716	69.2	10.6	26.7
127 x 152 x 19	4.61	2.67	132	43.8	27.2		78.0	42.7	0.606	14.9	0.718	47.4	7.38	23.6
102 x 152 x 17	4.82	2.15	118	42.3	19.0		75.8	30.0	0.656	15.8	0.749	36.8	6.10	20.9
102 x 152 x 14	4.84	2.08	100	37.4	15.3		67.4	24.2	0.673	18.7	0.756	25.2	3.70	17.9
102 x 152 x 13	4.88	1.97	85.0	34.8	12.1		63.4	19.4	0.702	21.7	0.766	20.4	2.39	15.8
146 x 127 x 22	3.54	3.52	130	33.2	46.0		59.6	70.5	0.195	10.6	0.613	64.9	11.9	27.4
146 x 127 x 19	3.52	3.48	115	28.5	39.0		50.7	59.7	0.233	12.2	0.616	41.0	7.67	23.6
146 x 127 x 16	3.61	3.36	97.4	26.2	30.6		46.1	47.1	0.376	14.8	0.623	24.5	4.28	19.8
102 x 127 x 14	3.92	2.22	85.5	28.3	17.5		50.5	27.4	0.608	13.7	0.720	21.0	4.78	18.0
102 x 127 x 13	3.95	2.15	75.3	26.2	14.6		46.9	23.0	0.629	15.7	0.727	15.9	3.21	16.0
102 x 127 x 11	3.99	2.06	64.5	24.1	11.7		43.4	18.6	0.655	18.2	0.735	12.0	2.07	14.0
133 x 102 x 15	2.84	3.17	73.1	18.8	28.7		33.5	44.1	–	10.7	0.569	21.7	5.15	19.1
133 x 102 x 13	2.00	3.10	62.4	16.2	23.1		28.7	35.5	–	12.8	0.572	12.6	2.98	16.0

(*) Note units are cm^6 and not dm^6.
– Indicates that no values of u and x are given, as lateral torsional buckling due to bending about the x-x axis is not possible, because the second moment of area about the y-y axis exceeds the second moment of area about the x-x axis.
FOR EXPLANATION OF TABLES SEE NOTES 2 AND 3

		STRUCTURAL TEES								
BS 5950-1: 2000		CUT FROM UNIVERSAL COLUMNS								
BS 4-1: 1993										

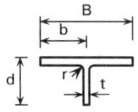
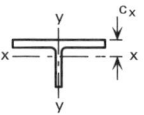

DIMENSIONS

Section Designation	Cut from Universal Column Section Designation	Mass per Metre	Width of Section	Depth of Section	Thickness		Root Radius	Ratios for Local Buckling		Dimension
					Web	Flange		Flange	Web	
			B	d	t	T	r	b/T	d/t	c_x
		kg/m	mm	mm	mm	mm	mm			cm
406 x 178 x 118	356 x 406 x 235	117.5	394.8	190.4	18.4	30.2	15.2	6.54	10.3	3.40
368 x 178 x 101	356 x 368 x 202	100.9	374.7	187.2	16.5	27.0	15.2	6.94	11.3	3.29
368 x 178 x 89	356 x 368 x 177	88.5	372.6	184.0	14.4	23.8	15.2	7.83	12.8	3.09
368 x 178 x 77	356 x 368 x 153	76.5	370.5	180.9	12.3	20.7	15.2	8.95	14.7	2.88
368 x 178 x 65	356 x 368 x 129	64.5	368.6	177.7	10.4	17.5	15.2	10.5	17.1	2.69
305 x 152 x 79	305 x 305 x 158	79.0	311.2	163.5	15.8	25.0	15.2	6.22	10.3	3.04
305 x 152 x 69	305 x 305 x 137	68.5	309.2	160.2	13.8	21.7	15.2	7.12	11.6	2.86
305 x 152 x 59	305 x 305 x 118	58.9	307.4	157.2	12.0	18.7	15.2	8.22	13.1	2.69
305 x 152 x 49	305 x 305 x 97	48.4	305.3	153.9	9.9	15.4	15.2	9.91	15.5	2.50
254 x 127 x 66	254 x 254 x 132	66.0	261.3	138.1	15.3	25.3	12.7	5.16	9.03	2.70
254 x 127 x 54	254 x 254 x 107	53.5	258.8	133.3	12.8	20.5	12.7	6.31	10.4	2.45
254 x 127 x 45	254 x 254 x 89	44.5	256.3	130.1	10.3	17.3	12.7	7.41	12.6	2.21
254 x 127 x 37	254 x 254 x 73	36.5	254.6	127.0	8.6	14.2	12.7	8.96	14.8	2.05
203 x 102 x 43	203 x 203 x 86	43.0	209.1	111.0	12.7	20.5	10.2	5.10	8.74	2.20
203 x 102 x 36	203 x 203 x 71	35.5	206.4	107.8	10.0	17.3	10.2	5.97	10.8	1.95
203 x 102 x 30	203 x 203 x 60	30.0	205.8	104.7	9.4	14.2	10.2	7.25	11.1	1.89
203 x 102 x 26	203 x 203 x 52	26.0	204.3	103.0	7.9	12.5	10.2	8.17	13.0	1.75
203 x 102 x 23	203 x 203 x 46	23.0	203.6	101.5	7.2	11.0	10.2	9.25	14.1	1.69
152 x 76 x 19	152 x 152 x 37	18.5	154.4	80.8	8.0	11.5	7.6	6.71	10.1	1.53
152 x 76 x 15	152 x 152 x 30	15.0	152.9	78.7	6.5	9.4	7.6	8.13	12.1	1.41
152 x 76 x 12	152 x 152 x 23	11.5	152.2	76.1	5.8	6.8	7.6	11.2	13.1	1.39

FOR EXPLANATION OF TABLES SEE NOTES 2 AND 3

BS 5950-1: 2000
BS 4-1: 1993

STRUCTURAL TEES
CUT FROM UNIVERSAL COLUMNS

PROPERTIES

Section Designation	Second Moment of Area		Radius of Gyration		Elastic Modulus			Plastic Modulus		Mono-symmetry Index	Warping Constant (*)	Torsional Constant	Area of Section
	Axis x-x	Axis y-y	Axis x-x	Axis y-y	Axis x-x		Axis y-y	Axis x-x	Axis y-y				
					Flange	Toe							
	cm^4	cm^4	cm	cm	cm^3	cm^3	cm^3	cm^3	cm^3	ψ	cm^6	cm^4	cm^2
406 x 178 x 118	2860	15500	4.37	10.2	843	183	785	367	1190	0.165	12700	405	150
368 x 178 x 101	2460	11800	4.38	9.60	749	160	632	312	960	0.216	7840	278	129
368 x 178 x 89	2090	10300	4.30	9.54	676	136	551	263	835	0.212	5270	190	113
368 x 178 x 77	1730	8780	4.22	9.49	601	114	474	216	717	0.209	3390	125	97.4
368 x 178 x 65	1420	7310	4.16	9.43	527	94.1	396	175	600	0.207	2010	76.2	82.2
305 x 152 x 79	1530	6290	3.90	7.90	503	115	404	225	615	0.268	3650	188	101
305 x 152 x 69	1290	5350	3.84	7.83	450	97.7	346	188	526	0.263	2340	124	87.2
305 x 152 x 59	1080	4530	3.79	7.77	401	82.8	295	156	448	0.262	1470	80.3	75.1
305 x 152 x 49	858	3650	3.73	7.69	343	66.5	239	123	363	0.258	806	45.5	61.7
254 x 127 x 66	871	3770	3.22	6.69	323	78.3	288	159	439	0.250	2200	159	84.1
254 x 127 x 54	676	2960	3.15	6.59	276	62.1	229	122	349	0.245	1150	85.9	68.2
254 x 127 x 45	524	2430	3.04	6.55	237	48.5	190	94.1	288	0.242	660	51.1	56.7
254 x 127 x 37	417	1950	2.99	6.48	204	39.2	153	74.1	233	0.236	359	28.8	46.5
203 x 102 x 43	373	1560	2.61	5.34	169	41.9	150	84.6	228	0.257	605	68.1	54.8
203 x 102 x 36	280	1270	2.49	5.30	143	31.8	123	63.6	187	0.254	343	40.0	45.2
203 x 102 x 30	244	1030	2.53	5.20	129	28.4	100	54.4	153	0.245	195	23.5	38.2
203 x 102 x 26	200	889	2.46	5.18	115	23.4	87.0	44.5	132	0.243	128	15.8	33.1
203 x 102 x 23	177	774	2.45	5.13	105	20.9	76.0	39.0	115	0.242	87.2	11.0	29.4
152 x 76 x 19	93.1	353	1.99	3.87	60.7	14.2	45.7	27.1	69.8	0.277	44.9	9.54	23.5
152 x 76 x 15	72.2	280	1.94	3.83	51.4	11.2	36.7	20.9	55.8	0.269	23.7	5.24	19.1
152 x 76 x 12	58.5	200	2.00	3.70	41.9	9.41	26.3	16.9	40.1	0.278	9.78	2.30	14.6

(*) Note units are cm^6 and not dm^6.
Values of u and x are not given, as lateral torsional buckling due to bending about the x-x axis is not possible, because the second moment of area about the y-y axis exceeds the second moment of area about the x-x axis.
FOR EXPLANATION OF TABLES SEE NOTES 2 AND 3

BS 5950-1: 2000
BS EN 10210-2: 1997

HOT-FINISHED CIRCULAR HOLLOW SECTIONS

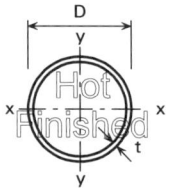

DIMENSIONS AND PROPERTIES

Section Designation		Mass per Metre	Area of Section	Ratio for Local Buckling	Second Moment of Area	Radius of Gyration	Elastic Modulus	Plastic Modulus	Torsional Constants		Surface Area	
Outside Diameter D mm	Thickness t mm	kg/m	A cm²	D/t	I cm⁴	r cm	Z cm³	S cm³	J cm⁴	C cm³	Per Metre m²	Per Tonne m²
26.9	3.2 ~	1.87	2.38	8.41	1.70	0.846	1.27	1.81	3.41	2.53	0.0845	45.2
42.4	3.2 ~	3.09	3.94	13.3	7.62	1.39	3.59	4.93	15.2	7.19	0.133	43.0
48.3	3.2 ~	3.56	4.53	15.1	11.6	1.60	4.80	6.52	23.2	9.59	0.152	42.7
	4.0 ~	4.37	5.57	12.1	13.8	1.57	5.70	7.87	27.5	11.4	0.152	34.8
	5.0 ~	5.34	6.80	9.66	16.2	1.54	6.69	9.42	32.3	13.4	0.152	28.5
60.3	3.2 ~	4.51	5.74	18.8	23.5	2.02	7.78	10.4	46.9	15.6	0.189	41.9
	5.0 ~	6.82	8.69	12.1	33.5	1.96	11.1	15.3	67.0	22.2	0.189	27.7
76.1	2.9 ^	5.24	6.67	26.2	44.7	2.59	11.8	15.5	89.0	23.5	0.239	45.6
	3.2 ~	5.75	7.33	23.8	48.8	2.58	12.8	17.0	97.6	25.6	0.239	41.6
	4.0 ~	7.11	9.06	19.0	59.1	2.55	15.5	20.8	118	31.0	0.239	33.6
	5.0 ~	8.77	11.2	15.2	70.9	2.52	18.6	25.3	142	37.3	0.239	27.3
88.9	3.2 ~	6.76	8.62	27.8	79.2	3.03	17.8	23.5	158	35.6	0.279	41.3
	4.0 ~	8.38	10.7	22.2	96.3	3.00	21.7	28.9	193	43.3	0.279	33.3
	5.0 ~	10.4	13.2	17.8	116	2.97	26.2	35.2	233	52.4	0.279	27.0
	6.3 ~	12.8	16.3	14.1	140	2.93	31.5	43.1	280	63.1	0.279	21.7
114.3	3.2 ~	8.77	11.2	35.7	172	3.93	30.2	39.5	345	60.4	0.359	40.9
	3.6	9.83	12.5	31.8	192	3.92	33.6	44.1	384	67.2	0.359	36.5
	5.0	13.5	17.2	22.9	257	3.87	45.0	59.8	514	89.9	0.359	26.6
	6.3	16.8	21.4	18.1	313	3.82	54.7	73.6	625	109	0.359	21.4
139.7	5.0	16.6	21.2	27.9	481	4.77	68.8	90.8	961	138	0.439	26.4
	6.3	20.7	26.4	22.2	589	4.72	84.3	112	1180	169	0.439	21.2
	8.0 ~	26.0	33.1	17.5	720	4.66	103	139	1440	206	0.439	16.9
	10.0 ~	32.0	40.7	14.0	862	4.60	123	169	1720	247	0.439	13.7
168.3	5.0 ~	20.1	25.7	33.7	856	5.78	102	133	1710	203	0.529	26.3
	6.3 ~	25.2	32.1	26.7	1050	5.73	125	165	2110	250	0.529	21.0
	8.0 ~	31.6	40.3	21.0	1300	5.67	154	206	2600	308	0.529	16.7
	10.0 ~	39.0	49.7	16.8	1560	5.61	186	251	3130	372	0.529	13.6
193.7	5.0 ~	23.3	29.6	38.7	1320	6.67	136	178	2640	273	0.609	26.1
	6.3 ~	29.1	37.1	30.7	1630	6.63	168	221	3260	337	0.609	20.9
	8.0 ~	36.6	46.7	24.2	2020	6.57	208	276	4030	416	0.609	16.6
	10.0 ~	45.3	57.7	19.4	2440	6.50	252	338	4880	504	0.609	13.4

~ Check availability in S275.
^ Check availability in S355.
FOR EXPLANATION OF TABLES SEE NOTES 2 AND 3

BS 5950-1: 2000
BS EN 10210-2: 1997

HOT-FINISHED
CIRCULAR HOLLOW SECTIONS

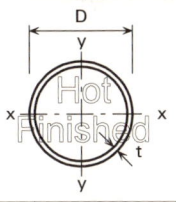

DIMENSIONS AND PROPERTIES

Section Designation		Mass per Metre	Area of Section	Ratio for Local Buckling	Second Moment of Area	Radius of Gyration	Elastic Modulus	Plastic Modulus	Torsional Constants		Surface Area	
Outside Diameter D mm	Thickness t mm	kg/m	A cm²	D/t	I cm⁴	r cm	Z cm³	S cm³	J cm⁴	C cm³	Per Metre m²	Per Tonne m²
219.1	5.0 ~	26.4	33.6	43.8	1930	7.57	176	229	3860	352	0.688	26.1
	6.3 ~	33.1	42.1	34.8	2390	7.53	218	285	4770	436	0.688	20.8
	8.0 ~	41.6	53.1	27.4	2960	7.47	270	357	5920	540	0.688	16.5
	10.0 ~	51.6	65.7	21.9	3600	7.40	328	438	7200	657	0.688	13.3
	12.5 ~	63.7	81.1	17.5	4350	7.32	397	534	8690	793	0.688	10.80
244.5	12.0 ~	68.8	87.7	20.4	5940	8.23	486	649	11900	972	0.768	11.16
273.0	5.0 ~	33.0	42.1	54.6	3780	9.48	277	359	7560	554	0.858	26.0
	6.3	41.4	52.8	43.3	4700	9.43	344	448	9390	688	0.858	20.7
	8.0 ~	52.3	66.6	34.1	5850	9.37	429	562	11700	857	0.858	16.4
	10.0	64.9	82.6	27.3	7150	9.31	524	692	14300	1050	0.858	13.2
	12.5	80.3	102	21.8	8700	9.22	637	849	17400	1270	0.858	10.68
	16.0 ~	101	129	17.1	10700	9.10	784	1060	21400	1570	0.858	8.46
323.9	6.3 ~	49.3	62.9	51.4	7930	11.2	490	636	15900	979	1.02	20.7
	8.0 ~	62.3	79.4	40.5	9910	11.2	612	799	19800	1220	1.02	16.4
	10.0 ~	77.4	98.6	32.4	12200	11.1	751	986	24300	1500	1.02	13.2
	12.5 ~	96.0	122	25.9	14800	11.0	917	1210	29700	1830	1.02	10.63
	16.0	122	155	20.2	18400	10.9	1140	1520	36800	2270	1.02	8.40
406.4	6.3 ~	62.2	79.2	64.5	15900	14.1	780	1010	31700	1560	1.28	20.6
	8.0 ~	78.6	100	50.8	19900	14.1	978	1270	39700	1960	1.28	16.3
	10.0 ~	97.8	125	40.6	24500	14.0	1210	1570	49000	2410	1.28	13.1
	12.5 ~	121	155	32.5	30000	13.9	1480	1940	60100	2960	1.28	10.54
	16.0	154	196	25.4	37500	13.8	1840	2440	74900	3690	1.28	8.31
457.0	8.0 ~	88.6	113	57.1	28500	15.9	1250	1610	56900	2490	1.44	16.3
	10.0 ~	110	140	45.7	35100	15.8	1540	2000	70200	3070	1.44	13.1
	12.5 ~	137	175	36.6	43100	15.7	1890	2470	86700	3780	1.44	10.51
	16.0	174	222	28.6	54000	15.6	2360	3110	108000	4730	1.44	8.28
508.0	8.0 ~	98.6	126	63.5	39300	17.7	1550	2000	78600	3090	1.60	16.2
	10.0 ~	123	156	50.8	48500	17.6	1910	2480	97000	3820	1.60	13.0
	12.5	153	195	40.0	59800	17.5	2350	3070	120000	4710	1.60	10.46
	16.0 ~	194	247	31.8	74900	17.4	2950	3870	150000	5900	1.60	8.25
	20.0 ~	241	307	25.4	91400	17.3	3600	4770	183000	7200	1.60	6.64

~ Check availability in S275.
FOR EXPLANATION OF TABLES SEE NOTES 2 AND 3

BS 5950-1: 2000
BS EN 10210-2: 1997

HOT-FINISHED
SQUARE HOLLOW SECTIONS

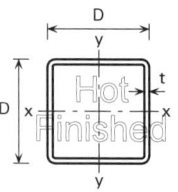

DIMENSIONS AND PROPERTIES

Section Designation Size D x D mm	Thickness t mm	Mass per Metre kg/m	Area of Section A cm²	Ratio for Local Buckling d/t (1)	Second Moment of Area I cm⁴	Radius of Gyration r cm	Elastic Modulus Z cm³	Plastic Modulus S cm³	Torsional Constants J cm⁴	Torsional Constants C cm³	Surface Area Per Metre m²	Surface Area Per Tonne m²
40 x 40	3.0 ~	3.41	4.34	10.3	9.78	1.50	4.89	5.97	15.7	7.10	0.152	44.6
	3.2 ~	3.61	4.60	9.50	10.2	1.49	5.11	6.28	16.5	7.42	0.152	42.1
	4.0	4.39	5.59	7.00	11.8	1.45	5.91	7.44	19.5	8.54	0.150	34.2
	5.0 ~	5.28	6.73	5.00	13.4	1.41	6.68	8.66	22.5	9.60	0.147	27.8
50 x 50	3.0 ~	4.35	5.54	13.7	20.2	1.91	8.08	9.70	32.1	11.8	0.192	44.1
	3.2 ~	4.62	5.88	12.6	21.2	1.90	8.49	10.2	33.8	12.4	0.192	41.6
	4.0	5.64	7.19	9.50	25.0	1.86	9.99	12.3	40.4	14.5	0.190	33.7
	5.0	6.85	8.73	7.00	28.9	1.82	11.6	14.5	47.6	16.7	0.187	27.3
	6.3 ~	8.31	10.6	4.94	32.8	1.76	13.1	17.0	55.2	18.8	0.184	22.1
60 x 60	3.0 ~	5.29	6.74	17.0	36.2	2.32	12.1	14.3	56.9	17.7	0.232	43.9
	3.2 ~	5.62	7.16	15.8	38.2	2.31	12.7	15.2	60.2	18.6	0.232	41.3
	4.0 ~	6.90	8.79	12.0	45.4	2.27	15.1	18.3	72.5	22.0	0.230	33.3
	5.0	8.42	10.7	9.00	53.3	2.23	17.8	21.9	86.4	25.7	0.227	27.0
	6.3 ~	10.3	13.1	6.52	61.6	2.17	20.5	26.0	102	29.6	0.224	21.7
	8.0 ~	12.5	16.0	4.50	69.7	2.09	23.2	30.4	118	33.4	0.219	17.5
70 x 70	3.6 ~	7.40	9.42	16.4	68.6	2.70	19.6	23.3	108	28.7	0.271	36.6
	5.0 ~	9.99	12.7	11.0	88.5	2.64	25.3	30.8	142	36.8	0.267	26.7
	6.3 ~	12.3	15.6	8.11	104	2.58	29.7	36.9	169	42.9	0.264	21.5
	8.0 ~	15.0	19.2	5.75	120	2.50	34.2	43.8	200	49.2	0.259	17.3
80 x 80	3.6 ~	8.53	10.9	19.2	105	3.11	26.2	31.0	164	38.5	0.311	36.5
	4.0 ~	9.41	12.0	17.0	114	3.09	28.6	34.0	180	41.9	0.310	32.9
	5.0 ~	11.6	14.7	13.0	137	3.05	34.2	41.1	217	49.8	0.307	26.6
	6.3	14.2	18.1	9.70	162	2.99	40.5	49.7	262	58.7	0.304	21.4
	8.0 ~	17.5	22.4	7.00	189	2.91	47.3	59.5	312	68.3	0.299	17.1
90 x 90	3.6 ~	9.66	12.3	22.0	152	3.52	33.8	39.7	237	49.7	0.351	36.3
	4.0 ~	10.7	13.6	19.5	166	3.50	37.0	43.6	260	54.2	0.350	32.7
	5.0 ~	13.1	16.7	15.0	200	3.45	44.4	53.0	316	64.8	0.347	26.5
	6.3 ~	16.2	20.7	11.3	238	3.40	53.0	64.3	382	77.0	0.344	21.2
	8.0 ~	20.1	25.6	8.25	281	3.32	62.6	77.6	459	90.5	0.339	16.9
100 x 100	4.0	11.9	15.2	22.0	232	3.91	46.4	54.4	361	68.2	0.390	32.8
	5.0	14.7	18.7	17.0	279	3.86	55.9	66.4	439	81.8	0.387	26.3
	6.3	18.2	23.2	12.9	336	3.80	67.1	80.9	534	97.8	0.384	21.1
	8.0 ~	22.6	28.8	9.50	400	3.73	79.9	98.2	646	116	0.379	16.8
	10.0	27.4	34.9	7.00	462	3.64	92.4	116	761	133	0.374	13.6
120 x 120	5.0 ~	17.8	22.7	21.0	498	4.68	83.0	97.6	777	122	0.467	26.2
	6.3	22.2	28.2	16.0	603	4.62	100	120	950	147	0.464	20.9
	8.0 ~	27.6	35.2	12.0	726	4.55	121	147	1160	176	0.459	16.6
	10.0 ~	33.7	42.9	9.00	852	4.46	142	175	1380	206	0.454	13.5
	12.5 ~	40.9	52.1	6.60	982	4.34	164	207	1620	236	0.448	11.0

~ Check availability in S275.
(1) For local buckling calculation d = D - 3t.
FOR EXPLANATION OF TABLES SEE NOTES 2 AND 3

HOT-FINISHED SQUARE HOLLOW SECTIONS

BS 5950-1: 2000
BS EN 10210-2: 1997

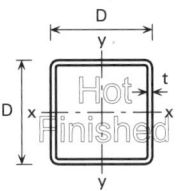

DIMENSIONS AND PROPERTIES

Section Designation Size D x D mm	Thickness t mm	Mass per Metre kg/m	Area of Section A cm²	Ratio for Local Buckling d/t [1]	Second Moment of Area I cm⁴	Radius of Gyration r cm	Elastic Modulus Z cm³	Plastic Modulus S cm³	Torsional Constants J cm⁴	Torsional Constants C cm³	Surface Area Per Metre m²	Surface Area Per Tonne m²
140 x 140	5.0 ~	21.0	26.7	25.0	807	5.50	115	135	1250	170	0.547	26.0
	6.3 ~	26.1	33.3	19.2	984	5.44	141	166	1540	206	0.544	20.8
	8.0 ~	32.6	41.6	14.5	1200	5.36	171	204	1890	249	0.539	16.5
	10.0 ~	40.0	50.9	11.0	1420	5.27	202	246	2270	294	0.534	13.4
	12.5 ~	48.7	62.1	8.20	1650	5.16	236	293	2700	342	0.528	10.8
150 x 150	5.0 ~	22.6	28.7	27.0	1000	5.90	134	156	1550	197	0.587	26.0
	6.3	28.1	35.8	20.8	1220	5.85	163	192	1910	240	0.584	20.8
	8.0 ~	35.1	44.8	15.8	1490	5.77	199	237	2350	291	0.579	16.5
	10.0	43.1	54.9	12.0	1770	5.68	236	286	2830	344	0.574	13.3
	12.5 ~	52.7	67.1	9.00	2080	5.57	277	342	3370	402	0.568	10.8
	16.0 ~	65.2	83.0	6.38	2430	5.41	324	411	4030	467	0.559	8.57
160 x 160	5.0 ~	24.1	30.7	29.0	1230	6.31	153	178	1890	226	0.627	26.0
	6.3 ~	30.1	38.3	22.4	1500	6.26	187	220	2330	275	0.624	20.7
	8.0 ~	37.6	48.0	17.0	1830	6.18	229	272	2880	335	0.619	16.5
	10.0 ~	46.3	58.9	13.0	2190	6.09	273	329	3480	398	0.614	13.3
	12.5 ~	56.6	72.1	9.80	2580	5.98	322	395	4160	467	0.608	10.7
180 x 180	6.3 ~	34.0	43.3	25.6	2170	7.07	241	281	3360	355	0.704	20.7
	8.0 ~	42.7	54.4	19.5	2660	7.00	296	349	4160	434	0.699	16.4
	10.0 ~	52.5	66.9	15.0	3190	6.91	355	424	5050	518	0.694	13.2
	12.5 ~	64.4	82.1	11.4	3790	6.80	421	511	6070	613	0.688	10.7
	16.0 ~	80.2	102	8.25	4500	6.64	500	621	7340	724	0.679	8.47
200 x 200	5.0 ~	30.4	38.7	37.0	2450	7.95	245	283	3760	362	0.787	25.9
	6.3 ~	38.0	48.4	28.7	3010	7.89	301	350	4650	444	0.784	20.6
	8.0 ~	47.7	60.8	22.0	3710	7.81	371	436	5780	545	0.779	16.3
	10.0	58.8	74.9	17.0	4470	7.72	447	531	7030	655	0.774	13.2
	12.5 ~	72.3	92.1	13.0	5340	7.61	534	643	8490	778	0.768	10.6
	16.0 ~	90.3	115	9.50	6390	7.46	639	785	10300	927	0.759	8.41
250 x 250	6.3 ~	47.9	61.0	36.7	6010	9.93	481	556	9240	712	0.984	20.5
	8.0 ~	60.3	76.8	28.3	7460	9.86	596	694	11500	880	0.979	16.2
	10.0 ~	74.5	94.9	22.0	9060	9.77	724	851	14100	1070	0.974	13.1
	12.5 ~	91.9	117	17.0	10900	9.66	873	1040	17200	1280	0.968	10.5
	16.0 ~	115	147	12.6	13300	9.50	1060	1280	21100	1550	0.959	8.31
300 x 300	6.3 ~	57.8	73.6	44.6	10500	12.0	703	809	16100	1040	1.18	20.4
	8.0 ~	72.8	92.8	34.5	13100	11.9	875	1010	20200	1290	1.18	16.2
	10.0 ~	90.2	115	27.0	16000	11.8	1070	1250	24800	1580	1.17	13.0
	12.5 ~	112	142	21.0	19400	11.7	1300	1530	30300	1900	1.17	10.5
	16.0 ~	141	179	15.8	23900	11.5	1590	1900	37600	2330	1.16	8.26
350 x 350	8.0 ~	85.4	109	40.8	21100	13.9	1210	1390	32400	1790	1.38	16.2
	10.0 ~	106	135	32.0	25900	13.9	1480	1720	39900	2190	1.37	12.9
	12.5 ~	131	167	25.0	31500	13.7	1800	2110	48700	2650	1.37	10.5
	16.0 ~	166	211	18.9	38900	13.6	2230	2630	61000	3260	1.36	8.19
400 x 400	10.0 ~	122	155	37.0	39100	15.9	1960	2260	60100	2900	1.57	12.9
	12.5	151	192	29.0	47800	15.8	2390	2780	73700	3530	1.57	10.4
	16.0 ~	191	243	22.0	59300	15.6	2970	3480	92400	4360	1.56	8.17
	20.0 ~	235	300	17.0	71500	15.4	3580	4250	113000	5240	1.55	6.60

~ Check availability in S275.
(1) For local buckling calculation d = D - 3t.
FOR EXPLANATION OF TABLES SEE NOTES 2 AND 3

BS 5950-1: 2000
BS EN 10210-2: 1997

HOT-FINISHED
RECTANGULAR HOLLOW SECTIONS

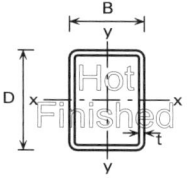

DIMENSIONS AND PROPERTIES

Section Designation		Mass per Metre	Area of Section	Ratios for Local Buckling		Second Moment of Area		Radius of Gyration		Elastic Modulus		Plastic Modulus		Torsional Constants		Surface Area	
Size	Thickness					Axis x-x	Axis y-y	Axis x-x	Axis y-y	Axis x-x	Axis y-y	Axis x-x	Axis y-y			Per Metre	Per Tonne
D x B mm	t mm	kg/m	A cm^2	$d/t^{(1)}$	$b/t^{(1)}$	cm^4	cm^4	cm	cm	cm^3	cm^3	cm^3	cm^3	J cm^4	C cm^3	m^2	m^2
50 x 30	3.2 ~	3.61	4.60	12.6	6.38	14.2	6.20	1.76	1.16	5.68	4.13	7.25	5.00	14.19	6.80	0.152	42.1
60 x 40	3.0 ~	4.35	5.54	17.0	10.3	26.5	13.9	2.18	1.58	8.82	6.95	10.9	8.19	29.2	11.2	0.192	44.1
	4.0 ~	5.64	7.19	12.0	7.00	32.8	17.0	2.14	1.54	10.9	8.52	13.8	10.3	36.7	13.7	0.190	33.7
	5.0 ~	6.85	8.73	9.00	5.00	38.1	19.5	2.09	1.50	12.7	9.77	16.4	12.2	43.0	15.7	0.187	27.3
80 x 40	3.2 ~	5.62	7.16	22.0	9.50	57.2	18.9	2.83	1.63	14.3	9.46	18.0	11.0	46.2	16.1	0.232	41.3
	4.0 ~	6.90	8.79	17.0	7.00	68.2	22.2	2.79	1.59	17.1	11.1	21.8	13.2	55.2	18.9	0.230	33.3
	5.0 ~	8.42	10.7	13.0	5.00	80.3	25.7	2.74	1.55	20.1	12.9	26.1	15.7	65.1	21.9	0.227	27.0
	6.3 ~	10.3	13.1	9.70	3.35	93.3	29.2	2.67	1.49	23.3	14.6	31.1	18.4	75.6	24.8	0.224	21.7
	8.0 ~	12.5	16.0	7.00	2.00	106	32.1	2.58	1.42	26.5	16.1	36.5	21.2	85.8	27.4	0.219	17.5
90 x 50	3.6 ~	7.40	9.42	22.0	10.9	98.3	38.7	3.23	2.03	21.8	15.5	27.2	18.0	89.4	25.9	0.271	36.6
	5.0 ~	9.99	12.7	15.0	7.00	127	49.2	3.16	1.97	28.3	19.7	36.0	23.5	116	32.9	0.267	26.7
	6.3 ~	12.3	15.6	11.3	4.94	150	57.0	3.10	1.91	33.3	22.8	43.2	28.0	138	38.1	0.264	21.5
100 x 50	3.0 ~	6.71	8.54	30.3	13.7	110	36.8	3.58	2.08	21.9	14.7	27.3	16.8	88.4	25.0	0.292	43.5
	3.2 ~	7.13	9.08	28.3	12.6	116	38.8	3.57	2.07	23.2	15.5	28.9	17.7	93.4	26.4	0.292	41.0
	5.0 ~	10.8	13.7	17.0	7.00	167	54.3	3.48	1.99	33.3	21.7	42.6	25.8	135	36.9	0.287	26.6
	6.3 ~	13.3	16.9	12.9	4.94	197	63.0	3.42	1.93	39.4	25.2	51.3	30.8	160	42.9	0.284	21.4
	8.0 ~	16.3	20.8	9.50	3.25	230	71.7	3.33	1.86	46.0	28.7	61.4	36.3	186	48.9	0.279	17.1
	10.0 ^	19.6	24.9	7.00	2.00	259	78.4	3.22	1.77	51.8	31.4	71.2	41.4	209	53.6	0.274	14.0
100 x 60	3.6 ~	8.53	10.9	24.8	13.7	145	64.8	3.65	2.44	28.9	21.6	35.6	24.9	142	35.6	0.311	36.5
	5.0 ~	11.6	14.7	17.0	9.00	189	83.6	3.58	2.38	37.8	27.9	47.4	32.9	188	45.9	0.307	26.5
	6.3 ~	14.2	18.1	12.9	6.52	225	98.1	3.52	2.33	45.0	32.7	57.3	39.5	224	53.8	0.304	21.4
	8.0 ~	17.5	22.4	9.50	4.50	264	113	3.44	2.25	52.8	37.8	68.7	47.1	265	62.2	0.299	17.1
120 x 60	3.6 ~	9.70	12.3	30.3	13.7	227	76.3	4.30	2.49	37.9	25.4	47.2	28.9	183	43.3	0.351	36.2
	5.0 ~	13.1	16.7	21.0	9.00	299	98.8	4.23	2.43	49.9	32.9	63.1	38.4	242	56.0	0.347	26.5
	6.3 ~	16.2	20.7	16.0	6.52	358	116	4.16	2.37	59.7	38.8	76.7	46.3	290	65.9	0.344	21.2
	8.0 ~	20.1	25.6	12.0	4.50	425	135	4.08	2.30	70.8	45.0	92.7	55.4	344	76.6	0.339	16.9
120 x 80	5.0 ~	14.7	18.7	21.0	13.0	365	193	4.42	3.21	60.9	48.2	74.6	56.1	401	77.9	0.387	26.3
	6.3	18.2	23.2	16.0	9.70	440	230	4.36	3.15	73.3	57.6	91.0	68.2	487	92.9	0.384	21.1
	8.0	22.6	28.8	12.0	7.00	525	273	4.27	3.08	87.5	68.1	111	82.6	587	110	0.379	16.8
	10.0 ^	27.4	34.9	9.00	5.00	609	313	4.18	2.99	102	78.1	131	97.3	688	126	0.374	13.6
150 x 100	5.0 ~	18.6	23.7	27.0	17.0	739	392	5.58	4.07	98.5	78.5	119	90.1	807	127	0.487	26.2
	6.3	23.1	29.5	20.8	12.9	898	474	5.52	4.01	120	94.8	147	110	986	153	0.484	21.0
	8.0 ~	28.9	36.8	15.8	9.50	1090	569	5.44	3.94	145	114	180	135	1200	183	0.479	16.6
	10.0 ~	35.3	44.9	12.0	7.00	1280	665	5.34	3.85	171	133	216	161	1430	214	0.474	13.4
	12.5 ~	42.8	54.6	9.00	5.00	1490	763	5.22	3.74	198	153	256	190	1680	246	0.468	10.9
160 x 80	4.0 ~	14.4	18.4	37.0	17.0	612	207	5.77	3.35	76.5	51.7	94.7	58.3	493	88.0	0.470	32.6
	5.0 ~	17.8	22.7	29.0	13.0	744	249	5.72	3.31	93.0	62.3	116	71.1	600	106	0.467	26.2
	6.3	22.2	28.2	22.4	9.70	903	299	5.66	3.26	113	74.8	142	86.8	730	127	0.464	20.9
	8.0	27.6	35.2	17.0	7.00	1090	356	5.57	3.18	136	89.0	175	106	883	151	0.459	16.6
	10.0 ~	33.7	42.9	13.0	5.00	1280	411	5.47	3.10	161	103	209	125	1040	175	0.454	13.5

~ Check availability in S275.
^ Check availability in S355.
(1) For local buckling calculation d = D - 3t and b = B -3t.
FOR EXPLANATION OF TABLES SEE NOTES 2 AND 3

BS 5950-1: 2000
BS EN 10210-2: 1997

HOT-FINISHED RECTANGULAR HOLLOW SECTIONS

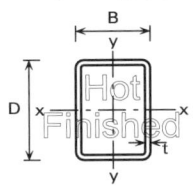

DIMENSIONS AND PROPERTIES

Section Designation Size D x B mm	Mass per Metre t mm	Area of Section A cm²	Ratios for Local Buckling d/t(1)	Ratios for Local Buckling b/t(1)	Second Moment of Area Axis x-x cm⁴	Second Moment of Area Axis y-y cm⁴	Radius of Gyration Axis x-x cm	Radius of Gyration Axis y-y cm	Elastic Modulus Axis x-x cm³	Elastic Modulus Axis y-y cm³	Plastic Modulus Axis x-x cm³	Plastic Modulus Axis y-y cm³	Torsional Constants J cm⁴	Torsional Constants C cm³	Surface Area Per Metre m²	Surface Area Per Tonne m²
200 x 100	5.0 ~	22.6	28.7	37.0	17.0	1500	505	7.21	4.19	149	101	185	114	1200	172	0.587	26.0
	6.3 ~	28.1	35.8	28.7	12.9	1830	613	7.15	4.14	183	123	228	140	1470	208	0.584	20.8
	8.0	35.1	44.8	22.0	9.50	2230	739	7.06	4.06	223	148	282	172	1800	251	0.579	16.5
	10.0	43.1	54.9	17.0	7.00	2660	869	6.96	3.98	266	174	341	206	2160	295	0.574	13.3
	12.5	52.7	67.1	13.0	5.20	3140	1000	6.84	3.87	314	201	408	245	2540	341	0.568	10.8
200 x 120	5.0 ~	24.1	30.7	37.0	21.0	1690	762	7.40	4.98	168	127	205	144	1650	210	0.627	26.0
	6.3 ~	30.1	38.3	28.7	16.0	2070	929	7.34	4.92	207	155	253	177	2030	255	0.624	20.7
	8.0 ~	37.6	48.0	22.0	12.0	2530	1130	7.26	4.85	253	188	313	218	2490	310	0.619	16.5
	10.0 ~	46.3	58.9	17.0	9.00	3030	1340	7.17	4.76	303	223	379	263	3000	367	0.614	13.3
200 x 150	8.0 ~	41.4	52.8	22.0	15.8	2970	1890	7.50	5.99	297	253	359	294	3640	398	0.679	16.4
	10.0 ~	51.1	64.9	17.0	12.0	3570	2260	7.41	5.91	357	302	436	356	4410	475	0.674	13.2
250 x 100	10.0 ~	51.0	64.9	22.0	7.00	4730	1070	8.54	4.06	379	214	491	251	2910	376	0.674	13.2
	12.5 ~	62.5	79.6	17.0	5.00	5620	1250	8.41	3.96	450	249	592	299	3440	438	0.668	10.7
250 x 150	5.0 ~	30.4	38.7	47.0	27.0	3360	1530	9.31	6.28	269	204	324	228	3280	337	0.787	25.9
	6.3	38.0	48.4	36.7	20.8	4140	1870	9.25	6.22	331	250	402	283	4050	413	0.784	20.6
	8.0 ~	47.7	60.8	28.3	15.8	5110	2300	9.17	6.15	409	306	501	350	5020	506	0.779	16.3
	10.0 ~	58.8	74.9	22.0	12.0	6170	2760	9.08	6.06	494	367	611	426	6090	605	0.774	13.2
	12.5 ~	72.3	92.1	17.0	9.00	7390	3270	8.96	5.96	591	435	740	514	7330	717	0.768	10.6
	16.0 ~	90.3	115	12.6	6.38	8880	3870	8.79	5.80	710	516	906	625	8870	849	0.759	8.41
300 x 100	8.0 ~	47.7	60.8	34.5	9.50	6310	1080	10.2	4.21	420	216	546	245	3070	387	0.779	16.3
	10.0 ~	58.8	74.9	27.0	7.00	7610	1280	10.1	4.13	508	255	666	296	3680	458	0.774	13.2
300 x 200	6.3 ~	47.9	61.0	44.6	28.7	7830	4190	11.3	8.29	522	419	624	472	8480	681	0.984	20.5
	8.0 ~	60.3	76.8	34.5	22.0	9720	5180	11.3	8.22	648	518	779	589	10600	840	0.979	16.2
	10.0 ~	74.5	94.9	27.0	17.0	11800	6280	11.2	8.13	788	628	956	721	12900	1020	0.974	13.1
	12.5 ~	91.9	117	21.0	13.0	14300	7540	11.0	8.02	952	754	1170	877	15700	1220	0.968	10.5
	16.0	115	147	15.8	9.50	17400	9110	10.9	7.87	1160	911	1440	1080	19300	1470	0.959	8.34
400 x 200	8.0 ~	72.8	92.8	47.0	22.0	19600	6660	14.5	8.47	978	666	1200	743	15700	1140	1.18	16.2
	10.0 ~	90.2	115	37.0	17.0	23900	8080	14.4	8.39	1200	808	1480	911	19300	1380	1.17	13.0
	12.5 ~	112	142	29.0	13.0	29100	9740	14.3	8.28	1450	974	1810	1110	23400	1660	1.17	10.5
	16.0 ~	141	179	22.0	9.50	35700	11800	14.1	0.13	1790	1180	2260	1370	28900	2010	1.16	8.26
450 x 250	8.0 ~	85.4	109	53.3	28.3	30100	12100	16.6	10.6	1340	971	1620	1080	27100	1630	1.38	16.2
	10.0 ~	106	135	42.0	22.0	36900	14800	16.5	10.5	1640	1190	2000	1330	33300	1990	1.37	12.9
	12.5 ~	131	167	33.0	17.0	45000	18000	16.4	10.4	2000	1440	2460	1630	40700	2410	1.37	10.5
	16.0	166	211	25.1	12.6	55700	22000	16.2	10.2	2480	1760	3070	2030	50500	2950	1.36	8.19
500 x 300	8.0 ~	98.0	125	59.5	34.5	43700	20000	18.7	12.6	1750	1330	2100	1480	42600	2200	1.58	16.1
	10.0 ~	122	155	47.0	27.0	53800	24400	18.6	12.6	2150	1630	2600	1830	52400	2700	1.57	12.9
	12.5 ~	151	192	37.0	21.0	66500	30000	18.5	12.5	2630	1990	3200	2240	64400	3280	1.57	10.4
	16.0 ~	191	243	28.3	15.8	81800	36800	18.3	12.3	3270	2450	4010	2800	80300	4040	1.56	8.17
	20.0 ~	235	300	22.0	12.0	98800	44100	18.2	12.1	3950	2940	4890	3410	97400	4840	1.55	6.60

~ Check availability in S275.
(1) For local buckling calculation d = D - 3t and b = B -3t.
FOR EXPLANATION OF TABLES SEE NOTES 2 AND 3

BS 5950-1: 2000
BS EN 10219-2: 1997

COLD-FORMED
CIRCULAR HOLLOW SECTIONS

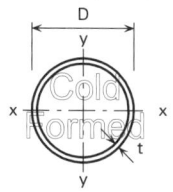

DIMENSIONS AND PROPERTIES

Section Designation		Mass per Metre	Area of Section	Ratio for Local Buckling	Second Moment of Area	Radius of Gyration	Elastic Modulus	Plastic Modulus	Torsional Constants		Surface Area	
Outside Diameter	Thickness										Per Metre	Per Tonne
D	t		A	D/t	I	r	Z	S	J	C		
mm	mm	kg/m	cm²		cm⁴	cm	cm³	cm³	cm⁴	cm³	m²	m²
26.9	2.0 ‡	1.23	1.56	13.5	1.22	0.883	0.907	1.24	2.44	1.81	0.0845	68.7
	2.5 ‡	1.50	1.92	10.8	1.44	0.867	1.07	1.49	2.88	2.14	0.0845	56.3
	3.0 ‡	1.77	2.25	8.97	1.63	0.852	1.21	1.72	3.27	2.43	0.0845	47.7
33.7	2.0 ‡	1.56	1.99	16.9	2.51	1.12	1.49	2.01	5.02	2.98	0.106	67.9
	2.5 ‡	1.92	2.45	13.5	3.00	1.11	1.78	2.44	6.00	3.56	0.106	55.2
	3.0 ‡	2.27	2.89	11.2	3.44	1.09	2.04	2.84	6.88	4.08	0.106	46.7
	4.0 ‡	2.93	3.73	8.43	4.19	1.06	2.49	3.55	8.38	4.97	0.106	36.2
	4.5 ‡	3.24	4.13	7.49	4.50	1.04	2.67	3.87	9.01	5.35	0.106	32.7
42.4	2.5 ‡	2.46	3.13	17.0	6.26	1.41	2.95	3.99	12.5	5.91	0.133	54.1
	3.0 ‡	2.91	3.71	14.1	7.25	1.40	3.42	4.67	14.5	6.84	0.133	45.7
	3.5 ‡	3.36	4.28	12.1	8.16	1.38	3.85	5.31	16.3	7.69	0.133	39.6
	4.0 ‡	3.79	4.83	10.6	8.99	1.36	4.24	5.92	18.0	8.48	0.133	35.1
48.3	2.5 ‡	2.82	3.60	19.3	9.46	1.62	3.92	5.25	18.9	7.83	0.152	53.9
	3.0 ‡	3.35	4.27	16.1	11.0	1.61	4.55	6.17	22.0	9.11	0.152	45.4
	3.5 ‡	3.87	4.93	13.8	12.4	1.59	5.15	7.04	24.9	10.3	0.152	39.3
	4.0 ‡	4.37	5.57	12.1	13.8	1.57	5.70	7.87	27.5	11.4	0.152	34.8
	5.0 ‡	5.34	6.80	9.66	16.2	1.54	6.69	9.42	32.3	13.4	0.152	28.5
60.3	2.5 ‡	3.56	4.54	24.1	19.0	2.05	6.30	8.36	38.0	12.6	0.189	53.1
	3.0 ‡	4.24	5.40	20.1	22.2	2.03	7.37	9.86	44.4	14.7	0.189	44.6
	3.5 ‡	4.90	6.25	17.2	25.3	2.01	8.39	11.3	50.6	16.8	0.189	38.6
	4.0 ‡	5.55	7.07	15.1	28.2	2.00	9.34	12.7	56.3	18.7	0.189	34.1
	5.0 ‡	6.82	8.69	12.1	33.5	1.96	11.1	15.3	67.0	22.2	0.189	27.7
76.1	2.5 ‡	4.54	5.78	30.4	39.2	2.60	10.3	13.5	78.4	20.6	0.239	52.6
	3.0 ‡	5.41	6.89	25.4	46.1	2.59	12.1	16.0	92.2	24.2	0.239	44.2
	3.5 ‡	6.27	7.98	21.7	52.7	2.57	13.9	18.5	105	27.7	0.239	38.1
	4.0 ‡	7.11	9.06	19.0	59.1	2.55	15.5	20.8	118	31.0	0.239	33.6
	5.0 ‡	8.77	11.2	15.2	70.9	2.52	18.6	25.3	142	37.3	0.239	27.3
88.9	3.0 ‡	6.36	8.10	29.6	74.8	3.04	16.8	22.1	150	33.6	0.279	43.9
	3.5 ‡	7.37	9.39	25.4	85.7	3.02	19.3	25.5	171	38.6	0.279	37.9
	4.0 ‡	8.38	10.7	22.2	96.3	3.00	21.7	28.9	193	43.3	0.279	33.3
	5.0 ‡	10.3	13.2	17.8	116	2.97	26.2	35.2	233	52.4	0.279	27.1
114.3	3.0 ‡	8.23	10.5	38.1	163	3.94	28.4	37.2	325	56.9	0.359	43.6
	3.5 ‡	9.56	12.2	32.7	187	3.92	32.7	43.0	374	65.5	0.359	37.6
	4.0 ‡	10.9	13.9	28.6	211	3.90	36.9	48.7	422	73.9	0.359	32.9
	5.0 ‡	13.5	17.2	22.9	257	3.87	45.0	59.8	514	89.9	0.359	26.6
	6.0 ‡	16.0	20.4	19.1	300	3.83	52.5	70.4	600	105	0.359	22.4
139.7	4.0 ‡	13.4	17.1	34.9	393	4.80	56.2	73.7	786	112	0.439	32.8
	5.0 ‡	16.6	21.2	27.9	481	4.77	68.8	90.8	961	138	0.439	26.4
	6.0 ‡	19.8	25.2	23.3	564	4.73	80.8	107	1130	162	0.439	22.2
	8.0 ‡	26.0	33.1	17.5	720	4.66	103	139	1440	206	0.439	16.9
	10.0 ‡	32.0	40.7	14.0	862	4.60	123	169	1720	247	0.439	13.7
	12.5 ‡	39.2	50.0	11.2	1020	4.52	146	203	2040	292	0.439	11.20

‡ Grade S275 not available from some leading producers. Check availability.
FOR EXPLANATION OF TABLES SEE NOTES 2 AND 3

		BS 5950-1: 2000
		BS EN 10219-2: 1997

COLD-FORMED CIRCULAR HOLLOW SECTIONS

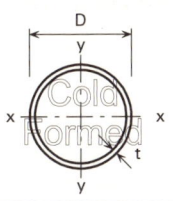

DIMENSIONS AND PROPERTIES

Section Designation		Mass per Metre	Area of Section	Ratio for Local Buckling	Second Moment of Area	Radius of Gyration	Elastic Modulus	Plastic Modulus	Torsional Constants		Surface Area	
Outside Diameter D mm	Thickness t mm	kg/m	A cm²	D/t	I cm⁴	r cm	Z cm³	S cm³	J cm⁴	C cm³	Per Metre m²	Per Tonne m²
168.3	4.0 ‡	16.2	20.6	42.1	697	5.81	82.8	108	1390	166	0.529	32.7
	5.0 ‡	20.1	25.7	33.7	856	5.78	102	133	1710	203	0.529	26.3
	6.0 ‡	24.0	30.6	28.1	1010	5.74	120	158	2020	240	0.529	22.0
	8.0 ‡	31.6	40.3	21.0	1300	5.67	154	206	2600	308	0.529	16.7
	10.0 ‡	39.0	49.7	16.8	1560	5.61	186	251	3130	372	0.529	13.6
	12.5 ‡	48.0	61.2	13.5	1870	5.53	222	304	3740	444	0.529	11.02
193.7	4.0 ‡	18.7	23.8	48.4	1070	6.71	111	144	2150	222	0.609	32.6
	4.5 ‡	21.0	26.7	43.0	1200	6.69	124	161	2400	247	0.609	29.0
	5.0 ‡	23.3	29.6	38.7	1320	6.67	136	178	2640	273	0.609	26.1
	6.0 ‡	27.8	35.4	32.3	1560	6.64	161	211	3120	322	0.609	21.9
	8.0 ‡	36.6	46.7	24.2	2020	6.57	208	276	4030	416	0.609	16.6
	10.0 ‡	45.3	57.7	19.4	2440	6.50	252	338	4880	504	0.609	13.4
	12.5 ‡	55.9	71.2	15.5	2930	6.42	303	411	5870	606	0.609	10.89
219.1	4.0 ‡	21.2	27.0	54.8	1560	7.61	143	185	3130	286	0.688	32.5
	4.5 ‡	23.8	30.3	48.7	1750	7.59	159	207	3490	319	0.688	28.9
	5.0 ‡	26.4	33.6	43.8	1930	7.57	176	229	3860	352	0.688	26.1
	6.0 ‡	31.5	40.2	36.5	2280	7.54	208	273	4560	417	0.688	21.8
	8.0 ‡	41.6	53.1	27.4	2960	7.47	270	357	5920	540	0.688	16.5
	10.0 ‡	51.6	65.7	21.9	3600	7.40	328	438	7200	657	0.688	13.3
	12.0 ‡	61.3	78.1	18.3	4200	7.33	383	515	8400	767	0.688	11.22
	12.5 ‡	63.7	81.1	17.5	4350	7.32	397	534	8690	793	0.688	10.80
	16.0 ‡	80.1	102	13.7	5300	7.20	483	661	10600	967	0.688	8.59
244.5	4.5 ‡	26.6	33.9	54.3	2440	8.49	200	259	4890	400	0.768	28.9
	5.0 ‡	29.5	37.6	48.9	2700	8.47	221	287	5400	441	0.768	26.0
	6.0 ‡	35.3	45.0	40.8	3200	8.43	262	341	6400	523	0.768	21.8
	8.0 ‡	46.7	59.4	30.6	4160	8.37	340	448	8320	681	0.768	16.4
	10.0 ‡	57.8	73.7	24.5	5070	8.30	415	550	10200	830	0.768	13.3
	12.0 ‡	68.8	87.7	20.4	5940	8.23	486	649	11900	972	0.768	11.16
	12.5 ‡	71.5	91.1	19.6	6150	8.21	503	673	12300	1010	0.768	10.74
	16.0 ‡	90.2	115	15.3	7530	8.10	616	837	15100	1230	0.768	8.51
273.0	4.0 ‡	26.5	33.8	68.3	3060	9.51	224	289	6120	448	0.858	32.4
	4.5 ‡	29.8	38.0	60.7	3420	9.49	251	324	6840	501	0.858	28.8
	5.0 ‡	33.0	42.2	54.6	3780	9.48	277	359	7560	554	0.858	26.0
	6.0 ‡	39.5	50.3	45.5	4490	9.44	329	428	8970	657	0.858	21.7
	8.0 ‡	52.3	66.6	34.1	5850	9.37	429	562	11700	857	0.858	16.4
	10.0 ‡	64.9	82.6	27.3	7150	9.31	524	692	14300	1050	0.858	13.2
	12.0 ‡	77.2	98.2	22.8	8400	9.24	615	818	16800	1230	0.858	11.11
	12.5 ‡	80.3	102	21.8	8700	9.22	637	849	17400	1270	0.858	10.68
	16.0 ‡	101	129	17.1	10700	9.10	784	1060	21400	1570	0.858	8.50

‡ Grade S275 not available from some leading producers. Check availability.
FOR EXPLANATION OF TABLES SEE NOTES 2 AND 3

BS 5950-1: 2000
BS EN 10219-2: 1997

COLD-FORMED
CIRCULAR HOLLOW SECTIONS

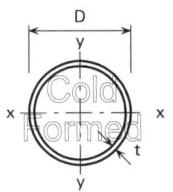

DIMENSIONS AND PROPERTIES

Section Designation		Mass per Metre	Area of Section	Ratio for Local Buckling	Second Moment of Area	Radius of Gyration	Elastic Modulus	Plastic Modulus	Torsional Constants		Surface Area	
Outside Diameter D	Thickness t		A	D/t	I	r	Z	S	J	C	Per Metre	Per Tonne
mm	mm	kg/m	cm²		cm⁴	cm	cm³	cm³	cm⁴	cm³	m²	m²
323.9	5.0 ‡	39.3	50.1	64.8	6370	11.3	393	509	12700	787	1.02	20.4
	6.0 ‡	47.0	59.9	54.0	7570	11.2	468	606	15100	935	1.02	17.0
	8.0 ‡	62.3	79.4	40.5	9910	11.2	612	799	19800	1220	1.02	12.8
	10.0 ‡	77.4	98.6	32.4	12200	11.1	751	986	24300	1500	1.02	10.3
	12.0 ‡	92.3	118	27.0	14300	11.0	884	1170	28600	1770	1.02	8.64
	12.5 ‡	96.0	122	25.9	14800	11.0	917	1210	29700	1830	1.02	8.36
	16.0 ‡	121	155	20.2	18400	10.9	1140	1520	36800	2270	1.02	6.58
355.6	5.0 ‡	43.2	55.1	71.1	8460	12.4	476	615	16900	952	1.12	20.3
	6.0 ‡	51.7	65.9	59.3	10100	12.4	566	733	20100	1130	1.12	17.0
	8.0 ‡	68.6	87.4	44.5	13200	12.3	742	967	26400	1490	1.12	12.8
	10.0 ‡	85.2	109	35.6	16200	12.2	912	1200	32400	1830	1.12	10.3
	12.0 ‡	102	130	29.6	19100	12.2	1080	1420	38300	2150	1.12	8.62
	12.5 ‡	106	135	28.4	19900	12.1	1120	1470	39700	2230	1.12	8.30
	16.0 ‡	134	171	22.2	24700	12.0	1390	1850	49300	2770	1.12	6.55
406.4	6.0 ‡	59.2	75.5	67.7	15100	14.2	745	962	30300	1490	1.28	17.0
	8.0 ‡	78.6	100	50.8	19900	14.1	978	1270	39700	1960	1.28	12.8
	10.0 ‡	97.8	125	40.6	24500	14.0	1210	1570	49000	2410	1.28	10.2
	12.0 ‡	117	149	33.9	28900	14.0	1420	1870	57900	2850	1.28	8.59
	12.5 ‡	121	155	32.5	30000	13.9	1480	1940	60100	2960	1.28	8.26
	16.0 ‡	154	196	25.4	37400	13.8	1840	2440	74900	3690	1.28	6.53
457.0	8.0 ‡	88.6	113	57.1	28400	15.9	1250	1610	56900	2490	1.44	12.7
	10.0 ‡	110	140	45.7	35100	15.8	1540	2000	70200	3070	1.44	10.3
	12.0 ‡	132	168	38.1	41600	15.7	1820	2380	83100	3640	1.44	8.57
	12.5 ‡	137	175	36.6	43100	15.7	1890	2470	86300	3780	1.44	8.23
	16.0 ‡	174	222	28.6	54000	15.6	2360	3110	108000	4720	1.44	6.49
508.0	8.0 ‡	98.6	126	63.5	39300	17.7	1550	2000	78600	3090	1.60	12.7
	10.0 ‡	123	156	50.8	48500	17.6	1910	2480	97000	3820	1.60	10.3
	12.0 ‡	147	187	42.3	57500	17.5	2270	2950	115000	4530	1.60	8.56
	12.5 ‡	153	195	40.6	59800	17.5	2350	3070	120000	4710	1.60	8.21
	16.0 ‡	194	247	31.8	74900	17.4	2950	3870	150000	5900	1.60	6.48

‡ Grade S275 not available from some leading producers. Check availability.
FOR EXPLANATION OF TABLES SEE NOTES 2 AND 3

BS 5950-1: 2000
BS EN 10219-2: 1997

COLD-FORMED SQUARE HOLLOW SECTIONS

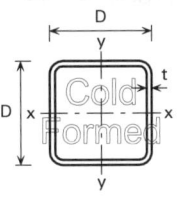

DIMENSIONS AND PROPERTIES

Section Designation		Mass per Metre	Area of Section	Ratio for Local Buckling	Second Moment of Area	Radius of Gyration	Elastic Modulus	Plastic Modulus	Torsional Constants		Surface Area	
Size D x D mm	Thickness t mm	kg/m	A cm^2	$d/t^{(1)}$	I cm^4	r cm	Z cm^3	S cm^3	J cm^4	C cm^3	Per Metre m^2	Per Tonne m^2
25 x 25	2.0 ‡	1.36	1.74	7.50	1.48	0.924	1.19	1.47	2.53	1.80	0.0931	68.5
	2.5 ‡	1.64	2.09	5.00	1.69	0.899	1.35	1.71	2.97	2.07	0.0914	55.7
	3.0 ‡	1.89	2.41	3.33	1.84	0.874	1.47	1.91	3.33	2.27	0.0897	47.5
30 x 30	2.0 ‡	1.68	2.14	10.0	2.72	1.13	1.81	2.21	4.54	2.75	0.113	67.3
	2.5 ‡	2.03	2.59	7.00	3.16	1.10	2.10	2.61	5.40	3.20	0.111	54.7
	3.0 ‡	2.36	3.01	5.00	3.50	1.08	2.34	2.96	6.15	3.58	0.110	46.6
40 x 40	2.0 ‡	2.31	2.94	15.0	6.94	1.54	3.47	4.13	11.3	5.23	0.153	66.2
	2.5 ‡	2.82	3.59	11.0	8.22	1.51	4.11	4.97	13.6	6.21	0.151	53.5
	3.0 ‡	3.30	4.21	8.33	9.32	1.49	4.66	5.72	15.8	7.07	0.150	45.5
	4.0 ‡	4.20	5.35	5.00	11.1	1.44	5.54	7.01	19.4	8.48	0.146	34.8
50 x 50	2.0 ‡	2.93	3.74	20.0	14.1	1.95	5.66	6.66	22.6	8.51	0.193	65.9
	2.5 ‡	3.60	4.59	15.0	16.9	1.92	6.78	8.07	27.5	10.2	0.191	53.1
	3.0 ‡	4.25	5.41	11.7	19.5	1.90	7.79	9.39	32.1	11.8	0.190	44.7
	4.0 ‡	5.45	6.95	7.50	23.7	1.85	9.49	11.7	40.4	14.4	0.186	34.1
	5.0 ‡	6.56	8.36	5.00	27.0	1.80	10.8	13.7	47.5	16.6	0.183	27.9
60 x 60	3.0 ‡	5.19	6.61	15.0	35.1	2.31	11.7	14.0	57.1	17.7	0.230	44.3
	4.0 ‡	6.71	8.55	10.0	43.6	2.26	14.5	17.6	72.6	22.0	0.226	33.7
	5.0 ‡	8.13	10.4	7.00	50.5	2.21	16.8	20.9	86.4	25.6	0.223	27.4
70 x 70	2.5 ‡	5.17	6.59	23.0	49.4	2.74	14.1	16.5	78.5	21.2	0.271	52.4
	3.0 ‡	6.13	7.81	18.3	57.5	2.71	16.4	19.4	92.4	24.7	0.270	44.0
	3.5 ‡	7.06	8.99	15.0	65.1	2.69	18.6	22.2	106	28.0	0.268	38.0
	4.0 ‡	7.97	10.1	12.5	72.1	2.67	20.6	24.8	119	31.1	0.266	33.4
	5.0 ‡	9.70	12.4	9.00	84.6	2.62	24.2	29.6	142	36.7	0.263	27.1
80 x 80	3.0 ‡	7.07	9.01	21.7	87.8	3.12	22.0	25.8	140	33.0	0.310	43.8
	3.5 ‡	8.16	10.4	17.9	99.8	3.10	25.0	29.5	161	37.6	0.308	37.7
	4.0 ‡	9.22	11.7	15.0	111	3.07	27.8	33.1	180	41.8	0.306	33.2
	5.0 ‡	11.3	14.4	11.0	131	3.03	32.9	39.7	218	49.7	0.303	26.8
	6.0 ‡	13.2	16.8	8.33	149	2.98	37.3	45.8	252	56.6	0.299	22.7
90 x 90	3.0 ‡	8.01	10.2	25.0	127	3.53	28.3	33.0	201	42.5	0.350	43.7
	3.5 ‡	9.26	11.8	20.7	145	3.51	32.2	37.9	232	40.5	0.348	37.6
	4.0 ‡	10.5	13.3	17.5	162	3.48	36.0	42.6	261	54.2	0.346	33.0
	5.0 ‡	12.8	16.4	13.0	193	3.43	42.9	51.4	316	64.7	0.343	26.8
	6.0 ‡	15.1	19.2	10.0	220	3.39	49.0	59.5	368	74.2	0.339	22.5
100 x 100	3.0 ‡	8.96	11.4	28.3	177	3.94	35.4	41.2	279	53.2	0.390	43.5
	4.0 ‡	11.7	14.9	20.0	226	3.89	45.3	53.3	362	68.1	0.386	33.0
	5.0 ‡	14.4	18.4	15.0	271	3.84	54.2	64.6	441	81.7	0.383	26.6
	6.0 ‡	17.0	21.6	11.7	311	3.79	62.3	75.1	514	94.1	0.379	22.3
	8.0 ‡	21.4	27.2	7.50	366	3.67	73.2	91.1	645	114	0.366	17.1
120 x 120	4.0 ‡	14.2	18.1	25.0	402	4.71	67.0	78.3	637	101	0.466	32.8
	5.0 ‡	17.5	22.4	19.0	485	4.66	80.9	95.4	778	122	0.463	26.5
	6.0 ‡	20.7	26.4	15.0	562	4.61	93.7	112	913	141	0.459	22.2
	8.0 ‡	26.4	33.6	10.0	677	4.49	113	138	1160	175	0.446	16.9
	10.0 ‡	31.8	40.6	7.00	777	4.38	129	162	1380	203	0.437	13.7

‡ Grade S275 not available from some leading producers. Check availability.
(1) For local buckling calculation d = D - 5t.
FOR EXPLANATION OF TABLES SEE NOTES 2 AND 3

BS 5950-1: 2000
BS EN 10219-2: 1997

COLD-FORMED
SQUARE HOLLOW SECTIONS

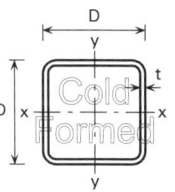

DIMENSIONS AND PROPERTIES

Section Designation Size D x D mm	Thickness t mm	Mass per Metre kg/m	Area of Section A cm²	Ratio for Local Buckling d/t (1)	Second Moment of Area I cm⁴	Radius of Gyration r cm	Elastic Modulus Z cm³	Plastic Modulus S cm³	Torsional Constants J cm⁴	Torsional Constants C cm³	Surface Area Per Metre m²	Surface Area Per Tonne m²
140 x 140	4.0 ‡	16.8	21.3	30.0	652	5.52	93.1	108	1020	140	0.546	32.5
	5.0 ‡	20.7	26.4	23.0	791	5.48	113	132	1260	170	0.543	26.2
	6.0 ‡	24.5	31.2	18.3	920	5.43	131	155	1480	198	0.539	22.0
	8.0 ‡	31.4	40.0	12.5	1130	5.30	161	194	1900	248	0.526	16.8
	10.0 ‡	38.1	48.6	9.00	1310	5.20	187	230	2270	291	0.517	13.6
150 x 150	4.0 ‡	18.0	22.9	32.5	808	5.93	108	125	1260	162	0.586	32.6
	5.0 ‡	22.3	28.4	25.0	982	5.89	131	153	1550	197	0.583	26.1
	6.0 ‡	26.4	33.6	20.0	1150	5.84	153	180	1830	230	0.579	21.9
	8.0 ‡	33.9	43.2	13.8	1410	5.71	188	226	2360	289	0.566	16.7
	10.0 ‡	41.3	52.6	10.0	1650	5.61	220	269	2840	341	0.557	13.5
160 x 160	4.0 ‡	19.3	24.5	35.0	987	6.34	123	143	1540	185	0.626	32.4
	5.0 ‡	23.8	30.4	27.0	1200	6.29	150	175	1900	226	0.623	26.2
	6.0 ‡	28.3	36.0	21.7	1410	6.25	176	206	2240	264	0.619	21.9
	8.0 ‡	36.5	46.4	15.0	1740	6.12	218	260	2900	334	0.606	16.6
	10.0 ‡	44.4	56.6	11.0	2050	6.02	256	311	3490	395	0.597	13.4
180 x 180	5.0 ‡	27.0	34.4	31.0	1740	7.11	193	224	2720	290	0.703	26.0
	6.0 ‡	32.1	40.8	25.0	2040	7.06	226	264	3220	340	0.699	21.8
	8.0 ‡	41.5	52.8	17.5	2550	6.94	283	336	4190	432	0.686	16.5
	10.0 ‡	50.7	64.6	13.0	3020	6.84	335	404	5070	515	0.677	13.4
	12.0 ‡	58.5	74.5	10.0	3320	6.68	369	454	5870	584	0.658	11.2
	12.5 ‡	60.5	77.0	9.40	3410	6.65	378	467	6050	600	0.656	10.8
200 x 200	5.0 ‡	30.1	38.4	35.0	2410	7.93	241	279	3760	362	0.783	26.0
	6.0 ‡	35.8	45.6	28.3	2830	7.88	283	330	4460	426	0.779	21.8
	8.0 ‡	46.5	59.2	20.0	3570	7.76	357	421	5820	544	0.766	16.5
	10.0 ‡	57.0	72.6	15.0	4250	7.65	425	508	7070	651	0.757	13.3
	12.0 ‡	66.0	84.1	11.7	4730	7.50	473	576	8230	743	0.738	11.2
	12.5 ‡	68.3	87.0	11.0	4860	7.47	486	594	8500	765	0.736	10.8
250 x 250	6.0 ‡	45.2	57.6	36.7	5670	9.92	454	524	8840	681	0.979	21.7
	8.0 ‡	59.1	75.2	26.3	7230	9.80	578	676	11600	878	0.966	16.3
	10.0 ‡	72.7	92.6	20.0	8710	9.70	697	822	14200	1060	0.957	13.2
	12.0 ‡	84.8	108	15.8	9860	9.55	789	944	16700	1230	0.938	11.1
	12.5 ‡	88.0	112	15.0	10200	9.52	813	975	17300	1270	0.936	10.6
300 x 300	8.0 ‡	71.6	91.2	32.5	12800	11.8	853	991	20300	1290	1.17	16.3
	10.0 ‡	88.4	113	25.0	15500	11.7	1030	1210	25000	1570	1.16	13.1
	12.0 ‡	104	132	20.0	17800	11.6	1180	1400	29200	1830	1.14	11.0
	12.5 ‡	108	137	19.0	18300	11.6	1220	1450	30600	1890	1.14	10.6

‡ Grade S275 not available from some leading producers. Check availability.
(1) For local buckling calculation d = D - 5t.
FOR EXPLANATION OF TABLES SEE NOTES 2 AND 3

BS 5950-1: 2000
BS EN 10219-2: 1997

COLD-FORMED RECTANGULAR HOLLOW SECTIONS

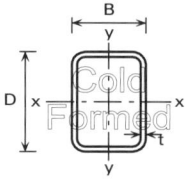

DIMENSIONS AND PROPERTIES

Section Designation Size D x B mm	Thickness t mm	Mass per Metre kg/m	Area of Section A cm²	Ratios for Local Buckling d/t (1)	Ratios for Local Buckling b/t (1)	Second Moment of Area Axis x-x cm⁴	Second Moment of Area Axis y-y cm⁴	Radius of Gyration Axis x-x cm	Radius of Gyration Axis y-y cm	Elastic Modulus Axis x-x cm³	Elastic Modulus Axis y-y cm³	Plastic Modulus Axis x-x cm³	Plastic Modulus Axis y-y cm³	Torsional Constants J cm⁴	Torsional Constants C cm³	Surface Area Per Metre m²	Surface Area Per Tonne m²
50 x 25	2.0 ‡	2.15	2.74	20.0	7.50	8.38	2.81	1.75	1.01	3.35	2.25	4.26	2.62	7.06	3.92	0.143	66.5
	2.5 ‡	2.62	3.34	15.0	5.00	9.89	3.28	1.72	0.991	3.95	2.62	5.11	3.12	8.43	4.60	0.141	53.8
	3.0 ‡	3.07	3.91	11.7	3.33	11.2	3.67	1.69	0.969	4.47	2.93	5.86	3.56	9.64	5.18	0.140	45.6
50 x 30	2.0 ‡	2.31	2.94	20.0	10.0	9.54	4.29	1.80	1.21	3.81	2.86	4.74	3.33	9.77	4.84	0.153	66.2
	2.5 ‡	2.82	3.59	15.0	7.00	11.3	5.05	1.77	1.19	4.52	3.37	5.70	3.98	11.7	5.72	0.151	53.5
	3.0 ‡	3.30	4.21	11.7	5.00	12.8	5.70	1.75	1.16	5.13	3.80	6.57	4.58	13.5	6.49	0.150	45.5
	4.0 ‡	4.20	5.35	7.50	2.50	15.3	6.69	1.69	1.12	6.10	4.46	8.05	5.58	16.5	7.71	0.146	34.8
60 x 30	3.0 ‡	3.77	4.81	15.0	5.00	20.5	6.80	2.06	1.19	6.83	4.53	8.82	5.39	17.5	7.95	0.170	45.1
	4.0 ‡	4.83	6.15	10.0	2.50	24.7	8.06	2.00	1.14	8.23	5.37	10.9	6.62	21.5	9.52	0.166	34.4
60 x 40	2.5 ‡	3.60	4.59	19.0	11.0	22.1	11.7	2.19	1.60	7.36	5.87	9.06	6.84	25.1	9.72	0.191	53.1
	3.0 ‡	4.25	5.41	15.0	8.33	25.4	13.4	2.17	1.58	8.46	6.72	10.5	7.94	29.3	11.2	0.190	44.7
	4.0 ‡	5.45	6.95	10.0	5.00	31.0	16.3	2.11	1.53	10.3	8.14	13.2	9.89	36.7	13.7	0.186	34.1
	5.0 ‡	6.56	8.36	7.00	3.00	35.3	18.4	2.06	1.48	11.8	9.21	15.4	11.5	42.8	15.6	0.183	27.9
70 x 40	3.0 ‡	4.72	6.01	18.3	8.33	37.3	15.5	2.49	1.61	10.7	7.75	13.4	9.05	36.5	13.2	0.210	44.5
	4.0 ‡	6.08	7.75	12.5	5.00	46.0	18.9	2.44	1.56	13.1	9.44	16.8	11.3	45.8	16.2	0.206	33.9
70 x 50	3.0 ‡	5.19	6.61	18.3	11.7	44.1	26.1	2.58	1.99	12.6	10.4	15.4	12.2	53.6	17.1	0.230	44.3
	4.0 ‡	6.71	8.55	12.5	7.50	54.7	32.2	2.53	1.94	15.6	12.9	19.5	15.4	68.1	21.2	0.226	33.7
80 x 40	3.0 ‡	5.19	6.61	21.7	8.33	52.3	17.6	2.81	1.63	13.1	8.78	16.5	10.2	43.9	15.3	0.230	44.3
	4.0 ‡	6.71	8.55	15.0	5.00	64.8	21.5	2.75	1.59	16.2	10.7	20.9	12.8	55.2	18.8	0.226	33.7
	5.0 ‡	8.13	10.4	11.0	3.00	75.1	24.6	2.69	1.54	18.8	12.3	24.7	15.0	65.0	21.7	0.223	27.4
80 x 50	3.0 ‡	5.66	7.21	21.7	11.7	61.4	29.4	2.91	2.02	15.3	11.8	18.8	13.6	65.0	19.7	0.250	44.2
	4.0 ‡	7.34	9.35	15.0	7.50	76.4	36.5	2.86	1.98	19.1	14.6	24.0	17.2	82.7	24.6	0.246	33.5
	5.0 ‡	8.91	11.4	11.0	5.00	89.2	42.3	2.80	1.93	22.3	16.9	28.5	20.5	98.4	28.7	0.243	27.3
80 x 60	3.0 ‡	6.13	7.81	21.7	15.0	70.0	44.9	3.00	2.40	17.5	15.0	21.2	17.4	88.3	24.1	0.270	44.0
	4.0 ‡	7.97	10.1	15.0	10.0	87.9	56.1	2.94	2.35	22.0	18.7	27.0	22.1	113	30.3	0.266	33.4
	5.0 ‡	9.70	12.4	11.0	7.00	103	65.7	2.89	2.31	25.8	21.9	32.2	26.4	136	35.7	0.263	27.1
90 x 50	3.0 ‡	6.13	7.81	25.0	11.7	81.9	32.7	3.24	2.05	18.2	13.1	22.6	15.0	76.7	22.4	0.270	44.0
	3.5 ‡	7.06	8.99	20.7	9.29	92.7	36.9	3.21	2.03	20.6	14.8	25.8	17.1	87.5	25.3	0.268	38.0
	4.0 ‡	7.97	10.1	17.5	7.50	103	40.7	3.18	2.00	22.8	16.3	28.8	19.1	97.7	28.0	0.266	33.4
	5.0 ‡	9.70	12.4	13.0	5.00	121	47.4	3.12	1.96	26.8	18.9	34.4	22.7	110	32.7	0.263	27.1
100 x 40	3.0 ‡	6.13	7.81	28.3	8.33	92.3	21.7	3.44	1.67	18.5	10.8	23.7	12.4	59.0	19.4	0.270	44.0
	4.0 ‡	7.97	10.1	20.0	5.00	116	26.7	3.38	1.62	23.1	13.3	30.3	15.7	74.5	24.0	0.266	33.4
	5.0 ‡	9.70	12.4	15.0	3.00	136	30.8	3.31	1.58	27.1	15.4	36.1	18.5	87.9	27.9	0.263	27.1
100 x 50	3.0 ‡	6.60	8.41	28.3	11.7	106	36.1	3.56	2.07	21.3	14.4	26.7	16.4	88.6	25.0	0.290	43.9
	4.0 ‡	8.59	10.9	20.0	7.50	134	44.9	3.50	2.03	26.8	18.0	34.1	20.9	113	31.3	0.286	33.3
	5.0 ‡	10.5	13.4	15.0	5.00	158	52.5	3.44	1.98	31.6	21.0	40.8	25.0	135	36.8	0.283	27.0
	6.0 ‡	12.3	15.6	11.7	3.33	179	58.7	3.38	1.94	35.8	23.5	46.9	28.5	154	41.4	0.279	22.7
100 x 60	3.0 ‡	7.07	9.01	28.3	15.0	121	54.6	3.66	2.46	24.1	18.2	29.6	20.8	122	30.6	0.310	43.8
	3.5 ‡	8.16	10.4	23.6	12.1	137	61.9	3.63	2.44	27.4	20.6	33.8	23.8	139	34.8	0.308	37.7
	4.0 ‡	9.22	11.7	20.0	10.0	153	68.7	3.60	2.42	30.5	22.9	37.9	26.6	156	38.7	0.306	33.2
	5.0 ‡	11.3	14.4	15.0	7.00	181	80.8	3.55	2.37	36.2	26.9	45.6	31.9	188	45.8	0.303	26.8
	6.0 ‡	13.2	16.8	11.7	5.00	205	91.2	3.49	2.33	41.1	30.4	52.5	36.6	216	51.9	0.299	22.7

‡ Grade S275 not available from some leading producers. Check availability.
(1) For local buckling calculation d = D - 5t and b = B -5t.
FOR EXPLANATION OF TABLES SEE NOTES 2 AND 3

BS 5950-1: 2000
BS EN 10219-2: 1997

COLD-FORMED RECTANGULAR HOLLOW SECTIONS

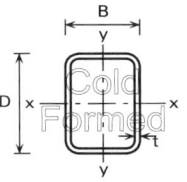

DIMENSIONS AND PROPERTIES

Section Designation Size D x B mm	Mass per Metre Thickness t mm	Area of Section A cm²	Ratios for Local Buckling d/t (1)	Ratios for Local Buckling b/t (1)	Second Moment of Area Axis x-x cm⁴	Second Moment of Area Axis y-y cm⁴	Radius of Gyration Axis x-x cm	Radius of Gyration Axis y-y cm	Elastic Modulus Axis x-x cm³	Elastic Modulus Axis y-y cm³	Plastic Modulus Axis x-x cm³	Plastic Modulus Axis y-y cm³	Torsional Constants J cm⁴	Torsional Constants C cm³	Surface Area Per Metre m²	Surface Area Per Tonne m²
100 x 80	3.0 ‡	8.01	10.2	28.3	21.7	149	106	3.82	3.22	29.8	26.4	35.4	30.4	196	41.9	0.350	43.7
	4.0 ‡	10.5	13.3	20.0	15.0	189	134	3.77	3.17	37.9	33.5	45.6	39.2	254	53.4	0.346	33.0
	5.0 ‡	12.8	16.4	15.0	11.0	226	160	3.72	3.12	45.2	39.9	55.1	47.2	308	63.7	0.343	26.8
120 x 40	3.0 ‡	7.07	9.01	35.0	8.33	148	25.8	4.05	1.69	24.7	12.9	32.2	14.6	74.6	23.5	0.310	43.8
	4.0 ‡	9.22	11.7	25.0	5.00	187	31.9	3.99	1.65	31.1	15.9	41.2	18.5	94.2	29.2	0.306	33.2
	5.0 ‡	11.3	14.4	19.0	3.00	221	36.9	3.92	1.60	36.8	18.5	49.4	22.0	111	34.1	0.303	26.8
120 x 60	3.0 ‡	8.01	10.2	35.0	15.0	189	64.4	4.30	2.51	31.5	21.5	39.2	24.2	156	37.1	0.350	43.7
	3.5 ‡	9.26	11.8	29.3	12.1	216	73.1	4.28	2.49	35.9	24.4	44.9	27.7	179	42.2	0.348	37.6
	4.0 ‡	10.5	13.3	25.0	10.0	241	81.2	4.25	2.47	40.1	27.1	50.5	31.1	201	47.0	0.346	33.0
	5.0 ‡	12.8	16.4	19.0	7.00	287	96.0	4.19	2.42	47.8	32.0	60.9	37.4	242	55.8	0.343	26.8
	6.0 ‡	15.1	19.2	15.0	5.00	328	109	4.13	2.38	54.7	36.3	70.6	43.1	280	63.6	0.339	22.5
120 x 80	3.0 ‡	8.96	11.4	35.0	21.7	230	123	4.49	3.29	38.4	30.9	46.2	35.0	255	50.8	0.390	43.5
	4.0 ‡	11.7	14.9	25.0	15.0	295	157	4.44	3.24	49.1	39.3	59.8	45.2	331	64.9	0.386	33.0
	5.0 ‡	14.4	18.4	19.0	11.0	353	188	4.39	3.20	58.9	46.9	72.4	54.7	402	77.8	0.383	26.6
	6.0 ‡	17.0	21.6	15.0	8.33	406	215	4.33	3.15	67.7	53.8	84.3	63.5	469	89.4	0.379	22.3
	8.0 ‡	21.4	27.2	10.0	5.00	476	252	4.18	3.04	79.3	62.9	102	76.9	584	108	0.366	17.1
140 x 80	3.0 ‡	9.90	12.6	41.7	21.7	334	141	5.15	3.35	47.8	35.3	58.2	39.6	317	59.7	0.430	43.4
	4.0 ‡	13.0	16.5	30.0	15.0	430	180	5.10	3.30	61.4	45.1	75.5	51.3	412	76.5	0.426	32.8
	5.0 ‡	16.0	20.4	23.0	11.0	517	216	5.04	3.26	73.9	54.0	91.8	62.2	501	91.8	0.423	26.4
	6.0 ‡	18.9	24.0	18.3	8.33	597	248	4.98	3.21	85.3	62.0	107	72.4	584	106	0.419	22.2
	8.0 ‡	23.9	30.4	12.5	5.00	708	293	4.82	3.10	101	73.3	131	88.4	731	129	0.406	17.0
	10.0 ‡	28.7	36.6	9.00	3.00	804	330	4.69	3.01	115	82.6	152	103	851	147	0.397	13.8
150 x 100	4.0 ‡	14.9	18.9	32.5	20.0	595	319	5.60	4.10	79.3	63.7	95.7	72.5	662	105	0.486	32.6
	5.0 ‡	18.3	23.4	25.0	15.0	719	384	5.55	4.05	95.9	76.8	117	88.3	809	127	0.483	26.4
	6.0 ‡	21.7	27.6	20.0	11.7	835	444	5.50	4.01	111	88.8	137	103	948	147	0.479	22.1
	8.0 ‡	27.7	35.2	13.8	7.50	1010	536	5.35	3.90	134	107	169	128	1210	182	0.466	16.8
	10.0 ‡	33.4	42.6	10.0	5.00	1160	614	5.22	3.80	155	123	199	150	1430	211	0.457	13.7
160 x 80	5.0 ‡	17.5	22.4	27.0	11.0	722	244	5.68	3.30	90.2	61.0	113	69.7	601	106	0.463	26.5
	6.0 ‡	20.7	26.4	21.7	8.33	836	281	5.62	3.26	105	70.2	132	81.3	702	122	0.459	22.2
	8.0 ‡	26.4	33.6	15.0	5.00	1000	335	5.46	3.16	125	83.7	163	100	882	150	0.446	16.9
180 x 80	4.0 ‡	15.5	19.7	40.0	15.0	802	227	6.37	3.39	89.1	56.7	112	63.5	578	99.6	0.506	32.6
	5.0 ‡	19.1	24.4	31.0	11.0	971	272	6.31	3.34	108	68.1	137	77.2	704	120	0.503	26.3
	6.0 ‡	22.6	28.8	25.0	8.33	1130	314	6.25	3.30	125	78.5	160	90.2	823	139	0.499	22.1
	8.0 ‡	28.9	36.8	17.5	5.00	1360	377	6.08	3.20	151	94.1	198	111	1040	170	0.486	16.8
	10.0 ‡	35.0	44.6	13.0	3.00	1570	429	5.94	3.10	174	107	234	131	1210	196	0.477	13.6
180 x 100	4.0 ‡	16.8	21.3	40.0	20.0	926	374	6.59	4.18	103	74.8	126	84.0	854	127	0.546	32.5
	5.0 ‡	20.7	26.4	31.0	15.0	1120	452	6.53	4.14	125	90.4	154	103	1040	154	0.543	26.2
	6.0 ‡	24.5	31.2	25.0	11.7	1310	524	6.48	4.10	146	105	181	120	1230	179	0.539	22.0
	8.0 ‡	31.4	40.0	17.5	7.50	1600	637	6.32	3.99	178	127	226	150	1570	222	0.526	16.8
	10.0 ‡	38.1	48.6	13.0	5.00	1860	736	6.19	3.89	207	147	268	177	1860	260	0.517	13.6

‡ Grade S275 not available from some leading producers. Check availability.
(1) For local buckling calculation d = D - 5t and b = B - 5t.
FOR EXPLANATION OF TABLES SEE NOTES 2 AND 3

BS 5950-1: 2000
BS EN 10219-2: 1997

COLD-FORMED
RECTANGULAR HOLLOW SECTIONS

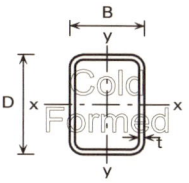

DIMENSIONS AND PROPERTIES

Section Designation Size D x B mm	Mass per Metre t mm	Area of Section A cm²	Ratios for Local Buckling d/t⁽¹⁾	Ratios for Local Buckling b/t⁽¹⁾	Second Moment of Area Axis x-x cm⁴	Second Moment of Area Axis y-y cm⁴	Radius of Gyration Axis x-x cm	Radius of Gyration Axis y-y cm	Elastic Modulus Axis x-x cm³	Elastic Modulus Axis y-y cm³	Plastic Modulus Axis x-x cm³	Plastic Modulus Axis y-y cm³	Torsional Constants J cm⁴	Torsional Constants C cm³	Surface Area Per Metre m²	Surface Area Per Tonne m²
200 x 100	4.0 ‡	18.0	22.9	45.0	20.0	1200	411	7.23	4.23	120	82.2	148	91.7	985	142	0.586	32.6
	5.0 ‡	22.3	28.4	35.0	15.0	1460	497	7.17	4.19	146	99.4	181	112	1210	172	0.583	26.1
	6.0 ‡	26.4	33.6	28.3	11.7	1700	577	7.12	4.14	170	115	213	132	1420	200	0.579	21.9
	8.0 ‡	33.9	43.2	20.0	7.50	2090	705	6.95	4.04	209	141	267	165	1810	250	0.566	16.7
	10.0 ‡	41.3	52.6	15.0	5.00	2440	818	6.82	3.94	244	164	318	195	2150	292	0.557	13.5
200 x 120	4.0 ‡	19.3	24.5	45.0	25.0	1350	618	7.43	5.02	135	103	164	115	1350	172	0.626	32.4
	5.0 ‡	23.8	30.4	35.0	19.0	1650	750	7.37	4.97	165	125	201	141	1650	210	0.623	26.2
	6.0 ‡	28.3	36.0	28.3	15.0	1930	874	7.32	4.93	193	146	237	166	1950	245	0.619	21.9
	8.0 ‡	36.5	46.4	20.0	10.0	2390	1080	7.17	4.82	239	180	298	209	2510	308	0.606	16.6
	10.0 ‡	44.4	56.6	15.0	7.00	2810	1260	7.04	4.72	281	210	356	250	3010	364	0.597	13.4
200 x 150	4.0 ‡	21.2	26.9	45.0	32.5	1580	1020	7.67	6.16	158	136	187	154	1940	219	0.686	32.4
	5.0 ‡	26.2	33.4	35.0	25.0	1930	1250	7.62	6.11	193	166	230	189	2390	267	0.683	26.1
	6.0 ‡	31.1	39.6	28.3	20.0	2270	1460	7.56	6.06	227	194	271	223	2830	313	0.679	21.8
	8.0 ‡	40.2	51.2	20.0	13.8	2830	1820	7.43	5.95	283	242	344	283	3680	396	0.666	16.6
	10.0 ‡	49.1	62.6	15.0	10.0	3350	2140	7.31	5.85	335	286	413	339	4430	471	0.657	13.4
250 x 150	5.0 ‡	30.1	38.4	45.0	25.0	3300	1510	9.28	6.27	264	201	320	225	3280	337	0.783	26.0
	6.0 ‡	35.8	45.6	36.7	20.0	3890	1770	9.23	6.23	311	236	378	266	3890	396	0.779	21.8
	8.0 ‡	46.5	59.2	26.3	13.8	4890	2220	9.08	6.12	391	296	482	340	5050	504	0.766	16.5
	10.0 ‡	57.0	72.6	20.0	10.0	5830	2630	8.96	6.02	466	351	582	409	6120	602	0.757	13.3
	12.0 ‡	66.0	84.1	15.8	7.50	6460	2930	8.77	5.90	517	390	658	463	7090	684	0.738	11.2
	12.5 ‡	68.3	87.0	15.0	7.00	6630	3000	8.73	5.87	531	400	678	477	7310	704	0.736	10.8
300 x 100	6.0 ‡	35.8	45.6	45.0	11.7	4780	842	10.2	4.30	318	168	411	188	2400	306	0.779	21.8
	8.0 ‡	46.5	59.2	32.5	7.50	5980	1040	10.0	4.20	399	209	523	238	3080	385	0.766	16.5
	10.0 ‡	57.0	72.6	25.0	5.00	7110	1220	9.90	4.11	474	245	631	285	3680	455	0.757	13.3
	12.0 ‡	66.0	84.1	20.0	3.33	7810	1340	9.64	4.00	521	269	710	321	4180	508	0.738	11.2
	12.5 ‡	68.3	87.0	19.0	3.00	8010	1370	9.59	3.97	534	275	732	330	4290	521	0.736	10.8
300 x 200	6.0 ‡	45.2	57.6	45.0	28.3	7370	3960	11.3	8.29	491	396	588	446	8120	651	0.979	21.7
	8.0 ‡	59.1	75.2	32.5	20.0	9390	5040	11.2	8.19	626	504	757	574	10600	838	0.966	16.3
	10.0 ‡	72.7	92.6	25.0	15.0	11300	6060	11.1	8.09	754	606	921	698	13000	1010	0.957	13.2
	12.0 ‡	84.8	108	20.0	11.7	12800	6850	10.9	7.96	853	685	1060	801	15200	1170	0.930	11.1
	12.5 ‡	88.0	112	19.0	11.0	13200	7060	10.8	7.94	879	706	1090	828	15800	1200	0.936	10.6
400 x 200	8.0 ‡	71.6	91.2	45.0	20.0	19000	6520	14.4	8.45	949	652	1170	728	15800	1130	1.17	16.3
	10.0 ‡	88.4	113	35.0	15.0	23000	7860	14.3	8.36	1150	786	1430	888	19400	1370	1.16	13.1
	12.0 ‡	104	132	28.3	11.7	26200	8980	14.1	8.24	1310	898	1660	1030	22800	1590	1.14	11.0
	12.5 ‡	108	137	27.0	11.0	27100	9260	14.1	8.22	1360	926	1710	1060	23600	1640	1.14	10.6

‡ Grade S275 not available from some leading producers. Check availability.
(1) For local buckling calculation d = D - 5t and b = B -5t.
FOR EXPLANATION OF TABLES SEE NOTES 2 AND 3

[BLANK PAGE]

MEMBER CAPACITIES

S275

BS 5950-1: 2000
BS 4-1: 1993

BENDING

UB SECTIONS
SUBJECT TO BENDING

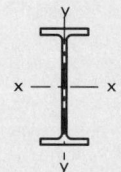

RESTRAINED BEAM CAPACITIES FOR S275

Section Designation	Section Classification	Shear Capacity P_v kN	Moment Capacity M_{cx} kNm	Ultimate U.D.L. Capacity (kN) for Restrained Beams for Lengths, L (m)											
				2.0	3.0	4.0	5.0	6.0	7.0	8.0	9.0	10.0	11.0	12.0	13.0	14.0
1016x305x487 # +	Plastic	4760	5920	9510	9510	8830	7970	7180	6460	5810	5260	4730	4300	3940	3640	3380
1016x305x437 # +	Plastic	4220	5290	8440	8440	7880	7110	6410	5770	5200	4710	4240	3850	3530	3260	3030
1016x305x393 # +	Plastic	3790	4730	7590	7590	7070	6370	5740	5160	4650	4200	3780	3440	3150	2910	2700
1016x305x349 # +	Plastic	3380	4400	6760	6760	6450	5840	5270	4760	4290	3910	3520	3200	2930	2710	2510
1016x305x314 # +	Plastic	3040	3940	6070	6070	5790	5240	4730	4260	3850	3500	3150	2860	2620	2420	2250
1016x305x272 # +	Plastic	2600	3400	5200	5200	4990	4510	4070	3670	3320	3020	2720	2470	2270	2090	1940
1016x305x249 # +	Plastic	2570	3010	5140	5140	4650	4170	3730	3330	3010	2670	2410	2190	2010	1850	1720
1016x305x222 # +	Plastic	2470	2600	4940	4720	4210	3740	3310	2930	2600	2310	2080	1890	1730	1600	1490
914 x 419 x 388 #	Plastic	3130	4680	6270	6270	6270	5940	5400	4910	4470	4070	3750	3410	3120	2880	2680
914 x 419 x 343 #	Plastic	2810	4100	5630	5630	5630	5260	4780	4330	3930	3580	3280	2980	2730	2520	2340
914 x 305 x 289 #	Plastic	2870	3330	5750	5750	5200	4650	4150	3700	3330	2960	2660	2420	2220	2050	1900
914 x 305 x 253 #	Plastic	2530	2900	5050	5050	4550	4060	3620	3230	2900	2580	2320	2110	1930	1780	1660
914 x 305 x 224 #	Plastic	2300	2530	4600	4540	4050	3600	3190	2830	2530	2250	2020	1840	1680	1550	1440
914 x 305 x 201 #	Plastic	2170	2210	4340	4130	3660	3230	2840	2520	2210	1970	1770	1610	1480	1360	1260
838 x 292 x 226 #	Plastic	2180	2430	4360	4360	3890	3460	3060	2720	2430	2160	1940	1760	1620	1490	1390
838 x 292 x 194 #	Plastic	1960	2020	3930	3800	3360	2960	2600	2310	2020	1800	1620	1470	1350	1250	1160
838 x 292 x 176 #	Plastic	1860	1800	3720	3490	3060	2680	2350	2060	1800	1600	1440	1310	1200	1110	1030
762 x 267 x 197	Plastic	1910	1900	3820	3670	3220	2820	2460	2170	1900	1690	1520	1380	1270	1170	1090
762 x 267 x 173	Plastic	1730	1640	3470	3250	2840	2470	2150	1880	1640	1460	1310	1190	1090	1010	939
762 x 267 x 147	Plastic	1530	1370	3070	2790	2420	2090	1820	1560	1370	1210	1090	994	911	841	781
762 x 267 x 134	Plastic	1490	1280	2970	2650	2280	1970	1700	1460	1280	1140	1020	929	851	786	730
686 x 254 x 170	Plastic	1600	1490	3190	3010	2610	2260	1960	1710	1490	1330	1190	1090	995	918	853
686 x 254 x 152	Plastic	1440	1330	2890	2700	2330	2020	1740	1510	1330	1180	1060	964	883	815	757
686 x 254 x 140	Plastic	1350	1210	2700	2490	2150	1850	1610	1380	1210	1070	966	878	805	743	690
686 x 254 x 125	Plastic	1260	1060	2520	2240	1920	1640	1410	1210	1060	941	847	770	706	651	605
610 x 305 x 238	Plastic	1860	1980	3720	3720	3340	2920	2550	2240	1980	1760	1590	1440	1320	1220	1130
610 x 305 x 179	Plastic	1390	1470	2780	2780	2470	2170	1900	1660	1470	1310	1180	1070	980	905	840
610 x 305 x 149	Plastic	1150	1220	2300	2300	2060	1800	1570	1380	1220	1080	974	885	812	749	696
610 x 229 x 140	Plastic	1290	1100	2570	2340	1990	1700	1460	1250	1100	976	878	798	732	675	627
610 x 229 x 125	Plastic	1160	974	2320	2090	1780	1510	1300	1110	974	866	779	708	649	599	557
610 x 229 x 113	Plastic	1070	869	2140	1890	1600	1360	1160	994	869	773	696	632	580	535	497
610 x 229 x 101	Plastic	1040	792	2090	1770	1490	1250	1060	905	792	704	634	576	528	488	453

+ Section is not given in BS 4-1: 1993.
# Check availability.
Section classification given applies to members subject to bending about the x-x axis only.
Loads given are the total Ultimate factored uniformaly distributed load supported over a beam span L assuming full lateral restraint to the compression flange. Self weight of the section has not been allowed for.
UDL values in **bold type** are governed by the shear capacity.
The unfactored imposed load is assumed to be 40% of the ultimate load given. Unless otherwise indicated, it is only necessary to perform deflection checks if the ratio of unfactored imposed load to dead load is greater than 1.556.
UDL values to the right of the zigzag line in *italic type* may be susceptible to serviceability deflections > L/360 and so should be checked under imposed loading. (This deflection limit is for beams which carry a brittle finish.)
FOR EXPLANATION OF TABLES SEE NOTE 5.1

BS 5950-1: 2000
BS 4-1: 1993

BENDING

UB SECTIONS
SUBJECT TO BENDING

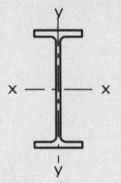

RESTRAINED BEAM CAPACITIES FOR S275

Section Designation	Section Classification	Shear Capacity P_v kN	Moment Capacity M_{cx} kNm	Ultimate U.D.L. Capacity (kN) for Restrained Beams for Lengths, L (m)											
				2.0	3.0	4.0	5.0	6.0	7.0	8.0	9.0	10.0	11.0	12.0	13.0	14.0
533 x 210 x 122	Plastic	1100	847	**2200**	1900	1590	1340	1130	968	847	753	678	616	565	*521*	*484*
533 x 210 x 109	Plastic	995	749	**1990**	1700	1420	1190	999	856	749	666	600	545	500	*461*	*428*
533 x 210 x 101	Plastic	922	692	**1840**	1570	1310	1100	923	791	692	615	554	503	*461*	*426*	*396*
533 x 210 x 92	Plastic	888	649	**1780**	1490	1240	1040	865	742	649	577	519	472	*433*	*399*	*371*
533 x 210 x 82	Plastic	837	566	1620	1340	1100	906	755	647	566	503	453	*412*	*377*	*348*	*324*
457 x 191 x 98	Plastic	847	591	**1690**	1390	1140	946	789	676	591	526	473	*430*	*394*	*364*	*338*
457 x 191 x 89	Plastic	774	534	1540	1260	1030	854	712	610	534	474	427	*388*	*356*	*328*	*305*
457 x 191 x 82	Plastic	751	504	1480	1200	980	806	671	575	504	448	403	*366*	*336*	*310*	*288*
457 x 191 x 74	Plastic	679	455	1330	1090	885	727	606	520	455	404	*364*	*331*	*303*	*280*	*260*
457 x 191 x 67	Plastic	636	405	1220	983	794	647	539	462	405	360	*324*	*294*	*270*	*249*	*231*
457 x 152 x 82	Plastic	778	480	1460	1170	946	768	640	548	480	427	*384*	*349*	*320*	*295*	*274*
457 x 152 x 74	Plastic	705	431	1320	1060	851	690	575	493	431	383	*345*	*314*	*287*	*265*	*246*
457 x 152 x 67	Plastic	680	400	1240	993	799	639	533	457	400	355	*320*	*291*	*266*	*246*	*228*
457 x 152 x 60	Plastic	608	354	1110	881	708	566	472	404	354	315	*283*	*257*	*236*	*218*	*202*
457 x 152 x 52	Plastic	564	301	978	767	603	482	402	344	301	268	*241*	*219*	*201*	*185*	*172*
406 x 178 x 74	Plastic	647	413	1260	1010	811	660	550	472	413	367	*330*	*300*	*275*	*254*	*236*
406 x 178 x 67	Plastic	594	370	1140	911	730	592	494	423	370	*329*	*296*	*269*	*247*	*228*	*212*
406 x 178 x 60	Plastic	530	330	1020	812	650	528	440	377	330	*293*	*264*	*240*	*220*	*203*	*188*
406 x 178 x 54	Plastic	512	290	930	731	580	464	387	332	290	*258*	*232*	*211*	*193*	*179*	*166*

Section classification given applies to members subject to bending about the x-x axis only.
Loads given are the total Ultimate factored uniformly distributed load supported over a beam span L assuming full lateral restraint to the compression flange. Self weight of the section has not been allowed for.
UDL values in **bold type** are governed by the shear capacity.
The unfactored imposed load is assumed to be 40% of the ultimate load given. Unless otherwise indicated, it is only necessary to perform deflection checks if the ratio of unfactored imposed load to dead load is greater than 1.556.
UDL values to the right of the zigzag line in *italic type* may be susceptible to serviceability deflections > L/360 and so should be checked under imposed loading. (This deflection limit is for beams which carry a brittle finish.)
FOR EXPLANATION OF TABLES SEE NOTE 5.1

BS 5950-1: 2000
BS 4-1: 1993

BENDING

UB SECTIONS
SUBJECT TO BENDING

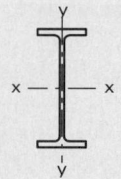

RESTRAINED BEAM CAPACITIES FOR S275

Section Designation	Section Classification	Shear Capacity P_v kN	Moment Capacity M_{cx} kNm	Ultimate U.D.L. Capacity (kN) for Restrained Beams for Lengths, L (m)												
				1.0	1.5	2.0	2.5	3.0	3.5	4.0	4.5	5.0	5.5	6.0	6.5	7.0
406 x 140 x 46	Plastic	452	244	**905**	902	799	705	623	550	488	434	391	355	326	301	279
406 x 140 x 39	Plastic	420	199	**841**	784	685	597	520	455	398	354	319	290	265	245	228
356 x 171 x 67	Plastic	546	333	**1090**	**1090**	1050	933	829	738	659	592	533	484	444	410	381
356 x 171 x 57	Plastic	478	278	**957**	**957**	895	791	700	620	556	494	444	404	370	342	317
356 x 171 x 51	Plastic	433	246	**867**	**867**	801	707	624	552	493	438	394	358	329	303	282
356 x 171 x 45	Plastic	406	213	**812**	**812**	715	626	549	482	426	379	341	310	284	262	244
356 x 127 x 39	Plastic	385	181	**770**	727	632	548	475	414	362	322	290	264	242	223	207
356 x 127 x 33	Plastic	346	149	**691**	623	536	460	398	341	299	265	239	217	199	184	171
305 x 165 x 54	Plastic	405	233	**809**	**809**	765	672	591	522	465	414	372	338	310	286	*266*
305 x 165 x 46	Plastic	339	198	**678**	**678**	648	569	501	443	396	352	317	288	264	244	*226*
305 x 165 x 40	Plastic	300	171	**601**	**601**	566	496	436	385	343	305	274	249	228	211	*196*
305 x 127 x 48	Plastic	462	196	**924**	835	712	607	521	447	391	348	313	284	261	241	*223*
305 x 127 x 42	Plastic	406	169	**811**	727	618	526	450	386	338	300	270	246	225	208	*193*
305 x 127 x 37	Plastic	357	148	**713**	639	543	462	395	339	296	264	237	216	198	182	*169*
305 x 102 x 33	Plastic	341	132	**681**	584	493	416	353	302	265	235	212	192	176	163	*151*
305 x 102 x 28	Plastic	306	111	**598**	503	421	355	296	253	222	197	177	161	148	*136*	*127*
305 x 102 x 25	Plastic	292	94.1	**539**	446	367	301	251	215	188	167	150	137	125	*116*	*107*
254 x 146 x 43	Plastic	308	156	**617**	**617**	544	469	406	356	311	277	249	226	*208*	*192*	*178*
254 x 146 x 37	Plastic	266	133	**532**	**532**	467	402	347	304	266	236	213	193	*177*	*163*	*152*
254 x 146 x 31	Plastic	249	108	**498**	468	396	336	288	247	216	192	173	157	*144*	*133*	*124*
254 x 102 x 28	Plastic	271	97.1	**541**	449	372	311	259	222	194	173	155	141	*129*	*119*	*111*
254 x 102 x 25	Plastic	255	84.2	**489**	401	328	269	224	192	168	150	135	*122*	*112*	*104*	*96.2*
254 x 102 x 22	Plastic	239	71.2	**434**	350	285	228	190	163	142	127	*114*	*104*	*95.0*	*87.7*	*81.4*
203 x 133 x 30	Plastic	218	86.4	**437**	395	327	273	230	197	173	154	*138*	*126*	*115*	*106*	*98.7*
203 x 133 x 25	Plastic	191	71.0	**382**	332	272	227	189	162	142	*126*	*114*	*103*	*94.6*	*87.3*	*81.1*
203 x 102 x 23	Plastic	181	64.4	**362**	305	249	206	172	147	129	*114*	*103*	*93.6*	*85.8*	*79.2*	*73.5*
178 x 102 x 19	Plastic	141	47.0	**282**	229	184	150	125	*107*	*94.1*	*83.6*	*75.2*	*68.4*	*62.7*	*57.9*	*53.7*
152 x 89 x 16	Plastic	113	33.8	220	171	135	108	90.2	*77.3*	*67.7*	*60.1*	*54.1*	*49.2*	*45.1*	*41.6*	*38.7*
127 x 76 x 13	Plastic	83.8	23.2	158	119	92.6	74.1	*61.7*	*52.9*	*46.3*	*41.2*	*37.0*	*33.7*	*30.9*	*28.5*	*26.5*

Section classification given applies to members subject to bending about the x-x axis only.
Loads given are the total Ultimate factored uniformaly distributed load supported over a beam span L assuming full lateral restraint to the compression flange. Self weight of the section has not been allowed for.
UDL values in **bold type** are governed by the shear capacity.
The unfactored imposed load is assumed to be 40% of the ultimate load given. Unless otherwise indicated, it is only necessary to perform deflection checks if the ratio of unfactored imposed load to dead load is greater than 1.556.
UDL values to the right of the zigzag line in *italic type* may be susceptible to serviceability deflections > L/360 and so should be checked under imposed loading. (This deflection limit is for beams which carry a brittle finish.)
Shaded UDL values may be susceptible to serviceability deflections > L/200 and so should be checked under imposed loading. (This deflection limit is for beams which do not carry a brittle finish.)
FOR EXPLANATION OF TABLES SEE NOTE 5.1

BS 5950-1: 2000
BS 4-1: 1993

BENDING

UC SECTIONS
SUBJECT TO BENDING

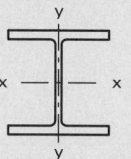

RESTRAINED BEAM CAPACITIES FOR S275

Section Designation	Section Classification	Shear Capacity P_v kN	Moment Capacity M_{cx} kNm	Ultimate U.D.L. Capacity (kN) for Restrained Beams for Lengths, L (m)											
				2.0	3.0	4.0	5.0	6.0	7.0	8.0	9.0	10.0	11.0	12.0	13.0	14.0
356 x 406 x 634 #	Plastic	3320	3400	6640	6640	6070	5230	4530	3890	3400	3030	2720	2480	2270	2100	1950
356 x 406 x 551 #	Plastic	2820	2930	5640	5640	5170	4450	3850	3350	2930	2600	2340	2130	1950	1800	1670
356 x 406 x 467 #	Plastic	2390	2550	4780	4780	4450	3830	3310	2890	2550	2270	2040	1850	1700	1570	1460
356 x 406 x 393 #	Plastic	1960	2100	3920	3920	3670	3150	2730	2380	2100	1860	1680	1520	1400	1290	1200
356 x 406 x 340 #	Plastic	1650	1780	3310	3310	3130	2680	2320	2020	1780	1590	1430	1300	1190	1100	1020
356 x 406 x 287 #	Plastic	1410	1540	2830	2830	2700	2310	2000	1750	1540	1370	1230	1120	1030	948	880
356 x 406 x 235 #	Plastic	1110	1240	2230	2230	2170	1860	1610	1410	1240	1100	994	903	828	764	710
356 x 368 x 202 #	Plastic	983	1050	1970	1970	1860	1590	1370	1190	1050	936	842	766	702	648	601
356 x 368 x 177 #	Plastic	843	916	1690	1690	1610	1380	1190	1040	916	814	732	666	610	563	523
356 x 368 x 153 #	Plastic	708	786	1420	1420	1380	1180	1020	890	786	698	629	571	524	484	449
356 x 368 x 129 #	Semi-compact	588	651	1180	1180	1150	980	846	738	651	579	521	474	434	401	372
305 x 305 x 283	Plastic	1500	1300	3000	2920	2430	2040	1740	1490	1300	1160	1040	947	868	801	744
305 x 305 x 240	Plastic	1290	1130	2580	2530	2100	1760	1500	1290	1130	1000	900	819	750	693	643
305 x 305 x 198	Plastic	1030	912	2060	2050	1700	1420	1220	1040	912	810	729	663	608	561	521
305 x 305 x 158	Plastic	822	710	1640	1620	1330	1110	947	812	710	631	568	517	473	437	406
305 x 305 x 137	Plastic	703	609	1410	1390	1140	955	812	696	609	541	487	443	406	375	348
305 x 305 x 118	Plastic	600	519	1200	1180	976	814	692	593	519	461	415	377	346	319	296
305 x 305 x 97	Compact	503	438	1010	1000	823	687	584	500	438	389	350	318	292	269	250
254 x 254 x 167	Plastic	883	642	1770	1550	1250	1030	856	734	642	571	514	467	428	395	367
254 x 254 x 132	Plastic	672	495	1340	1200	962	792	660	566	495	440	396	360	330	305	283
254 x 254 x 107	Plastic	543	393	1090	956	766	629	524	449	393	350	315	286	262	242	225
254 x 254 x 89	Plastic	426	324	853	780	628	516	432	371	324	288	259	236	216	200	185
254 x 254 x 73	Plastic	361	273	721	658	529	434	364	312	273	242	218	198	182	168	156
203 x 203 x 86	Plastic	449	259	881	665	518	414	345	296	259	230	207	188	173	159	148
203 x 203 x 71	Plastic	343	212	686	538	420	339	282	242	212	188	169	154	141	130	121
203 x 203 x 60	Plastic	325	180	625	467	361	289	241	206	180	160	144	131	120	111	103
203 x 203 x 52	Plastic	269	156	534	401	312	249	208	178	156	139	125	113	104	96.0	89.1
203 x 203 x 46	Compact	241	137	472	353	273	219	182	156	137	121	109	99.4	91.1	84.1	78.1
152 x 152 x 37	Plastic	214	85.0	324	227	170	136	113	97.1	85.0	75.5	68.0	61.8	56.7	52.3	48.6
152 x 152 x 30	Plastic	169	68.2	260	182	136	109	90.9	77.9	68.2	60.6	54.6	49.6	45.5	42.0	39.0
152 x 152 x 23	Semi-compact	146	48.5	190	129	96.9	77.6	64.6	55.4	48.5	43.1	38.8	35.3	32.3	29.8	27.7

\# Check availability.
Section classification given applies to members subject to bending about the x-x axis only.
M_{cx} values in *italic type* are governed by $1.2p_yZ$ and a higher value may be used in some circumstances.
Loads given are the total Ultimate factored uniformaly distributed load supported over a beam span L assuming full lateral restraint to the compression flange. Self weight of the section has not been allowed for.
UDL values in **bold type** are governed by the shear capacity.
The unfactored imposed load is assumed to be 40% of the ultimate load given. Unless otherwise indicated, it is only necessary to perform deflection checks if the ratio of unfactored imposed load to dead load is greater than 1.556.
UDL values to the right of the zigzag line in *italic type* may be susceptible to serviceability deflections > L/360 and so should be checked under imposed loading. (This deflection limit is for beams which carry a brittle finish.)
Shaded UDL values may be susceptible to serviceability deflections > L/200 and so should be checked under imposed loading. (This deflection limit is for beams which do not carry a brittle finish.)
FOR EXPLANATION OF TABLES SEE NOTE 5.1

BS 5950-1: 2000
BS 4-1: 1993

BENDING

JOISTS
SUBJECT TO BENDING

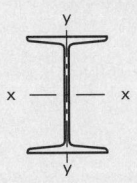

RESTRAINED BEAM CAPACITIES FOR S275

Section Designation	Section Classification	Shear Capacity P_v kN	Moment Capacity M_{cx} kNm	Ultimate U.D.L. Capacity (kN) for Restrained Beams for Lengths, L (m)											
				1.0	1.5	2.0	2.5	3.0	3.5	4.0	4.5	5.0	5.5	6.0	6.5	7.0
254 x 203 x 82 #	Plastic	412	285	**824**	**824**	**824**	794	703	626	560	504	457	415	*381*	*351*	*326*
254 x 114 x 37 #	Plastic	319	126	**637**	565	473	397	337	289	252	224	202	184	*168*	*155*	*144*
203 x 152 x 52 #	Plastic	288	143	**575**	**575**	513	438	376	328	287	255	229	209	*191*	*176*	*164*
152 x 127 x 37 #	Plastic	262	76.7	502	388	307	246	205	*175*	*153*	*136*	*123*	*112*	*102*	*94.4*	*87.7*
127 x 114 x 29 #	Plastic	214	49.8	357	262	199	159	*133*	*114*	*99.6*	*88.5*	*79.6*	*72.4*	*66.4*	*61.3*	*56.9*
127 x 114 x 27 #	Plastic	155	47.3	**310**	239	188	151	*126*	*108*	*94.6*	*84.1*	*75.7*	*68.8*	*63.1*	*58.2*	*54.1*
127 x 76 x 16 #	Plastic	117	28.6	203	150	114	91.5	76.3	65.4	57.2	50.8	45.8	41.6	*38.1*	*35.2*	*32.7*
114 x 114 x 27 #	Plastic	179	41.5	301	219	166	*133*	*111*	*94.9*	*83.1*	*73.8*	*66.4*	*60.4*	*55.4*	*51.1*	*47.5*
102 x 102 x 23 #	Plastic	159	31.1	236	166	124	99.4	82.9	71.0	62.2	55.2	49.7	45.2	*41.4*	*38.2*	*35.5*
102 x 44 x 7 #	Plastic	72.1	9.74	77.9	51.9	38.9	*31.2*	*26.0*	*22.3*	*19.5*	*17.3*	*15.6*	*14.2*	*13.0*	*12.0*	*11.1*
89 x 89 x 19 #	Plastic	139	22.7	179	121	91.0	72.8	60.6	52.0	45.5	40.4	36.4	33.1	*30.3*	*28.0*	*26.0*
76 x 76 x 15 #	Plastic	112	14.9	119	79.5	59.6	47.7	39.7	34.1	29.8	26.5	23.8	21.7	*19.9*	*18.3*	*17.0*
76 x 76 x 13 #	Plastic	64.1	13.4	102	71.4	53.6	42.9	35.7	30.6	26.8	23.8	21.4	19.5	*17.9*	*16.5*	*15.3*

\# Check availability.
Section classification given applies to members subject to bending about the x-x axis only.
Loads given are the total Ultimate factored uniformaly distributed load supported over a beam span L assuming full lateral restraint to the compression flange. Self weight of the section has not been allowed for.
UDL values in **bold type** are governed by the shear capacity.
The unfactored imposed load is assumed to be 40% of the ultimate load given. Unless otherwise indicated, it is only necessary to perform deflection checks if the ratio of unfactored imposed load to dead load is greater than 1.556.
UDL values to the right of the zigzag line in *italic type* may be susceptible to serviceability deflections > L/360 and so should be checked under imposed loading. (This deflection limit is for beams which carry a brittle finish.)
Shaded UDL values may be susceptible to serviceability deflections > L/200 and so should be checked under imposed loading. (This deflection limit is for beams which do not carry a brittle finish.)
FOR EXPLANATION OF TABLES SEE NOTE 5.1

BS 5950-1: 2000
BS 4-1: 1993

BENDING

PARALLEL FLANGE CHANNELS
SUBJECT TO BENDING

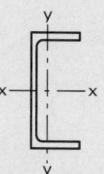

RESTRAINED BEAM CAPACITIES FOR S275

Section Designation	Section Classification	Shear Capacity P_v kN	Moment Capacity M_{cx} kNm	Ultimate U.D.L. Capacity (kN) for Restrained Beams for Lengths, L (m)												
				1.0	1.5	2.0	2.5	3.0	3.5	4.0	4.5	5.0	5.5	6.0	6.5	7.0
430 x 100 x 64	Plastic	752	324	**1500**	1320	1150	992	864	740	648	576	518	471	432	399	370
380 x 100 x 54	Plastic	574	247	**1150**	1020	883	760	659	565	494	440	396	360	330	304	283
300 x 100 x 46	Plastic	429	170	**859**	748	632	533	453	388	340	302	272	247	226	209	*194*
300 x 90 x 41	Plastic	446	156	861	720	599	500	417	357	312	278	250	227	208	*192*	*179*
260 x 90 x 35	Plastic	343	117	668	551	452	374	312	267	234	208	187	170	*156*	*144*	*134*
260 x 75 x 28	Plastic	300	90.2	547	442	356	289	241	206	180	160	144	*131*	*120*	*111*	*103*
230 x 90 x 32	Plastic	285	97.6	567	463	379	312	260	223	195	174	*156*	*142*	*130*	*120*	*112*
230 x 75 x 26	Plastic	247	76.5	464	374	301	245	204	175	153	136	*122*	*111*	*102*	*94.1*	*87.4*
200 x 90 x 30	Plastic	231	80.0	**462**	383	311	256	213	183	160	*142*	*128*	*116*	*107*	*98.5*	*91.5*
200 x 75 x 23	Plastic	198	62.4	384	307	246	200	166	143	125	*111*	*99.9*	*90.8*	*83.2*	*76.8*	*71.3*
180 x 90 x 26	Plastic	193	63.8	**386**	311	250	204	170	146	*128*	*113*	*102*	*92.8*	*85.1*	*78.5*	*72.9*
180 x 75 x 20	Plastic	178	48.4	320	247	194	155	129	*111*	*96.8*	*86.0*	*77.4*	*70.4*	*64.5*	*59.6*	*55.3*
150 x 90 x 24	Plastic	161	49.2	318	247	195	158	131	*113*	*98.5*	*87.5*	*78.8*	*71.6*	*65.6*	*60.6*	*56.3*
150 x 75 x 18	Plastic	136	36.3	246	187	145	116	*96.8*	*83.0*	*72.6*	*64.5*	*58.1*	*52.8*	*48.4*	*44.7*	*41.5*
125 x 65 x 15 #	Plastic	113	24.7	181	132	98.9	79.1	*65.9*	*56.5*	*49.4*	*44.0*	*39.6*	*36.0*	*33.0*	*30.4*	*28.3*
100 x 50 x 10 #	Plastic	82.5	13.4	105	71.7	53.8	*43.0*	*35.9*	*30.7*	*26.9*	*23.9*	*21.5*	*19.6*	*17.9*	*16.6*	*15.4*

\# Check availability.
Section classification given applies to members subject to bending about the x-x axis only.
Loads given are the total Ultimate factored uniformly distributed load supported over a beam span L assuming full lateral restraint to the compression flange. Self weight of the section has not been allowed for.
UDL values in **bold type** are governed by the shear capacity.
The unfactored imposed load is assumed to be 40% of the ultimate load given. Unless otherwise indicated, it is only necessary to perform deflection checks if the ratio of unfactored imposed load to dead load is greater than 1.556.
UDL values to the right of the zigzag line in *italic type* may be susceptible to serviceability deflections > L/360 and so should be checked under imposed loading. (This deflection limit is for beams which carry a brittle finish.)
Shaded UDL values may be susceptible to serviceability deflections > L/200 and so should be checked under imposed loading. (This deflection limit is for beams which do not carry a brittle finish.)
FOR EXPLANATION OF TABLES SEE NOTE 5.1

BS 5950-1: 2000
BS 4-1: 1993

WEB BEARING AND BUCKLING

UB SECTIONS

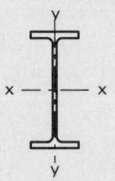

BEARING AND BUCKLING VALUES FOR UNSTIFFENED WEBS FOR S275

Section Designation	Web Thickness	Depth Between Fillets	Bearing				Buckling				Shear Capacity
			End Bearing		Continuous Over Bearing		End Bearing		Continuous Over Bearing		
			Beam Factor	Stiff Bearing Factor	Beam Factor	Stiff Bearing Factor	Buckling Factor	1.4d	Buckling Factor		
	t	d	C1	C2	C1	C2	C4		C4		P_v
	mm	mm	kN	kN/mm	kN	kN/mm	kN	mm	kN		kN
1016x305x487 # +	30.0	868	1290	7.65	3220	7.65	5350	1220	5350		4760
1016x305x437 # +	26.9	868	1080	6.86	2710	6.86	3850	1220	3850		4220
1016x305x393 # +	24.4	868	920	6.22	2300	6.22	2880	1220	2880		3790
1016x305x349 # +	21.1	868	783	5.59	1960	5.59	1860	1220	1860		3380
1016x305x314 # +	19.1	868	667	5.06	1670	5.06	1380	1220	1380		3040
1016x305x272 # +	16.5	868	533	4.37	1330	4.37	889	1220	889		2600
1016x305x249 # +	16.5	868	490	4.37	1220	4.37	889	1220	889		2570
1016x305x222 # +	16.0	868	433	4.24	1080	4.24	811	1220	811		2470
914 x 419 x 388 #	21.4	800	688	5.67	1720	5.67	2110	1120	2110		3130
914 x 419 x 343 #	19.4	800	577	5.14	1440	5.14	1570	1120	1570		2810
914 x 305 x 289 #	19.5	824	528	5.17	1320	5.17	1550	1150	1550		2870
914 x 305 x 253 #	17.3	824	431	4.58	1080	4.58	1080	1150	1080		2530
914 x 305 x 224 #	15.9	824	362	4.21	906	4.21	838	1150	838		2300
914 x 305 x 201 #	15.1	824	315	4.00	786	4.00	718	1150	718		2170
838 x 292 x 226 #	16.1	762	381	4.27	951	4.27	942	1070	942		2180
838 x 292 x 194 #	14.7	762	308	3.90	769	3.90	717	1070	717		1960
838 x 292 x 176 #	14.0	762	272	3.71	679	3.71	619	1070	619		1860
762 x 267 x 197	15.6	686	346	4.13	866	4.13	951	960	951		1910
762 x 267 x 173	14.3	686	289	3.79	722	3.79	733	960	733		1730
762 x 267 x 147	12.8	686	231	3.39	577	3.39	525	960	525		1530
762 x 267 x 134	12.0	686	211	3.30	528	3.30	433	960	433		1490
686 x 254 x 170	14.5	615	299	3.84	747	3.84	852	861	852		1600
686 x 254 x 152	13.2	615	253	3.50	633	3.50	643	861	643		1440
686 x 254 x 140	12.4	615	225	3.29	562	3.29	533	861	533		1350
686 x 254 x 125	11.7	615	195	3.10	487	3.10	448	861	448		1260
610 x 305 x 238	18.4	540	467	4.88	1170	4.88	1980	756	1980		1860
610 x 305 x 179	14.1	540	300	3.74	749	3.74	892	756	892		1390
610 x 305 x 149	11.8	540	226	3.13	566	3.13	523	756	523		1150
610 x 229 x 140	13.1	548	242	3.47	604	3.47	706	767	706		1290
610 x 229 x 125	11.9	548	204	3.15	509	3.15	529	767	529		1160
610 x 229 x 113	11.1	548	176	2.94	441	2.94	429	767	429		1070
610 x 229 x 101	10.5	548	159	2.89	397	2.89	363	767	363		1040

+ Section is not given in BS 4-1: 1993.
# Check availability.
Web bearing capacity, $P_w = C1 + b_1 C2$
Web buckling resistance, $P_x = K (C4\ P_w)^{0.5}$
For Continuous over bearing: K = 1.0
For End bearing: K = Min. { $(a_e / 1.4d) + 0.5$, 1.0 }
Where b_1 is the stiff bearing length and a_e is the distance from the load or reaction to the nearer end of the member.
FOR EXPLANATION OF TABLES SEE NOTE 6.1

BS 5950-1: 2000
BS 4-1: 1993

WEB BEARING AND BUCKLING

UB SECTIONS

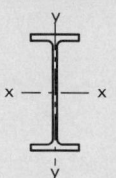

BEARING AND BUCKLING VALUES FOR UNSTIFFENED WEBS FOR S275

Section Designation	Web Thickness	Depth Between Fillets	Bearing				Buckling			Shear Capacity
			End Bearing		Continuous Over Bearing		End Bearing		Continuous Over Bearing	
			Beam Factor	Stiff Bearing Factor	Beam Factor	Stiff Bearing Factor	Buckling Factor	1.4d	Buckling Factor	
	t	d	C1	C2	C1	C2	C4		C4	P_v
	mm	mm	kN	kN/mm	kN	kN/mm	kN	mm	kN	kN
533 x 210 x 122	12.7	477	229	3.37	572	3.37	739	667	739	1100
533 x 210 x 109	11.6	477	194	3.07	484	3.07	563	667	563	995
533 x 210 x 101	10.8	477	172	2.86	431	2.86	454	667	454	922
533 x 210 x 92	10.1	477	157	2.78	393	2.78	372	667	372	888
533 x 210 x 82	9.6	477	137	2.64	342	2.64	319	667	319	837
457 x 191 x 98	11.4	408	180	3.02	450	3.02	625	571	625	847
457 x 191 x 89	10.5	408	155	2.78	388	2.78	488	571	488	774
457 x 191 x 82	9.9	408	143	2.72	357	2.72	409	571	409	751
457 x 191 x 74	9.0	408	122	2.48	306	2.48	307	571	307	679
457 x 191 x 67	8.5	408	107	2.34	268	2.34	259	571	259	636
457 x 152 x 82	10.5	408	162	2.78	405	2.78	488	571	488	778
457 x 152 x 74	9.6	408	138	2.54	346	2.54	373	571	373	705
457 x 152 x 67	9.0	408	125	2.48	312	2.48	307	571	307	680
457 x 152 x 60	8.1	408	105	2.23	262	2.23	224	571	224	608
457 x 152 x 52	7.6	408	88.2	2.09	220	2.09	185	571	185	564
406 x 178 x 74	9.5	360	137	2.61	342	2.61	409	505	409	647
406 x 178 x 67	8.8	360	119	2.42	296	2.42	325	505	325	594
406 x 178 x 60	7.9	360	99.9	2.17	250	2.17	235	505	235	530
406 x 178 x 54	7.7	360	89.4	2.12	223	2.12	218	505	218	512
406 x 140 x 46	6.8	360	80.0	1.87	200	1.87	150	505	150	452
406 x 140 x 39	6.4	360	66.2	1.76	165	1.76	125	505	125	420
356 x 171 x 67	9.1	312	130	2.50	324	2.50	416	436	416	546
356 x 171 x 57	8.1	312	103	2.23	258	2.23	293	436	293	478
356 x 171 x 51	7.4	312	88.3	2.04	221	2.04	224	436	224	433
356 x 171 x 45	7.0	312	76.6	1.93	192	1.93	189	436	189	406
356 x 127 x 39	6.6	312	75.9	1.82	190	1.82	159	436	159	385
356 x 127 x 33	6.0	312	61.7	1.65	154	1.65	119	436	119	346
305 x 165 x 54	7.9	265	98.2	2.17	245	2.17	320	371	320	405
305 x 165 x 46	6.7	265	76.3	1.84	191	1.84	195	371	195	339
305 x 165 x 40	6.0	265	63.0	1.65	158	1.65	140	371	140	300

Web bearing capacity, $P_w = C1 + b_1 C2$
Web buckling resistance, $P_x = K (C4 \, P_w)^{0.5}$
For Continuous over bearing: K = 1.0
For End bearing: K = Min. { $(a_e / 1.4d) + 0.5$, 1.0 }
Where b_1 is the stiff bearing length and a_e is the distance from the load or reaction to the nearer end of the member.
FOR EXPLANATION OF TABLES SEE NOTE 6.1

BS 5950-1: 2000
BS 4-1: 1993

WEB BEARING AND BUCKLING

UB SECTIONS

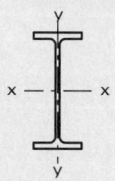

BEARING AND BUCKLING VALUES FOR UNSTIFFENED WEBS FOR S275

Section Designation	Web Thickness	Depth Between Fillets	Bearing				Buckling			Shear Capacity
			End Bearing		Continuous Over Bearing		End Bearing		Continuous Over Bearing	
			Beam Factor	Stiff Bearing Factor	Beam Factor	Stiff Bearing Factor	Buckling Factor	1.4d	Buckling Factor	
	t	d	C1	C2	C1	C2	C4		C4	P_v
	mm	mm	kN	kN/mm	kN	kN/mm	kN	mm	kN	kN
305 x 127 x 48	9.0	265	113	2.48	283	2.48	472	371	472	462
305 x 127 x 42	8.0	265	92.4	2.20	231	2.20	332	371	332	406
305 x 127 x 37	7.1	265	76.5	1.95	191	1.95	232	371	232	357
305 x 102 x 33	6.6	276	66.8	1.82	167	1.82	179	386	179	341
305 x 102 x 28	6.0	276	54.1	1.65	135	1.65	135	386	135	306
305 x 102 x 25	5.8	276	46.6	1.60	116	1.60	122	386	122	292
254 x 146 x 43	7.2	219	80.4	1.98	201	1.98	293	307	293	308
254 x 146 x 37	6.3	219	64.1	1.73	160	1.73	196	307	196	266
254 x 146 x 31	6.0	219	53.5	1.65	134	1.65	170	307	170	249
254 x 102 x 28	6.3	225	61.0	1.73	152	1.73	191	315	191	271
254 x 102 x 25	6.0	225	52.8	1.65	132	1.65	165	315	165	255
254 x 102 x 22	5.7	225	45.1	1.57	113	1.57	141	315	141	239
203 x 133 x 30	6.4	172	60.5	1.76	151	1.76	261	241	261	218
203 x 133 x 25	5.7	172	48.3	1.57	121	1.57	185	241	185	191
203 x 102 x 23	5.4	169	50.2	1.49	125	1.49	160	237	160	181
178 x 102 x 19	4.8	147	40.9	1.32	102	1.32	129	206	129	141
152 x 89 x 16	4.5	122	37.9	1.24	94.7	1.24	129	171	129	113
127 x 76 x 13	4.0	96.6	33.4	1.10	83.6	1.10	114	135	114	83.8

Web bearing capacity, $P_w = C1 + b_1 C2$
Web buckling resistance, $P_x = K (C4\ P_w)^{0.5}$
For Continuous over bearing: $K = 1.0$
For End bearing: $K = \text{Min.} \{ (a_e / 1.4d) + 0.5 , 1.0 \}$
Where b_1 is the stiff bearing length and a_e is the distance from the load or reaction to the nearer end of the member.
FOR EXPLANATION OF TABLES SEE NOTE 6.1

BS 5950-1: 2000
BS 4-1: 1993

WEB BEARING AND BUCKLING

UC SECTIONS

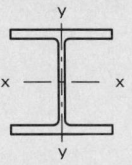

BEARING AND BUCKLING VALUES FOR UNSTIFFENED WEBS FOR S275

Section Designation	Web Thickness	Depth Between Fillets	Bearing				Buckling				Shear Capacity
			End Bearing		Continuous Over Bearing		End Bearing		Continuous Over Bearing		
			Beam Factor	Stiff Bearing Factor	Beam Factor	Stiff Bearing Factor	Buckling Factor	1.4d	Buckling Factor		
	t	d	C1	C2	C1	C2	C4		C4		P_v
	mm	mm	kN	kN/mm	kN	kN/mm	kN	mm	kN		kN
356 x 406 x 634 #	47.6	290	2150	11.7	5380	11.7	63900	406	63900		3320
356 x 406 x 551 #	42.1	290	1710	10.3	4270	10.3	44200	406	44200		2820
356 x 406 x 467 #	35.8	290	1340	9.13	3340	9.13	27200	406	27200		2390
356 x 406 x 393 #	30.6	290	1010	7.80	2510	7.80	17000	406	17000		1960
356 x 406 x 340 #	26.6	290	788	6.78	1970	6.78	11100	406	11100		1650
356 x 406 x 287 #	22.6	290	619	5.99	1550	5.99	6840	406	6840		1410
356 x 406 x 235 #	18.4	290	443	4.88	1110	4.88	3690	406	3690		1110
356 x 368 x 202 #	16.5	290	369	4.37	923	4.37	2660	406	2660		983
356 x 368 x 177 #	14.4	290	298	3.82	744	3.82	1770	406	1770		843
356 x 368 x 153 #	12.3	290	234	3.26	585	3.26	1100	406	1100		708
356 x 368 x 129 #	10.4	290	180	2.76	451	2.76	666	406	666		588
305 x 305 x 283	26.8	247	811	6.83	2030	6.83	13400	345	13400		1500
305 x 305 x 240	23.0	247	645	6.10	1610	6.10	8480	345	8480		1290
305 x 305 x 198	19.1	247	472	5.06	1180	5.06	4850	345	4850		1030
305 x 305 x 158	15.8	247	337	4.19	842	4.19	2750	345	2750		822
305 x 305 x 137	13.8	247	270	3.66	675	3.66	1830	345	1830		703
305 x 305 x 118	12.0	247	216	3.18	539	3.18	1200	345	1200		600
305 x 305 x 97	9.9	247	167	2.72	417	2.72	676	345	676		503
254 x 254 x 167	19.2	200	452	5.09	1130	5.09	6070	280	6070		883
254 x 254 x 132	15.3	200	308	4.05	770	4.05	3070	280	3070		672
254 x 254 x 107	12.8	200	225	3.39	563	3.39	1800	280	1800		543
254 x 254 x 89	10.3	200	164	2.73	409	2.73	938	280	938		426
254 x 254 x 73	8.6	200	127	2.36	318	2.36	546	280	546		361
203 x 203 x 86	12.7	161	207	3.37	517	3.37	2190	225	2190		449
203 x 203 x 71	10.0	161	146	2.65	364	2.65	1070	225	1070		343
203 x 203 x 60	9.4	161	126	2.59	315	2.59	888	225	888		325
203 x 203 x 52	7.9	161	98.6	2.17	247	2.17	527	225	527		269
203 x 203 x 46	7.2	161	84.0	1.98	210	1.98	399	225	399		241
152 x 152 x 37	8.0	124	84.0	2.20	210	2.20	712	173	712		214
152 x 152 x 30	6.5	124	60.8	1.79	152	1.79	382	173	382		169
152 x 152 x 23	5.8	124	45.9	1.60	115	1.60	271	173	271		146

\# Check availability.
Web bearing capacity, P_w = C1 + b_1 C2
Web buckling resistance, P_x = K $(C4\ P_w)^{0.5}$
For Continuous over bearing: K = 1.0
For End bearing: K = Min. { $(a_e / 1.4d) + 0.5$, 1.0 }
Where b_1 is the stiff bearing length and a_e is the distance from the load or reaction to the nearer end of the member.
FOR EXPLANATION OF TABLES SEE NOTE 6.1

BS 5950-1: 2000
BS 4-1: 1993

WEB BEARING AND BUCKLING

JOISTS

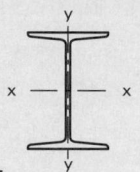

BEARING AND BUCKLING VALUES FOR UNSTIFFENED WEBS FOR S275

Section Designation	Web Thickness	Depth Between Fillets	Bearing				Buckling			Shear Capacity
			End Bearing		Continuous Over Bearing		End Bearing		Continuous Over Bearing	
			Beam Factor	Stiff Bearing Factor	Beam Factor	Stiff Bearing Factor	Buckling Factor	$1.4d$	Buckling Factor	
	t	d	C1	C2	C1	C2	C4		C4	P_v
	mm	mm	kN	kN/mm	kN	kN/mm	kN	mm	kN	kN
254 x 203 x 82 #	10.2	167	214	2.7	534	2.7	1090	233	1090	412
254 x 114 x 37 ‡	7.6	199	105	2.09	263	2.09	379	279	379	319
203 x 152 x 52 #	8.9	133	151	2.36	377	2.36	910	186	910	288
152 x 127 x 37 #	10.4	94.3	153	2.86	382	2.86	2050	132	2050	262
127 x 114 x 29 #	10.2	79.5	120	2.81	300	2.81	2290	111	2290	214
127 x 114 x 27 #	7.4	79.5	86.7	2.04	217	2.04	876	111	876	155
127 x 76 x 16 ‡	5.6	86.5	58.5	1.54	146	1.54	349	121	349	117
114 x 114 x 27 ‡	9.5	60.8	130	2.61	325	2.61	2420	85.1	2420	179
102 x 102 x 23 #	9.5	55.2	112	2.61	280	2.61	2670	77.3	2670	159
102 x 44 x 7 #	4.3	74.6	30.7	1.18	76.9	1.18	183	104	183	72.1
89 x 89 x 19 #	9.5	44.2	110	2.61	274	2.61	3330	61.9	3330	139
76 x 76 x 15 ‡	8.9	38.1	87.1	2.45	218	2.45	3180	53.3	3180	112
76 x 76 x 13 #	5.1	38.1	49.9	1.40	125	1.40	598	53.3	598	64.1

‡ Not available from some leading producers. Check availability.
# Check availability.
Web bearing capacity, $P_w = C1 + b_1 C2$
Web buckling resistance, $P_x = K (C4 \, P_w)^{0.5}$
For Continuous over bearing: $K = 1.0$
For End bearing: $K = \text{Min.} \{ (a_e / 1.4d) + 0.5 , 1.0 \}$
Where b_1 is the stiff bearing length and a_e is the distance from the load or reaction to the nearer end of the member.
FOR EXPLANATION OF TABLES SEE NOTE 6.1

BS 5950-1: 2000
BS 4-1: 1993

WEB BEARING AND BUCKLING

PARALLEL FLANGE CHANNELS

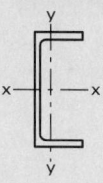

BEARING AND BUCKLING VALUES FOR UNSTIFFENED WEBS FOR S275

Section Designation	Web Thickness	Depth Between Fillets	Bearing					Buckling				Shear Capacity
			End Bearing			Continuous Over Bearing		End Bearing			Continuous Over Bearing	
			Beam Factor	Stiff Bearing Factor	Reduc. Factor	Beam Factor	Stiff Bearing Factor	Buckling Factor	1.4d	Reduc. Factor	Buckling Factor	
	t	d	C1	C2	K_b	C1	C2	C4		K_w	C4	P_v
	mm	mm	kN	kN/mm		kN	kN/mm	kN	mm		kN	kN
430 x 100 x 64	11.0	362	198	2.92	0.748	496	2.92	632	507	0.774	632	752
380 x 100 x 54	9.5	315	164	2.52	0.713	409	2.52	468	441	0.747	468	574
300 x 100 x 46	9.0	237	150	2.38	0.699	376	2.38	529	332	0.711	529	429
300 x 90 x 41	9.0	245	136	2.48	0.699	340	2.48	511	343	0.715	511	446
260 x 90 x 35	8.0	208	114	2.20	0.667	286	2.20	423	291	0.684	423	343
260 x 75 x 28	7.0	212	92.4	1.93	0.627	231	1.93	278	297	0.669	278	300
230 x 90 x 32	7.5	178	107	2.06	0.648	268	2.06	407	249	0.660	407	285
230 x 75 x 26	6.5	181	87.6	1.79	0.604	219	1.79	261	253	0.640	261	247
200 x 90 x 30	7.0	148	100	1.93	0.627	250	1.93	398	207	0.634	398	231
200 x 75 x 23	6.0	151	80.9	1.65	0.577	202	1.65	246	211	0.609	246	198
180 x 90 x 26	6.5	131	87.6	1.79	0.604	219	1.79	360	183	0.611	360	193
180 x 75 x 20	6.0	135	74.3	1.65	0.577	186	1.65	275	189	0.598	275	178
150 x 90 x 24	6.5	102	85.8	1.79	0.604	215	1.79	463	143	0.598	463	161
150 x 75 x 18	5.5	106	66.6	1.51	0.548	166	1.51	270	148	0.563	270	136
125 x 65 x 15 #	5.5	82.0	65.0	1.51	0.565	163	1.51	349	115	0.564	349	113
100 x 50 x 10 #	5.0	65.0	48.1	1.38	0.571	120	1.38	331	91.0	0.563	331	82.5

Check availability.
Web bearing capacity, $P_w = K_b (C1 + b_1 C2)$
Web buckling resistance, $P_x = K_w K (C4 \, P_w)^{0.5}$
For Continuous over bearing: K = 1.0
For End bearing: K = Min. { $(a_e / 1.4d) + 0.5$, 1.0 }
Where b_1 is the stiff bearing length and a_e is the distance from the load or reaction to the nearer end of the member.
Reduction factors K_b and K_w should be applied where appropriate to take account of the eccentricity due to the heel radius.
FOR EXPLANATION OF TABLES SEE NOTE 6.2

BS 5950-1: 2000
BS EN 10056-1: 1999

TIES

EQUAL ANGLES
SUBJECT TO AXIAL TENSION

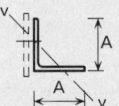

TENSION CAPACITY FOR S275

Section Designation		Mass per Metre	Radius of Gyration Axis v-v	Gross Area	Weld or Bolt Size	Holes Deducted From Angle		Equivalent Tension Area	Tension Capacity
A x A	t					No.	Diameter		P_t
mm	mm	kg	cm	cm^2			mm	cm^2	kN
200 x 200	24 #	71.1	3.90	90.6	Weld	0	-	77.8	2060
					M24	1	26	69.3	1840
					M24	2	26	63.9	1690
					M20	3	22	59.9	1590
	20	59.9	3.92	76.3	Weld	0	-	65.4	1730
					M24	1	26	58.2	1540
					M24	2	26	53.7	1420
					M20	3	22	50.3	1330
	18	54.3	3.90	69.1	Weld	0	-	59.2	1570
					M24	1	26	52.5	1390
					M24	2	26	48.5	1290
					M20	3	22	45.5	1210
	16	48.5	3.94	61.8	Weld	0	-	52.9	1450
					M24	1	26	46.9	1290
					M24	2	26	43.3	1190
					M20	3	22	40.6	1120
150 x 150	18 #	40.1	2.93	51.2	Weld	0	-	43.9	1160
					M24	1	26	38.9	1030
					M20	2	22	35.0	927
	15	33.8	2.93	43.0	Weld	0	-	36.9	1010
					M24	1	26	32.6	896
					M20	2	22	29.3	807
	12	27.3	2.95	34.8	Weld	0	-	29.8	818
					M24	1	26	26.3	722
					M20	2	22	23.7	651
	10	23.0	2.97	29.3	Weld	0	-	25.0	688
					M24	1	26	22.0	606
					M20	2	22	19.9	546
120 x 120	15 #	26.6	2.34	34.0	Weld	0	-	29.2	803
					M24	1	26	24.9	685
					M20	1	22	25.6	705
	12	21.6	2.35	27.5	Weld	0	-	23.6	648
					M24	1	26	20.1	552
					M20	1	22	20.7	568
	10	18.2	2.36	23.2	Weld	0	-	19.8	546
					M24	1	26	16.9	464
					M20	1	22	17.4	477
	8 #	14.7	2.38	18.8	Weld	0	-	16.0	441
					M24	1	26	13.6	375
					M20	1	22	14.0	385

Check availability.
FOR EXPLANATION OF TABLES SEE NOTE 7.1

BS 5950-1: 2000
BS EN 10056-1: 1999

TIES

EQUAL ANGLES
SUBJECT TO AXIAL TENSION

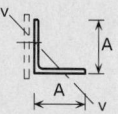

TENSION CAPACITY FOR S275

Section Designation		Mass per Metre	Radius of Gyration Axis v-v	Gross Area	Weld or Bolt Size	Holes Deducted From Angle		Equivalent Tension Area	Tension Capacity
A x A	t					No.	Diameter		P_t
mm	mm	kg	cm	cm^2			mm	cm^2	kN
100 x 100	15 #	21.9	1.94	28.0	Weld	0	-	24.1	663
					M24	1	26	19.8	545
					M20	1	22	20.5	565
	12	17.8	1.94	22.7	Weld	0	-	19.5	536
					M24	1	26	16.0	440
					M20	1	22	16.6	456
	10	15.0	1.95	19.2	Weld	0	-	16.4	452
					M24	1	26	13.5	371
					M20	1	22	14.0	384
	8	12.2	1.96	15.5	Weld	0	-	13.3	364
					M24	1	26	10.9	298
					M20	1	22	11.2	309
90 x 90	12 #	15.9	1.75	20.3	Weld	0	-	17.5	480
					M20	1	22	14.5	400
					M16	1	18	15.1	416
	10	13.4	1.75	17.1	Weld	0	-	14.7	403
					M20	1	22	12.2	336
					M16	1	18	12.7	349
	8	10.9	1.76	13.9	Weld	0	-	11.9	327
					M20	1	22	9.88	272
					M16	1	18	10.3	282
	7 #	9.61	1.77	12.2	Weld	0	-	10.4	287
					M20	1	22	8.66	238
					M16	1	18	9.0	247
80 x 80	10 ‡	11.9	1.55	15.1	Weld	0	-	13.0	357
					M20	1	22	10.5	289
					M16	1	18	11.0	302
	8 ‡	9.63	1.56	12.3	Weld	0	-	10.5	290
					M20	1	22	8.52	234
					M16	1	18	8.9	245
75 x 75	8 †	8.99	1.46	11.4	Weld	0	-	9.78	269
					M20	1	22	7.79	214
					M16	1	18	8.17	225
	6 ‡	6.85	1.47	8.73	Weld	0	-	7.46	205
					M20	1	22	5.93	163
					M16	1	18	6.22	171

‡ Not available from some leading producers. Check availability.
# Check availability.
FOR EXPLANATION OF TABLES SEE NOTE 7.1

BS 5950-1: 2000
BS EN 10056-1: 1999

TIES

EQUAL ANGLES
SUBJECT TO AXIAL TENSION

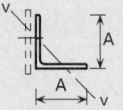

TENSION CAPACITY FOR S275

Section Designation		Mass per Metre	Radius of Gyration Axis v-v	Gross Area	Weld or Bolt Size	Holes Deducted From Angle		Equivalent Tension Area	Tension Capacity
A x A	t					No.	Diameter		P_t
mm	mm	kg	cm	cm²			mm	cm²	kN
70 x 70	7 ‡	7.38	1.36	9.40	Weld	0	-	8.05	221
					M20	1	22	6.28	173
					M16	1	18	6.62	182
	6 ‡	6.38	1.37	8.13	Weld	0	-	6.95	191
					M20	1	22	5.42	149
					M16	1	18	5.71	157
65 x 65	7 ‡	6.83	1.26	8.73	Weld	0	-	7.48	206
					M20	1	22	5.7	157
					M16	1	18	6.04	166
60 x 60	8 ‡	7.09	1.16	9.03	Weld	0	-	7.76	213
					M16	1	18	6.15	169
	6 ‡	5.42	1.17	6.91	Weld	0	-	5.92	163
					M16	1	18	4.68	129
	5 ‡	4.57	1.17	5.82	Weld	0	-	4.97	137
					M16	1	18	3.93	108
50 x 50	6 ‡	4.47	0.968	5.69	Weld	0	-	4.88	134
					M12	1	14	3.94	108
	5 ‡	3.77	0.973	4.80	Weld	0	-	4.11	113
					M12	1	14	3.31	91.0
	4 ‡	3.06	0.979	3.89	Weld	0	-	3.32	91.4
					M12	1	14	2.67	73.5
45 x 45	4.5 ‡	3.06	0.870	3.90	Weld	0	-	3.34	91.8
					M12	1	14	2.61	71.8
40 x 40	5 ‡	2.97	0.773	3.79	Weld	0	-	3.25	89.5
					M12	1	14	2.46	67.5
	4 ‡	2.42	0.777	3.08	Weld	0	-	2.64	72.5
					M12	1	14	1.99	54.7
35 x 35	4 ‡	2.09	0.678	2.67	Weld	0	-	2.29	62.9
					M12	1	14	1.64	45.2
30 x 30	4 ‡	1.78	0.577	2.27	Weld	0	-	1.95	53.6
					M12	1	14	1.3	35.8
	3 ‡	1.36	0.581	1.74	Weld	0	-	1.49	40.9
					M12	1	14	0.996	27.4
25 x 25	4 ‡	1.45	0.482	1.85	Weld	0	-	1.6	43.9
					M12	1	14	0.953	26.2
	3 ‡	1.12	0.484	1.42	Weld	0	-	1.22	33.5
					M12	1	14	0.731	20.1
20 x 20	3 ‡	0.882	0.383	1.12	Weld	0	-	0.964	26.5
					M12	1	14	0.476	13.1

‡ Not available from some leading producers. Check availability.
FOR EXPLANATION OF TABLES SEE NOTE 7.1

BS 5950-1: 2000
BS EN 10056-1: 1999

TIES

TWO EQUAL ANGLES BACK TO BACK SUBJECT TO AXIAL TENSION

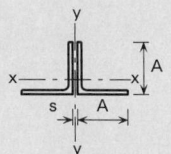

TENSION CAPACITY FOR S275

Composed Of Two Angles		Mass per Metre	Space Between Angles s	Radius of Gyration		Gross Area	Weld or Bolt Size	Holes Deducted From Angle		Gusset Between Angles		Gusset On Back of Angles	
A x A mm	t mm	kg	mm	Axis x-x cm	Axis y-y cm	cm²		No.	Diameter mm	Equivalent Tension Area cm²	Tension Capacity kN	Equivalent Tension Area cm²	Tension Capacity kN
200 x 200	24 #	142	15	6.06	8.95	181	Weld	0	-	168	4460	156	4120
							M24	1	26.0	160	4240	139	3670
							M24	2	26.0	149	3950	128	3390
							M20	3	22.0	141	3740	120	3170
	20	120	15	6.11	8.87	153	Weld	0	-	142	3760	131	3470
							M24	1	26.0	134	3560	116	3080
							M24	2	26.0	125	3330	107	2840
							M20	3	22.0	119	3150	101	2670
	18	109	15	6.13	8.83	138	Weld	0	-	128	3400	118	3140
							M24	1	26.0	122	3220	105	2790
							M24	2	26.0	114	3010	97.0	2570
							M20	3	22.0	108	2850	91.0	2410
	16	97.0	15	6.16	8.79	124	Weld	0	-	115	3150	106	2910
							M24	1	26.0	109	2990	93.8	2580
							M24	2	26.0	102	2790	86.6	2380
							M20	3	22.0	96.2	2640	81.3	2230
150 x 150	18 #	80.2	12	4.55	6.75	102	Weld	0	-	95.1	2520	87.9	2330
							M24	1	26.0	89.9	2380	77.8	2060
							M20	2	22.0	82.1	2180	70.0	1850
	15	67.6	12	4.57	6.66	86.0	Weld	0	-	79.9	2200	73.7	2030
							M24	1	26.0	75.4	2070	65.1	1790
							M20	2	22.0	68.9	1900	58.7	1610
	12	54.6	12	4.60	6.59	69.6	Weld	0	-	64.6	1780	59.5	1640
							M24	1	26.0	60.9	1680	52.5	1440
							M20	2	22.0	55.7	1530	47.3	1300
	10	46.0	12	4.62	6.54	58.6	Weld	0	-	54.3	1490	50.0	1380
							M24	1	26.0	51.2	1410	44.1	1210
							M20	2	22.0	46.9	1290	39.7	1090
120 x 120	15 #	53.2	12	3.63	5.49	68.0	Weld	0	-	63.2	1740	58.4	1610
							M24	1	26.0	57.8	1590	49.8	1370
							M20	1	22.0	59.3	1630	51.3	1410
	12	43.2	12	3.65	5.42	55.0	Weld	0	-	51.1	1400	47.1	1300
							M24	1	26.0	46.7	1280	40.2	1100
							M20	1	22.0	47.9	1320	41.3	1140
	10	36.4	12	3.67	5.36	46.4	Weld	0	-	43.0	1180	39.7	1090
							M24	1	26.0	39.4	1080	33.8	928
							M20	1	22.0	40.3	1110	34.7	955
	8 #	29.4	12	3.71	5.34	37.6	Weld	0	-	34.8	958	32.1	882
							M24	1	26.0	31.8	876	27.2	749
							M20	1	22.0	32.6	897	28.0	770

Check availability.
FOR EXPLANATION OF TABLES SEE NOTE 7.2

BS 5950-1: 2000
BS EN 10056-1: 1999

TIES

TWO EQUAL ANGLES
BACK TO BACK
SUBJECT TO AXIAL TENSION

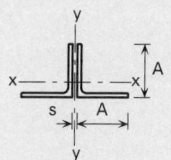

TENSION CAPACITY FOR S275

Composed Of Two Angles		Mass per Metre	Space Between Angles	Radius of Gyration		Gross Area	Weld or Bolt Size	Holes Deducted From Angle		Gusset Between Angles		Gusset On Back of Angles	
A x A	t		s	Axis x-x	Axis y-y			No.	Diameter	Equivalent Tension Area	Tension Capacity	Equivalent Tension Area	Tension Capacity
mm	mm	kg	mm	cm	cm	cm²			mm	cm²	kN	cm²	kN
100 x 100	15 #	43.8	10	2.99	4.62	56.0	Weld	0	-	52.1	1430	48.2	1330
							M24	1	26.0	46.1	1270	39.6	1090
							M20	1	22.0	47.6	1310	41.1	1130
	12	35.6	10	3.02	4.55	45.4	Weld	0	-	42.2	1160	39.0	1070
							M24	1	26.0	37.4	1030	32.0	880
							M20	1	22.0	38.5	1060	33.2	912
	10	30.0	10	3.04	4.50	38.4	Weld	0	-	35.6	980	32.9	904
							M24	1	26.0	31.6	868	27.0	741
							M20	1	22.0	32.5	894	27.9	768
	8	24.4	10	3.06	4.46	31.0	Weld	0	-	28.8	791	26.5	729
							M24	1	26.0	25.5	700	21.7	597
							M20	1	22.0	26.2	721	22.5	618
90 x 90	12 #	31.8	10	2.71	4.16	40.6	Weld	0	-	37.8	1040	34.9	960
							M20	1	22.0	33.8	930	29.1	800
							M16	1	18.0	35.0	962	30.2	831
	10	26.8	10	2.72	4.11	34.2	Weld	0	-	31.8	874	29.3	807
							M20	1	22.0	28.5	783	24.4	672
							M16	1	18.0	29.4	809	25.4	698
	8	21.8	10	2.74	4.06	27.8	Weld	0	-	25.8	709	23.8	654
							M20	1	22.0	23.1	635	19.8	543
							M16	1	18.0	23.9	657	20.5	564
	7 #	19.2	10	2.75	4.04	24.4	Weld	0	-	22.6	622	20.9	574
							M20	1	22.0	20.3	558	17.3	476
							M16	1	18.0	20.9	576	18.0	495
80 x 80	10 ‡	23.8	8	2.41	3.65	30.2	Weld	0	-	28.1	772	25.9	713
							M20	1	22.0	24.6	676	21.0	578
							M16	1	18.0	25.5	702	22.0	604
	8 ‡	19.3	8	2.43	3.60	24.6	Weld	0	-	22.8	628	21.1	579
							M20	1	22.0	20.0	550	17.0	468
							M16	1	18.0	20.8	571	17.8	490
75 x 75	8 ‡	18.0	8	2.27	3.41	22.8	Weld	0	-	21.2	582	19.6	538
							M20	1	22.0	18.3	503	15.6	428
							M16	1	18.0	19.0	524	16.3	449
	6 ‡	13.7	8	2.29	3.35	17.4	Weld	0	-	16.2	445	14.9	410
							M20	1	22.0	14.0	384	11.9	326
							M16	1	18.0	14.6	400	12.4	342

‡ Not available from some leading producers. Check availability.
# Check availability.
FOR EXPLANATION OF TABLES SEE NOTE 7.2

BS 5950-1: 2000
BS EN 10056-1: 1999

TIES

TWO EQUAL ANGLES BACK TO BACK SUBJECT TO AXIAL TENSION

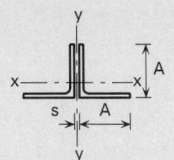

TENSION CAPACITY FOR S275

Composed Of Two Angles		Mass per Metre	Space Between Angles s	Radius of Gyration		Gross Area	Weld or Bolt Size	Holes Deducted From Angle		Gusset Between Angles		Gusset On Back of Angles	
A x A mm	t mm	kg	mm	Axis x-x cm	Axis y-y cm	cm²		No.	Diameter mm	Equivalent Tension Area cm²	Tension Capacity kN	Equivalent Tension Area cm²	Tension Capacity kN
70 x 70	7 ‡	14.8	8	2.12	3.18	18.8	Weld	0	-	17.5	480	16.1	443
							M20	1	22.0	14.8	407	12.6	346
							M16	1	18.0	15.5	426	13.2	364
	6 ‡	12.8	8	2.13	3.16	16.3	Weld	0	-	15.1	415	13.9	382
							M20	1	22.0	12.8	352	10.8	298
							M16	1	18.0	13.4	368	11.4	314
65 x 65	7 ‡	13.7	8	1.96	3.14	17.5	Weld	0	-	16.2	446	15.0	411
							M20	1	22.0	13.5	371	11.4	314
							M16	1	18.0	14.2	390	12.1	332
60 x 60	8 ‡	14.2	8	1.80	2.82	18.1	Weld	0	-	16.8	462	15.5	427
							M16	1	18.0	14.4	396	12.3	338
	6 ‡	10.8	8	1.82	2.77	13.8	Weld	0	-	12.8	353	11.8	325
							M16	1	18.0	11.0	303	9.36	257
	5 ‡	9.14	8	1.82	2.74	11.6	Weld	0	-	10.8	297	9.95	274
							M16	1	18.0	9.27	255	7.86	216
50 x 50	6 ‡	8.94	8	1.50	2.38	11.4	Weld	0	-	10.6	291	9.77	269
							M12	1	14.0	9.22	254	7.87	217
	5 ‡	7.54	8	1.51	2.35	9.60	Weld	0	-	8.91	245	8.22	226
							M12	1	14.0	7.77	214	6.62	182
	4 ‡	6.12	8	1.52	2.32	7.78	Weld	0	-	7.21	198	6.65	183
							M12	1	14.0	6.29	173	5.35	147

‡ Not available from some leading producers. Check availability.
FOR EXPLANATION OF TABLES SEE NOTE 7.2

BS 5950-1: 2000
BS EN 10056-1: 1999

TIES

UNEQUAL ANGLES
LONG LEG ATTACHED
SUBJECT TO AXIAL TENSION

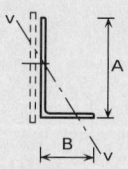

TENSION CAPACITY FOR S275

Section Designation A x B mm	t mm	Mass per Metre kg	Radius of Gyration Axis v-v cm	Gross Area cm^2	Weld or Bolt Size	Holes Deducted From Angle No.	Diameter mm	Equivalent Tension Area cm^2	Tension Capacity P_t kN
200 x 150	18 #	47.1	3.22	60.1	Weld	0	-	52.9	1400
					M24	1	26	48.1	1270
					M24	2	26	44.0	1170
					M20	3	22	41.0	1090
	15	39.6	3.23	50.5	Weld	0	-	44.4	1220
					M24	1	26	40.3	1110
					M24	2	26	36.9	1010
					M20	3	22	34.4	945
	12	32.0	3.25	40.8	Weld	0	-	35.8	983
					M24	1	26	32.4	891
					M24	2	26	29.7	817
					M20	3	22	27.7	762
200 x 100	15	33.8	2.12	43.0	Weld	0	-	39.1	1080
					M24	1	26	36.5	1000
					M24	2	26	33.1	911
					M20	3	22	30.6	842
	12	27.3	2.14	34.8	Weld	0	-	31.6	868
					M24	1	26	29.4	809
					M24	2	26	26.7	735
					M20	3	22	24.7	679
	10	23.0	2.15	29.2	Weld	0	-	26.4	727
					M24	1	26	24.6	677
					M24	2	26	22.4	615
					M20	3	22	20.7	569
150 x 90	15	26.6	1.93	33.9	Weld	0	-	30.5	838
					M24	1	26	28.0	771
					M20	2	22	24.8	681
	12	21.6	1.94	27.5	Weld	0	-	24.7	678
					M24	1	26	22.6	622
					M20	2	22	20.0	550
	10	18.2	1.95	23.2	Weld	0	-	20.7	570
					M24	1	26	19.0	522
					M20	2	22	16.8	463
150 x 75	15	24.8	1.58	31.7	Weld	0	-	28.9	796
					M24	1	26	26.9	740
					M20	2	22	23.7	651
	12	20.2	1.59	25.7	Weld	0	-	23.4	643
					M24	1	26	21.7	597
					M20	2	22	19.1	526
	10	17.0	1.60	21.7	Weld	0	-	19.7	541
					M24	1	26	18.2	501
					M20	2	22	16.1	442

Check availability.
FOR EXPLANATION OF TABLES SEE NOTE 7.1

BS 5950-1: 2000
BS EN 10056-1: 1999

TIES

UNEQUAL ANGLES
LONG LEG ATTACHED
SUBJECT TO AXIAL TENSION

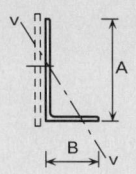

TENSION CAPACITY FOR S275

Section Designation		Mass per Metre	Radius of Gyration Axis v-v	Gross Area	Weld or Bolt Size	Holes Deducted From Angle		Equivalent Tension Area	Tension Capacity P_t
A x B mm	t mm	kg	cm	cm^2		No.	Diameter mm	cm^2	kN
125 x 75	12	17.8	1.61	22.7	Weld	0	-	20.4	561
					M24	1	26	18.1	498
					M20	2	22	15.5	427
	10	15.0	1.61	19.1	Weld	0	-	17.1	471
					M24	1	26	15.2	417
					M20	2	22	13.0	358
	8	12.2	1.63	15.5	Weld	0	-	13.9	381
					M24	1	26	12.3	337
					M20	2	22	10.5	289
100 x 75	12	15.4	1.59	19.7	Weld	0	-	17.4	478
					M24	1	26	14.5	399
					M20	1	22	15.1	415
	10	13.0	1.59	16.6	Weld	0	-	14.6	402
					M24	1	26	12.2	335
					M20	1	22	12.7	348
	8	10.6	1.60	13.5	Weld	0	-	11.9	326
					M24	1	26	9.85	271
					M20	1	22	10.2	282
100 x 65	10 #	12.3	1.39	15.6	Weld	0	-	13.9	383
					M24	1	26	11.7	321
					M20	1	22	12.2	334
	8 #	9.94	1.40	12.7	Weld	0	-	11.3	310
					M24	1	26	9.45	260
					M20	1	22	9.84	271
	7 #	8.77	1.40	11.2	Weld	0	-	9.94	273
					M24	1	26	8.32	229
					M20	1	22	8.65	238
100 x 50	8 ‡	8.97	1.06	11.4	Weld	0	-	10.4	285
					M24	1	26	8.8	242
					M20	1	22	9.19	253
	6 ‡	6.84	1.07	8.71	Weld	0	-	7.9	217
					M24	1	26	6.68	184
					M20	1	22	6.97	192
80 x 60	7 ‡	7.36	1.28	9.38	Weld	0	-	8.25	227
					M20	1	22	6.76	186
					M16	1	18	7.1	195

‡ Not available from some leading producers. Check availability.
# Check availability.
FOR EXPLANATION OF TABLES SEE NOTE 7.1

BS 5950-1: 2000
BS EN 10056-1: 1999

TIES

UNEQUAL ANGLES
LONG LEG ATTACHED
SUBJECT TO AXIAL TENSION

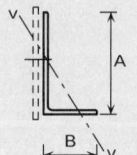

TENSION CAPACITY FOR S275

Section Designation		Mass per Metre	Radius of Gyration Axis	Gross Area	Weld or Bolt Size	Holes Deducted From Angle		Equivalent Tension Area	Tension Capacity
A x B	t		v-v			No.	Diameter		P_t
mm	mm	kg	cm	cm^2			mm	cm^2	kN
80 x 40	8 ‡	7.07	0.838	9.01	Weld	0	-	8.23	226
					M20	1	22	6.87	189
					M16	1	18	7.26	200
	6 ‡	5.41	0.845	6.89	Weld	0	-	6.26	172
					M20	1	22	5.22	144
					M16	1	18	5.51	151
75 x 50	8 ‡	7.39	1.07	9.41	Weld	0	-	8.39	231
					M20	1	22	6.79	187
					M16	1	18	7.18	197
	6 ‡	5.65	1.08	7.19	Weld	0	-	6.38	176
					M20	1	22	5.16	142
					M16	1	18	5.45	150
70 x 50	6 ‡	5.41	1.07	6.89	Weld	0	-	6.08	167
					M20	1	22	4.8	132
					M16	1	18	5.09	140
65 x 50	5 ‡	4.35	1.07	5.54	Weld	0	-	4.85	133
					M20	1	22	3.72	102
					M16	1	18	3.96	109
60 x 40	6 ‡	4.46	0.855	5.68	Weld	0	-	5.06	139
					M16	1	18	4.06	112
	5 ‡	3.76	0.860	4.79	Weld	0	-	4.25	117
					M16	1	18	3.42	93.9
60 x 30	5 ‡	3.36	0.633	4.28	Weld	0	-	3.9	107
					M16	1	18	3.16	86.9
50 x 30	5 ‡	2.96	0.639	3.78	Weld	0	-	3.4	93.4
					M12	1	14	2.8	77.0
45 x 30	4 ‡	2.25	0.640	2.87	Weld	0	-	2.55	70.1
					M12	1	14	2.02	55.6
40 x 25	4 ‡	1.93	0.534	2.46	Weld	0	-	2.2	60.6
					M12	1	14	1.68	46.1
40 x 20	4 ‡	1.77	0.417	2.26	Weld	0	-	2.06	56.7
					M12	1	14	1.58	43.4
30 x 20	4 ‡	1.46	0.421	1.86	Weld	0	-	1.66	45.7
					M12	1	14	1.1	30.2
	3 ‡	1.12	0.424	1.43	Weld	0	-	1.27	35.0
					M12	1	14	0.841	23.1

‡ Not available from some leading producers. Check availability.
FOR EXPLANATION OF TABLES SEE NOTE 7.1

BS 5950-1: 2000
BS EN 10056-1: 1999

TIES

TWO UNEQUAL ANGLES
LONG LEGS BACK TO BACK
SUBJECT TO AXIAL TENSION

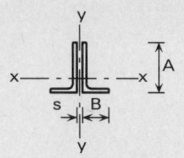

TENSION CAPACITY FOR S275

Composed Of Two Angles		Mass per Metre	Space Between Angles	Radius of Gyration		Gross Area	Weld or Bolt Size	Holes Deducted From Angle		Gusset Between Angles		Gusset On Back of Angles	
A x B	t		s	Axis x-x	Axis y-y			No.	Diameter	Equivalent Tension Area	Tension Capacity	Equivalent Tension Area	Tension Capacity
mm	mm	kg	mm	cm	cm	cm²			mm	cm²	kN	cm²	kN
200x150	18 #	94.2	15	6.30	6.36	120	Weld	0	-	113	2990	100	2660
							M24	1	26	108	2870	86.7	2300
							M20	2	22	104	2740	78.9	2090
	15	79.2	15	6.33	6.28	101	Weld	0	-	94.9	2610	84.2	2320
							M24	1	26	90.8	2500	72.6	2000
							M20	2	22	86.9	2390	66.2	1820
	12	64.0	15	6.36	6.22	81.6	Weld	0	-	76.6	2110	67.9	1870
							M24	1	26	73.2	2010	58.5	1610
							M20	2	22	70.1	1930	53.3	1470
200x100	15	67.5	15	6.40	3.97	86.0	Weld	0	-	82.1	2260	69.2	1900
							M24	1	26	79.5	2190	54.6	1500
							M20	1	22	79.5	2190	56.1	1540
	12	54.6	15	6.43	3.90	69.6	Weld	0	-	66.4	1820	55.9	1540
							M24	1	26	64.2	1770	44.1	1210
							M20	1	22	64.2	1770	45.3	1240
	10	46.0	15	6.46	3.85	58.4	Weld	0	-	55.6	1530	46.9	1290
							M24	1	26	53.8	1480	37.0	1020
							M20	1	22	53.8	1480	37.9	1040
150x90	15	53.2	12	4.74	3.749	67.8	Weld	0	-	64.4	1770	55.6	1530
							M20	1	22	62.1	1710	44.9	1230
							M16	1	18	62.1	1710	46.3	1270
	12	43.2	12	4.77	3.69	55.0	Weld	0	-	52.2	1430	45.0	1240
							M20	1	22	50.3	1380	36.3	998
							M16	1	18	50.3	1380	37.4	1030
	10	36.4	12	4.80	3.64	46.4	Weld	0	-	43.9	1210	37.9	1040
							M20	1	22	42.3	1160	30.5	839
							M16	1	18	42.3	1160	31.5	866
150x75	15	49.6	12	4.75	3.09	63.4	Weld	0	-	60.6	1670	51.1	1410
							M20	1	22	58.8	1620	39.5	1090
							M16	1	18	58.8	1620	41.0	1130
	12	40.4	12	4.78	3.02	51.4	Weld	0	-	49.1	1350	41.4	1140
							M20	1	22	47.6	1310	32.0	879
							M16	1	18	47.6	1310	33.1	911
	10	34.0	12	4.81	2.97	43.4	Weld	0	-	41.4	1140	34.9	959
							M20	1	22	40.1	1100	26.9	740
							M16	1	18	40.1	1100	27.9	767

Check availability.
FOR EXPLANATION OF TABLES SEE NOTE 7.2

BS 5950-1: 2000
BS EN 10056-1: 1999

TIES

TWO UNEQUAL ANGLES
LONG LEGS BACK TO BACK
SUBJECT TO AXIAL TENSION

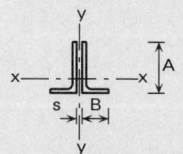

TENSION CAPACITY FOR S275

Composed Of Two Angles		Mass per Metre	Space Between Angles s	Radius of Gyration		Gross Area	Weld or Bolt Size	Holes Deducted From Angle		Gusset Between Angles		Gusset On Back of Angles	
A x B mm	t mm	kg	mm	Axis x-x cm	Axis y-y cm	cm²		No.	Diameter mm	Equivalent Tension Area cm²	Tension Capacity kN	Equivalent Tension Area cm²	Tension Capacity kN
125x75	12	35.6	12	3.95	3.19	45.4	Weld	0	-	43.1	1180	37.2	1020
							M20	1	22	41.2	1130	29.0	797
							M16	1	18	41.6	1140	30.1	828
	10	30.0	12	3.97	3.14	38.2	Weld	0	-	36.2	996	31.2	859
							M20	1	22	34.6	952	24.3	669
							M16	1	18	34.9	960	25.3	695
	8	24.4	12	4.00	3.09	31.0	Weld	0	-	29.4	807	25.3	696
							M20	1	22	28.0	771	19.7	541
							M16	1	18	28.3	777	20.4	562
100x75	12	30.8	10	3.10	3.31	39.4	Weld	0	-	37.1	1020	33.0	907
							M20	1	22	34.0	935	26.0	714
							M16	1	18	35.2	967	27.1	746
	10	26.0	10	3.12	3.27	33.2	Weld	0	-	31.2	859	27.7	763
							M20	1	22	28.6	787	21.8	600
							M16	1	18	29.6	813	22.8	626
	8	21.2	10	3.14	3.22	27.0	Weld	0	-	25.4	697	22.5	619
							M20	1	22	23.2	639	17.7	486
							M16	1	18	24.0	660	18.4	507
100x65	10 #	24.6	10	3.14	2.79	31.2	Weld	0	-	29.5	812	25.7	708
							M20	1	22	27.1	746	19.4	534
							M16	1	18	28.1	772	20.4	560
	8 #	19.9	10	3.16	2.74	25.4	Weld	0	-	24.0	660	20.9	575
							M20	1	22	22.0	606	15.8	433
							M16	1	18	22.8	627	16.5	454
	7 #	17.5	10	3.17	2.72	22.4	Weld	0	-	21.1	581	18.4	506
							M20	1	22	19.4	534	13.9	382
							M16	1	18	20.1	552	14.5	400
100x50	8 ‡	17.9	10	3.19	2.09	22.8	Weld	0	-	21.8	599	18.4	505
							M12	1	14	21.1	580	14.3	394
	6 ‡	13.7	10	3.21	2.04	17.4	Weld	0	-	16.6	457	14.0	385
							M12	1	14	16.1	442	10.9	300

‡ Not available from some leading producers. Check availability.
# Check availability.
FOR EXPLANATION OF TABLES SEE NOTE 7.2

BS 5950-1: 2000
BS EN 10056-1: 1999

TIES

TWO UNEQUAL ANGLES
LONG LEGS BACK TO BACK
SUBJECT TO AXIAL TENSION

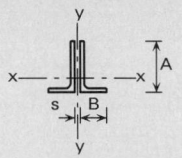

TENSION CAPACITY FOR S275

Composed Of Two Angles		Mass per Metre	Space Between Angles s	Radius of Gyration		Gross Area	Weld or Bolt Size	Holes Deducted From Angle		Gusset Between Angles		Gusset On Back of Angles	
A x B	t			Axis x-x	Axis y-y			No.	Diameter	Equivalent Tension Area	Tension Capacity	Equivalent Tension Area	Tension Capacity
mm	mm	kg	mm	cm	cm	cm^2			mm	cm^2	kN	cm^2	kN
80x60	7 ‡	14.7	8	2.51	2.59	18.8	Weld	0	-	17.6	485	15.7	430
							M16	1	18	16.1	442	12.2	336
80x40	8 ‡	14.1	8	2.53	1.71	18.0	Weld	0	-	17.2	474	14.5	400
							M12	1	14	16.6	456	10.8	297
	6 ‡	10.8	8	2.55	1.66	13.8	Weld	0	-	13.2	362	11.1	305
							M12	1	14	12.6	348	8.23	226
75x50	8 ‡	14.8	8	2.35	2.19	18.8	Weld	0	-	17.8	489	15.6	428
							M12	1	14	16.8	463	12.3	339
	6 ‡	11.3	8	2.37	2.14	14.4	Weld	0	-	13.6	373	11.9	326
							M12	1	14	12.8	353	9.37	258
70x50	6 ‡	10.8	8	2.20	2.19	13.8	Weld	0	-	13.0	357	11.4	315
							M12	1	14	12.1	333	9.07	250
65x50	5 ‡	8.70	8	2.05	2.21	11.1	Weld	0	-	10.4	286	9.26	255
							M12	1	14	9.56	263	7.36	202
60x40	6 ‡	8.92	8	1.88	1.80	11.4	Weld	0	-	10.7	295	9.39	258
							M12	1	14	9.74	268	7.02	193
	5 ‡	7.52	8	1.89	1.78	9.58	Weld	0	-	9.04	249	7.91	217
							M12	1	14	8.21	226	5.91	163

‡ Not available from some leading producers. Check availability.
FOR EXPLANATION OF TABLES SEE NOTE 7.2

BS 5950-1: 2000
BS 4-1: 1993

STRUTS

UB SECTIONS
SUBJECT TO AXIAL COMPRESSION

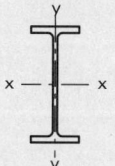

COMPRESSION RESISTANCE FOR S275

Section Designation	Axis	Compression resistance, P_{cx}, P_{cy} (kN) for Effective lengths, L_E (m)												
		2.0	3.0	4.0	5.0	6.0	7.0	8.0	9.0	10.0	11.0	12.0	13.0	14.0
1016x305x487 # +	P_{cx}	15800	15800	15800	15800	15800	15800	15700	15600	15400	15300	15100	15000	14800
	P_{cy}	14700	13200	11600	9930	8310	6890	5720	4780	4040	3450	2970	2580	2270
1016x305x437 # +	P_{cx}	14200	14200	14200	14200	14200	14200	14200	14100	14000	13900	13800	13700	13600	13500
	P_{cy}	13300	12200	10900	9370	7840	6470	5330	4430	3730	3160	2720	2350	2060
1016x305x393 # +	P_{cx}	12800	12800	12800	12800	12800	12800	12700	12600	12500	12400	12300	12200	12100
	P_{cy}	11900	10900	9680	8310	6930	5700	4690	3890	3270	2770	2380	2060	1800
* 1016x305x349 # +	P_{cx}	*11700*	*11700*	*11700*	*11700*	*11700*	*11700*	*11700*	*11600*	*11600*	*11500*	*11500*	*11400*	*11300*
	P_{cy}	*11200*	*10400*	*9350*	*8080*	*6730*	*5500*	*4490*	*3710*	*3090*	*2610*	*2230*	*1930*	*1680*
* 1016x305x314 # +	P_{cx}	*10100*	*10100*	*10100*	*10100*	*10100*	*10100*	*10100*	*10000*	*10000*	*9940*	*9880*	*9820*	*9760*
	P_{cy}	*9640*	*8960*	*8360*	*7200*	*5970*	*4860*	*3970*	*3270*	*2730*	*2300*	*1970*	*1700*	*1480*
* 1016x305x272 # +	P_{cx}	8340	8340	8340	8340	8340	8340	8310	8270	8230	8180	8140	8090	8040
	P_{cy}	7950	7410	6730	6090	5160	4200	3430	2820	2350	1990	1700	1470	1280
* 1016x305x249 # +	P_{cx}	7540	7540	7540	7540	7540	7540	7510	7470	7430	7390	7350	7310	7260
	P_{cy}	7160	6640	5980	5480	4470	3610	2930	2400	2000	1680	1440	1240	1080
* 1016x305x222 # +	P_{cx}	6580	6580	6580	6580	6580	6580	6550	6520	6480	6450	6410	6370	6330
	P_{cy}	6210	5740	5120	4670	3750	3000	2420	1980	1640	1380	1180	1010	883
914 x 419 x 388 #	P_{cx}	13100	13100	13100	13100	13100	13100	13000	12900	12800	12800	12700	12600	12500
	P_{cy}	12900	12400	11800	11100	10300	9420	8410	7400	6470	5640	4930	4330	3820
* 914 x 419 x 343 #	P_{cx}	*11500*	*11500*	*11500*	*11500*	*11500*	*11500*	*11400*	*11400*	*11400*	*11300*	*11200*	*11100*	*11100*
	P_{cy}	*11400*	*11000*	*10400*	*9820*	*9090*	*8250*	*7340*	*6400*	*5610*	*4890*	*4270*	*3740*	*3300*
* 914 x 305 x 289 #	P_{cx}	*9600*	*9600*	*9600*	*9600*	*9600*	*9570*	*9520*	*9460*	*9400*	*9350*	*9280*	*9220*	*9220*
	P_{cy}	9220	8610	7780	6740	5630	4620	3780	3120	2610	2200	1880	1630	1420
* 914 x 305 x 253 #	P_{cx}	8010	8010	8010	8010	8010	8010	8000	7950	7910	7860	7810	7760	7710	7660
	P_{cy}	7630	7110	6700	5850	4860	3970	3250	2680	2230	1890	1610	1390	1210
* 914 x 305 x 224 #	P_{cx}	6840	6840	6840	6840	6840	6820	6790	6750	6710	6670	6620	6580	6530
	P_{cy}	6510	6060	5490	5080	4190	3400	2770	2280	1900	1600	1370	1180	1030
* 914 x 305 x 201 #	P_{cx}	5950	5950	5950	5950	5950	5940	5900	5870	5840	5800	5760	5720	5680
	P_{cy}	5650	5240	4730	4200	3600	2900	2350	1930	1600	1350	1150	996	867
* 838 x 292 x 226 #	P_{cx}	7210	7210	7210	7210	7210	7170	7130	7080	7040	6990	6940	6890	6830
	P_{cy}	6840	6350	5990	5130	4230	3430	2800	2300	1920	1620	1380	1190	1040
* 838 x 292 x 194 #	P_{cx}	5910	5910	5910	5910	5910	5880	5850	5810	5770	5730	5690	5650	5600
	P_{cy}	5600	5190	4670	4250	3460	2790	2260	1850	1540	1300	1110	958	834
* 838 x 292 x 176 #	P_{cx}	5230	5230	5230	5230	5230	5200	5170	5140	5100	5070	5030	4990	4950
	P_{cy}	*4940*	*4570*	*4100*	*3700*	*3030*	*2430*	*1960*	*1610*	*1330*	*1120*	*958*	*826*	*720*
* 762 x 267 x 197	P_{cx}	6440	6440	6440	6440	6420	6380	6330	6290	6240	6190	6130	6070	6010
	P_{cy}	6020	5650	4920	4070	3250	2590	2080	1700	1410	1190	1010	871	758
* 762 x 267 x 173	P_{cx}	5440	5440	5440	5440	5420	5380	5350	5310	5270	5220	5180	5130	5080
	P_{cy}	5080	4640	4250	3480	2760	2190	1750	1430	1190	997	849	731	636
* 762 x 267 x 147	P_{cx}	4400	4400	4400	4400	4390	4360	4330	4290	4260	4230	4190	4150	4110
	P_{cy}	*4100*	*3740*	*3260*	*2850*	*2240*	*1760*	*1410*	*1150*	*950*	*798*	*679*	*585*	*508*
* 762 x 267 x 134	P_{cx}	4020	4020	4020	4020	4010	3980	3960	3930	3900	3870	3830	3800	3760
	P_{cy}	*3750*	*3410*	*2960*	*2580*	*2020*	*1580*	*1260*	*1020*	*844*	*708*	*602*	*518*	*450*

\# Check availability.
+ Section is not given in BS 4-1: 1993.
* Section may be slender under axial compression.
Values in *italic type* indicate that the section is slender when loaded to capacity and allowance has been made in calculating the capacity.
FOR EXPLANATION OF TABLES SEE NOTE 8.1

BS 5950-1: 2000
BS 4-1: 1993

STRUTS

UB SECTIONS
SUBJECT TO AXIAL COMPRESSION

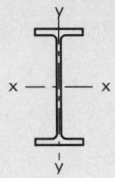

COMPRESSION RESISTANCE FOR S275

Section Designation	Axis	Compression resistance, P_{cx}, P_{cy} (kN) for Effective lengths, L_E (m)												
		2.0	3.0	4.0	5.0	6.0	7.0	8.0	9.0	10.0	11.0	12.0	13.0	14.0
* 686 x 254 x 170	P_{cx}	5660	5660	5660	5650	5610	5570	5520	5480	5430	5410	5400	5330	5260
	P_{cy}	5340	4830	4160	3390	2690	2130	1700	1390	1150	967	824	709	617
* 686 x 254 x 152	P_{cx}	4870	4870	4870	4870	4830	4800	4760	4720	4680	4630	4580	4530	4470
	P_{cy}	4530	4170	3690	2990	2360	1860	1490	1210	1010	845	719	619	539
* 686 x 254 x 140	P_{cx}	4360	4360	4360	4360	4320	4290	4260	4220	4180	4140	4100	4060	4010
	P_{cy}	4050	3680	3350	2700	2120	1670	1340	1090	901	757	644	555	482
* 686 x 254 x 125	P_{cx}	3780	3780	3780	3780	3760	3730	3700	3670	3640	3600	3560	3520	3480
	P_{cy}	3510	3180	2790	2340	1820	1430	1140	925	765	642	546	470	408
610 x 305 x 238	P_{cx}	8030	8030	8030	8000	7940	7870	7800	7730	7650	7560	7470	7360	7250
	P_{cy}	7720	7260	6700	5990	5180	4370	3650	3050	2570	2180	1870	1620	1420
610 x 305 x 179	P_{cx}	6040	6040	6040	6020	5970	5920	5860	5810	5740	5680	5600	5520	5440
	P_{cy}	5800	5440	5000	4440	3820	3200	2660	2210	1860	1580	1350	1170	1020
* 610 x 305 x 149	P_{cx}	4850	4850	4850	4830	4790	4750	4710	4660	4620	4560	4510	4440	4370
	P_{cy}	4660	4370	4150	3680	3150	2630	2180	1820	1520	1290	1110	960	838
* 610 x 229 x 140	P_{cx}	4670	4670	4670	4650	4610	4560	4540	4520	4470	4410	4350	4280	4210
	P_{cy}	4300	3810	3170	2480	1910	1490	1190	962	794	666	566	487	423
* 610 x 229 x 125	P_{cx}	4020	4020	4020	4000	3960	3930	3890	3850	3810	3770	3720	3660	3600
	P_{cy}	3670	3390	2800	2180	1680	1310	1040	841	694	582	494	425	369
* 610 x 229 x 113	P_{cx}	3540	3540	3540	3520	3490	3460	3430	3390	3360	3320	3270	3220	3170
	P_{cy}	3230	2870	2490	1930	1480	1150	910	737	608	509	433	372	323
* 610 x 229 x 101	P_{cx}	3180	3180	3180	3160	3140	3110	3080	3050	3010	2980	2940	2890	2840
	P_{cy}	2890	2550	2230	1690	1280	990	782	632	520	436	370	318	276
533 x 210 x 122	P_{cx}	4110	4110	4100	4060	4020	3980	3930	3880	3820	3760	3690	3610	3520
	P_{cy}	3690	3200	2570	1960	1480	1150	906	732	603	505	429	368	320
* 533 x 210 x 109	P_{cx}	3670	3670	3670	3640	3600	3560	3520	3480	3420	3370	3300	3230	3150
	P_{cy}	3300	2850	2270	1720	1300	1000	791	639	526	440	374	321	279
* 533 x 210 x 101	P_{cx}	3310	3310	3310	3280	3250	3210	3170	3130	3090	3040	3030	3000	2920
	P_{cy}	3030	2630	2090	1580	1190	918	725	586	482	403	342	294	255
* 533 x 210 x 92	P_{cx}	3020	3020	3010	2980	2950	2920	2890	2850	2810	2760	2710	2650	2590
	P_{cy}	2700	2440	1910	1420	1070	820	646	521	429	358	304	261	227
* 533 x 210 x 82	P_{cx}	2640	2640	2640	2610	2590	2560	2530	2500	2460	2420	2370	2320	2260
	P_{cy}	2360	2050	1660	1220	913	699	550	443	364	304	258	222	192
457 x 191 x 98	P_{cx}	3310	3310	3290	3250	3210	3170	3120	3060	3000	2930	2850	2750	2640
	P_{cy}	2920	2470	1910	1410	1060	810	638	514	422	353	300	257	223
457 x 191 x 89	P_{cx}	3020	3020	3000	2960	2930	2890	2840	2790	2730	2670	2590	2500	2400
	P_{cy}	2660	2240	1720	1270	948	726	572	461	379	316	268	230	200
* 457 x 191 x 82	P_{cx}	2830	2830	2800	2770	2740	2700	2680	2630	2580	2510	2430	2340	2240
	P_{cy}	2500	2080	1570	1150	852	651	511	411	338	282	239	205	178
* 457 x 191 x 74	P_{cx}	2480	2480	2460	2440	2410	2370	2340	2300	2250	2190	2150	2130	2030
	P_{cy}	2180	1880	1420	1030	766	585	459	369	303	253	215	184	160
* 457 x 191 x 67	P_{cx}	2190	2190	2180	2150	2120	2090	2060	2030	1980	1940	1880	1810	1770
	P_{cy}	1920	1680	1250	906	670	510	400	322	264	221	187	160	139

* Section may be slender under axial compression.
Values in *italic type* indicate that the section is slender when loaded to capacity and allowance has been made in calculating the capacity.
FOR EXPLANATION OF TABLES SEE NOTE 8.1

BS 5950-1: 2000
BS 4-1: 1993

STRUTS

UB SECTIONS
SUBJECT TO AXIAL COMPRESSION

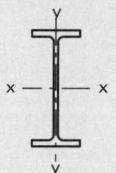

COMPRESSION RESISTANCE FOR S275

Section Designation	Axis	Compression resistance, P_{cx}, P_{cy} (kN) for Effective lengths, L_E (m)												
		1.0	1.5	2.0	2.5	3.0	3.5	4.0	4.5	5.0	5.5	6.0	6.5	7.0
457 x 152 x 82	P_{cx}	2780	2780	2780	2780	2780	2780	2760	2740	2730	2710	2690	2670	2650
	P_{cy}	2660	2480	2250	1980	1670	1380	1140	943	790	669	573	495	432
* 457 x 152 x 74	P_{cx}	2460	2460	2460	2460	2460	2460	2440	2430	2410	2400	2380	2370	2350
	P_{cy}	2350	2220	2020	1760	1480	1220	1010	832	696	589	504	436	380
* 457 x 152 x 67	P_{cx}	2240	2240	2240	2240	2240	2230	2220	2200	2190	2180	2160	2150	2130
	P_{cy}	2130	1980	1870	1620	1340	1100	896	738	616	520	445	384	335
* 457 x 152 x 60	P_{cx}	1910	1910	1910	1910	1910	1900	1890	1880	1870	1860	1850	1840	1820
	P_{cy}	1820	1700	1540	1420	1180	959	782	644	537	453	387	334	291
* 457 x 152 x 52	P_{cx}	1610	1610	1610	1610	1610	1610	1600	1590	1580	1570	1560	1550	1540
	P_{cy}	1540	1430	1290	1130	983	793	643	527	439	370	316	272	237
406 x 178 x 74	P_{cx}	2600	2600	2600	2600	2600	2580	2560	2550	2530	2510	2490	2470	2450
	P_{cy}	2530	2400	2240	2050	1830	1580	1350	1140	971	830	715	622	545
* 406 x 178 x 67	P_{cx}	2330	2330	2330	2330	2330	2310	2300	2280	2270	2270	2250	2230	2210
	P_{cy}	2270	2160	2020	1850	1640	1410	1200	1010	861	735	633	550	482
* 406 x 178 x 60	P_{cx}	2010	2010	2010	2010	2010	1990	1980	1970	1950	1940	1930	1910	1890
	P_{cy}	1950	1850	1740	1650	1460	1260	1070	901	764	652	562	488	427
* 406 x 178 x 54	P_{cx}	1790	1790	1790	1790	1780	1770	1760	1750	1740	1730	1710	1700	1680
	P_{cy}	1730	1640	1530	1460	1280	1090	922	775	655	559	480	417	364
* 406 x 140 x 46	P_{cx}	1450	1450	1450	1450	1450	1440	1430	1420	1410	1400	1390	1380	1370
	P_{cy}	1370	1270	1140	1020	836	671	542	444	369	310	265	228	199
* 406 x 140 x 39	P_{cx}	1180	1180	1180	1180	1180	1170	1170	1160	1150	1140	1130	1130	1120
	P_{cy}	1110	1030	911	773	659	523	420	342	284	238	203	175	152
356 x 171 x 67	P_{cx}	2350	2350	2350	2350	2340	2320	2300	2290	2270	2250	2230	2200	2180
	P_{cy}	2280	2160	2020	1850	1640	1410	1200	1010	861	735	633	550	482
356 x 171 x 57	P_{cx}	2000	2000	2000	2000	1980	1970	1950	1940	1920	1910	1890	1870	1840
	P_{cy}	1930	1830	1710	1550	1370	1170	991	835	707	603	519	451	394
* 356 x 171 x 51	P_{cx}	1750	1750	1750	1750	1740	1730	1720	1700	1690	1670	1660	1650	1650
	P_{cy}	1700	1630	1520	1380	1210	1030	870	732	619	528	454	394	344
* 356 x 171 x 45	P_{cx}	1510	1510	1510	1510	1500	1490	1480	1470	1460	1440	1430	1410	1400
	P_{cy}	1460	1380	1330	1200	1040	884	741	621	523	445	383	332	290
* 356 x 127 x 39	P_{cx}	1280	1280	1280	1280	1270	1270	1260	1250	1230	1220	1210	1200	1180
	P_{cy}	1190	1070	959	766	598	469	374	304	251	211	179	154	134
* 356 x 127 x 33	P_{cx}	1040	1040	1040	1040	1030	1030	1020	1010	1000	991	981	971	959
	P_{cy}	960	865	759	618	477	372	296	240	198	166	141	121	105
305 x 165 x 54	P_{cx}	1890	1890	1890	1880	1870	1850	1830	1820	1800	1780	1750	1730	1700
	P_{cy}	1830	1740	1620	1470	1300	1120	946	798	676	577	496	431	377
305 x 165 x 46	P_{cx}	1610	1610	1610	1610	1590	1580	1570	1550	1530	1520	1500	1470	1450
	P_{cy}	1560	1480	1380	1250	1100	945	799	673	569	486	418	363	317
* 305 x 165 x 40	P_{cx}	1370	1370	1370	1360	1350	1340	1330	1320	1300	1290	1270	1250	1230
	P_{cy}	1330	1260	1200	1090	955	816	688	579	489	417	359	311	272

* Section may be slender under axial compression.
Values in *italic type* indicate that the section is slender when loaded to capacity and allowance has been made in calculating the capacity.
FOR EXPLANATION OF TABLES SEE NOTE 8.1

BS 5950-1: 2000
BS 4-1: 1993

STRUTS

UB SECTIONS
SUBJECT TO AXIAL COMPRESSION

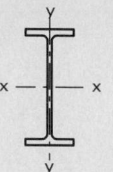

COMPRESSION RESISTANCE FOR S275

Section Designation	Axis	\multicolumn{11}{c}{Compression resistance, P_{cx}, P_{cy} (kN) for Effective lengths, L_E (m)}
		1.0	1.5	2.0	2.5	3.0	3.5	4.0	4.5	5.0	5.5	6.0	6.5	7.0
305 x 127 x 48	P_{cx}	1680	1680	1680	1670	1660	1640	1630	1610	1590	1570	1550	1520	1500
	P_{cy}	1560	1400	1200	967	759	597	477	388	321	270	229	197	172
305 x 127 x 42	P_{cx}	1470	1470	1470	1460	1450	1430	1420	1400	1390	1370	1350	1330	1300
	P_{cy}	1350	1220	1030	829	648	509	406	330	273	229	195	168	146
305 x 127 x 37	P_{cx}	1300	1300	1300	1290	1280	1270	1250	1240	1220	1210	1190	1170	1150
	P_{cy}	1200	1070	906	723	563	442	352	286	236	198	169	145	126
* 305 x 102 x 33	P_{cx}	*1130*	*1130*	*1130*	*1120*	*1110*	*1100*	*1090*	*1080*	1080	1070	1060	1040	1020
	P_{cy}	1010	846	645	473	352	269	212	170	140	117	99.2	85.1	73.9
* 305 x 102 x 28	P_{cx}	*928*	*928*	*928*	*923*	*915*	*906*	*898*	*888*	878	867	855	841	826
	P_{cy}	*815*	710	532	386	286	218	171	138	113	94.4	80	68.6	59.5
* 305 x 102 x 25	P_{cx}	*799*	*799*	*799*	*794*	*787*	*780*	*772*	*764*	*755*	*745*	*734*	*722*	*709*
	P_{cy}	*693*	*599*	*436*	*312*	*229*	*174*	*136*	*109*	*89.8*	*74.9*	*63.5*	*54.4*	*47.2*
254 x 146 x 43	P_{cx}	1510	1510	1500	1490	1470	1460	1440	1420	1400	1370	1350	1310	1280
	P_{cy}	1440	1350	1240	1100	934	777	643	534	448	380	325	281	246
254 x 146 x 37	P_{cx}	1300	1300	1290	1280	1270	1250	1240	1220	1200	1180	1160	1130	1100
	P_{cy}	1240	1160	1060	936	795	659	544	452	378	321	274	237	207
254 x 146 x 31	P_{cx}	1090	1090	1090	1080	1060	1050	1040	1020	1010	987	965	939	910
	P_{cy}	1040	969	879	766	643	529	434	358	300	253	217	187	163
254 x 102 x 28	P_{cx}	993	993	989	979	968	957	944	930	915	897	877	854	828
	P_{cy}	879	746	579	429	321	246	194	156	128	107	91	78.2	67.8
254 x 102 x 25	P_{cx}	880	880	876	867	857	846	835	822	808	792	773	752	727
	P_{cy}	773	648	494	362	270	206	162	130	107	89.6	75.9	65.2	56.5
254 x 102 x 22	P_{cx}	770	770	766	757	749	739	729	717	704	690	672	652	629
	P_{cy}	668	550	410	297	219	167	131	105	86.5	72.3	61.2	52.5	45.6
203 x 133 x 30	P_{cx}	1050	1050	1040	1020	1010	993	974	953	927	897	861	820	774
	P_{cy}	993	918	822	702	577	468	380	313	260	220	188	162	141
203 x 133 x 25	P_{cx}	880	879	868	857	844	830	814	795	772	746	715	678	638
	P_{cy}	829	764	680	576	470	379	307	252	210	177	151	130	113
203 x 102 x 23	P_{cx}	809	808	797	786	775	761	746	728	707	682	652	618	580
	P_{cy}	726	630	504	382	289	223	176	142	117	98.1	83.3	71.5	62.1
178 x 102 x 19	P_{cx}	668	664	654	644	632	618	601	581	556	527	493	456	418
	P_{cy}	601	522	418	318	241	186	147	119	97.6	81.7	69.4	59.6	51.8
152 x 89 x 16	P_{cx}	558	551	541	529	516	500	479	453	422	386	349	313	280
	P_{cy}	487	404	304	222	164	125	98.5	79.2	65.1	54.3	46.1	39.5	34.3
127 x 76 x 13	P_{cx}	452	443	432	420	403	382	354	320	284	250	219	192	169
	P_{cy}	379	294	207	145	106	80.3	62.7	50.3	41.2	34.3	29.1	24.9	21.6

* Section may be slender under axial compression.
Values in *italic type* indicate that the section is slender when loaded to capacity and allowance has been made in calculating the capacity.
FOR EXPLANATION OF TABLES SEE NOTE 8.1

BS 5950-1: 2000
BS 4-1: 1993

STRUTS

UC SECTIONS
SUBJECT TO AXIAL COMPRESSION

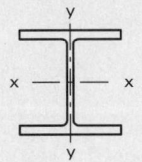

COMPRESSION RESISTANCE FOR S275

Section Designation	Axis	Compression resistance, P_{cx}, P_{cy} (kN) for Effective lengths, L_E (m)												
		2.0	3.0	4.0	5.0	6.0	7.0	8.0	9.0	10.0	11.0	12.0	13.0	14.0
356 x 406 x 634 #	P_{cx}	19800	19800	19400	18800	18200	17500	16900	16200	15500	14800	14100	13300	12600
	P_{cy}	19800	18300	16900	15600	14200	12900	11600	10500	9390	8420	7560	6800	6140
356 x 406 x 551 #	P_{cx}	17200	17200	16800	16300	15700	15100	14600	14000	13300	12700	12000	11400	10700
	P_{cy}	17200	15900	14700	13500	12300	11100	10000	9010	8080	7240	6490	5840	5260
356 x 406 x 467 #	P_{cx}	15200	15200	14700	14200	13700	13200	12700	12100	11500	10900	10300	9720	9120
	P_{cy}	15100	13900	12800	11700	10700	9610	8620	7700	6870	6130	5480	4910	4420
356 x 406 x 393 #	P_{cx}	12800	12800	12400	12100	11700	11300	10900	10500	10000	9550	9040	8520	7990
	P_{cy}	12700	11800	11000	10200	9290	8430	7580	6770	6030	5370	4780	4270	3830
356 x 406 x 340 #	P_{cx}	11000	11000	10700	10400	10100	9750	9390	9010	8600	8170	7720	7260	6790
	P_{cy}	10900	10200	9480	8740	7990	7230	6490	5800	5150	4580	4080	3640	3260
356 x 406 x 287 #	P_{cx}	9700	9670	9460	9230	8990	8730	8450	8130	7780	7400	7000	6570	6130
	P_{cy}	9590	9050	8490	7900	7260	6590	5920	5270	4670	4130	3660	3250	2900
356 x 406 x 235 #	P_{cx}	7920	7900	7720	7530	7330	7110	6880	6610	6320	6010	5660	5310	4940
	P_{cy}	7830	7380	6920	6430	5900	5350	4790	4260	3770	3330	2950	2610	2330
356 x 368 x 202 #	P_{cx}	6810	6780	6630	6460	6290	6100	5890	5660	5400	5120	4820	4510	4190
	P_{cy}	6680	6270	5840	5390	4890	4390	3890	3420	3000	2630	2310	2050	1820
356 x 368 x 177 #	P_{cx}	5990	5960	5820	5670	5520	5350	5160	4950	4720	4470	4200	3920	3640
	P_{cy}	5870	5510	5130	4720	4290	3840	3400	2980	2610	2290	2020	1780	1580
356 x 368 x 153 #	P_{cx}	5170	5140	5020	4890	4750	4610	4440	4260	4060	3840	3610	3360	3120
	P_{cy}	5060	4750	4420	4060	3690	3300	2910	2560	2240	1960	1730	1520	1350
356 x 368 x 129 #	P_{cx}	4350	4320	4210	4110	3990	3860	3720	3560	3390	3200	3000	2790	2590
	P_{cy}	4250	3990	3710	3410	3090	2760	2430	2130	1870	1640	1440	1270	1130
305 x 305 x 283	P_{cx}	9180	9070	8780	8480	8160	7830	7460	7070	6660	6230	5790	5360	4940
	P_{cy}	8770	8020	7250	6460	5680	4930	4260	3670	3170	2760	2410	2120	1880
305 x 305 x 240	P_{cx}	8110	8010	7800	7580	7340	7070	6770	6430	6060	5660	5250	4830	4430
	P_{cy}	7780	7200	6570	5890	5180	4480	3850	3300	2830	2450	2130	1870	1640
305 x 305 x 198	P_{cx}	6680	6590	6410	6220	6020	5790	5530	5240	4930	4590	4240	3890	3550
	P_{cy}	6400	5910	5380	4810	4220	3640	3120	2670	2290	1970	1720	1500	1320
305 x 305 x 158	P_{cx}	5330	5250	5100	4950	4780	4590	4370	4130	3870	3590	3310	3020	2760
	P_{cy}	5090	4690	4260	3800	3310	2850	2430	2070	1780	1530	1330	1160	1020
305 x 305 x 137	P_{cx}	4610	4540	4410	4270	4120	3950	3760	3550	3320	3070	2820	2570	2340
	P_{cy}	4400	4050	3680	3270	2850	2440	2080	1770	1520	1310	1130	991	873
305 x 305 x 118	P_{cx}	3980	3910	3800	3680	3550	3400	3230	3050	2840	2630	2410	2200	2000
	P_{cy}	3790	3490	3160	2800	2440	2090	1780	1510	1290	1110	965	844	743
305 x 305 x 97	P_{cx}	3380	3320	3220	3120	3000	2870	2720	2550	2370	2180	1990	1800	1630
	P_{cy}	3210	2950	2660	2350	2030	1730	1460	1240	1060	907	786	686	603

Check availability.
FOR EXPLANATION OF TABLES SEE NOTE 8.1

BS 5950-1: 2000
BS 4-1: 1993

STRUTS

UC SECTIONS
SUBJECT TO AXIAL COMPRESSION

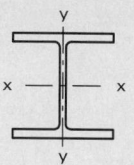

COMPRESSION RESISTANCE FOR S275

Section Designation	Axis	Compression resistance, P_{cx}, P_{cy} (kN) for Effective lengths, L_E (m)												
		1.0	1.5	2.0	2.5	3.0	3.5	4.0	4.5	5.0	5.5	6.0	6.5	7.0
254 x 254 x 167	P_{cx}	5640	5640	5640	5570	5480	5390	5300	5200	5090	4980	4860	4730	4590
	P_{cy}	5640	5500	5260	5020	4760	4490	4210	3910	3610	3320	3040	2770	2530
254 x 254 x 132	P_{cx}	4450	4450	4450	4390	4310	4240	4160	4080	3990	3900	3800	3690	3580
	P_{cy}	4450	4330	4140	3940	3730	3510	3280	3050	2810	2570	2350	2140	1950
254 x 254 x 107	P_{cx}	3600	3600	3600	3540	3480	3420	3360	3290	3210	3140	3050	2960	2860
	P_{cy}	3600	3500	3340	3180	3000	2820	2640	2440	2240	2050	1870	1700	1550
254 x 254 x 89	P_{cx}	2990	2990	2990	2940	2890	2840	2780	2730	2660	2600	2530	2450	2370
	P_{cy}	2990	2900	2770	2630	2490	2340	2180	2020	1860	1700	1540	1400	1270
254 x 254 x 73	P_{cx}	2560	2560	2550	2510	2470	2420	2370	2320	2270	2210	2150	2080	2000
	P_{cy}	2560	2470	2360	2240	2110	1980	1840	1700	1550	1410	1280	1160	1050
203 x 203 x 86	P_{cx}	2920	2920	2870	2810	2750	2690	2610	2540	2450	2360	2260	2160	2050
	P_{cy}	2890	2740	2580	2410	2220	2030	1840	1640	1460	1300	1160	1030	921
203 x 203 x 71	P_{cx}	2400	2400	2360	2310	2260	2200	2140	2080	2010	1930	1850	1760	1670
	P_{cy}	2380	2250	2110	1970	1820	1660	1500	1340	1190	1060	941	838	748
203 x 203 x 60	P_{cx}	2100	2100	2060	2020	1970	1920	1860	1800	1740	1670	1590	1500	1420
	P_{cy}	2080	1960	1840	1710	1570	1430	1280	1140	1010	888	786	698	622
203 x 203 x 52	P_{cx}	1820	1820	1790	1750	1710	1660	1620	1560	1510	1440	1370	1300	1220
	P_{cy}	1800	1700	1590	1480	1360	1230	1110	982	868	767	678	602	536
203 x 203 x 46	P_{cx}	1610	1610	1580	1550	1510	1470	1430	1380	1330	1270	1210	1140	1070
	P_{cy}	1590	1500	1410	1310	1200	1080	970	860	760	670	592	525	467
152 x 152 x 37	P_{cx}	1300	1270	1240	1200	1150	1110	1050	990	923	851	779	710	644
	P_{cy}	1230	1130	1020	904	782	667	565	479	409	351	304	266	234
152 x 152 x 30	P_{cx}	1050	1030	1000	972	936	896	850	799	743	684	625	567	514
	P_{cy}	999	918	828	730	630	536	453	384	327	281	243	212	187
152 x 152 x 23	P_{cx}	803	786	762	737	708	675	638	596	551	505	458	414	374
	P_{cy}	758	693	621	543	465	392	330	278	236	202	175	152	134

FOR EXPLANATION OF TABLES SEE NOTE 8.1

BS 5950-1: 2000
BS EN 10056-1: 1999

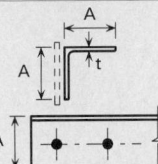

STRUTS

EQUAL ANGLES
SUBJECT TO AXIAL COMPRESSION

TWO OR MORE BOLTS IN LINE
OR EQUIVALENT WELDED AT EACH END

COMPRESSION RESISTANCE FOR STEEL S275

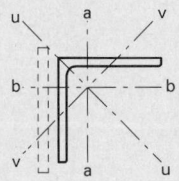

Section Designation		Area	Radius of Gyration			Compression resistance, P_c (kN) for Length between Intersections (m)											
A x A	t		Axis a-a	Axis b-b	Axis v-v	1.0	1.5	2.0	2.5	3.0	3.5	4.0	4.5	5.0	5.5	6.0	6.5	7.0
mm	mm	cm²	cm	cm	cm													
200 x 200	24.0 #	90.6	6.06	6.06	3.90	2060	1970	1880	1770	1620	1460	1310	1140	994	869	762	671	595
	20.0	76.3	6.11	6.11	3.92	1740	1660	1590	1490	1370	1240	1110	967	843	737	646	570	505
	18.0	69.1	6.13	6.13	3.90	1570	1510	1440	1350	1230	1120	997	870	758	662	581	512	454
	* 16.0	61.8	6.16	6.16	3.94	*1350*	*1300*	*1240*	*1170*	*1070*	*974*	*876*	*769*	*673*	*590*	*519*	*459*	*407*
150 x 150	18.0 #	51.2	4.55	4.55	2.93	1130	1060	972	856	740	617	515	432	365	312	269	234	205
	15.0	43.0	4.57	4.57	2.93	984	923	841	738	636	528	439	367	310	265	228	198	174
	* 12.0	34.8	4.60	4.60	2.95	*740*	*697*	*639*	*566*	*493*	*413*	*346*	*292*	*247*	*212*	*183*	*159*	*140*
	* 10.0	29.3	4.62	4.62	2.97	*447*	*424*	*396*	*360*	*324*	*282*	*244*	*211*	*183*	*159*	*139*	*123*	*109*
120 x 120	15.0 #	34.0	3.63	3.63	2.34	754	685	583	479	379	302	245	201	167	141	121	105	91.4
	12.0	27.5	3.65	3.65	2.35	610	555	473	389	309	246	199	164	136	115	98.7	85.3	74.5
	10.0	23.2	3.67	3.67	2.36	515	469	400	330	262	209	169	139	116	98.0	83.9	72.5	63.3
	* 8.0 #	18.8	3.71	3.71	2.38	*280*	*260*	*231*	*201*	*169*	*141*	*118*	*99.1*	*84.1*	*72.1*	*62.4*	*54.5*	*47.9*
100 x 100	15.0 #	28.0	2.99	2.99	1.94	599	512	411	310	237	184	147	119	98.8	83.0	70.8	61.0	53.1
	12.0	22.7	3.02	3.02	1.94	487	415	333	252	192	149	119	96.7	80.1	67.3	57.4	49.5	43.1
	10.0	19.2	3.04	3.04	1.95	412	352	283	214	163	127	101	82.5	68.3	57.5	49.0	42.2	36.8
	* 8.0	15.5	3.06	3.06	1.96	*310*	*268*	*219*	*168*	*129*	*101*	*81.0*	*66.1*	*54.9*	*46.2*	*39.4*	*34.0*	*29.7*
90 x 90	12.0 #	20.3	2.71	2.71	1.75	425	347	264	194	145	112	88.8	71.9	59.4	49.8	42.4	36.5	31.7
	10.0	17.1	2.72	2.72	1.75	358	293	222	163	122	94.4	74.8	60.6	50.0	42.0	35.7	30.7	26.7
	8.0	13.9	2.74	2.74	1.76	292	239	182	134	100	77.5	61.4	49.7	41.1	34.5	29.3	25.2	22.0
	* 7.0 #	12.2	2.75	2.75	1.77	*228*	*190*	*149*	*112*	*85.1*	*66.3*	*52.8*	*43.0*	*35.6*	*29.9*	*25.5*	*22.0*	*19.2*
80 x 80	10.0 ‡	15.1	2.41	2.41	1.55	303	235	167	119	88.2	67.5	53.1	42.9	35.3	29.6	25.1	21.6	18.8
	8.0 ‡	12.3	2.43	2.43	1.56	248	193	137	98.2	72.6	55.6	43.8	35.3	29.1	24.4	20.7	17.8	15.5
75 x 75	8.0 ‡	11.4	2.27	2.27	1.46	223	168	116	81.8	60.1	45.8	36.0	29.0	23.8	19.9	16.9	14.5	12.6
	* 6.0 ‡	8.73	2.29	2.29	1.47	*160*	*123*	*86.5*	*61.7*	*45.6*	*34.9*	*27.5*	*22.2*	*18.3*	*15.3*	*13.0*	*11.2*	*9.73*
70 x 70	7.0 ‡	9.40	2.12	2.12	1.36	177	128	85.8	59.9	43.8	33.2	26.1	21.0	17.2	14.4	12.2	10.5	9.10
	6.0 ‡	8.13	2.13	2.13	1.37	154	111	75.1	52.5	38.3	29.1	22.8	18.4	15.1	12.6	10.7	9.19	7.98
65 x 65	7.0 ‡	8.73	1.96	1.96	1.26	157	108	70.8	48.9	35.5	26.9	21.0	16.9	13.9	11.6	9.80	8.41	7.30
60 x 60	8.0 ‡	9.03	1.80	1.80	1.16	154	99.6	64.0	43.8	31.7	23.9	18.6	15.0	12.3	10.2	8.66	7.43	6.44
	6.0 ‡	6.91	1.82	1.82	1.17	118	77.1	49.7	34.0	24.6	18.6	14.5	11.6	9.53	7.95	6.74	5.78	5.01
	5.0 ‡	5.82	1.82	1.82	1.17	99.8	65.0	41.9	28.7	20.7	15.6	12.2	9.79	8.03	6.70	5.67	4.87	4.22
50 x 50	6.0 ‡	5.69	1.50	1.50	0.968	83.3	47.9	29.7	20.0	14.3	10.8	8.36	6.69	5.47	4.56	3.85	3.30	2.86
	5.0 ‡	4.80	1.51	1.51	0.973	70.4	40.7	25.3	17.0	12.2	9.16	7.12	5.70	4.66	3.88	3.28	2.81	2.44
	* 4.0 ‡	3.89	1.52	1.52	0.979	*54.8*	*32.4*	*20.3*	*13.7*	*9.88*	*7.43*	*5.79*	*4.63*	*3.79*	*3.16*	*2.68*	*2.29*	*1.99*

‡ Not available from some leading producers. Check availability.
# Check availability.
* Section is slender under axial compression and allowance has been made in calculating the compression resistance which is given in *italic type*.
FOR EXPLANATION OF TABLES SEE NOTE 8.2

BS 5950-1: 2000
BS EN 10056-1: 1999

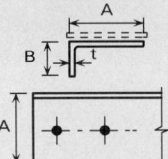

STRUTS

UNEQUAL ANGLES
LONG LEG ATTACHED
SUBJECT TO AXIAL COMPRESSION

TWO OR MORE BOLTS IN LINE
OR EQUIVALENT WELDED AT EACH END

COMPRESSION RESISTANCE FOR STEEL S275

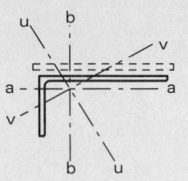

Section Designation		Area	Radius of Gyration			Compression resistance, P_c (kN) for Length between Intersections, L (m)												
			Axis a-a	Axis b-b	Axis v-v	1.0	1.5	2.0	2.5	3.0	3.5	4.0	4.5	5.0	5.5	6.0	6.5	7.0
A x B mm	t mm	cm²	cm	cm	cm													
200 x 150	18.0 #	60.1	4.38	6.30	3.22	1320	1240	1150	1060	943	813	688	584	498	428	371	324	285
	15.0	50.5	4.40	6.33	3.23	1150	1080	998	917	814	700	591	500	426	365	316	276	243
	* 12.0	40.8	4.44	6.36	3.25	653	618	582	546	499	445	389	339	296	259	228	201	179
200 x 100	15.0	43.0	2.64	6.40	2.12	894	781	668	537	416	327	262	214	178	150	128	110	96.3
	* 12.0	34.8	2.67	6.43	2.14	605	539	472	392	312	250	203	167	140	118	101	87.9	76.8
	* 10.0	29.2	2.68	6.46	2.15	368	335	301	260	216	179	149	125	106	91.1	78.8	68.8	60.5
150 x 90	15.0	33.9	2.46	4.74	1.93	692	596	494	373	284	221	176	143	118	99.6	84.9	73.1	63.7
	12.0	27.5	2.49	4.77	1.94	563	486	403	305	232	181	144	117	97.0	81.6	69.5	59.9	52.2
	10.0	23.2	2.51	4.80	1.95	476	412	342	259	198	154	123	99.7	82.6	69.5	59.2	51.0	44.4
150 x 75	15.0	31.7	1.94	4.75	1.58	599	486	360	258	191	147	115	93.2	76.8	64.3	54.6	47.0	40.8
	12.0	25.7	1.97	4.78	1.59	489	398	294	211	157	120	94.7	76.5	63.0	52.8	44.8	38.6	33.5
	10.0	21.7	1.99	4.81	1.60	414	338	251	180	134	103	80.9	65.3	53.8	45.1	38.3	32.9	28.6
125 x 75	12.0	22.7	2.05	3.95	1.61	438	361	265	191	141	108	85.5	69.1	56.9	47.7	40.5	34.9	30.3
	10.0	19.1	2.07	3.97	1.61	370	305	223	160	119	91.2	72.0	58.1	47.9	40.1	34.1	29.3	25.5
	* 8.0	15.5	2.09	4.00	1.63	282	236	176	129	96.4	74.3	58.8	47.6	39.3	33.0	28.0	24.1	21.0
100 x 75	12.0	19.7	2.14	3.10	1.59	386	314	226	162	120	92.1	72.6	58.6	48.3	40.4	34.4	29.6	25.7
	10.0	16.6	2.16	3.12	1.59	326	264	190	137	101	77.6	61.2	49.4	40.7	34.1	29.0	24.9	21.6
	8.0	13.5	2.18	3.14	1.60	266	216	156	112	83.2	63.8	50.3	40.6	33.5	28.0	23.8	20.5	17.8
100 x 65	10.0 #	15.6	1.81	3.14	1.39	287	217	147	103	75.5	57.4	45.0	36.2	29.8	24.9	21.1	18.1	15.7
	8.0 #	12.7	1.83	3.16	1.40	235	178	121	85.0	62.2	47.3	37.1	29.9	24.6	20.5	17.4	15.0	13.0
	7.0 #	11.2	1.83	3.17	1.40	207	157	107	74.9	54.9	41.7	32.7	26.4	21.7	18.1	15.4	13.2	11.5
100 x 50	8.0 ‡	11.4	1.31	3.19	1.06	176	110	69.5	47.1	33.9	25.5	19.9	15.9	13.0	10.9	9.20	7.88	6.83
	* 6.0 ‡	8.71	1.33	3.21	1.07	118	78.1	50.8	35.0	25.4	19.2	15.0	12.1	9.92	8.29	7.02	6.03	5.23
80 x 60	7.0 ‡	9.38	1.74	2.51	1.28	170	118	78.0	53.9	39.2	29.7	23.3	18.7	15.3	12.8	10.9	9.32	8.08
80 x 40	8.0 ‡	9.01	1.03	2.53	0.838	111	60.0	36.5	24.3	17.3	13.0	10.1	8.04	6.57	5.46	4.62	3.95	3.42
	6.0 ‡	6.89	1.05	2.55	0.845	85.6	46.6	28.3	18.9	13.5	10.1	7.82	6.25	5.10	4.25	3.59	3.07	2.66
75 x 50	8.0 ‡	9.41	1.40	2.35	1.07	151	92.3	58.3	39.6	28.5	21.4	16.7	13.4	11.0	9.13	7.73	6.63	5.74
	6.0 ‡	7.19	1.42	2.37	1.08	116	71.5	45.3	30.7	22.1	16.7	13.0	10.4	8.52	7.10	6.01	5.15	4.47
70 x 50	6.0 ‡	6.89	1.43	2.20	1.07	111	67.6	42.7	29.0	20.9	15.7	12.2	9.80	8.02	6.69	5.66	4.85	4.20
65 x 50	5.0 ‡	5.54	1.47	2.05	1.07	88.9	54.4	34.3	23.3	16.8	12.6	9.83	7.88	6.45	5.38	4.55	3.90	3.38
60 x 40	6.0 ‡	5.68	1.12	1.88	0.855	71.7	39.1	23.8	15.9	11.4	8.50	6.60	5.27	4.30	3.58	3.03	2.59	2.24
	5.0 ‡	4.79	1.13	1.89	0.860	60.9	33.3	20.3	13.6	9.68	7.25	5.63	4.49	3.67	3.05	2.58	2.21	1.91

‡ Not available from some leading producers. Check availability.
# Check availability.
* Section is slender under axial compression and allowance has been made in calculating the compression resistance which is given in *italic type*.
FOR EXPLANATION OF TABLES SEE NOTE 8.2

227

BS 5950-1: 2000
BS EN 10056-1: 1999

STRUTS

TWO EQUAL ANGLES BACK TO BACK CONNECTED TO ONE SIDE OF GUSSET OR MEMBER SUBJECT TO AXIAL COMPRESSION

TWO OR MORE BOLTS IN LINE ALONG EACH ANGLE OR EQUIVALENT WELDED AT EACH END

COMPRESSION RESISTANCE FOR STEEL S275

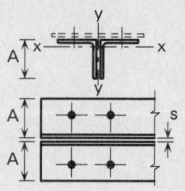

Section Designation		Space	Total Area	Radius of Gyration		Elastic Mod.	Compression resistance, P_c (kN) for Length between Intersections, L (m)												
				Axis x-x	Axis y-y	Axis x-x	1.0	1.5	2.0	2.5	3.0	3.5	4.0	4.5	5.0	5.5	6.0	6.5	7.0
A x A mm	t mm	s mm	cm²	cm	cm	cm³													
200 x 200	24.0 #	15	181	6.06	8.95	470	4120	3950	3760	3570	3380	3180	2980	2780	2600	2420	2250	2030	1830
	20.0	15	153	6.11	8.87	398	3480	3330	3170	3020	2850	2690	2520	2360	2200	2050	1910	1730	1560
	18.0	15	138	6.13	8.83	362	3150	3020	2880	2730	2590	2440	2290	2140	2000	1860	1730	1570	1420
	* 16.0	15	124	6.16	8.79	324	2700	2590	2480	2360	2230	2110	1990	1860	1740	1630	1520	1390	1250
150 x 150	18.0 #	12	102	4.55	6.75	200	2260	2130	1980	1830	1680	1540	1400	1270	1110	970	850	749	663
	15.0	12	86.0	4.57	6.66	167	1970	1850	1720	1590	1460	1330	1200	1090	954	832	728	640	567
	* 12.0	12	69.6	4.60	6.59	135	*1480*	*1390*	*1300*	*1210*	*1120*	*1020*	*934*	*851*	*751*	*658*	*578*	*510*	*453*
	* 10.0	12	58.6	4.62	6.54	114	*894*	*849*	*803*	*757*	*711*	*665*	*619*	*575*	*520*	*466*	*418*	*375*	*337*
120 x 120	15.0 #	12	68.0	3.63	5.49	106	1510	1380	1250	1120	995	880	747	629	533	**456**	**394**	**343**	**301**
	12.0	12	55.0	3.65	5.42	85.4	1220	1120	1010	909	808	715	609	513	435	**372**	**322**	**280**	**246**
	10.0	12	46.4	3.67	5.36	72.0	1030	946	858	769	684	606	518	436	370	317	**274**	**239**	**210**
	* 8.0 #	12	37.6	3.71	5.34	59.0	*559*	*523*	*486*	*449*	*413*	*377*	*335*	*292*	*255*	*223*	***196***	***173***	***154***
100 x 100	15.0 #	10	56.0	2.99	4.62	71.6	1200	1070	938	812	698	564	460	**380**	**318**	**270**	**231**	**200**	**175**
	12.0	10	45.4	3.02	4.55	58.2	973	870	765	663	572	464	379	313	**262**	**222**	**191**	**165**	**145**
	10.0	10	38.4	3.04	4.50	49.2	824	738	649	564	487	396	324	268	**224**	**190**	**163**	**142**	**124**
	* 8.0	10	31.0	3.06	4.46	39.8	*621*	*569*	*496*	*435*	*379*	*312*	*257*	*213*	*179*	***153***	***131***	***114***	***99.7***
90 x 90	12.0,#	10	40.6	2.71	4.16	47.0	850	746	642	545	443	353	285	**233**	**194**	**164**	**140**	**121**	**106**
	10.0	10	34.2	2.72	4.11	39.6	716	629	542	461	375	299	241	**198**	**165**	**139**	**119**	**103**	**89.9**
	8.0	10	27.8	2.74	4.06	32.2	583	513	442	377	308	246	198	**163**	**136**	**115**	**98.1**	**84.8**	**74.1**
	* 7.0 #	10	24.4	2.75	4.04	28.2	*455*	*405*	*354*	*306*	*255*	*206*	*168*	***139***	***116***	***98.4***	***84.5***	***73.2***	***64.0***
80 x 80	10.0 ‡	8	30.2	2.41	3.65	30.8	613	525	440	360	278	217	**174**	**142**	**117**	**98.9**	**84.4**	**72.8**	**63.4**
	8.0 ‡	8	24.6	2.43	3.60	25.2	501	430	361	297	229	179	**144**	**117**	**97.1**	**81.8**	**69.8**	**60.2**	**52.5**
75 x 75	8.0 ‡	8	22.8	2.27	3.41	22.0	455	384	318	251	191	**149**	**118**	**96.2**	**79.6**	**66.9**	**57.0**	**49.2**	**42.8**
	* 6.0 ‡	8	17.5	2.29	3.35	16.8	*326*	*279*	*234*	*187*	*144*	***113***	***90.2***	***73.6***	***61.0***	***51.4***	***43.9***	***37.8***	***33.0***
70 x 70	7.0 ‡	8	18.8	2.12	3.18	16.8	367	305	248	187	141	**109**	**86.6**	**70.2**	**58.0**	**48.7**	**41.4**	**35.7**	**31.1**
	6.0 ‡	8	16.3	2.13	3.16	14.5	318	264	215	163	123	**95.2**	**75.5**	**61.2**	**50.6**	**42.5**	**36.2**	**31.1**	**27.1**
65 x 65	7.0 ‡	8	17.5	1.96	3.14	14.4	331	269	213	155	**115**	**88.5**	**69.9**	**56.6**	**46.6**	**39.1**	**33.2**	**28.6**	**24.9**
60 x 60	8.0 ‡	8	18.1	1.80	2.82	13.8	331	263	196	140	**103**	**78.9**	**62.1**	**50.1**	**41.2**	**34.5**	**29.3**	**25.2**	**21.9**
	6.0 ‡	8	13.8	1.82	2.77	10.6	255	203	152	109	**80.5**	**61.5**	**48.5**	**39.1**	**32.2**	**27.0**	**22.9**	**19.7**	**17.1**
	5.0 ‡	8	11.6	1.82	2.74	8.90	214	171	128	91.7	**67.8**	**51.8**	**40.8**	**32.9**	**27.1**	**22.7**	**19.3**	**16.6**	**14.4**
50 x 50	6.0 ‡	8	11.4	1.50	2.38	7.22	191	143	94.0	**65.0**	**47.3**	**35.8**	**28.0**	**22.5**	**18.5**	**15.4**	**13.1**	**11.2**	**9.73**
	5.0 ‡	8	9.60	1.51	2.35	6.10	162	121	80.1	**55.5**	**40.3**	**30.6**	**23.9**	**19.2**	**15.8**	**13.2**	**11.2**	**9.58**	**8.32**
	* 4.0 ‡	8	7.78	1.52	2.32	4.92	*124*	*94.5*	*63.8*	***44.5***	***32.5***	***24.7***	***19.4***	***15.6***	***12.8***	***10.7***	***9.09***	***7.81***	***6.78***

‡ Not rolled by certain leading producers, check availability.
# Check availability.
* Section is slender under axial compression and allowance has been made in calculating the compression resistance which is given in *italic type*.
Values in **bold type** indicate that the section has been divided into more than 3 bays until $\lambda_c \leq 50$.
FOR EXPLANATION OF TABLES SEE NOTE 8.3

BS 5950-1: 2000
BS EN 10056-1: 1999

STRUTS

TWO EQUAL ANGLES BACK TO BACK
CONNECTED TO BOTH SIDES OF GUSSET OR MEMBER
SUBJECT TO AXIAL COMPRESSION

TWO OR MORE BOLTS IN LINE ALONG EACH ANGLE
AT EACH END

COMPRESSION RESISTANCE FOR STEEL S275

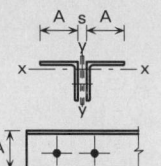

Section Designation		Space	Total Area	Radius of Gyration		Elastic Mod.	Compression resistance, P_c (kN) for Length between Intersections, L (m)											
				Axis x-x	Axis y-y	Axis x-x	1.0	1.5	2.0	2.5	3.0	3.5	4.0	4.5	5.0	5.5	6.0	6.5	7.0
A x A mm	t mm	s mm	cm²	cm	cm	cm³													
200 x 200	24.0 #	15	181	6.06	8.95	470	4120	3950	3760	3570	3380	3180	2980	2780	2600	2420	**2250**	**2090**	1940
	20.0	15	153	6.11	8.87	398	3480	3330	3170	3020	2850	2690	2520	2360	2200	2050	**1910**	**1770**	1650
	18.0	15	138	6.13	8.83	362	3150	3020	2880	2730	2590	2440	2290	2140	2000	1860	**1730**	**1610**	1500
	* 16.0	15	124	6.16	8.79	324	*2700*	*2590*	*2480*	*2360*	*2230*	*2110*	*1990*	*1860*	*1740*	*1630*	***1520***	***1410***	*1320*
150 x 150	18.0 #	12	102	4.55	6.75	200	2260	2130	1980	1830	1680	1540	1400	**1270**	**1150**	**1050**	953	869	794
	15.0	12	86.0	4.57	6.66	167	1970	1850	1720	1590	1460	1330	1200	**1090**	**990**	**898**	816	743	679
	* 12.0	12	69.6	4.60	6.59	135	*1480*	*1390*	*1300*	*1210*	*1120*	*1020*	*934*	***851***	***775***	***706***	*644*	*588*	*539*
	* 10.0	12	58.6	4.62	6.54	114	*894*	*849*	*803*	*757*	*711*	*665*	*619*	***575***	***533***	***494***	*457*	*424*	*392*
120 x 120	15.0 #	12	68.0	3.63	5.49	106	1510	1380	1250	1120	995	880	**777**	**688**	**610**	**544**	487	438	395
	12.0	12	55.0	3.65	5.42	85.4	1220	1120	1010	909	808	715	**632**	**559**	**497**	**443**	397	357	322
	10.0	12	46.4	3.67	5.36	72.0	1030	946	858	769	684	606	**536**	**475**	**422**	**376**	337	303	274
	* 8.0 #	12	37.6	3.71	5.34	59.0	*559*	*523*	*486*	*449*	*413*	*377*	***343***	***312***	***283***	***258***	*235*	*214*	*196*
100 x 100	15.0 #	10	56.0	2.99	4.62	71.6	1200	1070	938	812	**699**	**602**	**520**	**452**	**395**	**347**	307	268	235
	12.0	10	45.4	3.02	4.55	58.2	973	870	765	663	**572**	**493**	**426**	**371**	**324**	**286**	253	221	194
	10.0	10	38.4	3.04	4.50	49.2	824	738	649	564	**487**	**420**	**363**	**316**	**277**	**244**	216	189	166
	* 8.0	10	31.0	3.06	4.46	39.8	*621*	*559*	*496*	*435*	***379***	***329***	***286***	***250***	***220***	***194***	*172*	*151*	*133*
90 x 90	12.0 #	10	40.6	2.71	4.16	47.0	850	746	642	545	**462**	**392**	**335**	**289**	**251**	**219**	188	163	143
	10.0	10	34.2	2.72	4.11	39.6	716	629	542	461	**390**	**332**	**284**	**245**	**212**	**185**	159	138	121
	8.0	10	27.8	2.74	4.06	32.2	583	513	442	377	**320**	**272**	**233**	**201**	**174**	**153**	131	114	99.6
	* 7.0 #	10	24.4	2.75	4.04	28.2	*455*	*405*	*354*	*306*	***263***	***226***	***195***	***169***	***148***	***130***	*112*	*97.7*	*85.7*
80 x 80	10.0 ‡	8	30.2	2.41	3.65	30.8	613	525	440	**366**	**304**	**254**	**215**	**184**	**157**	**132**	113	98.1	85.7
	8.0 ‡	8	24.6	2.43	3.60	25.2	501	430	361	**300**	**250**	**209**	**177**	**151**	**129**	**109**	93.7	81.1	70.9
75 x 75	8.0 ‡	8	22.8	2.27	3.41	22.0	455	384	318	**261**	**215**	**178**	**150**	**127**	**106**	**89.9**	76.8	66.4	57.9
	* 6.0 ‡	8	17.5	2.29	3.35	16.8	*326*	*279*	*234*	***194***	***161***	***134***	***114***	***96.8***	***81.3***	***68.8***	*58.9*	*51.0*	*44.5*
70 x 70	7.0 ‡	8	18.8	2.12	3.18	16.8	367	305	248	**200**	**163**	**135**	**113**	**93.6**	**77.8**	**65.5**	55.9	48.3	42.1
	6.0 ‡	8	16.3	2.13	3.16	14.5	318	264	215	**174**	**142**	**117**	**98.1**	**81.7**	**67.8**	**57.2**	48.8	42.1	36.7
65 x 65	7.0 ‡	8	17.5	1.96	3.14	14.4	331	269	**215**	**171**	**138**	**113**	**93.1**	**75.7**	**62.7**	**52.8**	45.0	38.8	33.8
60 x 60	8.0 ‡	8	18.1	1.80	2.82	13.8	331	263	**205**	**161**	**128**	**104**	**83.0**	**67.3**	**55.6**	**46.7**	39.7	34.2	29.8
	6.0 ‡	8	13.8	1.82	2.77	10.6	255	203	**158**	**124**	**99.3**	**80.6**	**64.8**	**52.5**	**43.4**	**36.5**	31.0	26.7	23.3
	5.0 ‡	8	11.6	1.82	2.74	8.90	214	171	**133**	**105**	**83.6**	**67.9**	**54.5**	**44.2**	**36.6**	**30.7**	26.1	22.5	19.6
50 x 50	6.0 ‡	8	11.4	1.50	2.38	7.22	191	**143**	**106**	**80.6**	**62.8**	**48.0**	**37.7**	**30.4**	**25.0**	**20.9**	17.8	15.3	13.3
	5.0 ‡	8	9.60	1.51	2.35	6.10	162	**121**	**90.2**	**68.6**	**53.5**	**41.0**	**32.2**	**26.0**	**21.4**	**17.9**	15.2	13.1	11.3
	* 4.0 ‡	8	7.78	1.52	2.32	4.92	*124*	***94.5***	***71.3***	***54.7***	***42.9***	***33.0***	***26.0***	***21.0***	***17.3***	***14.5***	*12.4*	*10.6*	*9.24*

‡ Not rolled by certain leading producers, check availability.
\# Check availability.
* Section is slender under axial compression and allowance has been made in calculating the compression resistance which is given in *italic type*.
Values in **bold type** indicate that the section has been divided into more than 3 bays until $\lambda_c \leq 50$.
FOR EXPLANATION OF TABLES SEE NOTE 8.3

BS 5950-1: 2000
BS EN 10056-1: 1999

STRUTS

TWO UNEQUAL ANGLES LONG LEGS BACK TO BACK CONNECTED TO ONE SIDE OF GUSSET OR MEMBER SUBJECT TO AXIAL COMPRESSION

TWO OR MORE BOLTS IN LINE ALONG EACH ANGLE OR EQUIVALENT WELDED AT EACH END

COMPRESSION RESISTANCE FOR STEEL S275

Section Designation		Space	Total Area	Radius of Gyration		Elastic Mod.	Compression resistance, P_c (kN) for Length between Intersections, L (m)												
				Axis x-x	Axis y-y	Axis x-x													
A x A mm	t mm	s mm	cm²	cm	cm	cm³	1.0	1.5	2.0	2.5	3.0	3.5	4.0	4.5	5.0	5.5	6.0	6.5	7.0
200 x 150	18.0 #	15	120	6.30	6.36	350	2740	2630	2520	2390	2270	2140	2020	1890	**1770**	**1650**	**1540**	**1420**	**1280**
	15.0	15	101	6.33	6.28	294	2390	2290	2190	2080	1970	1860	1740	1630	**1530**	**1420**	**1330**	**1220**	**1100**
	* 12.0	15	81.6	6.36	6.22	238	*1350*	*1300*	*1250*	*1200*	*1150*	*1100*	*1050*	*998*	*948*	*898*	*850*	*798*	*735*
200 x 100	15.0	15	86.0	6.40	3.97	274	2030	1950	1840	1620	1390	**1300**	**1110**	**1010**	**868**	**781**	**682**	**615**	**543**
	* 12.0	15	69.6	6.43	3.90	222	*1360*	*1310*	*1240*	*1100*	*966*	*904*	*787*	*720*	*628*	*569*	*501*	*453*	*403*
	* 10.0	15	58.4	6.46	3.85	186	*813*	*786*	*747*	*678*	*609*	*577*	*516*	*479*	*428*	*393*	*352*	*322*	*290*
150 x 90	15.0	12	67.8	4.74	3.75	155	1560	1470	1370	1210	**1120**	**948**	**852**	**726**	**648**	**559**	**500**	**438**	**394**
	12.0	12	55.0	4.77	3.69	127	1260	1190	1110	976	**896**	**760**	**680**	**579**	**515**	**444**	**396**	**347**	**312**
	10.0	12	46.4	4.80	3.64	107	1070	1010	941	819	**749**	**634**	**566**	**481**	**427**	**368**	**328**	**287**	**258**
150 x 75	15.0	12	63.4	4.75	3.09	150	1460	1370	1160	**1030**	**845**	**736**	**633**	**529**	**459**	**392**	**344**	**304**	**267**
	12.0	12	51.4	4.78	3.02	123	1180	1110	929	**823**	**671**	**582**	**498**	**415**	**360**	**307**	**269**	**237**	**208**
	10.0	12	43.4	4.81	2.97	103	999	931	780	**687**	**559**	**483**	**397**	**343**	**297**	**253**	**221**	**195**	**171**
125 x 75	12.0	12	45.4	3.95	3.19	86.4	1020	944	845	**758**	**625**	**547**	**454**	**396**	**345**	**295**	**259**	**229**	**201**
	10.0	12	38.2	3.97	3.14	73.0	858	795	706	**631**	**519**	**453**	**375**	**327**	**284**	**243**	**213**	**188**	**165**
	* 8.0	12	31.0	4.00	3.09	59.2	*648*	*603*	*536*	***480***	***398***	***348***	***290***	***253***	***220***	***188***	***166***	***144***	***129***
100 x 75	12.0	10	39.4	3.10	3.31	56.0	849	762	673	**587**	**508**	**418**	**343**	**284**	**238**	**202**	**173**	**150**	**132**
	10.0	10	33.2	3.12	3.27	47.6	716	644	569	**497**	**430**	**356**	**292**	**242**	**203**	**172**	**148**	**128**	**112**
	8.0	10	27.0	3.14	3.22	38.6	583	525	464	**406**	**352**	**292**	**239**	**199**	**167**	**142**	**122**	**105**	**92.3**
100 x 65	10.0 #	10	31.2	3.14	2.79	46.4	674	606	519	**454**	**386**	**325**	**266**	**226**	**193**	**164**	**141**	**122**	**107**
	8.0 #	10	25.4	3.16	2.74	37.8	549	495	419	**364**	**309**	**248**	**211**	**180**	**154**	**130**	**114**	**99.5**	**87.0**
	7.0 #	10	22.4	3.17	2.72	33.2	485	437	367	**319**	**270**	**217**	**184**	**156**	**134**	**113**	**98.7**	**86.5**	**75.6**
100 x 50	8.0 ‡	10	22.8	3.19	2.09	36.4	494	381	**309**	**245**	**195**	**156**	**127**	**105**	**88.0**	**74.7**	**64.1**	**55.6**	**48.6**
	* 6.0 ‡	10	17.4	3.21	2.04	27.6	*313*	*248*	***205***	***166***	***133***	***108***	***88.4***	***73.5***	***61.8***	***52.6***	***45.3***	***39.3***	***34.5***
80 x 60	7.0 ‡	8	18.8	2.51	2.59	21.4	385	333	**282**	**236**	**183**	**144**	**116**	**94.5**	**78.5**	**66.1**	**56.5**	**48.7**	**42.5**
80 x 40	8.0 ‡	8	18.0	2.53	1.71	22.8	349	**267**	**198**	**148**	**116**	**90.1**	**71.9**	**58.7**	**48.7**	**41.2**	**35.2**	**30.4**	**26.5**
	6.0 ‡	8	13.8	2.55	1.66	17.5	264	**199**	**146**	**109**	**84.4**	**65.6**	**52.3**	**42.6**	**35.3**	**29.8**	**25.4**	**21.9**	**19.1**
75 x 50	8.0 ‡	8	18.8	2.35	2.19	20.8	379	323	**266**	**214**	**166**	**130**	**104**	**84.4**	**70.0**	**58.9**	**50.2**	**43.3**	**37.7**
	6.0 ‡	8	14.4	2.37	2.14	16.0	291	245	**200**	**160**	**127**	**101**	**80.4**	**65.5**	**54.3**	**45.7**	**39.0**	**33.6**	**29.3**
70 x 50	6.0 ‡	8	13.8	2.20	2.19	14.0	272	228	**187**	**145**	**110**	**85.2**	**67.7**	**55.0**	**45.5**	**38.2**	**32.5**	**28.0**	**24.4**
65 x 50	5.0 ‡	8	11.1	2.05	2.21	10.3	214	176	**142**	**105**	**78.8**	**60.7**	**48.1**	**38.9**	**32.1**	**27.0**	**22.9**	**19.7**	**17.2**
60 x 40	6.0 ‡	8	11.4	1.88	1.80	10.1	212	**170**	**131**	**94.2**	**69.9**	**53.6**	**42.2**	**34.1**	**28.1**	**23.5**	**20.0**	**17.2**	**15.0**
	5.0 ‡	8	9.58	1.89	1.78	8.50	179	**144**	**111**	**80.1**	**59.5**	**45.6**	**36.0**	**29.0**	**23.9**	**20.0**	**17.0**	**14.7**	**12.7**

‡ Not rolled by certain leading producers, check availability.
# Check availability.
* Section is slender under axial compression and allowance has been made in calculating the compression resistance which is given in *italic type*.
Values in **bold type** indicate that the section has been divided into more than 3 bays until $\lambda_c \leq 50$.
FOR EXPLANATION OF TABLES SEE NOTE 8.3

BS 5950-1: 2000
BS EN 10056-1: 1999

STRUTS

TWO UNEQUAL ANGLES LONG LEGS BACK TO BACK CONNECTED TO BOTH SIDES OF GUSSET OR MEMBER SUBJECT TO AXIAL COMPRESSION

TWO OR MORE BOLTS IN LINE ALONG EACH ANGLE AT EACH END

COMPRESSION RESISTANCE FOR STEEL S275

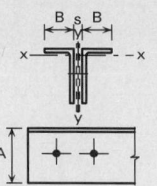

Section Designation		Space	Total Area	Radius of Gyration		Elastic Mod.	Compression resistance, P_c (kN) for Length between Intersections, L (m)												
				Axis x-x	Axis y-y	Axis x-x	1.0	1.5	2.0	2.5	3.0	3.5	4.0	4.5	5.0	5.5	6.0	6.5	7.0
A x A mm	t mm	s mm	cm²	cm	cm	cm³													
200 x 150	18.0 #	15	120	6.30	6.36	350	2740	2630	2520	2390	2270	2140	2000	1790	**1720**	**1550**	**1380**	**1300**	1170
	15.0	15	101	6.33	6.28	294	2390	2290	2190	2080	1970	1860	1710	1520	**1470**	**1310**	**1170**	**1100**	986
	*12.0	15	81.6	6.36	6.22	238	*1350*	*1300*	*1250*	*1200*	*1150*	*1100*	*1030*	*941*	*911*	*833*	*760*	*692*	*659*
200 x 100	15.0	15	86.0	6.40	3.97	274	2030	1950	1740	1490	1240	**1110**	**933**	**825**	**701**	**620**	**537**	**478**	420
	*12.0	15	69.6	6.43	3.90	222	*1360*	*1310*	*1170*	*1020*	*871*	*790*	*672*	*599*	*514*	*457*	*398*	*356*	*314*
	*10.0	15	58.4	6.46	3.85	186	*813*	*786*	*714*	*636*	*560*	*517*	*453*	*410*	*360*	*324*	*287*	*259*	*231*
150 x 90	15.0	12	67.8	4.74	3.75	155	1560	1470	1320	**1100**	**978**	**809**	**706**	**591**	**517**	**442**	**390**	**339**	302
	12.0	12	55.0	4.77	3.69	127	1260	1190	1060	**884**	**783**	**646**	**562**	**470**	**410**	**350**	**308**	**268**	239
	10.0	12	46.4	4.80	3.64	107	1070	1010	890	**740**	**653**	**537**	**466**	**390**	**339**	**290**	**255**	**222**	197
150 x 75	15.0	12	63.4	4.75	3.09	150	1460	1300	1060	**903**	**716**	**604**	**507**	**418**	**358**	**303**	**264**	**231**	202
	12.0	12	51.4	4.78	3.02	123	1180	1050	845	**717**	**566**	**475**	**398**	**327**	**279**	**237**	**206**	**180**	157
	10.0	12	43.4	4.81	2.97	103	999	878	707	**596**	**470**	**393**	**318**	**270**	**230**	**195**	**169**	**148**	129
125 x 75	12.0	12	45.4	3.95	3.19	86.4	1020	944	775	**667**	**533**	**451**	**368**	**314**	**269**	**229**	**199**	**175**	153
	10.0	12	38.2	3.97	3.14	73.0	858	791	647	**554**	**441**	**373**	**303**	**259**	**221**	**188**	**164**	**143**	125
	*8.0	12	31.0	4.00	3.09	59.2	*648*	*597*	*492*	*423*	*340*	*288*	*235*	*201*	*172*	*146*	*127*	*110*	*97.6*
100 x 75	12.0	10	39.4	3.10	3.31	56.0	849	762	673	**587**	**480**	**410**	**348**	**288**	**248**	**211**	**184**	**162**	142
	10.0	10	33.2	3.12	3.27	47.6	716	644	569	**497**	**399**	**340**	**287**	**238**	**204**	**174**	**152**	**133**	116
	8.0	10	27.0	3.14	3.22	38.6	583	525	463	**400**	**320**	**272**	**221**	**190**	**163**	**138**	**120**	**106**	92.4
100 x 65	10.0 #	10	31.2	3.14	2.79	46.4	674	599	466	**390**	**320**	**261**	**211**	**177**	**149**	**126**	**109**	**95.2**	83.6
	8.0 #	10	25.3	3.16	2.74	37.8	549	484	375	**312**	**255**	**201**	**167**	**140**	**118**	**99.7**	**86.2**	**75.1**	65.5
	7.0 #	10	22.4	3.17	2.72	33.2	485	425	329	**273**	**222**	**175**	**145**	**122**	**103**	**86.7**	**74.9**	**65.2**	56.9
100 x 50	8.0 ‡	10	22.8	3.19	2.09	36.4	471	342	**263**	**200**	**154**	**121**	**97.5**	**79.8**	**66.4**	**56.0**	**47.9**	**41.4**	36.1
	*6.0 ‡	10	17.4	3.21	2.04	27.6	*299*	*225*	*176*	*137*	*107*	*84.6*	*68.3*	*56.1*	*46.9*	*39.7*	*34.0*	*29.4*	*25.7*
80 x 60	7.0 ‡	8	18.8	2.51	2.59	21.4	385	333	**279**	**212**	**171**	**139**	**114**	**94.6**	**78.6**	**67.0**	**57.7**	**50.1**	43.7
80 x 40	8.0 ‡	8	18.0	2.53	1.71	22.8	324	**230**	**162**	**117**	**89.2**	**68.7**	**54.4**	**44.1**	**36.4**	**30.7**	**26.1**	**22.5**	19.5
	6.0 ‡	8	13.8	2.55	1.66	17.5	244	**171**	**119**	**85.6**	**64.9**	**49.9**	**39.4**	**31.9**	**26.3**	**22.2**	**18.8**	**16.2**	14.1
75 x 50	8.0 ‡	8	18.8	2.35	2.19	20.8	379	293	**228**	**176**	**137**	**108**	**87.0**	**71.4**	**59.5**	**50.3**	**43.0**	**37.2**	32.5
	6.0 ‡	8	14.4	2.37	2.14	16.0	291	221	**171**	**131**	**101**	**79.6**	**64.1**	**52.5**	**43.7**	**36.9**	**31.6**	**27.3**	23.7
70 x 50	6.0 ‡	8	13.8	2.20	2.19	14.0	272	214	**167**	**128**	**99.5**	**78.7**	**63.4**	**52.0**	**43.3**	**36.6**	**31.3**	**27.1**	23.7
65 x 50	5.0 ‡	8	11.1	2.05	2.21	10.3	214	173	**135**	**104**	**81.1**	**64.2**	**51.7**	**42.5**	**35.4**	**29.9**	**25.6**	**22.2**	19.3
60 x 40	6.0 ‡	8	11.4	1.88	1.80	10.1	211	**153**	**110**	**80.0**	**61.3**	**47.3**	**37.6**	**30.5**	**25.2**	**21.2**	**18.1**	**15.6**	13.6
	5.0 ‡	8	9.58	1.89	1.78	8.50	177	**128**	**91.0**	**66.3**	**49.8**	**39.1**	**31.0**	**25.1**	**20.8**	**17.5**	**14.9**	**12.8**	11.2

† Not rolled by certain loading producers, check availability.
# Check availability.
* Section is slender under axial compression and allowance has been made in calculating the compression resistance which is given in *italic type*.
Values in **bold type** indicate that the section has been divided into more than 3 bays until $\lambda_c \leq 50$.
FOR EXPLANATION OF TABLES SEE NOTE 8.3

| BS 5950-1: 2000 |
| BS 4-1: 1993 |

AXIAL LOAD & BENDING

UB SECTIONS SUBJECT TO AXIAL LOAD (COMPRESSION OR TENSION) AND BENDING

CROSS-SECTION CAPACITY CHECK

CAPACITIES FOR S275

Section Designation and Axial load Capacity P_z (kN)	F/P_z Limit Semi-Compact Compact		Moment Capacity M_{cx}, M_{cy} (kNm) and Reduced Moment Capacity M_{rx}, M_{ry} (kNm) for Ratios of Axial Load to Axial Load Capacity F/P_z										
		F/P_z	0.0	0.1	0.2	0.3	0.4	0.5	0.6	0.7	0.8	0.9	1.0
1016x305x487 # + $P_z = A_g p_y = 15800$	n/a 1.00	M_{cx} M_{cy} M_{rx} M_{ry}	5920 530 5920 530	5920 530 5830 530	5920 530 5590 530	5920 530 5180 530	5920 530 4610 530	5920 530 3900 530	5920 530 3150 530	5920 530 2390 528	5920 530 1610 397	5920 530 811 221	5920 530 0 0
1016x305x437 # + $P_z = A_g p_y = 14200$	n/a 1.00	M_{cx} M_{cy} M_{rx} M_{ry}	5290 470 5290 470	5290 470 5220 470	5290 470 5000 470	5290 470 4630 470	5290 470 4120 470	5290 470 3480 470	5290 470 2810 470	5290 470 2130 469	5290 470 1430 353	5290 470 722 197	5290 470 0 0
1016x305x393 # + $P_z = A_g p_y = 12800$	n/a 1.00	M_{cx} M_{cy} M_{rx} M_{ry}	4730 414 4730 414	4730 414 4660 414	4730 414 4470 414	4730 414 4140 414	4730 414 3680 414	4730 414 3110 414	4730 414 2510 414	4730 414 1900 414	4730 414 1270 314	4730 414 642 175	4730 414 0 0
1016x305x349 # + $P_z = A_g p_y = 11800$	0.986 0.405	M_{cx} M_{cy} M_{rx} M_{ry}	4400 389 4400 389	4400 389 4330 389	4400 389 4150 389	4400 389 3840 389	4400 389 3400 389	4380 389 - -	4370 389 - -	4360 389 - -	4350 389 - -	4310 389 - -	$ $ - -
1016x305x314 # + $P_z = A_g p_y = 10600$	0.845 0.343	M_{cx} M_{cy} M_{rx} M_{ry}	3940 344 3940 344	3940 344 3880 344	3940 344 3710 344	3940 344 3440 344	3800 344 - -	3750 344 - -	3710 344 - -	3650 344 - -	3520 344 - -	2740 230 - -	$ $ - -
1016x305x272 # + $P_z = A_g p_y = 9200$	0.662 0.258	M_{cx} M_{cy} M_{rx} M_{ry}	3400 297 3400 297	3400 297 3350 297	3400 297 3210 297	3300 297 - -	3140 297 - -	3090 297 - -	3020 297 - -	1780 148 - -	1780 148 - -	1780 148 - -	$ $ - -
1016x305x249 # + $P_z = A_g p_y = 8400$	0.662 0.282	M_{cx} M_{cy} M_{rx} M_{ry}	3010 249 3010 249	3010 249 2970 249	3010 249 2850 249	2960 249 - -	2790 249 - -	2720 249 - -	2660 249 - -	1560 125 - -	1560 125 - -	$ $ - -	$ $ - -

+ Section is not given in BS 4-1: 1993.
# Check availability.
F = Factored axial load.
$ For these values of F/P_z the section would be overloaded due to F alone even when M is zero, because F would exceed the local buckling resistance of the section.
- Not applicable for semi-compact and slender sections.
The values in this table are conservative for tension as the more onerous compression section classification limits have been used.
FOR EXPLANATION OF TABLES SEE NOTE 9.1

BS 5950-1: 2000
BS 4-1: 1993

AXIAL LOAD & BENDING

UB SECTIONS SUBJECT TO AXIAL COMPRESSION AND BENDING

MEMBER BUCKLING CHECK

RESISTANCES AND CAPACITIES FOR S275

Section Designation and Capacities (kN, kNm)	F/P_z Limit	Compression Resistance P_{cx}, P_{cy} (kN) and Buckling Resistance Moment M_b, M_{bs} (kNm) for Varying effective lengths L_E (m) within the limiting value of F/P_z												
		L_E (m)	2.0	3.0	4.0	5.0	6.0	7.0	8.0	9.0	10.0	11.0	12.0	13.0	14.0
1016x305x487 # +	1.00	P_{cx}	15800	15800	15800	15800	15800	15800	15700	15600	15400	15300	15100	15000	14800
$P_z = A_g p_y = 15800$		P_{cy}	14700	13200	11600	9930	8310	6890	5720	4780	4040	3450	2970	2580	2270
$p_y Z_x = 5030$	**1.00**	M_b	5920	5820	5280	4770	4300	3890	3530	3220	2960	2730	2540	2370	2220
$p_y Z_y = 442$		M_{bs}	5920	5920	5920	5800	5420	5020	4630	4230	3840	3480	3140	2840	2560
1016x305x437 # +	1.00	P_{cx}	14200	14200	14200	14200	14200	14200	14100	14000	13900	13800	13700	13600	13500
$P_z = A_g p_y = 14200$		P_{cy}	13300	12200	10900	9370	7840	6470	5330	4430	3730	3160	2720	2350	2060
$p_y Z_x = 4520$	**1.00**	M_b	5290	5170	4670	4190	3750	3360	3030	2750	2510	2310	2130	1990	1860
$p_y Z_y = 391$		M_{bs}	5290	5290	5290	5170	4820	4470	4110	3750	3400	3070	2770	2500	2260
1016x305x393 # +	1.00	P_{cx}	12700	12800	12700	12800	12800	12800	12700	12600	12500	12400	12300	12200	12100
$P_z = A_g p_y = 12800$		P_{cy}	11900	10900	9680	8310	6930	5700	4690	3890	3270	2770	2380	2060	1800
$p_y Z_x = 4050$	**1.00**	M_b	4730	4590	4120	3670	3260	2900	2590	2330	2120	1940	1780	1650	1540
$p_y Z_y = 345$		M_{bs}	4730	4730	4730	4590	4280	3960	3630	3310	2990	2700	2430	2190	1970
* 1016x305x349 # +	0.986	P_{cx}	11800	11800	11800	11800	11800	11800	11700	11700	11600	11500	11500	11400	11300
$P_z = A_g p_y = 11800$		P_{cy}	11200	10400	9350	8080	6730	5500	4490	3710	3090	2610	2230	1930	1680
$p_y Z_x = 3800$	0.986	M_b	3800	3750	3390	3040	2720	2420	2170	1950	1770	1620	1490	1380	1280
$p_y Z_y = 324$		M_{bs}	3800	3800	3800	3680	3430	3170	2900	2640	2390	2150	1930	1740	1570
	0.405	M_b	4400	4240	3780	3350	2940	2590	2300	2050	1850	1680	1540	1420	1320
		M_{bs}	4400	4400	4400	4250	3960	3660	3360	3050	2760	2490	2240	2010	1810
* 1016x305x314 # +	0.845	P_{cx}	10600	10600	10600	10600	10600	10600	10500	10500	10400	10400	10300	10200	10200
$P_z = A_g p_y = 10600$		P_{cy}	10000	9310	8360	7200	5970	4860	3970	3270	2730	2300	1970	1700	1480
$p_y Z_x = 3410$	0.845	M_b	3410	3350	3020	2690	2380	2110	1870	1670	1510	1370	1260	1160	1070
$p_y Z_y = 287$		M_{bs}	3410	3410	3410	3290	3060	2830	2590	2350	2120	1910	1710	1540	1380
	0.343	M_b	3940	3770	3350	2950	2570	2250	1980	1750	1570	1420	1300	1190	1100
		M_{bs}	3940	3940	3940	3790	3530	3260	2980	2710	2440	2200	1970	1770	1600
* 1016x305x272 # +	0.662	P_{cx}	9200	9200	9200	9200	9200	9200	9150	9100	9050	9000	8940	8890	8830
$P_z = A_g p_y = 9200$		P_{cy}	8710	8070	7240	6230	5160	4200	3430	2820	2350	1990	1700	1470	1280
$p_y Z_x = 2970$	0.662	M_b	2970	2900	2600	2300	2020	1770	1560	1380	1240	1120	1020	931	859
$p_y Z_y = 248$		M_{bs}	2970	2970	2970	2860	2660	2450	2240	2030	1830	1650	1480	1330	1200
	0.258	M_b	3400	3250	2870	2510	2170	1880	1640	1440	1280	1150	1050	956	880
		M_{bs}	3400	3400	3400	3270	3040	2810	2570	2330	2100	1890	1700	1530	1370
* 1016x305x249 # +	0.662	P_{cx}	8400	8400	8400	8400	8400	8390	8350	8300	8250	8200	8150	8100	8050
$P_z = A_g p_y = 8400$		P_{cy}	7910	7280	6470	5480	4470	3610	2930	2400	2000	1680	1440	1240	1080
$p_y Z_x = 2600$	0.662	M_b	2600	2520	2250	1970	1720	1490	1300	1140	1020	911	825	752	692
$p_y Z_y = 208$		M_{bs}	2600	2600	2600	2470	2290	2100	1900	1720	1540	1380	1230	1100	988
	0.282	M_b	3010	2840	2490	2160	1850	1580	1370	1190	1050	940	848	772	708
		M_{bs}	3010	3010	3010	2850	2640	2420	2200	1980	1780	1590	1420	1270	1140

+ Section is not given in BS 4-1: 1993.
# Check availability.
* The section can become slender under axial compression only. Under combined axial compression and bending the section becomes slender when the semi-compact F/P_z limit is exceeded.
Under combined axial compression and bending the capacities are only valid up to the given F/P_z limit. For higher values F/P_z the section would be overloaded due to F alone even when M is zero, because F would exceed the local buckling resistance of the section.
M_b is obtained using an equivalent slenderness = $u.v.L_E/r_y.\beta_w^{0.5}$
M_{bs} is obtained using an equivalent slenderness = $0.5 L/r_y$. Effective length $L_E = L$.
FOR EXPLANATION OF TABLES SEE NOTE 9.1

BS 5950-1: 2000
BS 4-1: 1993

AXIAL LOAD & BENDING

UB SECTIONS SUBJECT TO AXIAL LOAD (COMPRESSION OR TENSION) AND BENDING

CROSS-SECTION CAPACITY CHECK

CAPACITIES FOR S275

Section Designation and Axial load Capacity P_z (kN)	F/P_z Limit Semi-Compact Compact		Moment Capacity M_{cx}, M_{cy} (kNm) and Reduced Moment Capacity M_{rx}, M_{ry} (kNm) for Ratios of Axial Load to Axial Load Capacity F/P_z										
		F/P_z	0.0	0.1	0.2	0.3	0.4	0.5	0.6	0.7	0.8	0.9	1.0
1016x305x222 # + $P_z = A_g p_y = 7500$	0.627 **0.287**	M_{cx}	2600	2600	2600	2570	2400	2310	2250	1260	1260	$	$
		M_{cy}	202	202	202	202	202	202	202	95.1	95.1	$	$
		M_{rx}	2600	2570	2470	-	-	-	-	-	-	-	-
		M_{ry}	202	202	202	-	-	-	-	-	-	-	-
914 x 419 x 388 # $P_z = A_g p_y = 13100$	n/a **1.00**	M_{cx}	4680	4680	4680	4680	4680	4680	4680	4680	4680	4680	4680
		M_{cy}	687	687	687	687	687	687	687	687	687	687	687
		M_{rx}	4680	4610	4380	4000	3480	2920	2350	1770	1190	599	0
		M_{ry}	687	687	687	687	687	687	687	627	462	253	0
914 x 419 x 343 # $P_z = A_g p_y = 11600$	0.983 **0.348**	M_{cx}	4100	4100	4100	4100	4090	4080	4080	4070	4060	4020	$
		M_{cy}	595	595	595	595	595	595	595	595	595	595	$
		M_{rx}	4100	4040	3840	3520	-	-	-	-	-	-	-
		M_{ry}	595	595	595	595	-	-	-	-	-	-	-
914 x 305 x 289 # $P_z = A_g p_y = 9750$	0.946 **0.411**	M_{cx}	3330	3330	3330	3330	3330	3270	3260	3240	3200	3080	$
		M_{cy}	322	322	322	322	322	322	322	322	322	322	$
		M_{rx}	3330	3290	3150	2920	2590	-	-	-	-	-	-
		M_{ry}	322	322	322	322	322	-	-	-	-	-	-
914 x 305 x 253 # $P_z = A_g p_y = 8560$	0.783 **0.335**	M_{cx}	2900	2900	2900	2900	2780	2710	2680	2610	1840	1840	$
		M_{cy}	277	277	277	277	277	277	277	277	169	169	$
		M_{rx}	2900	2860	2740	2540	-	-	-	-	-	-	-
		M_{ry}	277	277	277	277	-	-	-	-	-	-	-
914 x 305 x 224 # $P_z = A_g p_y = 7580$	0.679 **0.295**	M_{cx}	2530	2530	2530	2520	2370	2300	2250	1350	1350	1350	$
		M_{cy}	235	235	235	235	235	235	235	121	121	121	$
		M_{rx}	2530	2490	2390	-	-	-	-	-	-	-	-
		M_{ry}	235	235	235	-	-	-	-	-	-	-	-

+ Section is not given in BS 4-1: 1993.
# Check availability.
F = Factored axial load.
$ For these values of F/P_z the section would be overloaded due to F alone even when M is zero, because F would exceed the local buckling resistance of the section.
- Not applicable for semi-compact and slender sections.
The values in this table are conservative for tension as the more onerous compression section classification limits have been used.
FOR EXPLANATION OF TABLES SEE NOTE 9.1

BS 5950-1: 2000
BS 4-1: 1993

AXIAL LOAD & BENDING

UB SECTIONS SUBJECT TO AXIAL COMPRESSION AND BENDING

MEMBER BUCKLING CHECK

RESISTANCES AND CAPACITIES FOR S275

Section Designation and Capacities (kN, kNm)	F/P_z Limit		Compression Resistance P_{cx}, P_{cy} (kN) and Buckling Resistance Moment M_b, M_{bs} (kNm) for Varying effective lengths L_E (m) within the limiting value of F/P_z											
		L_E (m)	2.0	3.0	4.0	5.0	6.0	7.0	8.0	9.0	10.0	11.0	12.0	13.0	14.0
* 1016x305x222 # +	0.627	P_{cx}	7500	7500	7500	7500	7500	7480	7440	7400	7360	7310	7260	7220	7170
$P_z = A_g p_y = 7500$		P_{cy}	7010	6410	5610	4670	3750	3000	2420	1980	1640	1380	1180	1010	883
$p_y Z_x = 2230$	0.627	M_b	2230	2130	1890	1640	1420	1220	1050	917	808	721	649	589	539
$p_y Z_y = 169$		M_{bs}	2230	2230	2230	2080	1910	1740	1570	1400	1250	1110	988	881	788
	0.287	M_b	2600	2420	2100	1800	1520	1290	1100	954	837	743	667	604	552
		M_{bs}	2600	2600	2600	2420	2230	2030	1830	1640	1460	1290	1150	1030	919
914 x 419 x 388 #	1.00	P_{cx}	13100	13100	13100	13100	13100	13100	13000	12900	12800	12800	12700	12600	12500
$P_z = A_g p_y = 13100$		P_{cy}	12900	12400	11800	11100	10300	9420	8410	7400	6470	5640	4930	4330	3820
$p_y Z_x = 4140$	1.00	M_b	4680	4680	4650	4320	3990	3680	3380	3100	2850	2620	2420	2240	2090
$p_y Z_y = 573$		M_{bs}	4680	4680	4680	4680	4680	4620	4420	4210	3990	3780	3560	3340	3130
* 914 x 419 x 343 #	0.983	P_{cx}	11600	11600	11600	11600	11600	11600	11500	11400	11400	11300	11200	11100	11100
$P_z = A_g p_y = 11600$		P_{cy}	11400	11000	10400	9820	9090	8250	7340	6440	5610	4890	4270	3740	3300
$p_y Z_x = 3640$	0.983	M_b	3640	3640	3640	3410	3170	2930	2690	2480	2270	2090	1930	1790	1660
$p_y Z_y = 496$		M_{bs}	3640	3640	3640	3640	3640	3580	3410	3250	3080	2910	2740	2560	2400
	0.348	M_b	4100	4100	4050	3750	3460	3170	2890	2640	2400	2200	2020	1860	1730
		M_{bs}	4100	4100	4100	4100	4100	4030	3850	3660	3470	3280	3080	2890	2700
* 914 x 305 x 289 #	0.946	P_{cx}	9750	9750	9750	9750	9750	9720	9670	9610	9550	9490	9430	9360	9290
$P_z = A_g p_y = 9750$		P_{cy}	9270	8610	7780	6740	5630	4620	3780	3120	2610	2200	1880	1630	1420
$p_y Z_x = 2880$	0.946	M_b	2880	2850	2580	2300	2050	1810	1610	1440	1300	1180	1080	990	917
$p_y Z_y = 269$		M_{bs}	2880	2880	2880	2800	2610	2420	2220	2020	1830	1650	1490	1340	1210
	0.411	M_b	3330	3220	2870	2530	2220	1940	1700	1510	1350	1220	1110	1020	944
		M_{bs}	3330	3330	3330	3230	3020	2790	2560	2340	2110	1910	1720	1550	1390
* 914 x 305 x 253 #	0.783	P_{cx}	8560	8560	8560	8560	8560	8530	8480	8430	8380	8330	8270	8210	8150
$P_z = A_g p_y = 8560$		P_{cy}	8120	7530	6780	5850	4860	3970	3250	2680	2230	1890	1610	1390	1210
$p_y Z_x = 2520$	0.783	M_b	2520	2480	2230	1980	1740	1530	1350	1190	1070	961	873	799	737
$p_y Z_y = 231$		M_{bs}	2520	2520	2520	2430	2270	2090	1920	1740	1580	1420	1280	1150	1030
	0.335	M_b	2900	2780	2470	2160	1880	1630	1420	1250	1110	994	900	822	756
		M_{bs}	2900	2900	2900	2800	2610	2410	2210	2010	1810	1630	1470	1320	1190
* 914 x 305 x 224 #	0.679	P_{cx}	7580	7580	7580	7580	7580	7550	7510	7460	7410	7370	7310	7260	7210
$P_z = A_g p_y = 7580$		P_{cy}	7170	6620	5930	5080	4190	3400	2770	2280	1900	1600	1370	1180	1030
$p_y Z_x = 2190$	0.679	M_b	2190	2140	1910	1690	1480	1280	1120	985	873	782	706	644	591
$p_y Z_y = 196$		M_{bs}	2190	2190	2190	2100	1950	1800	1640	1490	1340	1200	1080	966	869
	0.295	M_b	2530	2410	2130	1850	1590	1360	1180	1030	905	807	727	661	605
		M_{bs}	2530	2530	2530	2420	2250	2070	1890	1710	1540	1380	1240	1110	1000

+ Section is not given in BS 4-1: 1993.
# Check availability.
* The section can become slender under axial compression only. Under combined axial compression and bending the section becomes slender when the semi-compact F/P_z limit is exceeded.
Under combined axial compression and bending the capacities are only valid up to the given F/P_z limit. For higher values F/P_z the section would be overloaded due to F alone even when M is zero, because F would exceed the local buckling resistance of the section.
M_b is obtained using an equivalent slenderness = $u.v.L_E/r_y.\beta_w^{0.5}$
M_{bs} is obtained using an equivalent slenderness = $0.5 L/r_y$. Effective length $L_E = L$.
FOR EXPLANATION OF TABLES SEE NOTE 9.1

BS 5950-1: 2000
BS 4-1: 1993

AXIAL LOAD & BENDING

UB SECTIONS SUBJECT TO AXIAL LOAD (COMPRESSION OR TENSION) AND BENDING

CROSS-SECTION CAPACITY CHECK

CAPACITIES FOR S275

Section Designation and Axial load Capacity P_z (kN)	F/P_z Limit Semi-Compact Compact	\multicolumn{12}{c}{Moment Capacity M_{cx}, M_{cy} (kNm) and Reduced Moment Capacity M_{rx}, M_{ry} (kNm) for Ratios of Axial Load to Axial Load Capacity F/P_z}											
		F/P_z	0.0	0.1	0.2	0.3	0.4	0.5	0.6	0.7	0.8	0.9	1.0
914 x 305 x 201 # $P_z = A_g p_y = 6780$	0.620 0.281	M_{cx} M_{cy} M_{rx} M_{ry}	2210 197 2210 197	2210 197 2180 197	2210 197 2100 197	2180 197 - -	2050 197 - -	1970 197 - -	1920 197 - -	1060 91.7 - -	1060 91.7 - -	$ $ - -	$ $ - -
838 x 292 x 226 # $P_z = A_g p_y = 7660$	0.792 0.326	M_{cx} M_{cy} M_{rx} M_{ry}	2430 246 2430 246	2430 246 2390 246	2430 246 2290 246	2430 246 2120 246	2320 246 - -	2280 246 - -	2250 246 - -	2200 246 - -	1570 152 - -	1570 152 - -	$ $ - -
838 x 292 x 194 # $P_z = A_g p_y = 6550$	0.680 0.292	M_{cx} M_{cy} M_{rx} M_{ry}	2020 197 2020 197	2020 197 2000 197	2020 197 1910 197	2010 197 - -	1900 197 - -	1840 197 - -	1810 197 - -	1090 102 - -	1090 102 - -	1090 102 - -	$ $ - -
838 x 292 x 176 # $P_z = A_g p_y = 5940$	0.623 0.277	M_{cx} M_{cy} M_{rx} M_{ry}	1800 170 1800 170	1800 170 1780 170	1800 170 1710 170	1770 170 - -	1670 170 - -	1610 170 - -	1570 170 - -	876 79.5 - -	876 79.5 - -	$ $ - -	$ $ - -
762 x 267 x 197 $P_z = A_g p_y = 6650$	0.890 0.374	M_{cx} M_{cy} M_{rx} M_{ry}	1900 194 1900 194	1900 194 1870 194	1900 194 1790 194	1900 194 1660 194	1870 194 - -	1840 194 - -	1820 194 - -	1800 194 - -	1760 194 - -	1420 139 - -	$ $ - -

\# Check availability.
F = Factored axial load.
$ For these values of F/P_z the section would be overloaded due to F alone even when M is zero, because F would exceed the local buckling resistance of the section.
- Not applicable for semi-compact and slender sections.
The values in this table are conservative for tension as the more onerous compression section classification limits have been used.
FOR EXPLANATION OF TABLES SEE NOTE 9.1

BS 5950-1: 2000
BS 4-1: 1993

AXIAL LOAD & BENDING

UB SECTIONS SUBJECT TO AXIAL COMPRESSION AND BENDING

MEMBER BUCKLING CHECK

RESISTANCES AND CAPACITIES FOR S275

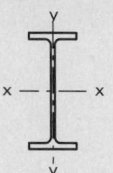

Section Designation and Capacities (kN, kNm)	F/P_z Limit		Compression Resistance P_{cx}, P_{cy} (kN) and Buckling Resistance Moment M_b, M_{bs} (kNm) for Varying effective lengths L_E (m) within the limiting value of F/P_z											
		L_E (m)	2.0	3.0	4.0	5.0	6.0	7.0	8.0	9.0	10.0	11.0	12.0	13.0	14.0
* 914 x 305 x 201 #	0.620	P_{cx}	6780	6780	6780	6780	6780	6750	6710	6670	6630	6580	6540	6490	6440
$P_z = A_g p_y = 6780$		P_{cy}	6380	5880	5210	4410	3600	2900	2350	1930	1600	1350	1150	996	867
$p_y Z_x = 1910$	0.620	M_b	1910	1850	1650	1440	1250	1080	932	813	717	638	573	520	475
$p_y Z_y = 165$		M_{bs}	1910	1910	1910	1810	1670	1530	1390	1260	1130	1010	899	804	721
	0.281	M_b	2210	2090	1830	1580	1340	1140	977	847	742	658	590	533	486
		M_{bs}	2210	2210	2210	2100	1940	1780	1610	1460	1300	1170	1040	932	836
* 838 x 292 x 226 #	0.792	P_{cx}	7660	7660	7660	7660	7660	7610	7560	7510	7460	7410	7350	7300	7230
$P_z = A_g p_y = 7660$		P_{cy}	7240	6690	5990	5130	4230	3430	2800	2300	1920	1620	1380	1190	1040
$p_y Z_x = 2120$	0.792	M_b	2120	2060	1850	1630	1430	1260	1100	979	875	790	718	659	608
$p_y Z_y = 205$		M_{bs}	2120	2120	2120	2030	1880	1740	1580	1430	1290	1160	1040	933	839
	0.326	M_b	2430	2310	2040	1780	1540	1330	1160	1020	907	815	739	676	623
		M_{bs}	2430	2430	2430	2330	2160	1990	1820	1640	1480	1330	1190	1070	962
* 838 x 292 x 194 #	0.680	P_{cx}	6550	6550	6550	6550	6540	6500	6460	6410	6370	6320	6270	6220	6170
$P_z = A_g p_y = 6550$		P_{cy}	6160	5670	5030	4250	3460	2790	2260	1850	1540	1300	1110	958	834
$p_y Z_x = 1760$	0.680	M_b	1760	1700	1510	1320	1150	993	863	757	670	600	542	493	453
$p_y Z_y = 164$		M_{bs}	1760	1760	1760	1670	1540	1410	1280	1160	1040	925	826	739	663
	0.292	M_b	2020	1910	1670	1440	1230	1050	904	787	693	618	556	506	463
		M_{bs}	2020	2020	2020	1920	1770	1630	1480	1330	1190	1060	951	851	763
* 838 x 292 x 176 #	0.623	P_{cx}	5940	5940	5940	5940	5930	5890	5850	5810	5770	5730	5680	5630	5580
$P_z = A_g p_y = 5940$		P_{cy}	5560	5100	4490	3750	3030	2430	1960	1610	1330	1120	958	826	720
$p_y Z_x = 1560$	0.623	M_b	1560	1500	1320	1150	992	852	735	640	564	502	451	409	374
$p_y Z_y = 142$		M_{bs}	1560	1560	1560	1460	1350	1230	1110	998	891	794	707	631	565
	0.277	M_b	1800	1680	1470	1250	1060	899	769	665	583	517	463	419	383
		M_{bs}	1800	1800	1800	1690	1560	1420	1290	1150	1030	917	817	729	653
* 762 x 267 x 197	0.890	P_{cx}	6650	6650	6650	6650	6620	6580	6530	6480	6430	6380	6320	6260	6190
$P_z = A_g p_y = 6650$		P_{cy}	6200	5650	4920	4070	3250	2590	2080	1700	1410	1190	1010	871	758
$p_y Z_x = 1650$	0.890	M_b	1650	1560	1380	1210	1050	911	799	708	634	573	523	481	445
$p_y Z_y = 162$		M_{bs}	1650	1650	1650	1530	1400	1270	1140	1020	907	805	715	636	569
	0.374	M_b	1900	1750	1520	1310	1120	961	836	736	656	591	537	493	455
		M_{bs}	1900	1900	1900	1760	1610	1460	1320	1170	1040	925	822	732	654

\# Check availability.
* The section can become slender under axial compression only. Under combined axial compression and bending the section becomes slender when the semi-compact F/P_a limit is exceeded.
Under combined axial compression and bending the capacities are only valid up to the given F/P_z limit. For higher values F/P_z the section would be overloaded due to F alone even when M is zero, because F would exceed the local buckling resistance of the section.
M_b is obtained using an equivalent slenderness = $u.v.L_E/r_y.\beta_w^{0.5}$
M_{bs} is obtained using an equivalent slenderness = $0.5 L/r_y$. Effective length $L_E = L$.
FOR EXPLANATION OF TABLES SEE NOTE 9.1

BS 5950-1: 2000
BS 4-1: 1993

AXIAL LOAD & BENDING

UB SECTIONS SUBJECT TO AXIAL LOAD (COMPRESSION OR TENSION) AND BENDING

CROSS-SECTION CAPACITY CHECK

CAPACITIES FOR S275

Section Designation and Axial load Capacity P_z (kN)	F/P_z Limit Semi-Compact **Compact**		Moment Capacity M_{cx}, M_{cy} (kNm) and Reduced Moment Capacity M_{rx}, M_{ry} (kNm) for Ratios of Axial Load to Axial Load Capacity F/P_z										
		F/P_z	0.0	0.1	0.2	0.3	0.4	0.5	0.6	0.7	0.8	0.9	1.0
762 x 267 x 173 $P_z = A_g p_y = 5830$	0.774 **0.334**	M_{cx} M_{cy} M_{rx} M_{ry}	1640 163 1640 163	1640 163 1620 163	1640 163 1550 163	1640 163 1440 163	1580 163 - -	1530 163 - -	1510 163 - -	1480 163 - -	1030 98.3 - -	1030 98.3 - -	$ $ - -
762 x 267 x 147 $P_z = A_g p_y = 4960$	0.640 **0.282**	M_{cx} M_{cy} M_{rx} M_{ry}	1370 131 1370 131	1370 131 1350 131	1370 131 1290 131	1350 131 - -	1270 131 - -	1230 131 - -	1200 131 - -	685 63 - -	685 63 - -	$ $ - -	$ $ - -
762 x 267 x 134 $P_z = A_g p_y = 4700$	0.550 **0.240**	M_{cx} M_{cy} M_{rx} M_{ry}	1280 119 1280 119	1280 119 1260 119	1280 119 1210 119	1220 119 - -	1150 119 - -	1120 119 - -	541 48.7 - -	541 48.7 - -	541 48.7 - -	$ $ - -	$ $ - -
686 x 254 x 170 $P_z = A_g p_y = 5750$	0.941 **0.384**	M_{cx} M_{cy} M_{rx} M_{ry}	1490 165 1490 165	1490 165 1470 165	1490 165 1410 165	1490 165 1300 165	1480 165 - -	1470 165 - -	1460 165 - -	1450 165 - -	1430 165 - -	1380 165 - -	$ $ - -
686 x 254 x 152 $P_z = A_g p_y = 5140$	0.812 **0.331**	M_{cx} M_{cy} M_{rx} M_{ry}	1330 145 1330 145	1330 145 1310 145	1330 145 1250 145	1330 145 1150 145	1270 145 - -	1250 145 - -	1240 145 - -	1220 145 - -	1170 145 - -	886 92.2 - -	$ $ - -
686 x 254 x 140 $P_z = A_g p_y = 4720$	0.732 **0.301**	M_{cx} M_{cy} M_{rx} M_{ry}	1210 130 1210 130	1210 130 1190 130	1210 130 1140 130	1210 130 1060 130	1140 130 - -	1120 130 - -	1100 130 - -	1070 130 - -	713 73.1 - -	713 73.1 - -	$ $ - -

F = Factored axial load.
$ For these values of F/Pz the section would be overloaded due to F alone even when M is zero, because F would exceed the local buckling resistance of the section.
- Not applicable for semi-compact and slender sections.
The values in this table are conservative for tension as the more onerous compression section classification limits have been used.
FOR EXPLANATION OF TABLES SEE NOTE 9.1

BS 5950-1: 2000
BS 4-1: 1993

AXIAL LOAD & BENDING

UB SECTIONS SUBJECT TO AXIAL COMPRESSION AND BENDING

MEMBER BUCKLING CHECK

RESISTANCES AND CAPACITIES FOR S275

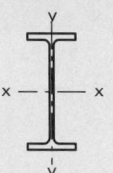

Section Designation and Capacities (kN, kNm)	F/P_z Limit	Compression Resistance P_{cx}, P_{cy} (kN) and Buckling Resistance Moment M_b, M_{bs} (kNm) for Varying effective lengths L_E (m) within the limiting value of F/P_z												
		L_E (m)	2.0	3.0	4.0	5.0	6.0	7.0	8.0	9.0	10.0	11.0	12.0	13.0	14.0
* 762 x 267 x 173	0.774	P_{cx}	5830	5830	5830	5830	5800	5760	5720	5680	5630	5580	5530	5480	5420
$P_z = A_g p_y = 5830$		P_{cy}	5420	4910	4250	3480	2760	2190	1750	1430	1190	997	849	731	636
$p_y Z_x = 1430$	0.774	M_b	1430	1340	1180	1020	873	751	652	573	509	457	415	379	349
$p_y Z_y = 136$		M_{bs}	1430	1430	1420	1310	1200	1080	969	861	762	674	598	531	474
	0.334	M_b	1640	1500	1290	1100	928	790	680	594	525	470	425	388	357
		M_{bs}	1640	1640	1630	1510	1380	1250	1110	991	877	776	688	611	546
* 762 x 267 x 147	0.640	P_{cx}	4960	4960	4960	4960	4930	4890	4860	4820	4780	4740	4690	4640	4590
$P_z = A_g p_y = 4960$		P_{cy}	4580	4130	3530	2850	2240	1760	1410	1150	950	798	679	585	508
$p_y Z_x = 1180$	0.640	M_b	1180	1100	955	816	691	587	504	438	386	344	310	281	258
$p_y Z_y = 109$		M_{bs}	1180	1180	1160	1070	974	876	780	689	607	535	473	420	374
	0.282	M_b	1370	1230	1050	880	733	616	524	453	397	353	317	288	263
		M_{bs}	1370	1370	1340	1230	1120	1010	899	795	700	617	546	484	431
* 762 x 267 x 134	0.550	P_{cx}	4700	4700	4700	4700	4670	4640	4600	4570	4530	4490	4440	4390	4340
$P_z = A_g p_y = 4700$		P_{cy}	4320	3870	3260	2590	2020	1580	1260	1020	844	708	602	518	450
$p_y Z_x = 1100$	0.550	M_b	1100	1010	873	738	619	521	443	383	335	297	266	241	220
$p_y Z_y = 99.6$		M_{bs}	1100	1100	1070	984	891	796	704	619	543	477	420	371	330
	0.240	M_b	1280	1130	959	794	655	545	460	395	344	304	272	246	224
		M_{bs}	1280	1280	1240	1140	1030	920	814	715	627	551	485	429	382
* 686 x 254 x 170	0.941	P_{cx}	5750	5750	5750	5750	5700	5660	5610	5560	5510	5460	5400	5330	5260
$P_z = A_g p_y = 5750$		P_{cy}	5340	4830	4160	3390	2690	2130	1700	1390	1150	967	824	709	617
$p_y Z_x = 1300$	0.941	M_b	1300	1220	1070	930	805	700	614	545	489	443	405	373	345
$p_y Z_y = 137$		M_{bs}	1300	1300	1290	1190	1090	981	877	778	688	608	538	478	427
	0.384	M_b	1490	1360	1170	1000	856	736	641	565	505	456	416	382	353
		M_{bs}	1490	1490	1480	1360	1240	1120	1000	891	788	697	617	548	489
* 686 x 254 x 152	0.812	P_{cx}	5140	5140	5140	5140	5100	5060	5010	4970	4920	4870	4820	4760	4700
$P_z = A_g p_y = 5140$		P_{cy}	4760	4300	3690	2990	2360	1860	1490	1210	1010	845	719	619	539
$p_y Z_x = 1160$	0.812	M_b	1160	1080	940	810	694	598	520	458	409	368	335	307	283
$p_y Z_y = 121$		M_{bs}	1160	1160	1140	1050	959	864	771	683	603	532	470	417	372
	0.331	M_b	1330	1190	1030	870	735	627	541	474	421	378	343	314	289
		M_{bs}	1330	1330	1310	1200	1100	988	881	780	689	608	538	477	425
* 686 x 254 x 140	0.732	P_{cx}	4720	4720	4720	4710	4670	4640	4600	4560	4510	4470	4420	4360	4300
$P_z = A_g p_y = 4720$		P_{cy}	4360	3930	3350	2700	2120	1670	1340	1090	901	757	644	555	482
$p_y Z_x = 1060$	0.732	M_b	1060	975	848	726	618	529	458	401	356	319	289	264	243
$p_y Z_y = 108$		M_{bs}	1060	1060	1040	954	868	780	694	613	540	476	421	373	332
	0.301	M_b	1210	1080	926	779	654	553	475	414	366	327	296	270	248
		M_{bs}	1210	1210	1190	1090	992	892	793	701	618	544	481	427	380

* The section can become slender under axial compression only. Under combined axial compression and bending the section becomes slender when the semi-compact F/P_z limit is exceeded.
Under combined axial compression and bending the capacities are only valid up to the given F/P_z limit. For higher values F/P_z the section would be overloaded due to F alone even when M is zero, because F would exceed the local buckling resistance of the section.
M_b is obtained using an equivalent slenderness = $u.v.L_E/r_y.\beta_w^{0.5}$
M_{bs} is obtained using an equivalent slenderness = $0.5 L/r_y$. Effective length $L_E = L$.
FOR EXPLANATION OF TABLES SEE NOTE 9.1

BS 5950-1: 2000
BS 4-1: 1993

AXIAL LOAD & BENDING

UB SECTIONS SUBJECT TO AXIAL LOAD (COMPRESSION OR TENSION) AND BENDING

CROSS-SECTION CAPACITY CHECK

CAPACITIES FOR S275

Section Designation and Axial load Capacity P_z (kN)	F/P_z Limit Semi-Compact Compact		Moment Capacity M_{cx}, M_{cy} (kNm) and Reduced Moment Capacity M_{rx}, M_{ry} (kNm) for Ratios of Axial Load to Axial Load Capacity F/P_z										
		F/P_z	0.0	0.1	0.2	0.3	0.4	0.5	0.6	0.7	0.8	0.9	1.0
686 x 254 x 125 $P_z = A_g p_y = 4210$	0.663 0.283	M_{cx}	1060	1060	1060	1040	987	961	941	554	554	$	$
		M_{cy}	110	110	110	110	110	110	110	55.1	55.1	$	$
		M_{rx}	1060	1040	1000	-	-	-	-	-	-	-	-
		M_{ry}	110	110	110	-	-	-	-	-	-	-	-
610 x 305 x 238 $P_z = A_g p_y = 8030$	n/a 1.00	M_{cx}	1980	1980	1980	1980	1980	1980	1980	1980	1980	1980	1980
		M_{cy}	323	323	323	323	323	323	323	323	323	323	323
		M_{rx}	1980	1950	1850	1690	1460	1230	990	748	503	253	0
		M_{ry}	323	323	323	323	323	323	323	288	211	115	0
610 x 305 x 179 $P_z = A_g p_y = 6040$	n/a 1.00	M_{cx}	1470	1470	1470	1470	1470	1470	1470	1470	1470	1470	1470
		M_{cy}	236	236	236	236	236	236	236	236	236	236	236
		M_{rx}	1470	1450	1370	1250	1080	909	732	552	370	186	0
		M_{ry}	236	236	236	236	236	236	236	213	156	85.5	0
610 x 305 x 149 $P_z = A_g p_y = 5040$	0.836 0.274	M_{cx}	1220	1220	1220	1200	1180	1170	1160	1140	1110	864	$
		M_{cy}	194	194	194	194	194	194	194	194	194	128	$
		M_{rx}	1220	1200	1140	-	-	-	-	-	-	-	-
		M_{ry}	194	194	194	-	-	-	-	-	-	-	-
610 x 229 x 140 $P_z = A_g p_y = 4720$	0.962 0.386	M_{cx}	1100	1100	1100	1100	1090	1090	1080	1080	1070	1040	$
		M_{cy}	124	124	124	124	124	124	124	124	124	124	$
		M_{rx}	1100	1080	1030	953	-	-	-	-	-	-	-
		M_{ry}	124	124	124	124	-	-	-	-	-	-	-
610 x 229 x 125 $P_z = A_g p_y = 4210$	0.828 0.332	M_{cx}	974	974	974	974	937	927	918	902	870	669	$
		M_{cy}	109	109	109	109	109	109	109	109	109	71.3	$
		M_{rx}	974	960	918	847	-	-	-	-	-	-	-
		M_{ry}	109	109	109	109	-	-	-	-	-	-	-

F = Factored axial load.
$ For these values of F/Pz the section would be overloaded due to F alone even when M is zero, because F would exceed the local buckling resistance of the section.
- Not applicable for semi-compact and slender sections.
The values in this table are conservative for tension as the more onerous compression section classification limits have been used.
FOR EXPLANATION OF TABLES SEE NOTE 9.1

BS 5950-1: 2000
BS 4-1: 1993

AXIAL LOAD & BENDING

UB SECTIONS SUBJECT TO AXIAL COMPRESSION AND BENDING

MEMBER BUCKLING CHECK

RESISTANCES AND CAPACITIES FOR S275

Section Designation and Capacities (kN, kNm)	F/P_z Limit	Compression Resistance P_{cx}, P_{cy} (kN) and Buckling Resistance Moment M_b, M_{bs} (kNm) for Varying effective lengths L_E (m) within the limiting value of F/P_z												
		L_E (m)	2.0	3.0	4.0	5.0	6.0	7.0	8.0	9.0	10.0	11.0	12.0	13.0	14.0
* 686 x 254 x 125 $P_z = A_g p_y = 4210$ $p_y Z_x = 922$ $p_y Z_y = 91.7$	0.663	P_{cx}	4210	4210	4210	4210	4170	4140	4100	4070	4030	3980	3940	3890	3830
		P_{cy}	3870	3470	2930	2340	1820	1430	1140	925	765	642	546	470	408
	0.663	M_b	922	843	728	617	520	441	378	329	290	258	233	212	194
		M_{bs}	922	922	898	822	745	666	589	518	455	400	352	312	277
	0.283	M_b	1060	938	795	661	549	460	392	339	298	265	238	216	198
		M_{bs}	1060	1060	1030	944	854	764	676	595	522	459	404	358	318
610 x 305 x 238 $P_z = A_g p_y = 8030$ $p_y Z_x = 1750$ $p_y Z_y = 270$	1.00	P_{cx}	8030	8030	8030	8000	7940	7870	7800	7730	7650	7560	7470	7360	7250
		P_{cy}	7720	7260	6700	5990	5180	4370	3650	3050	2570	2180	1870	1620	1420
	1.00	M_b	1980	1980	1810	1640	1480	1340	1220	1110	1020	942	874	815	763
		M_{bs}	1980	1980	1980	1980	1870	1760	1640	1510	1390	1270	1160	1060	960
610 x 305 x 179 $P_z = A_g p_y = 6040$ $p_y Z_x = 1310$ $p_y Z_y = 197$	1.00	P_{cx}	6040	6040	6040	6020	5970	5920	5860	5810	5740	5680	5600	5520	5440
		P_{cy}	5800	5440	5000	4440	3820	3200	2660	2210	1860	1580	1350	1170	1020
	1.00	M_b	1470	1450	1310	1170	1040	924	823	737	665	605	554	511	475
		M_{bs}	1470	1470	1470	1460	1380	1290	1200	1100	1010	921	837	760	690
* 610 x 305 x 149 $P_z = A_g p_y = 5040$ $p_y Z_x = 1090$ $p_y Z_y = 162$	0.836	P_{cx}	5040	5040	5040	5010	4970	4930	4880	4830	4780	4730	4660	4600	4520
		P_{cy}	4830	4520	4150	3680	3150	2630	2180	1820	1520	1290	1110	960	838
	0.836	M_b	1090	1090	985	884	787	698	619	552	496	449	409	376	347
		M_{bs}	1090	1090	1090	1080	1020	949	880	811	742	676	614	556	505
	0.274	M_b	1220	1190	1070	953	838	736	648	574	513	462	420	385	355
		M_{bs}	1220	1220	1220	1210	1140	1060	984	906	829	755	686	622	564
* 610 x 229 x 140 $P_z = A_g p_y = 4720$ $p_y Z_x = 960$ $p_y Z_y = 104$	0.962	P_{cx}	4720	4720	4720	4690	4650	4610	4560	4520	4470	4410	4350	4280	4210
		P_{cy}	4300	3810	3170	2480	1910	1490	1190	962	794	666	566	487	423
	0.962	M_b	960	864	747	639	548	474	415	368	331	300	275	254	235
		M_{bs}	960	960	921	839	754	669	588	514	448	392	345	304	270
	0.386	M_b	1100	957	813	684	578	495	431	381	341	308	282	259	240
		M_{bs}	1100	1100	1050	959	862	765	672	587	513	449	394	348	309
* 610 x 229 x 125 $P_z = A_g p_y = 4210$ $p_y Z_x = 854$ $p_y Z_y = 90.9$	0.828	P_{cx}	4210	4210	4210	4190	4150	4120	4080	4030	3990	3940	3880	3820	3750
		P_{cy}	3830	3390	2800	2180	1680	1310	1040	841	694	582	494	425	369
	0.828	M_b	854	762	654	554	470	403	350	309	275	249	227	208	193
		M_{bs}	854	854	816	742	665	589	516	450	392	343	301	266	236
	0.332	M_b	974	842	709	591	494	420	362	318	283	255	232	213	197
		M_{bs}	974	974	931	846	759	672	589	514	448	391	343	303	269

* The section can become slender under axial compression only. Under combined axial compression and bending the section becomes slender when the semi-compact F/P_z limit is exceeded.
Under combined axial compression and bending the capacities are only valid up to the given F/Pz limit. For higher values F/Pz the section would be overloaded due to F alone even when M is zero, because F would exceed the local buckling resistance of the section.
M_b is obtained using an equivalent slenderness = $u.v.L_E/r_y.\beta_w^{0.5}$
M_{bs} is obtained using an equivalent slenderness = $0.5 L/r_y$. Effective length $L_E = L$.
FOR EXPLANATION OF TABLES SEE NOTE 9.1

		AXIAL LOAD & BENDING

BS 5950-1: 2000
BS 4-1: 1993

AXIAL LOAD & BENDING

UB SECTIONS SUBJECT TO AXIAL LOAD (COMPRESSION OR TENSION) AND BENDING

CROSS-SECTION CAPACITY CHECK

CAPACITIES FOR S275

Section Designation and Axial load Capacity P_z (kN)	F/P_z Limit Semi-Compact Compact		Moment Capacity M_{cx}, M_{cy} (kNm) and Reduced Moment Capacity M_{rx}, M_{ry} (kNm) for Ratios of Axial Load to Axial Load Capacity F/P_z										
		F/P_z	0.0	0.1	0.2	0.3	0.4	0.5	0.6	0.7	0.8	0.9	1.0
610 x 229 x 113 $P_z = A_g p_y = 3820$	0.739 0.300	M_{cx} M_{cy} M_{rx} M_{ry}	869 95.7 869 95.7	869 95.7 857 95.7	869 95.7 820 95.7	869 95.7 - -	822 95.7 - -	808 95.7 - -	795 95.7 - -	774 95.7 - -	520 54.4 - -	520 54.4 - -	$ $ - -
610 x 229 x 101 $P_z = A_g p_y = 3550$	0.650 0.273	M_{cx} M_{cy} M_{rx} M_{ry}	792 84.5 792 84.5	792 84.5 781 84.5	792 84.5 749 84.5	776 84.5 - -	736 84.5 - -	718 84.5 - -	703 84.5 - -	407 41.4 - -	407 41.4 - -	$ $ - -	$ $ - -
533 x 210 x 122 $P_z = A_g p_y = 4110$	n/a 1.00	M_{cx} M_{cy} M_{rx} M_{ry}	847 102 847 102	847 102 834 102	847 102 797 102	847 102 734 102	847 102 646 102	847 102 540 102	847 102 435 102	847 102 329 96.9	847 102 221 72.1	847 102 111 39.8	847 102 0 0
533 x 210 x 109 $P_z = A_g p_y = 3680$	0.988 0.392	M_{cx} M_{cy} M_{rx} M_{ry}	749 88.7 749 88.7	749 88.7 738 88.7	749 88.7 705 88.7	749 88.7 650 88.7	747 88.7 - -	747 88.7 - -	746 88.7 - -	745 88.7 - -	743 88.7 - -	738 88.7 - -	$ $ - -
533 x 210 x 101 $P_z = A_g p_y = 3420$	0.885 0.348	M_{cx} M_{cy} M_{rx} M_{ry}	692 81.4 692 81.4	692 81.4 682 81.4	692 81.4 651 81.4	692 81.4 600 81.4	673 81.4 - -	670 81.4 - -	665 81.4 - -	658 81.4 - -	642 81.4 - -	518 57.9 - -	$ $ - -
533 x 210 x 92 $P_z = A_g p_y = 3220$	0.772 0.307	M_{cx} M_{cy} M_{rx} M_{ry}	649 75.2 649 75.2	649 75.2 640 75.2	649 75.2 612 75.2	649 75.2 565 75.2	617 75.2 - -	609 75.2 - -	601 75.2 - -	587 75.2 - -	410 45.1 - -	410 45.1 - -	$ $ - -

F = Factored axial load.
$ For these values of F/Pz the section would be overloaded due to F alone even when M is zero, because F would exceed the local buckling resistance of the section.
- Not applicable for semi-compact and slender sections.
The values in this table are conservative for tension as the more onerous compression section classification limits have been used.
FOR EXPLANATION OF TABLES SEE NOTE 9.1

BS 5950-1: 2000
BS 4-1: 1993

AXIAL LOAD & BENDING

UB SECTIONS SUBJECT TO AXIAL COMPRESSION AND BENDING

MEMBER BUCKLING CHECK

RESISTANCES AND CAPACITIES FOR S275

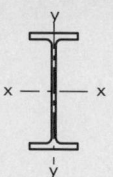

Section Designation and Capacities (kN, kNm)	F/P_z Limit	Compression Resistance P_{cx}, P_{cy} (kN) and Buckling Resistance Moment M_b, M_{bs} (kNm) for Varying effective lengths L_E (m) within the limiting value of F/P_z												
		L_E (m)	2.0	3.0	4.0	5.0	6.0	7.0	8.0	9.0	10.0	11.0	12.0	13.0	14.0
* 610 x 229 x 113	0.739	P_{cx}	3820	3820	3820	3790	3760	3720	3690	3650	3610	3560	3510	3450	3390
$P_z = A_g p_y = 3820$		P_{cy}	3460	3040	2490	1930	1480	1150	910	737	608	509	433	372	323
$p_y Z_x = 762$	0.739	M_b	762	673	573	481	404	343	296	259	230	207	187	172	158
$p_y Z_y = 79.8$		M_{bs}	762	762	723	656	586	517	451	392	341	298	261	230	204
	0.300	M_b	868	743	621	512	424	357	305	266	236	211	191	175	161
		M_{bs}	869	869	825	748	669	590	515	448	390	340	298	263	233
* 610 x 229 x 101	0.650	P_{cx}	3550	3550	3550	3520	3490	3460	3420	3380	3340	3300	3240	3190	3120
$P_z = A_g p_y = 3550$		P_{cy}	3190	2770	2230	1690	1280	990	782	632	520	436	370	318	276
$p_y Z_x = 692$	0.650	M_b	692	600	504	417	345	289	247	214	189	168	152	139	127
$p_y Z_y = 70.4$		M_{bs}	692	692	646	582	516	452	391	338	292	254	222	195	173
	0.273	M_b	781	663	545	442	361	300	254	220	193	172	155	141	130
		M_{bs}	792	792	740	667	592	517	448	387	335	291	254	223	198
533 x 210 x 122	1.00	P_{cx}	4110	4110	4100	4060	4020	3980	3930	3880	3820	3760	3690	3610	3520
$P_z = A_g p_y = 4110$		P_{cy}	3690	3200	2570	1960	1480	1150	906	732	603	505	429	368	320
$p_y Z_x = 740$	1.00	M_b	836	715	602	504	427	368	322	286	257	234	215	199	185
$p_y Z_y = 84.8$		M_{bs}	847	847	791	712	631	551	477	412	357	310	271	238	211
* 533 x 210 x 109	0.988	P_{cx}	3680	3680	3680	3640	3600	3560	3520	3480	3420	3370	3300	3230	3150
$P_z = A_g p_y = 3680$		P_{cy}	3300	2850	2270	1720	1300	1000	791	639	526	440	374	321	279
$p_y Z_x = 656$	0.988	M_b	656	568	481	405	343	295	258	229	205	186	171	157	146
$p_y Z_y = 73.9$		M_{bs}	656	656	609	547	483	421	363	313	271	235	205	180	159
	0.392	M_b	736	625	520	430	360	307	267	235	211	191	174	161	149
		M_{bs}	749	749	696	624	551	480	415	358	309	268	234	206	182
* 533 x 210 x 101	0.885	P_{cx}	3420	3420	3410	3380	3350	3310	3270	3230	3180	3130	3070	3000	2920
$P_z = A_g p_y = 3420$		P_{cy}	3060	2630	2090	1580	1190	918	725	586	482	403	342	294	255
$p_y Z_x = 607$	0.885	M_b	607	522	440	368	309	264	230	203	181	164	150	138	128
$p_y Z_y = 67.8$		M_{bs}	607	607	562	504	445	387	334	287	248	215	188	165	146
	0.348	M_b	678	573	474	389	324	274	237	208	186	168	153	141	130
		M_{bs}	692	692	641	574	507	441	380	327	283	245	214	188	166
* 533 x 210 x 92	0.772	P_{cx}	3220	3220	3210	3180	3140	3110	3070	3030	2980	2930	2870	2800	2720
$P_z = A_g p_y = 3220$		P_{cy}	2860	2440	1910	1420	1070	820	646	521	429	358	304	261	227
$p_y Z_x = 570$	0.772	M_b	564	481	401	331	275	232	200	175	156	140	128	117	108
$p_y Z_y = 62.7$		M_{bs}	570	570	521	465	408	353	303	260	223	193	168	148	130
	0.307	M_b	629	528	430	349	286	240	206	180	159	143	130	119	110
		M_{bs}	649	649	594	530	465	402	346	290	254	220	192	168	148

* The section can become slender under axial compression only. Under combined axial compression and bending the section becomes slender when the semi-compact F/P_z limit is exceeded.
Under combined axial compression and bending the capacities are only valid up to the given F/P_z limit. For higher values F/P_z the section would be overloaded due to F alone even when M is zero, because F would exceed the local buckling resistance of the section.
M_b is obtained using an equivalent slenderness = $u.v.L_E/r_y.\beta_w^{0.5}$
M_{bs} is obtained using an equivalent slenderness = $0.5 L/r_y$. Effective length $L_E = L$.
FOR EXPLANATION OF TABLES SEE NOTE 9.1

BS 5950-1: 2000
BS 4-1: 1993

AXIAL LOAD & BENDING

UB SECTIONS SUBJECT TO AXIAL LOAD (COMPRESSION OR TENSION) AND BENDING

CROSS-SECTION CAPACITY CHECK

CAPACITIES FOR S275

Section Designation and Axial load Capacity P_z (kN)	F/P_z Limit Semi-Compact Compact		Moment Capacity M_{cx}, M_{cy} (kNm) and Reduced Moment Capacity M_{rx}, M_{ry} (kNm) for Ratios of Axial Load to Axial Load Capacity F/P_z										
		F/P_z	0.0	0.1	0.2	0.3	0.4	0.5	0.6	0.7	0.8	0.9	1.0
533 x 210 x 82 $P_z = A_g p_y = 2890$	0.709 **0.295**	M_{cx} M_{cy} M_{rx} M_{ry}	566 63.4 566 63.4	566 63.4 558 63.4	566 63.4 535 63.4	564 63.4 - -	533 63.4 - -	521 63.4 - -	512 63.4 - -	497 63.4 - -	321 34.3 - -	321 34.3 - -	$ $ - -
457 x 191 x 98 $P_z = A_g p_y = 3310$	n/a **1.00**	M_{cx} M_{cy} M_{rx} M_{ry}	591 77.3 591 77.3	591 77.3 582 77.3	591 77.3 555 77.3	591 77.3 510 77.3	591 77.3 445 77.3	591 77.3 373 77.3	591 77.3 301 77.3	591 77.3 227 72	591 77.3 153 53.3	591 77.3 76.8 29.3	591 77.3 0 0
457 x 191 x 89 $P_z = A_g p_y = 3020$	n/a **1.00**	M_{cx} M_{cy} M_{rx} M_{ry}	534 69.3 534 69.3	534 69.3 526 69.3	534 69.3 501 69.3	534 69.3 460 69.3	534 69.3 404 69.3	534 69.3 339 69.3	534 69.3 273 69.3	534 69.3 206 65.1	534 69.3 138 48.2	534 69.3 69.5 26.6	534 69.3 0 0
457 x 191 x 82 $P_z = A_g p_y = 2860$	0.957 **0.370**	M_{cx} M_{cy} M_{rx} M_{ry}	504 64.7 504 64.7	504 64.7 496 64.7	504 64.7 473 64.7	504 64.7 436 64.7	498 64.7 - -	497 64.7 - -	496 64.7 - -	494 64.7 - -	490 64.7 - -	477 64.7 - -	$ $ - -
457 x 191 x 74 $P_z = A_g p_y = 2600$	0.825 **0.312**	M_{cx} M_{cy} M_{rx} M_{ry}	455 58.1 455 58.1	455 58.1 448 58.1	455 58.1 427 58.1	455 58.1 393 58.1	436 58.1 - -	433 58.1 - -	429 58.1 - -	422 58.1 - -	407 58.1 - -	313 37.8 - -	$ $ - -
457 x 191 x 67 $P_z = A_g p_y = 2350$	0.751 **0.293**	M_{cx} M_{cy} M_{rx} M_{ry}	405 50.5 405 50.5	405 50.5 399 50.5	405 50.5 381 50.5	403 50.5 - -	382 50.5 - -	378 50.5 - -	373 50.5 - -	364 50.5 - -	248 29.3 - -	248 29.3 - -	$ $ - -

F = Factored axial load.
$ For these values of F/Pz the section would be overloaded due to F alone even when M is zero, because F would exceed the local buckling resistance of the section.
- Not applicable for semi-compact and slender sections.
The values in this table are conservative for tension as the more onerous compression section classification limits have been used.
FOR EXPLANATION OF TABLES SEE NOTE 9.1

BS 5950-1: 2000
BS 4-1: 1993

AXIAL LOAD & BENDING

UB SECTIONS SUBJECT TO AXIAL COMPRESSION AND BENDING

MEMBER BUCKLING CHECK

RESISTANCES AND CAPACITIES FOR S275

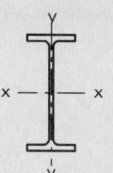

Section Designation and Capacities (kN, kNm)	F/P_z Limit	Compression Resistance P_{cx}, P_{cy} (kN) and Buckling Resistance Moment M_b, M_{bs} (kNm) for Varying effective lengths L_E (m) within the limiting value of F/P_z												
		L_E (m)	2.0	3.0	4.0	5.0	6.0	7.0	8.0	9.0	10.0	11.0	12.0	13.0	14.0
* 533 x 210 x 82 $P_z = A_g p_y = 2890$ $p_y Z_x = 495$ $p_y Z_y = 52.8$	0.709	P_{cx}	2890	2890	2880	2850	2820	2790	2750	2710	2670	2620	2560	2500	2420
		P_{cy}	2550	2150	1660	1220	913	699	550	443	364	304	258	222	192
	0.709	M_b	487	412	340	276	227	190	162	141	125	112	101	92.4	85.1
		M_{bs}	495	495	447	397	346	297	254	217	186	160	139	122	108
	0.295	M_b	545	453	364	291	236	196	167	144	127	114	103	93.9	86.5
		M_{bs}	566	566	512	454	396	340	290	248	212	183	159	140	123
457 x 191 x 98 $P_z = A_g p_y = 3310$ $p_y Z_x = 519$ $p_y Z_y = 64.4$	1.00	P_{cx}	3310	3310	3290	3250	3210	3170	3120	3060	3000	2930	2850	2750	2640
		P_{cy}	2920	2470	1910	1410	1060	810	638	514	422	353	300	257	223
	1.00	M_b	570	481	401	334	283	245	215	192	173	158	145	135	125
		M_{bs}	591	591	535	475	414	356	304	260	223	193	168	147	130
457 x 191 x 89 $P_z = A_g p_y = 3020$ $p_y Z_x = 469$ $p_y Z_y = 57.8$	1.00	P_{cx}	3020	3020	3000	2960	2930	2890	2840	2790	2730	2670	2590	2500	2400
		P_{cy}	2660	2240	1720	1270	948	726	572	461	379	316	268	230	200
	1.00	M_b	512	429	353	291	245	209	183	162	146	133	122	112	105
		M_{bs}	534	534	481	426	370	318	271	231	199	171	149	131	115
* 457 x 191 x 82 $P_z = A_g p_y = 2860$ $p_y Z_x = 443$ $p_y Z_y = 53.9$	0.957	P_{cx}	2860	2860	2840	2800	2770	2730	2680	2630	2580	2510	2430	2340	2240
		P_{cy}	2500	2080	1570	1150	852	651	511	411	338	282	239	205	178
	0.957	M_b	430	363	301	249	208	178	155	137	123	112	102	94.3	87.6
		M_{bs}	443	439	394	347	300	256	217	184	158	136	118	103	90.8
	0.370	M_b	478	396	322	261	217	184	159	141	126	114	104	96	89.1
		M_{bs}	504	499	448	395	341	291	247	210	179	154	134	117	103
* 457 x 191 x 74 $P_z = A_g p_y = 2600$ $p_y Z_x = 401$ $p_y Z_y = 48.4$	0.825	P_{cx}	2600	2600	2580	2550	2520	2480	2440	2400	2340	2280	2210	2130	2030
		P_{cy}	2270	1880	1420	1030	766	585	459	369	303	253	215	184	160
	0.825	M_b	388	326	267	218	181	154	133	117	104	94.3	86.1	79.2	73.4
		M_{bs}	401	396	355	313	270	230	195	165	141	121	105	92.2	81.2
	0.312	M_b	430	354	284	229	188	158	136	120	106	96.1	87.6	80.5	74.6
		M_{bs}	455	449	403	354	306	260	221	187	160	138	119	104	92.1
* 457 x 191 x 67 $P_z = A_g p_y = 2350$ $p_y Z_x = 356$ $p_y Z_y = 42.1$	0.751	P_{cx}	2350	2350	2330	2300	2270	2240	2200	2160	2110	2050	1990	1910	1820
		P_{cy}	2040	1680	1250	906	670	510	400	322	264	221	187	160	139
	0.751	M_b	343	286	232	187	154	129	111	96.9	86.1	77.4	70.4	64.6	59.7
		M_{bs}	356	350	313	274	235	200	169	143	122	105	90.7	79.3	69.8
	0.293	M_b	380	310	246	196	159	133	113	99	87.7	78.8	71.6	65.6	60.6
		M_{bs}	405	397	355	311	267	226	191	162	138	119	103	90	79.3

* The section can become slender under axial compression only. Under combined axial compression and bending the section becomes slender when the semi-compact F/P_z limit is exceeded.
Under combined axial compression and bending the capacities are only valid up to the given F/P_z limit. For higher values F/P_z the section would be overloaded due to F alone even when M is zero, because F would exceed the local buckling resistance of the section.
M_b is obtained using an equivalent slenderness = $u.v.L_E/r_y.\beta_w^{0.5}$
M_{bs} is obtained using an equivalent slenderness = $0.5 L/r_y$. Effective length $L_E = L$.
FOR EXPLANATION OF TABLES SEE NOTE 9.1

BS 5950-1: 2000
BS 4-1: 1993

AXIAL LOAD & BENDING

UB SECTIONS SUBJECT TO AXIAL LOAD (COMPRESSION OR TENSION) AND BENDING

CROSS-SECTION CAPACITY CHECK

CAPACITIES FOR S275

Section Designation and Axial load Capacity P_z (kN)	F/P_z Limit Semi-Compact Compact		Moment Capacity M_{cx}, M_{cy} (kNm) and Reduced Moment Capacity M_{rx}, M_{ry} (kNm) for Ratios of Axial Load to Axial Load Capacity F/P_z										
		F/P_z	0.0	0.1	0.2	0.3	0.4	0.5	0.6	0.7	0.8	0.9	1.0
457 x 152 x 82 $P_z = A_g p_y = 2780$	n/a 1.00	M_{cx}	480	480	480	480	480	480	480	480	480	480	480
		M_{cy}	48.7	48.7	48.7	48.7	48.7	48.7	48.7	48.7	48.7	48.7	48.7
		M_{rx}	480	473	452	417	369	312	252	190	128	64.3	0
		M_{ry}	48.7	48.7	48.7	48.7	48.7	48.7	48.7	47.4	35.5	19.7	0
457 x 152 x 74 $P_z = A_g p_y = 2500$	0.940 0.386	M_{cx}	431	431	431	431	428	423	422	419	413	397	$
		M_{cy}	43.2	43.2	43.2	43.2	43.2	43.2	43.2	43.2	43.2	43.2	$
		M_{rx}	431	425	407	376	-	-	-	-	-	-	-
		M_{ry}	43.2	43.2	43.2	43.2	-	-	-	-	-	-	-
457 x 152 x 67 $P_z = A_g p_y = 2350$	0.825 0.345	M_{cx}	400	400	400	400	386	379	375	368	353	271	$
		M_{cy}	39.3	39.3	39.3	39.3	39.3	39.3	39.3	39.3	39.3	25.5	$
		M_{rx}	400	394	377	349	-	-	-	-	-	-	-
		M_{ry}	39.3	39.3	39.3	39.3	-	-	-	-	-	-	-
457 x 152 x 60 $P_z = A_g p_y = 2100$	0.692 0.285	M_{cx}	354	354	354	350	331	324	318	195	195	195	$
		M_{cy}	34.3	34.3	34.3	34.3	34.3	34.3	34.3	18.1	18.1	18.1	$
		M_{rx}	354	349	334	-	-	-	-	-	-	-	-
		M_{ry}	34.3	34.3	34.3	-	-	-	-	-	-	-	-
457 x 152 x 52 $P_z = A_g p_y = 1830$	0.619 0.268	M_{cx}	301	301	301	294	278	269	263	145	145	$	$
		M_{cy}	27.9	27.9	27.9	27.9	27.9	27.9	27.9	12.9	12.9	$	$
		M_{rx}	301	297	285	-	-	-	-	-	-	-	-
		M_{ry}	27.9	27.9	27.9	-	-	-	-	-	-	-	-
406 x 178 x 74 $P_z = A_g p_y = 2600$	n/a 1.00	M_{cx}	413	413	413	413	413	413	413	413	413	413	413
		M_{cy}	56.8	56.8	56.8	56.8	56.8	56.8	56.8	56.8	56.8	56.8	56.8
		M_{rx}	413	406	387	355	310	260	209	158	106	53.3	0
		M_{ry}	56.8	56.8	56.8	56.8	56.8	56.8	56.8	52.7	39	21.4	0

F = Factored axial load.
$ For these values of F/Pz the section would be overloaded due to F alone even when M is zero, because F would exceed the local buckling resistance of the section.
- Not applicable for semi-compact and slender sections.
The values in this table are conservative for tension as the more onerous compression section classification limits have been used.
FOR EXPLANATION OF TABLES SEE NOTE 9.1

BS 5950-1: 2000
BS 4-1: 1993

AXIAL LOAD & BENDING

UB SECTIONS SUBJECT TO AXIAL COMPRESSION AND BENDING

MEMBER BUCKLING CHECK

RESISTANCES AND CAPACITIES FOR S275

Section Designation and Capacities (kN, kNm)	F/P_z Limit		Compression Resistance P_{cx}, P_{cy} (kN) and Buckling Resistance Moment M_b, M_{bs} (kNm) for Varying effective lengths L_E (m) within the limiting value of F/P_z											
			L_E (m)	1.0	1.5	2.0	2.5	3.0	3.5	4.0	4.5	5.0	5.5	6.0	6.5	7.0
457 x 152 x 82	1.00	P_{cx}	2780	2780	2780	2780	2780	2780	2760	2740	2730	2710	2690	2670	2650
$P_z = A_g p_y = 2780$		P_{cy}	2660	2480	2250	1980	1670	1380	1140	943	790	669	573	495	432
$p_y Z_x = 416$	**1.00**	M_b	480	469	422	376	333	295	263	235	212	193	177	164	152
$p_y Z_y = 40.5$		M_{bs}	480	480	480	471	441	410	378	347	315	286	258	233	211
* 457 x 152 x 74	0.940	P_{cx}	2500	2500	2500	2500	2500	2500	2480	2470	2450	2440	2420	2400	2390
$P_z = A_g p_y = 2500$		P_{cy}	2390	2220	2020	1760	1480	1220	1010	832	696	589	504	436	380
$p_y Z_x = 375$	0.940	M_b	375	372	337	303	270	241	215	193	174	158	145	134	124
$p_y Z_y = 36$		M_{bs}	375	375	375	367	343	318	293	268	243	220	199	179	162
	0.386	M_b	431	419	375	333	293	257	227	202	182	164	150	138	128
		M_{bs}	431	431	431	422	394	366	337	308	280	253	229	206	186
* 457 x 152 x 67	0.825	P_{cx}	2350	2350	2350	2350	2350	2340	2330	2320	2300	2290	2270	2260	2240
$P_z = A_g p_y = 2350$		P_{cy}	2230	2080	1870	1620	1340	1100	896	738	616	520	445	384	335
$p_y Z_x = 347$	0.825	M_b	347	341	308	274	242	213	188	167	150	135	123	113	105
$p_y Z_y = 32.7$		M_{bs}	347	347	347	336	313	290	266	242	218	197	177	159	143
	0.345	M_b	400	384	342	300	261	227	198	175	156	140	127	116	107
		M_{bs}	400	400	400	386	360	333	306	278	251	226	204	183	165
* 457 x 152 x 60	0.692	P_{cx}	2100	2100	2100	2100	2100	2090	2070	2060	2050	2040	2020	2010	1990
$P_z = A_g p_y = 2100$		P_{cy}	1990	1840	1650	1420	1180	959	782	644	537	453	387	334	291
$p_y Z_x = 309$	0.692	M_b	309	302	271	240	210	184	161	142	126	114	103	94	86.5
$p_y Z_y = 28.6$		M_{bs}	309	309	309	297	277	255	234	212	192	172	155	139	125
	0.285	M_b	354	338	300	262	226	195	169	148	131	117	106	96.4	88.5
		M_{bs}	354	354	354	341	317	293	268	244	220	198	177	159	143
* 457 x 152 x 52	0.619	P_{cx}	1830	1830	1830	1830	1830	1820	1810	1800	1790	1780	1760	1750	1740
$P_z = A_g p_y = 1830$		P_{cy}	1730	1590	1420	1200	983	793	643	527	439	370	316	272	237
$p_y Z_x = 261$	0.619	M_b	261	253	226	198	172	149	129	113	99.4	88.6	79.8	72.4	66.3
$p_y Z_y = 23.3$		M_{bs}	261	261	261	248	230	211	192	174	156	139	125	112	100
	0.268	M_b	301	285	250	216	184	157	135	117	103	91.3	81.9	74.2	67.8
		M_{bs}	301	301	301	286	266	244	222	200	180	161	144	129	116
406 x 178 x 74	1.00	P_{cx}	2600	2600	2600	2600	2600	2580	2560	2550	2530	2510	2490	2470	2450
$P_z = A_g p_y = 2600$		P_{cy}	2530	2400	2240	2050	1830	1580	1350	1140	971	830	715	622	545
$p_y Z_x = 364$	**1.00**	M_b	413	413	386	351	318	286	257	232	210	192	176	162	150
$p_y Z_y = 47.3$		M_{bs}	413	413	413	413	403	381	359	336	313	290	267	246	226

* The section can become slender under axial compression only. Under combined axial compression and bending the section becomes slender when the semi-compact F/P_z limit is exceeded.
Under combined axial compression and bending the capacities are only valid up to the given F/P_z limit. For higher values F/P_z the section would be overloaded due to F alone even when M is zero, because F would exceed the local buckling resistance of the section.
M_b is obtained using an equivalent slenderness = $u.v.L_E/r_y.\beta_w^{0.5}$
M_{bs} is obtained using an equivalent slenderness = $0.5 L/r_y$. Effective length $L_E = L$.
FOR EXPLANATION OF TABLES SEE NOTE 9.1

BS 5950-1: 2000
BS 4-1: 1993

AXIAL LOAD & BENDING

UB SECTIONS SUBJECT TO AXIAL LOAD (COMPRESSION OR TENSION) AND BENDING

CROSS-SECTION CAPACITY CHECK

CAPACITIES FOR S275

Section Designation and Axial load Capacity P_z (kN)	F/P_z Limit Semi-Compact Compact		Moment Capacity M_{cx}, M_{cy} (kNm) and Reduced Moment Capacity M_{rx}, M_{ry} (kNm) for Ratios of Axial Load to Axial Load Capacity F/P_z										
		F/P_z	0.0	0.1	0.2	0.3	0.4	0.5	0.6	0.7	0.8	0.9	1.0
406 x 178 x 67 $P_z = A_g p_y = 2350$	0.965 **0.357**	M_{cx}	370	370	370	370	367	367	366	365	362	355	$
		M_{cy}	50.5	50.5	50.5	50.5	50.5	50.5	50.5	50.5	50.5	50.5	$
		M_{rx}	370	364	347	319	-	-	-	-	-	-	-
		M_{ry}	50.5	50.5	50.5	50.5	-	-	-	-	-	-	-
406 x 178 x 60 $P_z = A_g p_y = 2100$	0.815 **0.296**	M_{cx}	330	330	330	329	316	314	311	306	295	225	$
		M_{cy}	44.6	44.6	44.6	44.6	44.6	44.6	44.6	44.6	44.6	28.5	$
		M_{rx}	330	325	309	-	-	-	-	-	-	-	-
		M_{ry}	44.6	44.6	44.6	-	-	-	-	-	-	-	-
406 x 178 x 54 $P_z = A_g p_y = 1900$	0.782 **0.305**	M_{cx}	290	290	290	290	276	273	270	264	187	187	$
		M_{cy}	38	38	38	38	38	38	38	38	23.1	23.1	$
		M_{rx}	290	286	273	252	-	-	-	-	-	-	-
		M_{ry}	38	38	38	38	-	-	-	-	-	-	-
406 x 140 x 46 $P_z = A_g p_y = 1610$	0.632 **0.247**	M_{cx}	244	244	244	235	225	221	216	122	122	$	$
		M_{cy}	25	25	25	25	25	25	25	11.9	11.9	$	$
		M_{rx}	244	241	230	-	-	-	-	-	-	-	-
		M_{ry}	25	25	25	-	-	-	-	-	-	-	-
406 x 140 x 39 $P_z = A_g p_y = 1370$	0.565 **0.240**	M_{cx}	199	199	199	190	181	176	87.3	87.3	87.3	$	$
		M_{cy}	19.1	19.1	19.1	19.1	19.1	19.1	8.02	8.02	8.02	$	$
		M_{rx}	199	196	188	-	-	-	-	-	-	-	-
		M_{ry}	19.1	19.1	19.1	-	-	-	-	-	-	-	-
356 x 171 x 67 $P_z = A_g p_y = 2350$	n/a **1.00**	M_{cx}	333	333	333	333	333	333	333	333	333	333	333
		M_{cy}	51.8	51.8	51.8	51.8	51.8	51.8	51.8	51.8	51.8	51.8	51.8
		M_{rx}	333	328	311	283	246	206	166	126	84.3	42.4	0
		M_{ry}	51.8	51.8	51.8	51.8	51.8	51.8	51.8	46.7	34.3	18.8	0
356 x 171 x 57 $P_z = A_g p_y = 2000$	n/a **1.00**	M_{cx}	278	278	278	278	278	278	278	278	278	278	278
		M_{cy}	42.6	42.6	42.6	42.6	42.6	42.6	42.6	42.6	42.6	42.6	42.6
		M_{rx}	278	273	260	237	207	173	140	105	70.6	35.5	0
		M_{ry}	42.6	42.6	42.6	42.6	42.6	42.6	39	28.8	15.8	0	0

F = Factored axial load.
$ For these values of F/P_z the section would be overloaded due to F alone even when M is zero, because F would exceed the local buckling resistance of the section.
- Not applicable for semi-compact and slender sections.
The values in this table are conservative for tension as the more onerous compression section classification limits have been used.
FOR EXPLANATION OF TABLES SEE NOTE 9.1

BS 5950-1: 2000
BS 4-1: 1993

AXIAL LOAD & BENDING

UB SECTIONS SUBJECT TO AXIAL COMPRESSION AND BENDING

MEMBER BUCKLING CHECK

RESISTANCES AND CAPACITIES FOR S275

Section Designation and Capacities (kN, kNm)	F/P_z Limit		Compression Resistance P_{cx}, P_{cy} (kN) and Buckling Resistance Moment M_b, M_{bs} (kNm) for Varying effective lengths L_E (m) within the limiting value of F/P_z											
			L_E (m)	1.0	1.5	2.0	2.5	3.0	3.5	4.0	4.5	5.0	5.5	6.0	6.5	7.0
* 406 x 178 x 67	0.965	P_{cx}	2350	2350	2350	2350	2350	2330	2320	2300	2290	2270	2250	2230	2210
$P_z = A_g p_y = 2350$		P_{cy}	2280	2160	2020	1850	1640	1410	1200	1010	861	735	633	550	482
$p_y Z_x = 327$	0.965	M_b	327	327	311	285	259	234	211	191	173	158	145	133	123
$p_y Z_y = 42.1$		M_{bs}	327	327	327	327	318	301	283	264	246	227	209	192	176
	0.357	M_b	370	370	344	312	281	251	224	201	181	164	150	138	127
		M_{bs}	370	370	370	370	360	340	320	299	278	257	237	217	199
* 406 x 178 x 60	0.815	P_{cx}	2100	2100	2100	2100	2100	2090	2070	2060	2050	2030	2010	2000	1980
$P_z = A_g p_y = 2100$		P_{cy}	2040	1930	1810	1650	1460	1260	1070	901	764	652	562	488	427
$p_y Z_x = 292$	0.815	M_b	292	292	277	253	229	206	185	166	149	135	123	113	104
$p_y Z_y = 37.1$		M_{bs}	292	292	292	292	284	268	252	235	219	202	186	171	157
	0.296	M_b	330	330	305	276	247	220	195	174	156	140	128	117	107
		M_{bs}	330	330	330	330	320	302	284	266	247	228	210	193	177
* 406 x 178 x 54	0.782	P_{cx}	1900	1900	1900	1900	1890	1880	1870	1860	1840	1830	1810	1800	1780
$P_z = A_g p_y = 1900$		P_{cy}	1840	1730	1610	1460	1280	1090	922	775	655	559	480	417	364
$p_y Z_x = 256$	0.782	M_b	256	256	240	218	196	176	156	139	125	112	102	93	85.4
$p_y Z_y = 31.6$		M_{bs}	256	256	256	256	246	232	217	202	187	172	158	144	132
	0.305	M_b	290	290	266	239	212	188	165	146	130	117	105	95.7	87.7
		M_{bs}	290	290	290	290	279	263	246	229	212	195	179	164	149
* 406 x 140 x 46	0.632	P_{cx}	1610	1610	1610	1610	1610	1600	1590	1580	1560	1550	1540	1530	1510
$P_z = A_g p_y = 1610$		P_{cy}	1510	1390	1230	1030	836	671	542	444	369	310	265	228	199
$p_y Z_x = 214$	0.632	M_b	214	204	181	158	137	118	103	90	79.8	71.5	64.6	59	54.2
$p_y Z_y = 20.8$		M_{bs}	214	214	214	201	186	170	154	139	124	110	98.5	88	78.8
	0.247	M_b	244	228	199	171	146	124	107	93.3	82.3	73.5	66.3	60.3	55.3
		M_{bs}	244	244	244	230	212	194	176	158	141	126	112	100	90
* 406 x 140 x 39	0.565	P_{cx}	1370	1370	1370	1370	1360	1350	1340	1330	1320	1310	1300	1290	1280
$P_z = A_g p_y = 1370$		P_{cy}	1270	1160	1010	827	659	523	420	342	284	238	203	175	152
$p_y Z_x = 173$	0.565	M_b	173	163	143	123	105	89.2	76.5	66.3	58.2	51.7	46.4	42	38.4
$p_y Z_y = 15.9$		M_{bs}	173	173	172	159	146	133	119	106	94	83.2	73.8	65.7	58.6
	0.240	M_b	199	182	157	133	111	93.7	79.6	68.6	60	53	47.5	42.9	39.1
		M_{bs}	199	199	198	184	168	153	137	122	108	95.8	85	75.6	67.5
356 x 171 x 67	1.00	P_{cx}	2350	2350	2350	2350	2340	2320	2300	2290	2270	2250	2230	2200	2180
$P_z = A_g p_y = 2350$		P_{cy}	2280	2160	2020	1850	1640	1410	1200	1010	861	735	633	550	482
$p_y Z_x = 295$	1.00	M_b	333	333	310	283	257	232	210	191	174	159	147	136	127
$p_y Z_y = 43.2$		M_{bs}	333	333	333	333	324	306	288	269	250	231	213	196	179
356 x 171 x 57	1.00	P_{cx}	2000	2000	2000	2000	1980	1970	1950	1940	1920	1910	1890	1870	1840
$P_z = A_g p_y = 2000$		P_{cy}	1930	1830	1710	1550	1370	1170	991	835	707	603	519	451	394
$p_y Z_x = 246$	1.00	M_b	278	278	256	232	208	187	167	150	135	123	112	103	95.7
$p_y Z_y = 35.5$		M_{bs}	278	278	278	278	268	253	238	222	205	189	174	159	146

* The section can become slender under axial compression only. Under combined axial compression and bending the section becomes slender when the semi-compact F/P_z limit is exceeded.
Under combined axial compression and bending the capacities are only valid up to the given F/P_z limit. For higher values F/P_z the section would be overloaded due to F alone even when M is zero, because F would exceed the local buckling resistance of the section.
M_b is obtained using an equivalent slenderness = $u.v.L_E/r_y.\beta_w^{0.5}$
M_{bs} is obtained using an equivalent slenderness = $0.5 L/r_y$. Effective length $L_E = L$.
FOR EXPLANATION OF TABLES SEE NOTE 9.1

| BS 5950-1: 2000 |
| BS 4-1: 1993 |

AXIAL LOAD & BENDING

UB SECTIONS SUBJECT TO AXIAL LOAD (COMPRESSION OR TENSION) AND BENDING

CROSS-SECTION CAPACITY CHECK

CAPACITIES FOR S275

Section Designation and Axial load Capacity P_z (kN)	F/P_z Limit Semi-Compact Compact	\multicolumn{12}{c}{Moment Capacity M_{cx}, M_{cy} (kNm) and Reduced Moment Capacity M_{rx}, M_{ry} (kNm) for Ratios of Axial Load to Axial Load Capacity F/P_z}
		F/P_z	0.0	0.1	0.2	0.3	0.4	0.5	0.6	0.7	0.8	0.9	1.0
356 x 171 x 51 $P_z = A_g p_y = 1780$	0.925 0.326	M_{cx} M_{cy} M_{rx} M_{ry}	246 37.3 246 37.3	246 37.3 242 37.3	246 37.3 231 37.3	246 37.3 211 37.3	242 37.3 - -	242 37.3 - -	241 37.3 - -	239 37.3 - -	236 37.3 - -	226 37.3 - -	$ $ - -
356 x 171 x 45 $P_z = A_g p_y = 1580$	0.848 0.316	M_{cx} M_{cy} M_{rx} M_{ry}	213 31.3 213 31.3	213 31.3 210 31.3	213 31.3 200 31.3	213 31.3 184 31.3	206 31.3 - -	205 31.3 - -	203 31.3 - -	200 31.3 - -	194 31.3 - -	153 21.1 - -	$ $ - -
356 x 127 x 39 $P_z = A_g p_y = 1370$	0.771 0.308	M_{cx} M_{cy} M_{rx} M_{ry}	181 18.7 181 18.7	181 18.7 179 18.7	181 18.7 171 18.7	181 18.7 158 18.7	172 18.7 - -	170 18.7 - -	167 18.7 - -	163 18.7 - -	114 11.2 - -	114 11.2 - -	$ $ - -
356 x 127 x 33 $P_z = A_g p_y = 1160$	0.655 0.274	M_{cx} M_{cy} M_{rx} M_{ry}	149 14.8 149 14.8	149 14.8 147 14.8	149 14.8 141 14.8	146 14.8 - -	139 14.8 - -	135 14.8 - -	132 14.8 - -	77.2 7.29 - -	77.2 7.29 - -	$ $ - -	$ $ - -
305 x 165 x 54 $P_z = A_g p_y = 1890$	n/a 1.00	M_{cx} M_{cy} M_{rx} M_{ry}	233 41.9 233 41.9	233 41.9 229 41.9	233 41.9 216 41.9	233 41.9 196 41.9	233 41.9 169 41.9	233 41.9 142 41.9	233 41.9 114 41.9	233 41.9 86.3 36.7	233 41.9 57.9 26.8	233 41.9 29.2 14.6	233 41.9 0 0
305 x 165 x 46 $P_z = A_g p_y = 1610$	n/a 1.00	M_{cx} M_{cy} M_{rx} M_{ry}	198 35.6 198 35.6	198 35.6 194 35.6	198 35.6 184 35.6	198 35.6 166 35.6	198 35.6 143 35.6	198 35.6 120 35.6	198 35.6 96.7 35.6	198 35.6 73 31.1	198 35.6 48.9 22.7	198 35.6 24.6 12.4	198 35.6 0 0
305 x 165 x 40 $P_z = A_g p_y = 1410$	0.857 0.261	M_{cx} M_{cy} M_{rx} M_{ry}	171 30.6 171 30.6	171 30.6 168 30.6	171 30.6 159 30.6	168 30.6 - -	166 30.6 - -	166 30.6 - -	165 30.6 - -	163 30.6 - -	159 30.6 - -	126 20.9 - -	$ $ - -

F = Factored axial load.
$ For these values of F/Pz the section would be overloaded due to F alone even when M is zero, because F would exceed the local buckling resistance of the section.
- Not applicable for semi-compact and slender sections.
The values in this table are conservative for tension as the more onerous compression section classification limits have been used.
FOR EXPLANATION OF TABLES SEE NOTE 9.1

BS 5950-1: 2000
BS 4-1: 1993

AXIAL LOAD & BENDING

UB SECTIONS SUBJECT TO AXIAL COMPRESSION AND BENDING

MEMBER BUCKLING CHECK

RESISTANCES AND CAPACITIES FOR S275

Section Designation and Capacities (kN, kNm)	F/P_z Limit		Compression Resistance P_{cx}, P_{cy} (kN) and Buckling Resistance Moment M_b, M_{bs} (kNm) for Varying effective lengths L_E (m) within the limiting value of F/P_z												
		L_E (m)	1.0	1.5	2.0	2.5	3.0	3.5	4.0	4.5	5.0	5.5	6.0	6.5	7.0
* 356 x 171 x 51	0.925	P_{cx}	1780	1780	1780	1780	1770	1760	1750	1730	1720	1700	1680	1670	1650
$P_z = A_g p_y = 1780$		P_{cy}	1730	1630	1520	1380	1210	1030	870	732	619	528	454	394	344
$p_y Z_x = 219$	0.925	M_b	219	219	205	187	169	151	136	122	110	99.9	91.2	83.8	77.4
$p_y Z_y = 31.1$		M_{bs}	219	219	219	219	211	198	186	173	160	148	135	124	113
	0.326	M_b	246	246	226	203	182	161	143	128	115	103	94.1	86.2	79.5
		M_{bs}	246	246	246	246	237	223	209	195	180	166	152	139	127
* 356 x 171 x 45	0.848	P_{cx}	1580	1580	1580	1580	1560	1550	1540	1530	1510	1500	1480	1470	1450
$P_z = A_g p_y = 1580$		P_{cy}	1520	1430	1330	1200	1040	884	741	621	523	445	383	332	290
$p_y Z_x = 189$	0.848	M_b	189	189	176	159	143	127	113	101	90.1	81.2	73.6	67.3	61.9
$p_y Z_y = 26.1$		M_{bs}	189	189	189	189	180	169	158	147	135	124	114	104	94.4
	0.316	M_b	213	213	193	173	153	135	119	105	93.5	83.9	75.9	69.1	63.4
		M_{bs}	213	213	213	213	203	191	178	166	153	140	128	117	107
* 356 x 127 x 39	0.771	P_{cx}	1370	1370	1370	1370	1360	1350	1340	1330	1310	1300	1290	1270	1260
$P_z = A_g p_y = 1370$		P_{cy}	1260	1130	959	766	598	469	374	304	251	211	179	154	134
$p_y Z_x = 158$	0.771	M_b	158	145	126	108	92	79	68.5	60.3	53.7	48.3	43.9	40.3	37.2
$p_y Z_y = 15.6$		M_{bs}	158	158	154	142	129	115	102	89.9	79	69.4	61.2	54.2	48.2
	0.308	M_b	181	161	138	116	97.2	82.5	71.1	62.2	55.2	49.6	45	41.2	38
		M_{bs}	181	181	177	162	147	132	117	103	90.3	79.4	70	62	55.1
* 356 x 127 x 33	0.655	P_{cx}	1160	1160	1160	1160	1150	1140	1130	1120	1110	1100	1080	1070	1060
$P_z = A_g p_y = 1160$		P_{cy}	1060	942	786	618	477	372	296	240	198	166	141	121	105
$p_y Z_x = 130$	0.655	M_b	130	117	101	85	71.4	60.3	51.7	45	39.7	35.4	32	29.1	26.8
$p_y Z_y = 12.3$		M_{bs}	130	130	125	114	103	91.6	80.7	70.6	61.7	54	47.5	41.9	37.2
	0.274	M_b	149	130	110	90.9	75.2	62.9	53.5	46.3	40.7	36.3	32.7	29.7	27.3
		M_{bs}	149	149	144	131	118	105	92.6	81	70.8	62	54.5	48.1	42.7
305 x 165 x 54	1.00	P_{cx}	1890	1890	1890	1880	1870	1850	1830	1820	1800	1780	1750	1730	1700
$P_z = A_g p_y = 1890$		P_{cy}	1830	1740	1620	1470	1300	1120	946	798	676	577	496	431	377
$p_y Z_x = 207$	1.00	M_b	233	233	216	196	178	161	146	132	121	111	102	94.9	88.6
$p_y Z_y = 34.9$		M_{bs}	233	233	233	233	225	212	199	186	173	159	147	134	123
305 x 165 x 46	1.00	P_{cx}	1610	1610	1610	1610	1590	1580	1570	1550	1530	1520	1500	1470	1450
$P_z = A_g p_y = 1610$		P_{cy}	1560	1480	1380	1250	1100	945	799	673	569	486	418	363	317
$p_y Z_x = 178$	1.00	M_b	198	198	182	165	148	133	119	107	96.8	88.2	80.8	74.5	69.2
$p_y Z_y = 29.7$		M_{bs}	198	198	198	198	191	180	169	158	146	135	124	113	104
* 305 x 165 x 40	0.857	P_{cx}	1410	1410	1410	1400	1390	1380	1370	1350	1340	1320	1310	1290	1260
$P_z = A_g p_y = 1410$		P_{cy}	1360	1290	1200	1090	955	816	688	579	489	417	359	311	272
$p_y Z_x = 154$	0.857	M_b	154	154	144	131	118	106	94.6	85	76.7	69.7	63.7	58.5	54.1
$p_y Z_y = 25.5$		M_{bs}	154	154	154	154	148	140	131	122	113	104	95.2	87.1	79.6
	0.261	M_b	171	171	156	141	126	112	99.3	88.6	79.5	71.9	65.5	60.1	55.4
		M_{bs}	171	171	171	171	165	155	145	135	125	116	106	96.9	88.6

* The section can become slender under axial compression only. Under combined axial compression and bending the section becomes slender when the semi-compact F/P_z limit is exceeded.
Under combined axial compression and bending the capacities are only valid up to the given F/P_z limit. For higher values F/P_z the section would be overloaded due to F alone even when M is zero, because F would exceed the local buckling resistance of the section.
M_b is obtained using an equivalent slenderness = $u.v.L_E/r_y.\beta_w^{0.5}$
M_{bs} is obtained using an equivalent slenderness = $0.5 L/r_y$. Effective length $L_E = L$.
FOR EXPLANATION OF TABLES SEE NOTE 9.1

BS 5950-1: 2000
BS 4-1: 1993

AXIAL LOAD & BENDING

UB SECTIONS SUBJECT TO AXIAL LOAD (COMPRESSION OR TENSION) AND BENDING

CROSS-SECTION CAPACITY CHECK

CAPACITIES FOR S275

Section Designation and Axial load Capacity P_z (kN)	F/P_z Limit Semi-Compact Compact		Moment Capacity M_{cx}, M_{cy} (kNm) and Reduced Moment Capacity M_{rx}, M_{ry} (kNm) for Ratios of Axial Load to Axial Load Capacity F/P_z										
		F/P_z	0.0	0.1	0.2	0.3	0.4	0.5	0.6	0.7	0.8	0.9	1.0
305 x 127 x 48 $P_z = A_g p_y = 1680$	n/a 1.00	M_{cx}	196	196	196	196	196	196	196	196	196	196	196
		M_{cy}	24.3	24.3	24.3	24.3	24.3	24.3	24.3	24.3	24.3	24.3	24.3
		M_{rx}	196	193	184	170	150	126	101	76.7	51.5	26	0
		M_{ry}	24.3	24.3	24.3	24.3	24.3	24.3	24.3	23.4	17.4	9.62	0
305 x 127 x 42 $P_z = A_g p_y = 1470$	n/a 1.00	M_{cx}	169	169	169	169	169	169	169	169	169	169	169
		M_{cy}	20.7	20.7	20.7	20.7	20.7	20.7	20.7	20.7	20.7	20.7	20.7
		M_{rx}	169	166	159	147	130	109	87.7	66.2	44.5	22.4	0
		M_{ry}	20.7	20.7	20.7	20.7	20.7	20.7	20.7	20.1	15	8.32	0
305 x 127 x 37 $P_z = A_g p_y = 1300$	n/a 1.00	M_{cx}	148	148	148	148	148	148	148	148	148	148	148
		M_{cy}	18	18	18	18	18	18	18	18	18	18	18
		M_{rx}	148	146	140	129	114	95.7	77	58.1	39	19.6	0
		M_{ry}	18	18	18	18	18	18	18	17.6	13.2	7.29	0
305 x 102 x 33 $P_z = A_g p_y = 1150$	0.935 0.404	M_{cx}	132	132	132	132	132	130	129	128	126	121	$
		M_{cy}	12.5	12.5	12.5	12.5	12.5	12.5	12.5	12.5	12.5	12.5	$
		M_{rx}	132	130	125	116	103	-	-	-	-	-	-
		M_{ry}	12.5	12.5	12.5	12.5	12.5	-	-	-	-	-	-
305 x 102 x 28 $P_z = A_g p_y = 987$	0.805 0.361	M_{cx}	111	111	111	111	108	104	103	101	96	72.4	$
		M_{cy}	10.1	10.1	10.1	10.1	10.1	10.1	10.1	10.1	10.1	6.35	$
		M_{rx}	111	109	105	97.5	-	-	-	-	-	-	-
		M_{ry}	10.1	10.1	10.1	10.1	-	-	-	-	-	-	-
305 x 102 x 25 $P_z = A_g p_y = 869$	0.761 0.372	M_{cx}	94.1	94.1	94.1	94.1	92.1	87	85.3	82.9	56.8	56.8	$
		M_{cy}	7.99	7.99	7.99	7.99	7.99	7.99	7.99	7.99	4.71	4.71	$
		M_{rx}	94.1	92.9	89.3	83.4	-	-	-	-	-	-	-
		M_{ry}	7.99	7.99	7.99	7.99	-	-	-	-	-	-	-
254 x 146 x 43 $P_z = A_g p_y = 1510$	n/a 1.00	M_{cx}	156	156	156	156	156	156	156	156	156	156	156
		M_{cy}	30.4	30.4	30.4	30.4	30.4	30.4	30.4	30.4	30.4	30.4	30.4
		M_{rx}	156	153	144	130	112	94.3	76	57.4	38.6	19.4	0
		M_{ry}	30.4	30.4	30.4	30.4	30.4	30.4	30.4	26	18.9	10.3	0
254 x 146 x 37 $P_z = A_g p_y = 1300$	n/a 1.00	M_{cx}	133	133	133	133	133	133	133	133	133	133	133
		M_{cy}	25.7	25.7	25.7	25.7	25.7	25.7	25.7	25.7	25.7	25.7	25.7
		M_{rx}	133	130	123	111	95.9	80.5	64.8	48.9	32.8	16.5	0
		M_{ry}	25.7	25.7	25.7	25.7	25.7	25.7	25.7	22.2	16.2	8.8	0

F = Factored axial load.
$ For these values of F/P_z the section would be overloaded due to F alone even when M is zero, because F would exceed the local buckling resistance of the section.
- Not applicable for semi-compact and slender sections.
The values in this table are conservative for tension as the more onerous compression section classification limits have been used.
FOR EXPLANATION OF TABLES SEE NOTE 9.1

BS 5950-1: 2000
BS 4-1: 1993

AXIAL LOAD & BENDING

UB SECTIONS SUBJECT TO AXIAL COMPRESSION AND BENDING

MEMBER BUCKLING CHECK

RESISTANCES AND CAPACITIES FOR S275

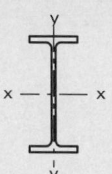

Section Designation and Capacities (kN, kNm)	F/P_z Limit	\multicolumn{13}{c}{Compression Resistance P_{cx}, P_{cy} (kN) and Buckling Resistance Moment M_b, M_{bs} (kNm) for Varying effective lengths L_E (m) within the limiting value of F/P_z}												
		L_E (m)	1.0	1.5	2.0	2.5	3.0	3.5	4.0	4.5	5.0	5.5	6.0	6.5	7.0
305 x 127 x 48	1.00	P_{cx}	1680	1680	1680	1670	1660	1640	1630	1610	1590	1570	1550	1520	1500
$P_z = A_g p_y = 1680$		P_{cy}	1560	1400	1200	967	759	597	477	388	321	270	229	197	172
$p_y Z_x = 169$	1.00	M_b	196	178	156	136	118	104	92.4	83	75.3	68.9	63.5	59	55
$p_y Z_y = 20.2$		M_{bs}	196	196	192	177	161	145	129	114	100	88.4	78	69.2	61.6
305 x 127 x 42	1.00	P_{cx}	1470	1470	1470	1460	1450	1430	1420	1400	1390	1370	1350	1330	1300
$P_z = A_g p_y = 1470$		P_{cy}	1350	1220	1030	829	648	509	406	330	273	229	195	168	146
$p_y Z_x = 147$	1.00	M_b	169	152	132	113	97.4	84.7	74.5	66.4	59.8	54.4	49.9	46.1	42.9
$p_y Z_y = 17.2$		M_{bs}	169	169	165	152	138	124	110	96.7	85	74.8	66	58.4	52
305 x 127 x 37	1.00	P_{cx}	1300	1300	1300	1290	1280	1270	1250	1240	1220	1210	1190	1170	1150
$P_z = A_g p_y = 1300$		P_{cy}	1200	1070	906	723	563	442	352	286	236	198	169	145	126
$p_y Z_x = 130$	1.00	M_b	148	132	114	96.6	82.2	70.7	61.7	54.5	48.8	44.2	40.4	37.2	34.5
$p_y Z_y = 15$		M_{bs}	148	148	144	132	120	107	95.2	83.8	73.5	64.6	56.9	50.4	44.8
* 305 x 102 x 33	0.935	P_{cx}	1150	1150	1150	1140	1130	1120	1110	1100	1090	1070	1060	1040	1020
$P_z = A_g p_y = 1150$		P_{cy}	1010	846	645	473	352	269	212	170	140	117	99.2	85.1	73.9
$p_y Z_x = 114$	0.935	M_b	112	95.7	79.9	66.4	55.8	47.8	41.6	36.8	33	29.9	27.4	25.3	23.5
$p_y Z_y = 10.4$		M_{bs}	114	114	103	90.6	78.6	67.3	57.2	48.8	41.7	36	31.3	27.4	24.1
	0.404	M_b	127	106	86.4	70.5	58.5	49.6	43	37.9	33.9	30.6	28	25.8	23.9
		M_{bs}	132	132	119	105	90.9	77.8	66.2	56.4	48.3	41.6	36.1	31.6	27.9
* 305 x 102 x 28	0.805	P_{cx}	987	987	987	980	972	962	953	942	931	918	904	889	871
$P_z = A_g p_y = 987$		P_{cy}	859	710	532	386	286	218	171	138	113	94.4	80	68.6	59.5
$p_y Z_x = 95.7$	0.805	M_b	93	78.4	64.3	52.5	43.4	36.6	31.5	27.6	24.5	22.1	20.1	18.5	17.1
$p_y Z_y = 8.39$		M_{bs}	95.7	94.3	84.4	74.1	63.8	54.2	45.8	38.9	33.2	28.5	24.8	21.6	19.1
	0.361	M_b	105	86.6	69.3	55.4	45.2	37.9	32.4	28.3	25.1	22.6	20.5	18.8	17.4
		M_{bs}	111	109	97.8	85.8	73.9	62.8	53.1	45	38.4	33.1	28.7	25.1	22.1
* 305 x 102 x 25	0.761	P_{cx}	869	869	869	862	854	846	837	827	817	805	792	778	761
$P_z = A_g p_y = 869$		P_{cy}	744	599	436	312	229	174	136	109	89.8	74.9	63.5	54.4	47.2
$p_y Z_x = 80.3$	0.761	M_b	77.1	64.1	51.7	41.5	33.8	28.2	24	20.9	18.5	16.6	15	13.7	12.7
$p_y Z_y = 6.65$		M_{bs}	80.3	77.8	68.9	59.7	50.7	42.6	35.7	30.1	25.6	21.9	19	16.6	14.6
	0.372	M_b	87.8	71.1	55.7	43.8	35.2	29.1	24.7	21.4	18.9	16.9	15.3	14	12.9
		M_{bs}	94.1	91.1	80.8	70	59.4	49.9	41.8	35.3	30	25.7	22.2	19.4	17.1
254 x 146 x 43	1.00	P_{cx}	1510	1510	1500	1490	1470	1460	1440	1420	1400	1370	1350	1310	1280
$P_z = A_g p_y = 1510$		P_{cy}	1440	1350	1240	1100	934	777	643	534	448	380	325	281	246
$p_y Z_x = 139$	1.00	M_b	156	153	139	125	113	102	92	83.6	76.5	70.4	65.2	60.7	56.8
$p_y Z_y = 25.3$		M_{bs}	156	156	156	154	145	135	125	115	105	95.7	86.8	78.6	71.3
254 x 146 x 37	1.00	P_{cx}	1300	1300	1290	1280	1270	1250	1240	1220	1200	1180	1160	1130	1100
$P_z = A_g p_y = 1300$		P_{cy}	1240	1160	1060	936	795	659	544	452	378	321	274	237	207
$p_y Z_x = 119$	1.00	M_b	133	130	117	105	93.5	83.3	74.5	67.1	60.9	55.7	51.2	47.5	44.2
$p_y Z_y = 21.5$		M_{bs}	133	133	133	131	123	115	106	97.4	88.8	80.7	73.1	66.1	59.9

* The section can become slender under axial compression only. Under combined axial compression and bending the section becomes slender when the semi-compact F/P_z limit is exceeded.
Under combined axial compression and bending the capacities are only valid up to the given F/P_z limit. For higher values F/P_z the section would be overloaded due to F alone even when M is zero, because F would exceed the local buckling resistance of the section.
M_b is obtained using an equivalent slenderness = $u.v.L_E/r_y.\beta_w^{0.5}$
M_{bs} is obtained using an equivalent slenderness = $0.5 L/r_y$. Effective length $L_E = L$.
FOR EXPLANATION OF TABLES SEE NOTE 9.1

| BS 5950-1: 2000 |
| BS 4-1: 1993 |

AXIAL LOAD & BENDING

UB SECTIONS SUBJECT TO AXIAL LOAD (COMPRESSION OR TENSION) AND BENDING

CROSS-SECTION CAPACITY CHECK

CAPACITIES FOR S275

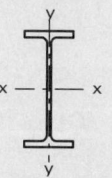

Section Designation and Axial load Capacity P_z (kN)	F/P_z Limit Semi-Compact Compact		Moment Capacity M_{cx}, M_{cy} (kNm) and Reduced Moment Capacity M_{rx}, M_{ry} (kNm) for Ratios of Axial Load to Axial Load Capacity F/P_z										
		F/P_z	0.0	0.1	0.2	0.3	0.4	0.5	0.6	0.7	0.8	0.9	1.0
254 x 146 x 31 $P_z = A_g p_y = 1090$	n/a 1.00	M_{cx} M_{cy} M_{rx} M_{ry}	108 20.2 108 20.2	108 20.2 106 20.2	108 20.2 101 20.2	108 20.2 91.8 20.2	108 20.2 79.7 20.2	108 20.2 66.8 20.2	108 20.2 53.7 20.2	108 20.2 40.5 20.2	108 20.2 27.1 18.3	108 20.2 13.6 13.4	108 20.2 0 7.35
254 x 102 x 28 $P_z = A_g p_y = 993$	n/a 1.00	M_{cx} M_{cy} M_{rx} M_{ry}	97.1 11.5 97.1 11.5	97.1 11.5 95.7 11.5	97.1 11.5 91.4 11.5	97.1 11.5 84.3 11.5	97.1 11.5 74.3 11.5	97.1 11.5 62.4 11.5	97.1 11.5 50.3 11.5	97.1 11.5 38.0 11.5	97.1 11.5 25.5 11.2	97.1 11.5 12.8 8.35	97.1 11.5 0 4.62
254 x 102 x 25 $P_z = A_g p_y = 880$	n/a 1.00	M_{cx} M_{cy} M_{rx} M_{ry}	84.2 9.64 84.2 9.64	84.2 9.64 83.0 9.64	84.2 9.64 79.5 9.64	84.2 9.64 73.6 9.64	84.2 9.64 65.4 9.64	84.2 9.64 54.9 9.64	84.2 9.64 44.2 9.64	84.2 9.64 33.3 9.64	84.2 9.64 22.4 9.64	84.2 9.64 11.2 7.29	84.2 9.64 0 4.06
254 x 102 x 22 $P_z = A_g p_y = 770$	n/a 1.00	M_{cx} M_{cy} M_{rx} M_{ry}	71.2 7.76 71.2 7.76	71.2 7.76 70.3 7.76	71.2 7.76 67.4 7.76	71.2 7.76 62.7 7.76	71.2 7.76 56.1 7.76	71.2 7.76 47.6 7.76	71.2 7.76 38.3 7.76	71.2 7.76 28.9 7.76	71.2 7.76 19.3 7.76	71.2 7.76 9.73 6.24	71.2 7.76 0 3.52
203 x 133 x 30 $P_z = A_g p_y = 1050$	n/a 1.00	M_{cx} M_{cy} M_{rx} M_{ry}	86.4 19.0 86.4 19.0	86.4 19.0 84.8 19.0	86.4 19.0 80.1 19.0	86.4 19.0 72.2 19.0	86.4 19.0 62.5 19.0	86.4 19.0 52.4 19.0	86.4 19.0 42.2 19.0	86.4 19.0 31.9 16.4	86.4 19.0 21.4 12.0	86.4 19.0 10.8 6.51	86.4 19.0 0 0
203 x 133 x 25 $P_z = A_g p_y = 880$	n/a 1.00	M_{cx} M_{cy} M_{rx} M_{ry}	71.0 15.2 71.0 15.2	71.0 15.2 69.7 15.2	71.0 15.2 66.0 15.2	71.0 15.2 59.8 15.2	71.0 15.2 51.7 15.2	71.0 15.2 43.4 15.2	71.0 15.2 34.9 15.2	71.0 15.2 26.3 15.2	71.0 15.2 17.7 13.5	71.0 15.2 8.89 9.92	71.0 15.2 0 5.41
203 x 102 x 23 $P_z = A_g p_y = 809$	n/a 1.00	M_{cx} M_{cy} M_{rx} M_{ry}	64.4 10.6 64.4 10.6	64.4 10.6 63.2 10.6	64.4 10.6 59.9 10.6	64.4 10.6 54.4 10.6	64.4 10.6 47.2 10.6	64.4 10.6 39.6 10.6	64.4 10.6 31.9 10.6	64.4 10.6 24.1 9.47	64.4 10.6 16.2 6.95	64.4 10.6 8.16 3.80	64.4 10.6 0 0

F = Factored axial load.
$ For these values of F/Pz the section would be overloaded due to F alone even when M is zero, because F would exceed the local buckling resistance of the section.
- Not applicable for semi-compact and slender sections.
The values in this table are conservative for tension as the more onerous compression section classification limits have been used.
FOR EXPLANATION OF TABLES SEE NOTE 9.1

BS 5950-1: 2000
BS 4-1: 1993

AXIAL LOAD & BENDING

UB SECTIONS SUBJECT TO AXIAL COMPRESSION AND BENDING

MEMBER BUCKLING CHECK

RESISTANCES AND CAPACITIES FOR S275

Section Designation and Capacities (kN, kNm)	F/P_z Limit		Compression Resistance P_{cx}, P_{cy} (kN) and Buckling Resistance Moment M_b, M_{bs} (kNm) for Varying effective lengths L_E (m) within the limiting value of F/P_z											
		L_E (m)	1.0	1.5	2.0	2.5	3.0	3.5	4.0	4.5	5.0	5.5	6.0	6.5	7.0
254 x 146 x 31	1.00	P_{cx}	1090	1090	1090	1080	1060	1050	1040	1020	1010	987	965	939	910
$P_z = A_g p_y = 1090$		P_{cy}	1040	969	879	766	643	529	434	358	300	253	217	187	163
$p_y Z_x = 96.5$	1.00	M_b	108	105	93.5	82.7	72.6	63.7	56.1	49.9	44.8	40.5	37.0	34.0	31.4
$p_y Z_y = 16.9$		M_{bs}	108	108	108	105	98.6	91.5	84.3	77.0	69.9	63.2	57.0	51.4	46.4
254 x 102 x 28	1.00	P_{cx}	993	993	989	979	968	957	944	930	915	897	877	854	828
$P_z = A_g p_y = 993$		P_{cy}	879	746	579	429	321	246	194	156	128	107	91.0	78.2	67.8
$p_y Z_x = 84.7$	1.00	M_b	94.0	79.4	65.9	54.7	46.0	39.5	34.5	30.7	27.6	25.1	23.0	21.3	19.8
$p_y Z_y = 9.6$		M_{bs}	97.1	97.1	88.2	78.5	68.6	59.1	50.6	43.3	37.2	32.1	27.9	24.5	21.6
254 x 102 x 25	1.00	P_{cx}	880	880	876	867	857	846	835	822	808	792	773	752	727
$P_z = A_g p_y = 880$		P_{cy}	773	648	494	362	270	206	162	130	107	89.6	75.9	65.2	56.5
$p_y Z_x = 73.2$	1.00	M_b	80.7	67.4	55.0	44.9	37.3	31.6	27.4	24.1	21.6	19.5	17.9	16.4	15.3
$p_y Z_y = 8.03$		M_{bs}	84.2	83.7	75.4	66.7	57.8	49.5	42.1	35.9	30.7	26.5	23.0	20.1	17.7
254 x 102 x 22	1.00	P_{cx}	770	770	766	757	749	739	729	717	704	690	672	652	629
$P_z = A_g p_y = 770$		P_{cy}	668	550	410	297	219	167	131	105	86.5	72.3	61.2	52.5	45.6
$p_y Z_x = 61.6$	1.00	M_b	67.4	55.5	44.5	35.6	29.2	24.5	21.0	18.4	16.3	14.7	13.4	12.3	11.4
$p_y Z_y = 6.46$		M_{bs}	71.2	70.0	62.6	54.8	47.0	39.9	33.7	28.5	24.3	20.9	18.1	15.8	14.0
203 x 133 x 30	1.00	P_{cx}	1050	1050	1040	1020	1010	993	974	953	927	897	861	820	774
$P_z = A_g p_y = 1050$		P_{cy}	993	918	822	702	577	468	380	313	260	220	188	162	141
$p_y Z_x = 77$	1.00	M_b	86.4	82.6	74.1	66.1	58.9	52.6	47.3	42.8	39.1	35.9	33.2	30.9	28.9
$p_y Z_y = 15.8$		M_{bs}	86.4	86.4	86.4	82.6	76.8	70.7	64.5	58.4	52.6	47.1	42.2	37.9	34.0
203 x 133 x 25	1.00	P_{cx}	880	879	868	857	844	830	814	795	772	746	715	678	638
$P_z = A_g p_y = 880$		P_{cy}	829	764	680	576	470	379	307	252	210	177	151	130	113
$p_y Z_x = 63.3$	1.00	M_b	71.0	67.1	59.6	52.5	46.0	40.5	35.9	32.1	29.0	26.5	24.3	22.5	20.9
$p_y Z_y = 12.7$		M_{bs}	71.0	71.0	71.0	67.4	62.4	57.3	52.1	47.0	42.2	37.7	33.7	30.2	27.1
203 x 102 x 23	1.00	P_{cx}	809	808	797	786	775	761	746	728	707	682	652	618	580
$P_z = A_g p_y = 809$		P_{cy}	726	630	504	382	289	223	176	142	117	98.1	83.3	71.5	62.1
$p_y Z_x = 56.9$	1.00	M_b	63.4	54.6	46.5	39.6	34.1	29.8	26.4	23.7	21.5	19.7	18.1	16.8	15.7
$p_y Z_y = 8.86$		M_{bs}	64.4	64.4	60.0	54.0	47.8	41.7	36.1	31.2	26.9	23.4	20.4	18.0	15.9

* The section can become slender under axial compression only. Under combined axial compression and bending the section becomes slender when the semi-compact F/P_z limit is exceeded.
Under combined axial compression and bending the capacities are only valid up to the given F/P_z limit. For higher values F/P_z the section would be overloaded due to F alone even when M is zero, because F would exceed the local buckling resistance of the section.
M_b is obtained using an equivalent slenderness = $u.v.L_E/r_y.\beta_w^{0.5}$
M_{bs} is obtained using an equivalent slenderness = $0.5 L/r_y$. Effective length $L_E = L$.
FOR EXPLANATION OF TABLES SEE NOTE 9.1

BS 5950-1: 2000
BS 4-1: 1993

AXIAL LOAD & BENDING

UB SECTIONS SUBJECT TO AXIAL LOAD (COMPRESSION OR TENSION) AND BENDING

CROSS-SECTION CAPACITY CHECK

CAPACITIES FOR S275

Section Designation and Axial load Capacity P_z (kN)	F/P_z Limit Semi-Compact Compact		\multicolumn{11}{c}{Moment Capacity M_{cx}, M_{cy} (kNm) and Reduced Moment Capacity M_{rx}, M_{ry} (kNm) for Ratios of Axial Load to Axial Load Capacity F/P_z}
		F/P_z	0.0	0.1	0.2	0.3	0.4	0.5	0.6	0.7	0.8	0.9	1.0
178 x 102 x 19 $P_z = A_g p_y = 668$	n/a 1.00	M_{cx}	47.0	47.0	47.0	47.0	47.0	47.0	47.0	47.0	47.0	47.0	47.0
		M_{cy}	8.91	8.91	8.91	8.91	8.91	8.91	8.91	8.91	8.91	8.91	8.91
		M_{rx}	47.0	46.2	43.6	39.4	34.2	28.7	23.1	17.5	11.7	5.90	0
		M_{ry}	8.91	8.91	8.91	8.91	8.91	8.91	8.91	7.83	5.73	3.12	0
152 x 89 x 16 $P_z = A_g p_y = 558$	n/a 1.00	M_{cx}	33.8	33.8	33.8	33.8	33.8	33.8	33.8	33.8	33.8	33.8	33.8
		M_{cy}	6.67	6.67	6.67	6.67	6.67	6.67	6.67	6.67	6.67	6.67	6.67
		M_{rx}	33.8	33.2	31.3	28.2	24.4	20.5	16.5	12.5	8.38	4.22	0
		M_{ry}	6.67	6.67	6.67	6.67	6.67	6.67	6.67	5.77	4.22	2.29	0
127 x 76 x 13 $P_z = A_g p_y = 454$	n/a 1.00	M_{cx}	23.2	23.2	23.2	23.2	23.2	23.2	23.2	23.2	23.2	23.2	23.2
		M_{cy}	4.85	4.85	4.85	4.85	4.85	4.85	4.85	4.85	4.85	4.85	4.85
		M_{rx}	23.2	22.7	21.3	19.0	16.4	13.8	11.1	8.42	5.66	2.86	0
		M_{ry}	4.85	4.85	4.85	4.85	4.85	4.85	4.85	4.06	2.96	1.60	0

F = Factored axial load.
$ For these values of F/Pz the section would be overloaded due to F alone even when M is zero, because F would exceed the local buckling resistance of the section.
- Not applicable for semi-compact and slender sections.
The values in this table are conservative for tension as the more onerous compression section classification limits have been used.
FOR EXPLANATION OF TABLES SEE NOTE 9.1

BS 5950-1: 2000
BS 4-1: 1993

AXIAL LOAD & BENDING

UB SECTIONS SUBJECT TO AXIAL COMPRESSION AND BENDING

MEMBER BUCKLING CHECK

RESISTANCES AND CAPACITIES FOR S275

Section Designation and Capacities (kN, kNm)	F/P_z Limit		Compression Resistance P_{cx}, P_{cy} (kN) and Buckling Resistance Moment M_b, M_{bs} (kNm) for Varying effective lengths L_E (m) within the limiting value of F/P_z												
		L_E (m)	1.0	1.5	2.0	2.5	3.0	3.5	4.0	4.5	5.0	5.5	6.0	6.5	7.0
178 x 102 x 19	1.00	P_{cx}	668	664	654	644	632	618	601	581	556	527	493	456	418
$P_z = A_g p_y = 668$		P_{cy}	601	522	418	318	241	186	147	119	97.6	81.7	69.4	59.6	51.8
$p_y Z_x = 42.1$	1.00	M_b	46.4	40.0	34.0	29.0	25.0	21.8	19.3	17.3	15.7	14.4	13.3	12.3	11.5
$p_y Z_y = 7.43$		M_{bs}	47.0	47.0	43.9	39.5	35.1	30.6	26.5	22.9	19.8	17.2	15.0	13.2	11.7
152 x 89 x 16	1.00	P_{cx}	558	551	541	529	516	500	479	453	422	386	349	313	280
$P_z = A_g p_y = 558$		P_{cy}	487	404	304	222	164	125	98.5	79.2	65.1	54.3	46.1	39.5	34.3
$p_y Z_x = 30$	1.00	M_b	32.3	27.5	23.2	19.8	17.2	15.1	13.5	12.2	11.1	10.2	9.44	8.79	8.23
$p_y Z_y = 5.56$		M_{bs}	33.8	33.4	30.0	26.4	22.7	19.4	16.4	13.9	11.9	10.2	8.89	7.78	6.85
127 x 76 x 13	1.00	P_{cx}	452	443	432	420	403	382	354	320	284	250	219	192	169
$P_z = A_g p_y = 454$		P_{cy}	379	294	207	145	106	80.3	62.7	50.3	41.2	34.3	29.1	24.9	21.6
$p_y Z_x = 20.5$	1.00	M_b	21.3	18.0	15.3	13.2	11.6	10.3	9.22	8.37	7.67	7.08	6.57	6.14	5.76
$p_y Z_y = 4.04$		M_{bs}	23.2	21.9	19.1	16.3	13.6	11.3	9.34	7.82	6.61	5.64	4.87	4.24	3.72

* The section can become slender under axial compression only. Under combined axial compression and bending the section becomes slender when the semi-compact F/P_z limit is exceeded.
Under combined axial compression and bending the capacities are only valid up to the given F/P_z limit. For higher values F/P_z the section would be overloaded due to F alone even when M is zero, because F would exceed the local buckling resistance of the section.
M_b is obtained using an equivalent slenderness = $u.v.L_E/r_y.\beta_w^{0.5}$
M_{bs} is obtained using an equivalent slenderness = $0.5 L/r_y$. Effective length $L_E = L$.
FOR EXPLANATION OF TABLES SEE NOTE 9.1

BS 5950-1: 2000
BS 4-1: 1993

AXIAL LOAD & BENDING

UC SECTIONS SUBJECT TO AXIAL LOAD (COMPRESSION OR TENSION) AND BENDING

CROSS-SECTION CAPACITY CHECK

CAPACITIES FOR S275

Section Designation and Axial load Capacity P_z (kN)	F/P_z Limit Semi-Compact Compact	Moment Capacity M_{cx}, M_{cy} (kNm) and Reduced Moment Capacity M_{rx}, M_{ry} (kNm) for Ratios of Axial Load to Axial Load Capacity F/P_z											
		F/P_z	0.0	0.1	0.2	0.3	0.4	0.5	0.6	0.7	0.8	0.9	1.0
356 x 406 x 634 # $P_z = A_g p_y = 19800$	n/a 1.00	M_{cx}	3400	3400	3400	3400	3400	3400	3400	3400	3400	3400	3400
		M_{cy}	1360	1360	1360	1360	1360	1360	1360	1360	1360	1360	1360
		M_{rx}	3400	3400	3150	2830	2480	2110	1730	1320	902	460	0
		M_{ry}	1360	1360	1360	1360	1360	1360	1360	1260	1030	735	394	0
356 x 406 x 551 # $P_z = A_g p_y = 17200$	n/a 1.00	M_{cx}	2930	2930	2930	2930	2930	2930	2930	2930	2930	2930	2930
		M_{cy}	1160	1160	1160	1160	1160	1160	1160	1160	1160	1160	1160
		M_{rx}	2930	2890	2670	2390	2090	1780	1450	1110	755	385	0
		M_{ry}	1160	1160	1160	1160	1160	1160	1160	1080	878	630	338	0
356 x 406 x 467 # $P_z = A_g p_y = 15200$	n/a 1.00	M_{cx}	2550	2550	2550	2550	2550	2550	2550	2550	2550	2550	2550
		M_{cy}	1010	1010	1010	1010	1010	1010	1010	1010	1010	1010	1010
		M_{rx}	2550	2490	2300	2050	1790	1520	1240	944	641	326	0
		M_{ry}	1010	1010	1010	1010	1010	1010	1010	940	763	548	293	0
356 x 406 x 393 # $P_z = A_g p_y = 12800$	n/a 1.00	M_{cx}	2100	2100	2100	2100	2100	2100	2100	2100	2100	2100	2100
		M_{cy}	833	833	833	833	833	833	833	833	833	833	833
		M_{rx}	2100	2040	1890	1680	1460	1240	1010	768	520	264	0
		M_{ry}	833	833	833	833	833	833	833	780	634	455	244	0
356 x 406 x 340 # $P_z = A_g p_y = 11000$	n/a 1.00	M_{cx}	1780	1780	1780	1780	1780	1780	1780	1780	1780	1780	1780
		M_{cy}	711	711	711	711	711	711	711	711	711	711	711
		M_{rx}	1780	1740	1610	1430	1240	1050	850	646	437	221	0
		M_{ry}	711	711	711	711	711	711	711	667	542	389	209	0
356 x 406 x 287 # $P_z = A_g p_y = 9700$	n/a 1.00	M_{cx}	1540	1540	1540	1540	1540	1540	1540	1540	1540	1540	1540
		M_{cy}	617	617	617	617	617	617	617	617	617	617	617
		M_{rx}	1540	1500	1380	1230	1070	899	728	553	373	189	0
		M_{ry}	617	617	617	617	617	617	617	579	471	338	181	0
356 x 406 x 235 # $P_z = A_g p_y = 7920$	n/a 1.00	M_{cx}	1240	1240	1240	1240	1240	1240	1240	1240	1240	1240	1240
		M_{cy}	499	499	499	499	499	499	499	499	499	499	499
		M_{rx}	1240	1210	1110	983	852	717	580	439	296	149	0
		M_{ry}	499	499	499	499	499	499	499	469	381	274	147	0
356 x 368 x 202 # $P_z = A_g p_y = 6810$	n/a 1.00	M_{cx}	1050	1050	1050	1050	1050	1050	1050	1050	1050	1050	1050
		M_{cy}	402	402	402	402	402	402	402	402	402	402	402
		M_{rx}	1050	1030	947	836	723	609	492	372	250	126	0
		M_{ry}	402	402	402	402	402	402	381	310	223	119	0
356 x 368 x 177 # $P_z = A_g p_y = 5990$	n/a 1.00	M_{cx}	916	916	916	916	916	916	916	916	916	916	916
		M_{cy}	350	350	350	350	350	350	350	350	350	350	350
		M_{rx}	916	892	822	727	629	529	426	323	217	109	0
		M_{ry}	350	350	350	350	350	350	333	271	195	104	0

\# Check availability.
F = Factored axial load.
- Not applicable for semi-compact and slender sections.
The values in this table are conservative for tension as the more onerous compression section classification limits have been used.
FOR EXPLANATION OF TABLES SEE NOTE 9.1

BS 5950-1: 2000
BS 4-1: 1993

AXIAL LOAD & BENDING

UC SECTIONS SUBJECT TO AXIAL COMPRESSION AND BENDING

MEMBER BUCKLING CHECK

RESISTANCES AND CAPACITIES FOR S275

Section Designation and Capacities (kN, kNm)	F/P_z Limit		Compression Resistance P_{cx}, P_{cy} (kN) and Buckling Resistance Moment M_b, M_{bs} (kNm) for Varying effective lengths L_E (m) within the limiting value of F/P_z											
		L_E (m)	2.0	3.0	4.0	5.0	6.0	7.0	8.0	9.0	10.0	11.0	12.0	13.0	14.0
356 x 406 x 634 #	1.00	P_{cx}	19800	19800	19400	18800	18200	17500	16900	16200	15500	14800	14100	13300	12600
$P_z = A_g p_y = 19800$		P_{cy}	19800	18300	16900	15600	14200	12900	11600	10500	9390	8420	7560	6800	6140
$p_y Z_x = 2840$	1.00	M_b	3400	3400	3400	3400	3400	3400	3400	3400	3400	3350	3290	3240	3190
$p_y Z_y = 1130$		M_{bs}	3400	3400	3400	3400	3400	3400	3400	3360	3220	3090	2950	2810	2670
356 x 406 x 551 #	1.00	P_{cx}	17200	17200	16800	16300	15700	15100	14600	14000	13300	12700	12000	11400	10700
$P_z = A_g p_y = 17200$		P_{cy}	17200	15900	14700	13500	12300	11100	10000	9010	8080	7240	6490	5840	5260
$p_y Z_x = 2440$	1.00	M_b	2930	2930	2930	2930	2930	2930	2930	2900	2840	2790	2740	2690	2640
$p_y Z_y = 968$		M_{bs}	2930	2930	2930	2930	2930	2930	2930	2840	2720	2610	2490	2370	2250
356 x 406 x 467 #	1.00	P_{cx}	15200	15200	14700	14200	13700	13200	12700	12100	11500	10900	10300	9720	9120
$P_z = A_g p_y = 15200$		P_{cy}	15100	13900	12800	11700	10700	9610	8620	7700	6870	6130	5480	4910	4420
$p_y Z_x = 2140$	1.00	M_b	2550	2550	2550	2550	2550	2550	2550	2490	2430	2370	2320	2280	2230	2180
$p_y Z_y = 839$		M_{bs}	2550	2550	2550	2550	2550	2550	2510	2410	2310	2210	2100	2000	1890
356 x 406 x 393 #	1.00	P_{cx}	12800	12800	12400	12100	11700	11300	10900	10500	10000	9550	9040	8520	7990
$P_z = A_g p_y = 12800$		P_{cy}	12700	11800	11000	10200	9290	8430	7580	6770	6030	5370	4780	4270	3830
$p_y Z_x = 1780$	1.00	M_b	2100	2100	2100	2100	2100	2100	2050	2000	1950	1900	1850	1810	1760	1720
$p_y Z_y = 694$		M_{bs}	2100	2100	2100	2100	2100	2100	2050	1970	1880	1800	1710	1620	1530
356 x 406 x 340 #	1.00	P_{cx}	11000	11000	10700	10400	10100	9750	9390	9010	8600	8170	7720	7260	6790
$P_z = A_g p_y = 11000$		P_{cy}	10900	10200	9480	8740	7990	7230	6490	5800	5150	4580	4080	3640	3260
$p_y Z_x = 1540$	1.00	M_b	1780	1780	1780	1780	1780	1780	1720	1670	1620	1570	1530	1490	1450	1410
$p_y Z_y = 593$		M_{bs}	1780	1780	1780	1780	1780	1780	1740	1670	1600	1520	1450	1370	1290
356 x 406 x 287 #	1.00	P_{cx}	9700	9670	9460	9230	8990	8730	8450	8130	7780	7400	7000	6570	6130
$P_z = A_g p_y = 9700$		P_{cy}	9590	9050	8490	7900	7260	6590	5920	5270	4670	4130	3660	3250	2900
$p_y Z_x = 1340$	1.00	M_b	1540	1540	1540	1540	1500	1440	1390	1350	1300	1260	1220	1190	1150
$p_y Z_y = 514$		M_{bs}	1540	1540	1540	1540	1540	1490	1430	1360	1300	1230	1160	1100
356 x 406 x 235 #	1.00	P_{cx}	7920	7900	7720	7530	7330	7110	6880	6610	6320	6010	5660	5310	4940
$P_z = A_g p_y = 7920$		P_{cy}	7830	7380	6920	6430	5900	5350	4790	4260	3770	3330	2950	2610	2330
$p_y Z_x = 1100$	1.00	M_b	1240	1240	1240	1240	1190	1140	1090	1050	1010	969	934	900	869
$p_y Z_y = 416$		M_{bs}	1240	1240	1240	1240	1240	1200	1150	1090	1040	985	930	876
356 x 368 x 202 #	1.00	P_{cx}	6810	6780	6630	6460	6290	6100	5890	5660	5400	5120	4820	4510	4190
$P_z = A_g p_y = 6810$		P_{cy}	6680	6270	5840	5390	4890	4390	3890	3420	3000	2630	2310	2050	1820
$p_y Z_x = 938$	1.00	M_b	1050	1050	1050	1020	973	926	883	842	804	769	736	705	676
$p_y Z_y = 335$		M_{bs}	1050	1050	1050	1050	1050	1040	993	946	898	849	800	751	703
356 x 368 x 177 #	1.00	P_{cx}	5990	5960	5820	5670	5520	5350	5160	4950	4720	4470	4200	3920	3640
$P_z = A_g p_y = 5990$		P_{cy}	5870	5510	5130	4720	4290	3840	3400	2980	2610	2290	2020	1780	1580
$p_y Z_x = 822$	1.00	M_b	916	916	916	902	835	791	750	711	676	643	612	584	558
$p_y Z_y = 292$		M_{bs}	916	916	916	916	916	902	862	821	778	736	693	650	608

Check availability.
Under combined axial compression and bending the capacities are only valid up to the given F/P_z limit. For higher values F/P_z the section would be overloaded due to F alone even when M is zero, because F would exceed the local buckling resistance of the section.
M_b is obtained using an equivalent slenderness = $u.v.L_E/r_y.\beta_w^{0.5}$
M_{bs} is obtained using an equivalent slenderness = $0.5 L/r_y$. Effective length $L_E = L$.
FOR EXPLANATION OF TABLES SEE NOTE 9.1

BS 5950-1: 2000
BS 4-1: 1993

AXIAL LOAD & BENDING

UC SECTIONS SUBJECT TO AXIAL LOAD (COMPRESSION OR TENSION) AND BENDING

CROSS-SECTION CAPACITY CHECK

CAPACITIES FOR S275

Section Designation and Axial load Capacity P_z (kN)	F/P_z Limit Semi-Compact Compact		Moment Capacity M_{cx}, M_{cy} (kNm) and Reduced Moment Capacity M_{rx}, M_{ry} (kNm) for Ratios of Axial Load to Axial Load Capacity F/P_z										
		F/P_z	0.0	0.1	0.2	0.3	0.4	0.5	0.6	0.7	0.8	0.9	1.0
356 x 368 x 153 # $P_z = A_g p_y = 5170$	n/a 1.00	M_{cx} M_{cy} M_{rx} M_{ry}	786 301 786 301	786 301 765 301	786 301 704 301	786 301 621 301	786 301 537 301	786 301 451 301	786 301 363 301	786 301 274 286	786 301 184 232	786 301 92.9 167	786 301 0 89.6
356 x 368 x 129 # $P_z = A_g p_y = 4350$	1.00 0.00	M_{cx} M_{cy} M_{rx} M_{ry}	651 252 - -	651 252 - -	651 252 - -	651 252 - -	651 252 - -	651 252 - -	651 252 - -	651 252 - -	651 252 - -	651 252 - -	651 252 - -
305 x 305 x 283 $P_z = A_g p_y = 9180$	n/a 1.00	M_{cx} M_{cy} M_{rx} M_{ry}	1300 468 1300 468	1300 468 1270 468	1300 468 1180 468	1300 468 1050 468	1300 468 914 468	1300 468 774 468	1300 468 630 442	1300 468 480 359	1300 468 325 258	1300 468 165 139	1300 468 0 0
305 x 305 x 240 $P_z = A_g p_y = 8110$	n/a 1.00	M_{cx} M_{cy} M_{rx} M_{ry}	1130 406 1130 406	1130 406 1100 406	1130 406 1020 406	1130 406 905 406	1130 406 787 406	1130 406 666 406	1130 406 541 385	1130 406 411 313	1130 406 278 225	1130 406 141 121	1130 406 0 0
305 x 305 x 198 $P_z = A_g p_y = 6680$	n/a 1.00	M_{cx} M_{cy} M_{rx} M_{ry}	912 330 912 330	912 330 890 330	912 330 823 330	912 330 729 330	912 330 633 330	912 330 534 330	912 330 433 313	912 330 328 255	912 330 222 183	912 330 112 98.3	912 330 0 0
305 x 305 x 158 $P_z = A_g p_y = 5330$	n/a 1.00	M_{cx} M_{cy} M_{rx} M_{ry}	710 257 710 257	710 257 693 257	710 257 642 257	710 257 568 257	710 257 492 257	710 257 414 257	710 257 335 246	710 257 254 200	710 257 171 144	710 257 86.3 77.5	710 257 0 0
305 x 305 x 137 $P_z = A_g p_y = 4610$	n/a 1.00	M_{cx} M_{cy} M_{rx} M_{ry}	609 220 609 220	609 220 594 220	609 220 551 220	609 220 485 220	609 220 420 220	609 220 353 220	609 220 285 211	609 220 216 172	609 220 145 124	609 220 73.2 66.7	609 220 0 0
305 x 305 x 118 $P_z = A_g p_y = 3980$	n/a 1.00	M_{cx} M_{cy} M_{rx} M_{ry}	519 187 519 187	519 187 506 187	519 187 469 187	519 187 414 187	519 187 358 187	519 187 300 187	519 187 242 181	519 187 183 147	519 187 123 106	519 187 62.0 57.1	519 187 0 0
305 x 305 x 97 $P_z = A_g p_y = 3380$	n/a 1.00	M_{cx} M_{cy} M_{rx} M_{ry}	438 158 438 158	438 158 427 158	438 158 396 158	438 158 348 158	438 158 300 158	438 158 252 158	438 158 203 153	438 158 153 125	438 158 103 89.8	438 158 51.7 48.3	438 158 0 0

\# Check availability.
F = Factored axial load.
- Not applicable for semi-compact and slender sections.
The values in this table are conservative for tension as the more onerous compression section classification limits have been used.
FOR EXPLANATION OF TABLES SEE NOTE 9.1

BS 5950-1: 2000
BS 4-1: 1993

AXIAL LOAD & BENDING

UC SECTIONS SUBJECT TO AXIAL COMPRESSION AND BENDING

MEMBER BUCKLING CHECK

RESISTANCES AND CAPACITIES FOR S275

Section Designation and Capacities (kN, kNm)	F/P_z Limit		Compression Resistance P_{cx}, P_{cy} (kN) and Buckling Resistance Moment M_b, M_{bs} (kNm) for Varying effective lengths L_E (m) within the limiting value of F/P_z											
		L_E (m)	2.0	3.0	4.0	5.0	6.0	7.0	8.0	9.0	10.0	11.0	12.0	13.0	14.0
356 x 368 x 153 #	1.00	P_{cx}	5170	5140	5020	4890	4750	4610	4440	4260	4060	3840	3610	3360	3120
$P_z = A_g p_y = 5170$		P_{cy}	5060	4750	4420	4060	3690	3300	2910	2560	2240	1960	1730	1520	1350
$p_y Z_x = 711$	1.00	M_b	786	786	786	750	707	666	627	592	559	528	500	475	451
$p_y Z_y = 251$		M_{bs}	786	786	786	786	786	773	738	703	666	629	592	555	519
356 x 368 x 129 #	1.00	P_{cx}	4350	4320	4210	4110	3990	3860	3720	3560	3390	3200	3000	2790	2590
$P_z = A_g p_y = 4350$		P_{cy}	4250	3990	3710	3410	3090	2760	2430	2130	1870	1640	1440	1270	1130
$p_y Z_x = 600$	1.00	M_b	600	600	600	576	542	510	479	450	424	399	376	356	337
$p_y Z_y = 210$		M_{bs}	600	600	600	600	600	589	562	535	507	479	450	422	394
305 x 305 x 283	1.00	P_{cx}	9180	9070	8780	8480	8160	7830	7460	7070	6660	6230	5790	5360	4940
$P_z = A_g p_y = 9180$		P_{cy}	8770	8020	7250	6460	5680	4930	4260	3670	3170	2760	2410	2120	1880
$p_y Z_x = 1100$	1.00	M_b	1300	1300	1300	1290	1250	1210	1170	1130	1100	1070	1030	1010	977
$p_y Z_y = 390$		M_{bs}	1300	1300	1300	1300	1290	1230	1160	1090	1020	953	885	819	756
305 x 305 x 240	1.00	P_{cx}	8110	8010	7800	7580	7340	7070	6770	6430	6060	5660	5250	4830	4430
$P_z = A_g p_y = 8110$		P_{cy}	7780	7200	6570	5890	5180	4480	3850	3300	2830	2450	2130	1870	1640
$p_y Z_x = 965$	1.00	M_b	1130	1130	1130	1090	1050	1010	971	936	904	873	843	816	789
$p_y Z_y = 338$		M_{bs}	1130	1130	1130	1130	1110	1050	990	929	867	805	745	687	632
305 x 305 x 198	1.00	P_{cx}	6680	6590	6410	6220	6020	5790	5530	5240	4930	4590	4240	3890	3550
$P_z = A_g p_y = 6680$		P_{cy}	6400	5910	5380	4810	4220	3640	3120	2670	2290	1970	1720	1500	1320
$p_y Z_x = 794$	1.00	M_b	912	912	912	866	826	789	755	724	694	667	641	617	594
$p_y Z_y = 275$		M_{bs}	912	912	912	912	894	846	796	746	695	645	596	548	504
305 x 305 x 158	1.00	P_{cx}	5330	5250	5100	4950	4780	4590	4370	4130	3870	3590	3310	3020	2760
$P_z = A_g p_y = 5330$		P_{cy}	5090	4690	4260	3800	3310	2850	2430	2070	1780	1530	1330	1160	1020
$p_y Z_x = 628$	1.00	M_b	710	710	697	657	620	587	556	528	502	478	456	435	417
$p_y Z_y = 214$		M_{bs}	710	710	710	710	692	654	615	575	535	495	456	419	384
305 x 305 x 137	1.00	P_{cx}	4610	4540	4410	4270	4120	3950	3760	3550	3320	3070	2820	2570	2340
$P_z = A_g p_y = 4610$		P_{cy}	4400	4050	3680	3270	2850	2440	2080	1770	1520	1310	1130	991	873
$p_y Z_x = 543$	1.00	M_b	609	609	591	554	519	488	459	433	409	387	367	349	332
$p_y Z_y = 183$		M_{bs}	609	609	609	609	592	559	525	490	455	421	387	356	326
305 x 305 x 118	1.00	P_{cx}	3980	3910	3800	3680	3550	3400	3230	3050	2840	2630	2410	2200	2000
$P_z = A_g p_y = 3980$		P_{cy}	3790	3490	3160	2800	2440	2090	1780	1510	1290	1110	965	844	743
$p_y Z_x = 466$	1.00	M_b	519	519	499	464	433	403	376	352	330	311	293	277	263
$p_y Z_y = 156$		M_{bs}	519	519	519	519	503	475	445	416	386	356	328	301	275
305 x 305 x 97	1.00	P_{cx}	3380	3320	3220	3120	3000	2870	2720	2550	2370	2180	1990	1800	1630
$P_z = A_g p_y = 3380$		P_{cy}	3210	2950	2660	2350	2030	1730	1460	1240	1060	907	786	686	603
$p_y Z_x = 397$	1.00	M_b	438	438	413	381	351	323	297	275	255	237	221	208	195
$p_y Z_y = 132$		M_{bs}	438	438	438	438	421	396	371	345	320	294	270	247	225

\# Check availability.
Under combined axial compression and bending the capacities are only valid up to the given F/P_z limit. For higher values F/P_z the section would be overloaded due to F alone even when M is zero, because F would exceed the local buckling resistance of the section.
M_b is obtained using an equivalent slenderness = $u.v.L_E/r_y.\beta_w^{0.5}$
M_{bs} is obtained using an equivalent slenderness = $0.5 L/r_y$. Effective length $L_E = L$.
FOR EXPLANATION OF TABLES SEE NOTE 9.1

BS 5950-1: 2000
BS 4-1: 1993

AXIAL LOAD & BENDING

UC SECTIONS SUBJECT TO AXIAL LOAD (COMPRESSION OR TENSION) AND BENDING

CROSS-SECTION CAPACITY CHECK

CAPACITIES FOR S275

Section Designation and Axial load Capacity P_z (kN)	F/P_z Limit Semi-Compact Compact		Moment Capacity M_{cx}, M_{cy} (kNm) and Reduced Moment Capacity M_{rx}, M_{ry} (kNm) for Ratios of Axial Load to Axial Load Capacity F/P_z										
		F/P_z	0.0	0.1	0.2	0.3	0.4	0.5	0.6	0.7	0.8	0.9	1.0
254 x 254 x 167 $P_z = A_g p_y = 5640$	n/a 1.00	M_{cx}	642	642	642	642	642	642	642	642	642	642	642
		M_{cy}	237	237	237	237	237	237	237	237	237	237	237
		M_{rx}	642	627	580	516	449	380	308	235	159	80.5	0
		M_{ry}	237	237	237	237	237	237	224	182	131	70.1	0
254 x 254 x 132 $P_z = A_g p_y = 4450$	n/a 1.00	M_{cx}	495	495	495	495	495	495	495	495	495	495	495
		M_{cy}	183	183	183	183	183	183	183	183	183	183	183
		M_{rx}	495	483	446	395	343	290	235	178	120	60.8	0
		M_{ry}	183	183	183	183	183	174	141	102	54.5	0	
254 x 254 x 107 $P_z = A_g p_y = 3600$	n/a 1.00	M_{cx}	393	393	393	393	393	393	393	393	393	393	393
		M_{cy}	146	146	146	146	146	146	146	146	146	146	146
		M_{rx}	393	384	355	313	271	228	185	140	94.2	47.6	0
		M_{ry}	146	146	146	146	146	139	113	81.3	43.6	0	
254 x 254 x 89 $P_z = A_g p_y = 2990$	n/a 1.00	M_{cx}	324	324	324	324	324	324	324	324	324	324	324
		M_{cy}	121	121	121	121	121	121	121	121	121	121	121
		M_{rx}	324	316	292	257	222	187	151	114	76.6	38.6	0
		M_{ry}	121	121	121	121	121	121	114	93.1	67.0	35.9	0
254 x 254 x 73 $P_z = A_g p_y = 2560$	n/a 1.00	M_{cx}	273	273	273	273	273	273	273	273	273	273	273
		M_{cy}	101	101	101	101	101	101	101	101	101	101	101
		M_{rx}	273	266	245	216	187	157	126	95.5	64.1	32.3	0
		M_{ry}	101	101	101	101	101	101	96.8	78.9	56.8	30.5	0
203 x 203 x 86 $P_z = A_g p_y = 2920$	n/a 1.00	M_{cx}	259	259	259	259	259	259	259	259	259	259	259
		M_{cy}	95.1	95.1	95.1	95.1	95.1	95.1	95.1	95.1	95.1	95.1	95.1
		M_{rx}	259	253	234	208	181	152	123	93.7	63.2	32.0	0
		M_{ry}	95.1	95.1	95.1	95.1	95.1	95.1	90.6	73.8	53.1	28.5	0
203 x 203 x 71 $P_z = A_g p_y = 2400$	n/a 1.00	M_{cx}	212	212	212	212	212	212	212	212	212	212	212
		M_{cy}	78.2	78.2	78.2	78.2	78.2	78.2	78.2	78.2	78.2	78.2	78.2
		M_{rx}	212	206	190	168	146	123	99.2	75.2	50.6	25.6	0
		M_{ry}	78.2	78.2	78.2	78.2	78.2	78.2	73.9	60.1	43.2	23.2	0
203 x 203 x 60 $P_z = A_g p_y = 2100$	n/a 1.00	M_{cx}	180	180	180	180	180	180	180	180	180	180	180
		M_{cy}	66.3	66.3	66.3	66.3	66.3	66.3	66.3	66.3	66.3	66.3	66.3
		M_{rx}	180	176	163	145	125	105	85.0	64.3	43.3	21.8	0
		M_{ry}	66.3	66.3	66.3	66.3	66.3	63.9	52.1	37.6	20.2	0	

F = Factored axial load.
- Not applicable for semi-compact and slender sections.
The values in this table are conservative for tension as the more onerous compression section classification limits have been used.
FOR EXPLANATION OF TABLES SEE NOTE 9.1

BS 5950-1: 2000
BS 4-1: 1993

AXIAL LOAD & BENDING

UC SECTIONS SUBJECT TO AXIAL COMPRESSION AND BENDING

MEMBER BUCKLING CHECK

RESISTANCES AND CAPACITIES FOR S275

Section Designation and Capacities (kN, kNm)	F/P_z Limit	Compression Resistance P_{cx}, P_{cy} (kN) and Buckling Resistance Moment M_b, M_{bs} (kNm) for Varying effective lengths L_E (m) within the limiting value of F/P_z												
		L_E (m)	1.0	1.5	2.0	2.5	3.0	3.5	4.0	4.5	5.0	5.5	6.0	6.5	7.0
254 x 254 x 167	1.00	P_{cx}	5640	5640	5640	5570	5480	5390	5300	5200	5090	4980	4860	4730	4590
$P_z = A_g p_y = 5640$		P_{cy}	5640	5500	5260	5020	4760	4490	4210	3910	3610	3320	3040	2770	2530
$p_y Z_x = 550$	1.00	M_b	642	642	642	642	642	642	632	617	602	589	576	563	551
$p_y Z_y = 197$		M_{bs}	642	642	642	642	642	642	642	642	633	613	593	573	552
254 x 254 x 132	1.00	P_{cx}	4450	4450	4450	4390	4310	4240	4160	4080	3990	3900	3800	3690	3580
$P_z = A_g p_y = 4450$		P_{cy}	4450	4330	4140	3940	3730	3510	3280	3050	2810	2570	2350	2140	1950
$p_y Z_x = 432$	1.00	M_b	495	495	495	495	495	490	476	462	449	437	425	414	403
$p_y Z_y = 153$		M_{bs}	495	495	495	495	495	495	495	495	485	470	454	438	422
254 x 254 x 107	1.00	P_{cx}	3600	3600	3600	3540	3480	3420	3360	3290	3210	3140	3050	2960	2860
$P_z = A_g p_y = 3600$		P_{cy}	3600	3500	3340	3180	3000	2820	2640	2440	2240	2050	1870	1700	1550
$p_y Z_x = 348$	1.00	M_b	393	393	393	393	393	382	369	357	345	334	323	313	303
$p_y Z_y = 121$		M_{bs}	393	393	393	393	393	393	393	393	383	371	358	345	332
254 x 254 x 89	1.00	P_{cx}	2990	2990	2990	2940	2890	2840	2780	2730	2660	2600	2530	2450	2370
$P_z = A_g p_y = 2990$		P_{cy}	2990	2900	2770	2630	2490	2340	2180	2020	1860	1700	1540	1400	1270
$p_y Z_x = 290$	1.00	M_b	324	324	324	324	324	311	299	287	276	266	256	247	238
$p_y Z_y = 100$		M_{bs}	324	324	324	324	324	324	324	324	316	305	295	284	273
254 x 254 x 73	1.00	P_{cx}	2560	2560	2550	2510	2470	2420	2370	2320	2270	2210	2150	2080	2000
$P_z = A_g p_y = 2560$		P_{cy}	2560	2470	2360	2240	2110	1980	1840	1700	1550	1410	1280	1160	1050
$p_y Z_x = 247$	1.00	M_b	273	273	273	273	268	256	245	234	224	214	204	195	187
$p_y Z_y = 84.4$		M_{bs}	273	273	273	273	273	273	273	272	263	254	245	236	226
203 x 203 x 86	1.00	P_{cx}	2920	2920	2870	2810	2750	2690	2610	2540	2450	2360	2260	2160	2050
$P_z = A_g p_y = 2920$		P_{cy}	2890	2740	2580	2410	2220	2030	1840	1640	1460	1300	1160	1030	921
$p_y Z_x = 225$	1.00	M_b	259	259	259	259	252	243	235	227	220	213	206	200	194
$p_y Z_y = 79.2$		M_{bs}	259	259	259	259	259	259	253	243	233	222	211	201	190
203 x 203 x 71	1.00	P_{cx}	2400	2400	2360	2310	2260	2200	2140	2080	2010	1930	1850	1760	1670
$P_z = A_g p_y = 2400$		P_{cy}	2380	2250	2110	1970	1820	1660	1500	1340	1190	1060	941	838	748
$p_y Z_x = 187$	1.00	M_b	212	212	212	212	203	194	187	179	173	166	160	154	149
$p_y Z_y = 65.2$		M_{bs}	212	212	212	212	212	212	207	198	190	181	172	163	154
203 x 203 x 60	1.00	P_{cx}	2100	2100	2060	2020	1970	1920	1860	1800	1740	1670	1590	1500	1420
$P_z = A_g p_y = 2100$		P_{cy}	2080	1960	1840	1710	1570	1430	1280	1140	1010	888	786	698	622
$p_y Z_x = 161$	1.00	M_b	180	180	180	177	169	161	153	146	139	133	127	122	117
$p_y Z_y = 55.3$		M_{bs}	180	180	180	180	180	180	174	167	159	151	144	136	128

Under combined axial compression and bending the capacities are only valid up to the given F/P_z limit. For higher values F/P, the section would be overloaded due to F alone even when M is zero, because F would exceed the local buckling resistance of the section.
M_b is obtained using an equivalent slenderness = $u.v.L_E/r_y.\beta_w^{0.5}$
M_{bs} is obtained using an equivalent slenderness = $0.5 L/r_y$. Effective length $L_E = L$.
FOR EXPLANATION OF TABLES SEE NOTE 9.1

BS 5950-1: 2000
BS 4-1: 1993

AXIAL LOAD & BENDING

UC SECTIONS SUBJECT TO AXIAL LOAD (COMPRESSION OR TENSION) AND BENDING

CROSS-SECTION CAPACITY CHECK

CAPACITIES FOR S275

Section Designation and Axial load Capacity P_z (kN)	F/P_z Limit Semi-Compact Compact	Moment Capacity M_{cx}, M_{cy} (kNm) and Reduced Moment Capacity M_{rx}, M_{ry} (kNm) for Ratios of Axial Load to Axial Load Capacity F/P_z										
		F/P_z	0.0	0.1	0.2	0.3	0.4	0.5	0.6	0.7	0.8	0.9	1.0
203 x 203 x 52 $P_z = A_g p_y = 1820$	n/a 1.00	M_{cx}	156	156	156	156	156	156	156	156	156	156	156
		M_{cy}	57.4	57.4	57.4	57.4	57.4	57.4	57.4	57.4	57.4	57.4	57.4
		M_{rx}	156	152	141	124	107	90.3	72.8	55.1	37.0	18.6	0
		M_{ry}	57.4	57.4	57.4	57.4	57.4	57.4	55.2	45.0	32.4	17.4	0
203 x 203 x 46 $P_z = A_g p_y = 1610$	n/a 1.00	M_{cx}	137	137	137	137	137	137	137	137	137	137	137
		M_{cy}	50.2	50.2	50.2	50.2	50.2	50.2	50.2	50.2	50.2	50.2	50.2
		M_{rx}	137	133	124	109	94.2	79.1	63.7	48.2	32.3	16.3	0
		M_{ry}	50.2	50.2	50.2	50.2	50.2	50.2	48.5	39.6	28.6	15.4	0
152 x 152 x 37 $P_z = A_g p_y = 1300$	n/a 1.00	M_{cx}	85.0	85.0	85.0	85.0	85.0	85.0	85.0	85.0	85.0	85.0	85.0
		M_{cy}	30.2	30.2	30.2	30.2	30.2	30.2	30.2	30.2	30.2	30.2	30.2
		M_{rx}	85.0	83.1	77.3	68.5	59.3	49.9	40.3	30.5	20.6	10.4	0
		M_{ry}	30.2	30.2	30.2	30.2	30.2	30.2	29.4	24.0	17.3	9.34	0
152 x 152 x 30 $P_z = A_g p_y = 1050$	n/a 1.00	M_{cx}	68.2	68.2	68.2	68.2	68.2	68.2	68.2	68.2	68.2	68.2	68.2
		M_{cy}	24.2	24.2	24.2	24.2	24.2	24.2	24.2	24.2	24.2	24.2	24.2
		M_{rx}	68.2	66.6	62.0	54.9	47.4	39.8	32.1	24.3	16.3	8.23	0
		M_{ry}	24.2	24.2	24.2	24.2	24.2	24.2	23.6	19.3	14.0	7.52	0
152 x 152 x 23 $P_z = A_g p_y = 803$	1.00 0.00	M_{cx}	48.5	48.5	48.5	48.5	48.5	48.5	48.5	48.5	48.5	48.5	48.5
		M_{cy}	17.4	17.4	17.4	17.4	17.4	17.4	17.4	17.4	17.4	17.4	17.4
		M_{rx}	-	-	-	-	-	-	-	-	-	-	-
		M_{ry}	-	-	-	-	-	-	-	-	-	-	-

F = Factored axial load.
- Not applicable for semi-compact and slender sections.
The values in this table are conservative for tension as the more onerous compression section classification limits have been used.
FOR EXPLANATION OF TABLES SEE NOTE 9.1

BS 5950-1: 2000
BS 4-1: 1993

AXIAL LOAD & BENDING

UC SECTIONS SUBJECT TO AXIAL COMPRESSION AND BENDING

MEMBER BUCKLING CHECK

RESISTANCES AND CAPACITIES FOR S275

Section Designation and Capacities (kN, kNm)	F/P_z Limit		Compression Resistance P_{cx}, P_{cy} (kN) and Buckling Resistance Moment M_b, M_{bs} (kNm) for Varying effective lengths L_E (m) within the limiting value of F/P_z											
			L_E (m)	1.0	1.5	2.0	2.5	3.0	3.5	4.0	4.5	5.0	5.5	6.0	6.5	7.0
203 x 203 x 52	1.00	P_{cx}	1820	1820	1790	1750	1710	1660	1620	1560	1510	1440	1370	1300	1220
$P_z = A_g p_y = 1820$		P_{cy}	1800	1700	1590	1480	1360	1230	1110	982	868	767	678	602	536
$p_y Z_x = 140$	1.00	M_b	156	156	156	152	144	137	130	123	117	111	106	101	96.0
$p_y Z_y = 47.9$		M_{bs}	156	156	156	156	156	156	150	144	137	131	124	117	110
203 x 203 x 46	1.00	P_{cx}	1610	1610	1580	1550	1510	1470	1430	1380	1330	1270	1210	1140	1070
$P_z = A_g p_y = 1610$		P_{cy}	1590	1500	1410	1310	1200	1080	970	860	760	670	592	525	467
$p_y Z_x = 124$	1.00	M_b	137	137	137	132	125	118	111	105	99.1	93.7	88.7	84.1	79.8
$p_y Z_y = 41.8$		M_{bs}	137	137	137	137	137	137	131	126	120	114	108	102	95.8
152 x 152 x 37	1.00	P_{cx}	1300	1270	1240	1200	1150	1110	1050	990	923	851	779	710	644
$P_z = A_g p_y = 1300$		P_{cy}	1230	1130	1020	904	782	667	565	479	409	351	304	266	234
$p_y Z_x = 75.1$	1.00	M_b	85.0	85.0	82.3	77.2	72.6	68.3	64.4	60.8	57.6	54.6	51.9	49.4	47.1
$p_y Z_y = 25.2$		M_{bs}	85.0	85.0	85.0	85.0	81.8	77.1	72.3	67.3	62.3	57.4	52.7	48.2	44.1
152 x 152 x 30	1.00	P_{cx}	1050	1030	1000	972	936	896	850	799	743	684	625	567	514
$P_z = A_g p_y = 1050$		P_{cy}	999	918	828	730	630	536	453	384	327	281	243	212	187
$p_y Z_x = 61.1$	1.00	M_b	68.2	68.2	65.0	60.4	56.2	52.3	48.8	45.6	42.7	40.1	37.8	35.7	33.8
$p_y Z_y = 20.2$		M_{bs}	68.2	68.2	68.2	68.2	65.4	61.6	57.7	53.7	49.6	45.6	41.8	38.2	34.9
152 x 152 x 23	1.00	P_{cx}	803	786	762	737	708	675	638	596	551	505	458	414	374
$P_z = A_g p_y = 803$		P_{cy}	758	693	621	543	465	392	330	278	236	202	175	152	134
$p_y Z_x = 45.1$	1.00	M_b	45.1	45.1	42.9	39.7	36.6	33.7	31.1	28.7	26.7	24.8	23.2	21.8	20.5
$p_y Z_y = 14.5$		M_{bs}	45.1	45.1	45.1	45.1	42.7	40.1	37.4	34.6	31.9	29.2	26.6	24.3	22.1

Under combined axial compression and bending the capacities are only valid up to the given F/P_z limit. For higher values F/P_z the section would be overloaded due to F alone even when M is zero, because F would exceed the local buckling resistance of the section.
M_b is obtained using an equivalent slenderness = $u.v.L_E/r_y.\beta_w^{0.5}$
M_{bs} is obtained using an equivalent slenderness = $0.5 L/r_y$. Effective length $L_E = L$.
FOR EXPLANATION OF TABLES SEE NOTE 9.1

BS 5950-1: 2000
BS 4190: 2001

BOLT CAPACITIES

NON-PRELOADED ORDINARY BOLTS

GRADE 4.6 BOLTS IN S275

Diameter of Bolt	Tensile Stress Area	Tension Capacity		Shear Capacity		Bearing Capacity in kN (Minimum of P_{bb} and P_{bs}) End distance equal to 2 x bolt diameter.								
		Nominal $0.8A_t p_t$ P_{nom}	Exact $A_t p_t$ P_t	Single Shear P_s	Double Shear $2P_s$	Thickness in mm of ply passed through.								
mm	mm²	kN	kN	kN	kN	5	6	7	8	9	10	12	15	20	25	30
12	84.3	16.2	20.2	13.5	27.0	*27.6*	*33.1*	*38.6*	*44.2*	*49.7*	*55.2*	*66.2*	*82.8*	*110*	*138*	*166*
16	157	30.1	37.7	25.1	50.2	36.8	44.2	*51.5*	*58.9*	*66.2*	*73.6*	*88.3*	*110*	*147*	*184*	*221*
20	245	47.0	58.8	39.2	78.4	46.0	55.2	64.4	73.6	*82.8*	*92.0*	*110*	*138*	*184*	*230*	*276*
22	303	58.2	72.7	48.5	97.0	50.6	60.7	70.8	81.0	91.1	*101*	*121*	*152*	*202*	*253*	*304*
24	353	67.8	84.7	56.5	113	**55.2**	66.2	77.5	88.3	99.4	110	*132*	*166*	*221*	*276*	*331*
27	459	88.1	110	73.4	147	**62.1**	74.5	86.9	99.4	112	124	149	*186*	*248*	*311*	*373*
30	561	108	135	89.8	180	**69.0**	**82.8**	96.6	110	124	138	166	*207*	*276*	*345*	*414*

Values in **bold** are less than the single shear capacity of the bolt.
Values in *italic* are greater than the double shear capacity of the bolt.
Bearing values assume standard clearance holes.
If oversize or short slotted holes are used, bearing values should be multiplied by 0.7.
If long slotted or kidney shaped holes are used, bearing values should be multiplied by 0.5.
If appropriate, shear capacity must be reduced for large packings, large grip lengths and long joints.
FOR EXPLANATION OF TABLES SEE NOTE 10.1

GRADE 8.8 BOLTS IN S275

Diameter of Bolt	Tensile Stress Area	Tension Capacity		Shear Capacity		Bearing Capacity in kN (Minimum of P_{bb} and P_{bs}) End distance equal to 2 x bolt diameter.								
		Nominal $0.8A_t p_t$ P_{nom}	Exact $A_t p_t$ P_t	Single Shear P_s	Double Shear $2P_s$	Thickness in mm of ply passed through.								
mm	mm²	kN	kN	kN	kN	5	6	7	8	9	10	12	15	20	25	30
12	84.3	37.8	47.2	31.6	63.2	**27.6**	33.1	38.6	44.2	49.7	55.2	66.2	*82.8*	*110*	*138*	*166*
16	157	70.3	87.9	58.9	118	**36.8**	**44.2**	**51.5**	58.9	66.2	73.6	88.3	110	*147*	*184*	*221*
20	245	110	137	91.9	184	46.0	55.2	64.4	73.6	82.8	92.0	110	138	*184*	*230*	*276*
22	303	136	170	114	227	50.6	60.7	70.8	81.0	91.1	101	121	152	*202*	*253*	*304*
24	353	158	198	132	265	**55.2**	66.2	77.3	88.3	99.4	110	132	166	221	*276*	*331*
27	459	206	257	172	344	**62.1**	74.5	86.9	99.4	112	124	149	186	248	*311*	*373*
30	561	251	314	210	421	**69.0**	**82.8**	96.6	110	124	138	166	207	276	345	414

Values in **bold** are less than the single shear capacity of the bolt.
Values in *italic* are greater than the double shear capacity of the bolt.
Bearing values assume standard clearance holes.
If oversize or short slotted holes are used, bearing values should be multiplied by 0.7.
If long slotted or kidney shaped holes are used, bearing values should be multiplied by 0.5.
If appropriate, shear capacity must be reduced for large packings, large grip lengths and long joints.
FOR EXPLANATION OF TABLES SEE NOTE 10.1

BS 5950-1: 2000
BS 4190: 2001

BOLT CAPACITIES

NON-PRELOADED ORDINARY BOLTS

GRADE 10.9 BOLTS IN S275

Diameter of Bolt	Tensile Stress Area	Tension Capacity		Shear Capacity		Bearing Capacity in kN (Minimum of P_{bb} and P_{bs}) End distance equal to 2 x bolt diameter.									
		Nominal $0.8A_tp_t$ P_{nom}	Exact A_tp_t P_t	Single Shear P_s	Double Shear $2P_s$	Thickness in mm of ply passed through.									
mm	mm²	kN	kN	kN	kN	5	6	7	8	9	10	12	15	20	25	30
12	84.3	47.2	59.0	33.7	67.4	**27.6**	**33.1**	38.6	44.2	49.7	55.2	66.2	82.8	*110*	*138*	*166*
16	157	87.9	110	62.8	126	**36.8**	**44.2**	**51.5**	**58.9**	66.2	73.6	88.3	110	*147*	*184*	*221*
20	245	137	172	98.0	196	**46.0**	**55.2**	**64.4**	**73.6**	**82.8**	**92.0**	110	138	184	*230*	*276*
22	303	170	212	121	242	**50.6**	**60.7**	**70.8**	**81.0**	**91.1**	**101**	121	152	202	*253*	*304*
24	353	198	247	141	282	**55.2**	**66.2**	**77.3**	**88.3**	**99.4**	**110**	**132**	166	221	276	*331*
27	459	257	321	184	367	**62.1**	**74.5**	**86.9**	**99.4**	**112**	**124**	**149**	186	248	311	373
30	561	314	393	224	449	**69.0**	**82.8**	**96.6**	**110**	**124**	**138**	**166**	**207**	276	345	414

Values in **bold** are less than the single shear capacity of the bolt.
Values in *italic* are greater than the double shear capacity of the bolt.
Bearing values assume standard clearance holes.
If oversize or short slotted holes are used, bearing values should be multiplied by 0.7.
If long slotted or kidney shaped holes are used, bearing values should be multiplied by 0.5.
If appropriate, shear capacity must be reduced for large packings, large grip lengths and long joints.
FOR EXPLANATION OF TABLES SEE NOTE 10.1

BS 5950-1: 2000
BS 4190: 2001
BS 4933: 1973

BOLT CAPACITIES

NON-PRELOADED COUNTERSUNK BOLTS

GRADE 4.6 COUNTERSUNK BOLTS IN S275

Diameter of Bolt	Tensile Stress Area	Tension Capacity		Shear Capacity		Bearing Capacity in kN (Minimum of P_{bb} and P_{bs}) End distance equal to 2 x bolt diameter.									
		Nominal $0.8A_t p_t$ P_{nom}	Exact $A_t p_t$ P_t	Single Shear P_s	Double Shear $2P_s$	Thickness in mm of ply passed through.									
mm	A_t mm²	kN	kN	kN	kN	5	6	7	8	9	10	12	15	20	25	30
12	84.3	16.2	20.2	13.5	27.0	**11.0**	16.6	22.1	*27.6*	*33.1*	*38.6*	*49.7*	*66.2*	*93.8*	*121*	*149*
16	157	30.1	37.7	25.1	50.2	**7.36**	**14.7**	22.1	29.4	36.8	44.2	58.9	81.0	*118*	*155*	*191*
20	245	47.0	58.8	39.2	78.4	0	**9.20**	**18.4**	27.6	36.8	46.0	64.4	92.0	138	*184*	*230*
22	303	58.2	72.7	48.5	97.0	0	**5.06**	**15.2**	25.3	35.4	45.5	65.8	96.1	147	*197*	*248*
24	353	67.8	84.7	56.5	113	0	0	**11.0**	22.1	33.1	44.2	66.2	99.4	155	*210*	*265*
27	459	88.1	110	73.4	147	0	0	**3.11**	**15.5**	27.9	40.4	65.2	102	*165*	*227*	*289*
30	561	108	135	89.8	180	0	0	0	**6.90**	**20.7**	34.5	62.8	104	*173*	*242*	*311*

Values in **bold** are less than the single shear capacity of the bolt.
Values in *italic* are greater than the double shear capacity of the bolt.
Bearing values assume standard clearance holes.
If oversize or short slotted holes are used, bearing values should be multiplied by 0.7.
If long slotted or kidney shaped holes are used, bearing values should be multiplied by 0.5.
Depth of countersink is taken as half the bolt diameter.
FOR EXPLANATION OF TABLES SEE NOTE 10.1

GRADE 8.8 COUNTERSUNK BOLTS IN S 275

Diameter of Bolt	Tensile Stress Area	Tension Capacity		Shear Capacity		Bearing Capacity in kN (Minimum of P_{bb} and P_{bs}) End distance equal to 2 x bolt diameter.									
		Nominal $0.8A_t p_t$ P_{nom}	Exact $A_t p_t$ P_t	Single Shear P_s	Double Shear $2P_s$	Thickness in mm of ply passed through.									
mm	A_t mm²	kN	kN	kN	kN	5	6	7	8	9	10	12	15	20	25	30
12	84.3	37.8	47.2	31.6	63.2	**11.0**	**16.6**	**22.1**	**27.6**	33.1	38.6	49.7	66.2	*93.8*	*121*	*149*
16	157	70.3	87.9	58.9	118	**7.36**	**14.7**	**22.1**	29.4	36.8	44.2	58.9	81.0	*118*	*155*	*191*
20	245	110	137	91.9	184	0	**9.20**	**18.4**	27.6	36.8	46.0	64.4	92.0	138	*184*	*230*
22	303	136	170	114	227	0	**5.06**	**15.2**	25.3	35.4	45.5	65.8	96.1	147	197	*248*
24	353	158	198	132	265	0	0	**11.0**	22.1	33.1	44.2	66.2	99.4	155	210	*265*
27	459	206	257	172	344	0	0	**3.11**	**15.5**	27.9	40.4	65.2	102	165	227	289
30	561	251	314	210	421	0	0	0	**6.90**	**20.7**	34.5	62.8	104	173	242	311

Values in **bold** are less than the single shear capacity of the bolt.
Values in *italic* are greater than the double shear capacity of the bolt.
Bearing values assume standard clearance holes.
If oversize or short slotted holes are used, bearing values should be multiplied by 0.7.
If long slotted or kidney shaped holes are used, bearing values should be multiplied by 0.5.
Depth of countersink is taken as half the bolt diameter.
FOR EXPLANATION OF TABLES SEE NOTE 10.1

BS 5950-1: 2000
BS 4190: 2001
BS 4933: 1973

BOLT CAPACITIES

NON-PRELOADED COUNTERSUNK BOLTS

GRADE 10.9 COUNTERSUNK BOLTS IN S 275

Diameter of Bolt	Tensile Stress Area	Tension Capacity		Shear Capacity		Bearing Capacity in kN (Minimum of P_{bb} and P_{bs}) End distance equal to 2 x bolt diameter. Thickness in mm of ply passed through.									
		Nominal $0.8A_t p_t$ P_{nom}	Exact $A_t p_t$ P_t	Single Shear P_s	Double Shear $2P_s$										
mm	A_t mm²	kN	kN	kN	kN	5	6	7	8	9	10	12	15	20	25	30
12	84.3	47.2	59.0	33.7	67.4	**11.0**	**16.6**	**22.1**	**27.6**	**33.1**	38.6	49.7	66.2	*93.8*	*121*	*149*
16	157	87.9	110	62.8	126	**7.36**	**14.7**	**22.1**	**29.4**	**36.8**	**44.2**	**58.9**	81.0	118	*155*	*191*
20	245	137	172	98.0	196	**0**	**9.20**	**18.4**	**27.6**	**36.8**	**46.0**	**64.4**	**92.0**	138	184	*230*
22	303	170	212	121	242	**0**	**5.06**	**15.2**	**25.3**	**35.4**	**45.5**	**65.8**	**96.1**	147	197	*248*
24	353	198	247	141	282	**0**	**0**	**11.0**	**22.1**	**33.1**	**44.2**	**66.2**	**99.4**	155	210	265
27	459	257	321	184	367	**0**	**0**	**3.11**	**15.5**	**27.9**	**40.4**	**65.2**	**102**	165	227	289
30	561	314	393	224	449	**0**	**0**	**0**	**6.90**	**20.7**	**34.5**	**62.1**	**104**	173	242	311

Values in **bold** are less than the single shear capacity of the bolt.
Values in *italic* are greater than the double shear capacity of the bolt.
Bearing values assume standard clearance holes.
If oversize or short slotted holes are used, bearing values should be multiplied by 0.7.
If long slotted or kidney shaped holes are used, bearing values should be multiplied by 0.5.
Depth of countersink is taken as half the bolt diameter.
FOR EXPLANATION OF TABLES SEE NOTE 10.1

BS 5950-1: 2000
BS 4395: 1969

BOLT CAPACITIES

NON-PRELOADED HSFG BOLTS

GENERAL GRADE HSFG BOLTS IN S275

Diameter of Bolt	Tensile Stress Area	Tension Capacity		Shear Capacity		Bearing Capacity in kN (Minimum of P_{bb} and P_{bs}) End distance equal to 2 x bolt diameter. Thickness in mm of ply passed through.								
		Nominal $0.8A_t p_t$ P_{nom}	Exact $A_t p_t$ P_t	Single Shear P_s	Double Shear $2P_s$									
mm	mm²	kN	kN	kN	kN	5	6	7	8	9	10	12	15	20	25	30
12	84.3	39.8	49.7	33.7	67.4	**27.6**	**33.1**	38.6	44.2	49.7	55.2	66.2	*82.8*	*110*	*138*	*166*
16	157	74.1	92.6	62.8	126	**36.8**	**44.2**	**51.5**	**58.9**	66.2	73.6	88.3	110	*147*	*184*	*221*
20	245	116	145	98.0	196	**46.0**	**55.2**	**64.4**	**73.6**	**82.8**	**92.0**	110	138	184	*230*	*276*
22	303	143	179	121	242	**50.6**	**60.7**	**70.8**	**81.0**	**91.1**	101	121	152	202	*253*	*304*
24	353	167	208	141	282	**55.2**	**66.2**	**77.3**	**88.3**	**99.4**	110	**132**	166	221	276	*331*
27	459	189	236	161	321	**62.1**	**74.5**	**86.9**	**99.4**	112	124	149	186	248	311	*373*
30	561	231	289	196	393	**69.0**	**82.8**	**96.6**	110	124	138	166	207	276	345	414

Values in **bold** are less than the single shear capacity of the bolt.
Values in *italic* are greater than the double shear capacity of the bolt.
Bearing values assume standard clearance holes.
If oversize or short slotted holes are used, bearing values should be multiplied by 0.7.
If long slotted or kidney shaped holes are used, bearing values should be multiplied by 0.5.
If appropriate, shear capacity must be reduced for large packings, large grip lengths and long joints.
FOR EXPLANATION OF TABLES SEE NOTE 10.1

HIGHER GRADE HSFG BOLTS IN S 275

Diameter of Bolt	Tensile Stress Area	Tension Capacity		Shear Capacity		Bearing Capacity in kN (Minimum of P_{bb} and P_{bs}) End distance equal to 2 x bolt diameter. Thickness in mm of ply passed through.								
		Nominal $0.8A_t p_t$ P_{nom}	Exact $A_t p_t$ P_t	Single Shear P_s	Double Shear $2P_s$									
mm	mm²	kN	kN	kN	kN	5	6	7	8	9	10	12	15	20	25	30
16	157	87.9	110	62.8	126	**36.8**	**44.2**	**51.5**	**58.9**	66.2	73.6	88.3	110	*147*	*184*	*221*
20	245	137	172	98.0	196	**46.0**	**55.2**	**64.4**	**73.6**	**82.8**	**92.0**	110	138	184	*230*	*276*
22	303	170	212	121	242	**50.6**	**60.7**	**70.8**	**81.0**	**91.1**	101	121	152	202	*253*	*304*
24	353	198	247	141	282	**55.2**	**66.2**	**77.3**	**88.3**	**99.4**	110	**132**	166	221	276	*331*
27	459	257	321	184	367	**62.1**	**74.5**	**86.9**	**99.4**	112	124	149	186	248	311	*373*
30	561	314	393	224	449	**69.0**	**82.8**	**96.6**	110	124	138	166	**207**	276	345	414

Values in **bold** are less than the single shear capacity of the bolt.
Values in *italic* are greater than the double shear capacity of the bolt.
Bearing values assume standard clearance holes.
If oversize or short slotted holes are used, bearing values should be multiplied by 0.7.
If long slotted or kidney shaped holes are used, bearing values should be multiplied by 0.5.
If appropriate, shear capacity must be reduced for large packings, large grip lengths and long joints.
FOR EXPLANATION OF TABLES SEE NOTE 10.1

BS 5950-1: 2000
BS 4395: 1969
BS 4604: 1970

BOLT CAPACITIES

PRELOADED HSFG BOLTS: NON-SLIP IN SERVICE

GENERAL GRADE HSFG BOLTS IN S275

Diameter of Bolt	Min. Shank Tension	Tension		Shear Capacity		Slip Resistance for $\mu = 0.5$		Bearing Capacity, P_{bg} in kN End distance equal to 3 x bolt diameter. Thickness in mm of ply passed through.									
				Single Shear	Double Shear	Single Shear	Double Shear										
mm	P_o kN	$1.1P_o$ kN	$A_t p_t$ kN	kN	kN	kN	kN	5	6	7	8	9	10	12	15	20	25	30
12	49.4	54.3	49.7	33.7	67.4	27.2	54.3	41.4	49.7	58.0	66.2	74.5	82.8	99.4	*124*	*166*	*207*	*248*
16	92.1	101	92.6	62.8	126	50.7	101	**55.2**	66.2	77.3	88.3	99.4	110	*132*	*166*	*221*	*276*	*331*
20	144	158	145	98.0	196	79.2	158	**69.0**	**82.8**	**96.6**	110	124	138	166	*207*	*276*	*345*	*414*
22	177	195	179	121	242	97.4	195	**75.9**	**91.1**	**106**	121	137	152	182	*228*	*304*	*380*	*455*
24	207	228	208	141	282	114	228	**82.8**	**99.4**	**116**	**132**	149	166	199	248	*331*	*414*	*497*
27	234	257	236	161	321	129	257	**93.2**	**112**	**130**	**149**	168	186	224	279	*373*	*466*	*559*
30	286	315	289	196	393	157	315	**104**	**124**	**145**	**166**	**186**	207	248	311	*414*	*518*	*621*

Values in **bold** are less than the single shear capacity of the bolt.
Values in *italic* are greater than the double shear capacity of the bolt.
Shading indicates that the ply thickness is not suitable for an outer ply.
FOR EXPLANATION OF TABLES SEE NOTE 10.1

HIGHER GRADE HSFG BOLTS IN S275

Diameter of Bolt	Min. Shank Tension	Tension		Shear Capacity		Slip Resistance for $\mu = 0.5$		Bearing Capacity, P_{bg} in kN End distance equal to 3 x bolt diameter. Thickness in mm of ply passed through.									
				Single Shear	Double Shear	Single Shear	Double Shear										
mm	P_o kN	$1.1P_o$ kN	$A_t p_t$ kN	kN	kN	kN	kN	5	6	7	8	9	10	12	15	20	25	30
16	104	114	110	62.8	126	57.1	114	**55.2**	66.2	77.3	88.3	99.4	110	*132*	*166*	*221*	*276*	*331*
20	162	178	172	98.0	196	89.0	178	**69.0**	**82.8**	**96.6**	110	124	138	166	*207*	*276*	*345*	*414*
22	200	220	212	121	242	110	220	**75.9**	**91.1**	**106**	121	137	152	182	*228*	*304*	*380*	*455*
24	233	257	247	141	282	128	257	**82.8**	**99.4**	**116**	**132**	149	166	199	248	*331*	*414*	*497*
27	303	333	321	184	367	167	333	**93.2**	**112**	**130**	**149**	**168**	186	224	279	*373*	*466*	*559*
30	370	407	393	224	449	204	407	**104**	**124**	**145**	**166**	**186**	**207**	248	311	*414*	*518*	*621*

Values in **bold** are less than the single shear capacity of the bolt.
Values in *italic* are greater than the double shear capacity of the bolt.
Shading indicates that the ply thickness is not suitable for an outer ply.
FOR EXPLANATION OF TABLES SEE NOTE 10.1

BS 5950-1: 2000
BS 4395: 1969
BS 4604: 1970

BOLT CAPACITIES

PRELOADED HSFG BOLTS: NON-SLIP UNDER FACTORED LOADS

GENERAL GRADE HSFG BOLTS IN S275

Diameter of Bolt	Min. Shank Tension	Bolt Tension Capacity	Slip Resistance P_{sL}							
			$\mu = 0.2$		$\mu = 0.3$		$\mu = 0.4$		$\mu = 0.5$	
			Single Shear	Double Shear	Single Shear	Double Shear	Single Shear	Double Shear	Single Shear	Double Shear
mm	P_o kN	$0.9 P_o$ kN	kN	kN	kN	kN	kN	kN	kN	kN
12	49.4	44.5	8.89	17.8	13.3	26.7	17.8	35.6	22.2	44.5
16	92.1	82.9	16.6	33.2	24.9	49.7	33.2	66.3	41.4	82.9
20	144	130	25.9	51.8	38.9	77.8	51.8	104	64.8	130
22	177	159	31.9	63.7	47.8	95.6	63.7	127	79.7	159
24	207	186	37.3	74.5	55.9	112	74.5	149	93.2	186
27	234	211	42.1	84.2	63.2	126	84.2	168	105	211
30	286	257	51.5	103	77.2	154	103	206	129	257

FOR EXPLANATION OF TABLES SEE NOTE 10.1

HIGHER GRADE HSFG BOLTS IN S 275

Diameter of Bolt	Min. Shank Tension	Bolt Tension Capacity	Slip Resistance P_{sL}							
			$\mu = 0.2$		$\mu = 0.3$		$\mu = 0.4$		$\mu = 0.5$	
			Single Shear	Double Shear	Single Shear	Double Shear	Single Shear	Double Shear	Single Shear	Double Shear
mm	P_o kN	$0.9 P_o$ kN	kN	kN	kN	kN	kN	kN	kN	kN
16	104	93.5	18.7	37.4	28.1	56.1	37.4	74.8	46.8	93.5
20	162	146	29.1	58.2	43.7	87.4	58.2	116	72.8	146
22	200	180	36.0	72.1	54.1	108	72.1	144	90.1	180
24	233	210	42.0	84.0	63.0	126	84.0	168	105	210
27	303	273	54.5	109	81.8	164	109	218	136	273
30	370	333	66.6	133	100	200	133	266	167	333

FOR EXPLANATION OF TABLES SEE NOTE 10.1

BS 5950-1: 2000
BS 4395: 1969
BS 4604: 1970
BS 4933: 1973

BOLT CAPACITIES

PRELOADED HSFG BOLTS: NON-SLIP IN SERVICE

GENERAL GRADE COUNTERSUNK HSFG BOLTS IN S275

Diameter of Bolt	Min. Shank Tension	Tension		Shear Capacity		Slip Resistance for $\mu = 0.5$		Bearing Capacity, P_{bg} in kN — End distance equal to 3 × bolt diameter. Thickness in mm of ply passed through.									
				Single Shear	Double Shear	Single Shear	Double Shear										
mm	P_o kN	$1.1P_o$ kN	$A_t p_t$ kN	kN	kN	kN	kN	5	6	7	8	9	10	12	15	20	25	30
12	49.4	54.3	49.7	33.7	67.4	27.2	54.3	**16.6**	**24.8**	**33.1**	41.4	49.7	58.0	*74.5*	*99.4*	*141*	*182*	*224*
16	92.1	101	92.6	62.8	126	50.7	101	**11.0**	**22.1**	**33.1**	**44.2**	**55.2**	66.2	88.3	121	*177*	*232*	*287*
20	144	158	145	98.0	196	79.2	158	**0**	**13.8**	**27.6**	**41.4**	**55.2**	**69.0**	**96.6**	138	*207*	*276*	*345*
22	177	195	179	121	242	97.4	195	**0**	**7.59**	**22.8**	**38.0**	**53.1**	**68.3**	**98.7**	144	220	*296*	*372*
24	207	228	208	141	282	114	228	**0**	**0**	**16.6**	**33.1**	**49.7**	**66.2**	**99.4**	149	232	*315*	*397*
27	234	257	236	161	321	129	257	**0**	**0**	**4.66**	**23.3**	**41.9**	**60.5**	**97.8**	154	247	*340*	*433*
30	286	315	289	196	393	157	315	**0**	**0**	**0**	**10.4**	**31.1**	**51.8**	**93.2**	155	259	362	*466*

Values in **bold** are less than the single shear capacity of the bolt.
Values in *italic* are greater than the double shear capacity of the bolt.
Shading indicates that the ply thickness is not suitable for an outer ply.
FOR EXPLANATION OF TABLES SEE NOTE 10.1

HIGHER GRADE COUNTERSUNK HSFG BOLTS IN S275

Diameter of Bolt	Min. Shank Tension	Tension		Shear Capacity		Slip Resistance for $\mu = 0.5$		Bearing Capacity, P_{bg} in kN — End distance equal to 3 × bolt diameter. Thickness in mm of ply passed through.									
				Single Shear	Double Shear	Single Shear	Double Shear										
mm	P_o kN	$1.1P_o$ kN	$A_t p_t$ kN	kN	kN	kN	kN	5	6	7	8	9	10	12	15	20	25	30
16	104	114	110	62.8	126	57.1	114	**11.0**	**22.1**	**33.1**	**44.2**	**55.2**	66.2	88.3	121	*177*	*232*	*287*
20	162	178	172	98.0	196	89.0	178	**0**	**13.8**	**27.6**	**41.4**	**55.2**	**69.0**	**96.6**	138	*207*	*276*	*345*
22	200	220	212	121	242	110	220	**0**	**7.59**	**22.8**	**38.0**	**53.1**	**68.3**	**98.7**	144	220	*296*	*372*
24	233	257	247	141	282	128	257	**0**	**0**	**16.6**	**33.1**	**49.7**	**66.2**	**99.4**	149	232	*315*	*397*
27	303	333	321	184	367	167	333	**0**	**0**	**4.66**	**23.3**	**41.9**	**60.5**	**97.8**	154	247	340	*433*
30	370	407	393	224	449	204	407	**0**	**0**	**0**	**10.4**	**31.1**	**51.8**	**93.2**	155	259	362	*466*

Values in **bold** are less than the single shear capacity of the bolt.
Values in *italic* are greater than the double shear capacity of the bolt.
Shading indicates that the ply thickness is not suitable for an outer ply.
FOR EXPLANATION OF TABLES SEE NOTE 10.1

	BS 5950-1: 2000
	BS 4395: 1969
	BS 4604: 1970
	BS 4933: 1973

BOLT CAPACITIES

PRELOADED HSFG BOLTS: NON-SLIP UNDER FACTORED LOADS

GENERAL GRADE COUNTERSUNK HSFG BOLTS IN S275

Diameter of Bolt	Min. Shank Tension	Bolt Tension Capacity	Slip Resistance P_{sL}							
			$\mu = 0.2$		$\mu = 0.3$		$\mu = 0.4$		$\mu = 0.5$	
			Single Shear	Double Shear	Single Shear	Double Shear	Single Shear	Double Shear	Single Shear	Double Shear
mm	P_o kN	$0.9P_o$ kN	kN	kN	kN	kN	kN	kN	kN	kN
12	49.4	44.5	8.89	17.8	13.3	26.7	17.8	35.6	22.2	44.5
16	92.1	82.9	16.6	33.2	24.9	49.7	33.2	66.3	41.4	82.9
20	144	130	25.9	51.8	38.9	77.8	51.8	104	64.8	130
22	177	159	31.9	63.7	47.8	95.6	63.7	127	79.7	159
24	207	186	37.3	74.5	55.9	112	74.5	149	93.2	186
27	234	211	42.1	84.2	63.2	126	84.2	168	105	211
30	286	257	51.5	103	77.2	154	103	206	129	257

FOR EXPLANATION OF TABLES SEE NOTE 10.1

HIGHER GRADE COUNTERSUNK HSFG BOLTS IN S 275

Diameter of Bolt	Min. Shank Tension	Bolt Tension Capacity	Slip Resistance P_{sL}							
			$\mu = 0.2$		$\mu = 0.3$		$\mu = 0.4$		$\mu = 0.5$	
			Single Shear	Double Shear	Single Shear	Double Shear	Single Shear	Double Shear	Single Shear	Double Shear
mm	P_o kN	$0.9P_o$ kN	kN	kN	kN	kN	kN	kN	kN	kN
16	104	93.5	18.7	37.4	28.1	56.1	37.4	74.8	46.8	93.5
20	162	146	29.1	58.2	43.7	87.4	58.2	116	72.8	146
22	200	180	36.0	72.1	54.1	108	72.1	144	90.1	180
24	233	210	42.0	84.0	63.0	126	84.0	168	105	210
27	303	273	54.5	109	81.8	164	109	218	136	273
30	370	333	66.6	133	100	200	133	266	167	333

FOR EXPLANATION OF TABLES SEE NOTE 10.1

BS 5950-1 :2000
BS EN 440
BS EN 499
BS EN 756
BS EN 758
BS EN 1668

FILLET WELDS

WELD CAPACITIES WITH E35 ELECTRODE WITH S275

Leg Length s mm	Throat Thickness a mm	Longitudinal Capacity P_L kN/mm	Transverse Capacity P_T kN/mm
3.0	2.1	0.462	0.577
4.0	2.8	0.616	0.770
5.0	3.5	0.770	0.963
6.0	4.2	0.924	1.155
8.0	5.6	1.232	1.540
10.0	7.0	1.540	1.925
12.0	8.4	1.848	2.310
15.0	10.5	2.310	2.888
18.0	12.6	2.772	3.465
20.0	14.0	3.080	3.850
22.0	15.4	3.388	4.235
25.0	17.5	3.850	4.813

Welds are between two elements at 90° to each other.
$P_L = p_w a$
$P_T = K p_w a$
$p_w = 220$ N/mm^2
$K = 1.25$ for elements at 90° to each other.
FOR EXPLANATION OF TABLES SEE NOTE 10.2

[BLANK PAGE]

[BLANK PAGE]

[BLANK PAGE]

MEMBER CAPACITIES

S355

BS 5950-1: 2000
BS 4-1: 1993

BENDING

UB SECTIONS SUBJECT TO BENDING

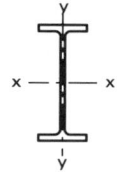

RESTRAINED BEAM CAPACITIES FOR S355

Section Designation	Section Classification	Shear Capacity P_v kN	Moment Capacity M_{cx} kNm	Ultimate U.D.L. Capacity (kN) for Restrained Beams for Lengths, L (m)												
				2.0	3.0	4.0	5.0	6.0	7.0	8.0	9.0	10.0	11.0	12.0	13.0	14.0
1016x305x487 # +	Plastic	6250	7770	####	####	####	####	9430	8490	7640	6910	6220	5650	5180	4780	4440
1016x305x437 # +	Plastic	5550	6950	####	####	####	9350	8420	7580	6830	6180	5560	5060	4640	4280	3970
1016x305x393 # +	Plastic	4980	6210	9970	9970	9280	8370	7540	6780	6100	5520	4970	4520	4140	3820	3550
1016x305x349 # +	Plastic	4400	5720	8810	8810	8400	7600	6860	6190	5590	5090	4580	4160	3820	3520	3270
1016x305x314 # +	Plastic	3950	5120	7910	7910	7540	6820	6150	5550	5010	4550	4100	3730	3420	3150	2930
1016x305x272 # +	Plastic	3380	4430	6760	6760	6490	5870	5300	4780	4320	3930	3540	3220	2950	2720	2530
1016x305x249 # +	Plastic	3350	3920	6700	6700	6050	5420	4850	4340	3920	3480	3130	2850	2610	2410	2240
1016x305x222 # +	Plastic	3210	3380	6430	6140	5480	4860	4310	3810	3380	3010	2710	2460	2260	2080	1930
914 x 419 x 388 #	Plastic	4080	6100	8160	8160	8160	7740	7030	6390	5810	5300	4880	4430	4060	3750	3480
914 x 419 x 343 #	Plastic	3660	5340	7320	7320	7320	6850	6220	5640	5120	4660	4270	3880	3560	3290	3050
914 x 305 x 289 #	Plastic	3740	4340	7480	7480	6770	6050	5400	4820	4340	3850	3470	3150	2890	2670	2480
914 x 305 x 253 #	Plastic	3290	3770	6580	6580	5930	5290	4720	4200	3770	3350	3020	2740	2520	2320	2160
914 x 305 x 224 #	Plastic	3000	3290	5990	5910	5270	4690	4160	3690	3290	2920	2630	2390	2190	2020	1880
914 x 305 x 201 #	Plastic	2820	2880	5650	5370	4760	4200	3700	3290	2880	2560	2300	2100	1920	1770	1650
838 x 292 x 226 #	Plastic	2840	3160	5670	5670	5070	4500	3990	3540	3160	2810	2530	2300	2110	1940	1800
838 x 292 x 194 #	Plastic	2560	2640	5120	4950	4370	3850	3390	3010	2640	2340	2110	1920	1760	1620	1510
838 x 292 x 176 #	Plastic	2420	2350	4840	4540	3990	3490	3050	2680	2350	2090	1880	1710	1570	1450	1340
762 x 267 x 197	Plastic	2490	2470	4970	4780	4190	3670	3210	2830	2470	2200	1980	1800	1650	1520	*1410*
762 x 267 x 173	Plastic	2260	2140	4510	4230	3690	3210	2800	2440	2140	1900	1710	1560	1430	*1320*	*1220*
762 x 267 x 147	Plastic	2000	1780	4000	3630	3140	2720	2370	2030	1780	1580	1420	1290	1190	*1090*	*1020*
762 x 267 x 134	Compact	1920	1650	3830	3420	2950	2540	2200	1880	1650	1470	1320	1200	1100	*1010*	*942*
686 x 254 x 170	Plastic	2080	1940	4160	3920	3400	2940	2550	2220	1940	1730	1550	1410	*1300*	*1200*	*1110*
686 x 254 x 152	Plastic	1880	1730	3760	3510	3040	2620	2270	1970	1730	1530	1380	1250	*1150*	*1060*	*986*
686 x 254 x 140	Plastic	1750	1570	3510	3240	2790	2410	2100	1800	1570	1400	1260	1140	*1050*	*968*	*899*
686 x 254 x 125	Plastic	1640	1380	3280	2920	2500	2140	1840	1570	1380	1220	1100	1000	*919*	*848*	*787*
610 x 305 x 238	Plastic	2420	2580	4840	4840	4350	3800	3320	2920	2580	2300	2070	1880	*1720*	*1590*	*1480*
610 x 305 x 179	Plastic	1810	1910	3620	3620	3240	2830	2470	2170	1910	1700	1530	*1390*	*1280*	*1180*	*1090*
610 x 305 x 149	Plastic	1500	1580	2990	2990	2690	2340	2040	1790	1580	1410	1270	*1150*	*1060*	*975*	*906*
610 x 229 x 140	Plastic	1670	1430	3350	3040	2600	2210	1910	1630	1430	1270	1140	*1040*	*953*	*879*	*817*
610 x 229 x 125	Plastic	1510	1270	3020	2720	2320	1970	1690	1450	1270	1130	1010	*922*	*845*	*780*	*725*
610 x 229 x 113	Plastic	1400	1130	2790	2470	2090	1770	1510	1290	1130	1010	906	*823*	*755*	*697*	*647*
610 x 229 x 101	Plastic	1350	1020	2700	2290	1920	1620	1360	1170	1020	909	818	*744*	*682*	*629*	*584*

+ Section is not given in BS 4-1: 1993.
# Check availability.
Section classification given applies to members subject to bending about the x-x axis only.
Loads given are the total Ultimate factored uniformaly distributed load supported over a beam span L assuming full lateral restraint to the compression flange. Self weight of the section has not been allowed for.
UDL values in **bold type** are governed by the shear capacity.
The unfactored imposed load is assumed to be 40% of the ultimate load given. Unless otherwise indicated, it is only necessary to perform deflection checks if the ratio of unfactored imposed load to dead load is greater than 1.556.
UDL values to the right of the zigzag line in *italic type* may be susceptible to serviceability deflections > L/360 and so should be checked under imposed loading. (This deflection limit is for beams which carry a brittle finish.)
FOR EXPLANATION OF TABLES SEE NOTE 5.1

| BS 5950-1: 2000 |
| BS 4-1: 1993 |

BENDING

UB SECTIONS
SUBJECT TO BENDING

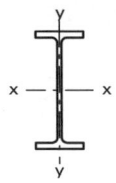

RESTRAINED BEAM CAPACITIES FOR S355

Section Designation	Section Classification	Shear Capacity P_v kN	Moment Capacity M_{cx} kNm	Ultimate U.D.L. Capacity (kN) for Restrained Beams for Lengths, L (m)											
				2.0	3.0	4.0	5.0	6.0	7.0	8.0	9.0	10.0	11.0	12.0	13.0	14.0
533 x 210 x 122	Plastic	1430	1100	**2860**	2480	2070	1740	1470	1260	1100	980	882	802	735	679	630
533 x 210 x 109	Plastic	1300	976	**2590**	2210	1850	1540	1300	1120	976	867	781	710	650	600	558
533 x 210 x 101	Plastic	1200	901	**2400**	2050	1710	1430	1200	1030	901	801	721	655	601	555	515
533 x 210 x 92	Plastic	1150	838	**2290**	1920	1600	1340	1120	957	838	745	670	609	559	516	479
533 x 210 x 82	Plastic	1080	731	2090	1720	1420	1170	975	835	731	650	585	532	487	450	418
457 x 191 x 98	Plastic	1100	770	**2200**	1810	1490	1230	1030	880	770	684	616	560	513	474	440
457 x 191 x 89	Plastic	1010	695	**2010**	1640	1350	1110	926	794	695	618	556	505	463	428	397
457 x 191 x 82	Plastic	970	650	1910	1550	1270	1040	867	743	650	578	520	473	433	400	371
457 x 191 x 74	Plastic	876	587	1720	1400	1140	939	782	671	587	522	469	427	391	361	335
457 x 191 x 67	Plastic	821	522	1570	1270	1030	836	696	597	522	464	418	380	348	321	298
457 x 152 x 82	Plastic	1010	625	1900	1530	1230	1000	833	714	625	555	500	454	417	384	357
457 x 152 x 74	Plastic	918	561	1710	1380	1110	898	748	642	561	499	449	408	374	345	321
457 x 152 x 67	Plastic	878	516	1600	1280	1030	825	688	590	516	459	413	375	344	317	295
457 x 152 x 60	Plastic	784	457	1430	1140	914	731	609	522	457	406	366	332	305	281	261
457 x 152 x 52	Plastic	728	389	1260	990	778	623	519	445	389	346	311	283	259	239	222
406 x 178 x 74	Plastic	835	533	1620	1300	1050	853	710	609	533	474	426	388	355	328	304
406 x 178 x 67	Plastic	767	478	1470	1180	942	765	637	546	478	425	382	348	319	294	273
406 x 178 x 60	Plastic	684	426	1310	1050	839	681	568	486	426	378	341	310	284	262	243
406 x 178 x 54	Compact	660	375	1200	944	749	599	499	428	375	333	300	272	250	230	214

Section classification given applies to members subject to bending about the x-x axis only.
Loads given are the total Ultimate factored uniformily distributed load supported over a beam span L assuming full lateral restraint to the compression flange. Self weight of the section has not been allowed for.
UDL values in **bold type** are governed by the shear capacity.
The unfactored imposed load is assumed to be 40% of the ultimate load given. Unless otherwise indicated, it is only necessary to perform deflection checks if the ratio of unfactored imposed load to dead load is greater than 1.556.
UDL values to the right of the zigzag line in *italic type* may be susceptible to serviceability deflections > L/360 and so should be checked under imposed loading. (This deflection limit is for beams which carry a brittle finish.)
Shaded UDL values may be susceptible to serviceability deflections > L/200 and so should be checked under imposed loading. (This deflection limit is for beams which do not carry a brittle finish.)
FOR EXPLANATION OF TABLES SEE NOTE 5.1

BENDING

UB SECTIONS
SUBJECT TO BENDING

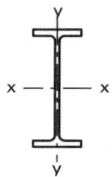

RESTRAINED BEAM CAPACITIES FOR S355

Section Designation	Section Classification	Shear Capacity P_v kN	Moment Capacity M_{cx} kNm	Ultimate U.D.L. Capacity (kN) for Restrained Beams for Lengths, L (m)											
				1.0	1.5	2.0	2.5	3.0	3.5	4.0	4.5	5.0	5.5	6.0	6.5	7.0
406 x 140 x 46	Plastic	584	315	**1170**	1160	1030	911	804	710	630	560	504	459	420	388	360
406 x 140 x 39	Compact	543	257	**1090**	1010	885	771	672	587	514	457	411	374	343	316	294
356 x 171 x 67	Plastic	704	430	**1410**	**1410**	1360	1200	1070	952	851	764	688	625	573	529	491
356 x 171 x 57	Plastic	618	359	**1240**	**1240**	1160	1020	904	801	717	637	574	522	478	441	410
356 x 171 x 51	Plastic	560	318	**1120**	**1120**	1030	912	806	713	636	565	509	463	424	391	364
356 x 171 x 45	Semi-compact	524	275	**1050**	**1050**	922	808	708	628	550	489	440	400	367	338	314
356 x 127 x 39	Plastic	497	234	**994**	939	815	707	613	535	468	416	374	340	312	288	267
356 x 127 x 33	Plastic	446	193	**892**	805	692	594	514	441	386	343	308	280	257	237	220
305 x 165 x 54	Plastic	522	300	**1040**	**1040**	987	867	763	674	601	534	481	437	400	370	343
305 x 165 x 46	Plastic	438	256	**875**	**875**	837	735	647	572	511	454	409	372	341	315	292
305 x 165 x 40	Compact	388	221	**775**	**775**	731	641	563	497	442	393	354	322	295	272	253
305 x 127 x 48	Plastic	596	252	**1190**	1080	919	783	673	577	505	449	404	367	337	311	288
305 x 127 x 42	Plastic	523	218	**1050**	938	798	679	581	498	436	388	349	317	291	268	249
305 x 127 x 37	Plastic	460	191	**921**	825	701	596	510	437	383	340	306	278	255	236	219
305 x 102 x 33	Plastic	440	171	**879**	754	637	537	455	390	342	304	273	248	228	210	195
305 x 102 x 28	Plastic	395	143	**772**	650	543	458	382	327	286	254	229	208	191	176	164
305 x 102 x 25	Plastic	377	121	**696**	576	473	389	324	278	243	216	194	177	162	149	139
254 x 146 x 43	Plastic	398	201	**796**	**796**	703	606	524	459	402	357	321	292	268	247	230
254 x 146 x 37	Plastic	344	171	**687**	**687**	603	518	448	392	343	305	274	249	229	211	196
254 x 146 x 31	Compact	321	140	**643**	605	511	433	372	319	279	248	223	203	186	172	159
254 x 102 x 28	Plastic	349	125	**699**	580	480	401	334	286	251	223	201	182	167	154	143
254 x 102 x 25	Plastic	329	109	**631**	517	423	348	290	248	217	193	174	158	145	134	124
254 x 102 x 22	Plastic	308	91.9	**561**	452	368	294	245	210	184	163	147	134	123	113	105
203 x 133 x 30	Plastic	282	111	**564**	510	422	352	297	255	223	198	178	162	149	137	127
203 x 133 x 25	Compact	247	91.6	**493**	428	351	293	244	209	183	163	147	133	122	113	105
203 x 102 x 23	Plastic	234	83.1	**467**	394	321	266	222	190	166	148	133	121	111	102	94.9
178 x 102 x 19	Plastic	182	60.7	**364**	296	238	194	162	139	121	108	97.1	88.3	80.9	74.7	69.4
152 x 89 x 16	Plastic	146	43.7	**284**	220	175	140	116	99.8	87.3	77.6	69.9	63.5	58.2	53.7	49.9
127 x 76 x 13	Plastic	108	29.9	**204**	154	120	95.7	79.7	68.3	59.8	53.1	47.8	43.5	39.9	36.8	34.2

Section classification given applies to members subject to bending about the x-x axis only.
Loads given are the total Ultimate factored uniformally distributed load supported over a beam span L assuming full lateral restraint to the compression flange. Self weight of the section has not been allowed for.
UDL values in **bold type** are governed by the shear capacity.
The unfactored imposed load is assumed to be 40% of the ultimate load given. Unless otherwise indicated, it is only necessary to perform deflection checks if the ratio of unfactored imposed load to dead load is greater than 1.556.
UDL values to the right of the zigzag line in *italic type* may be susceptible to serviceability deflections > L/360 and so should be checked under imposed loading. (This deflection limit is for beams which carry a brittle finish.)
Shaded UDL values may be susceptible to serviceability deflections > L/200 and so should be checked under imposed loading. (This deflection limit is for beams which do not carry a brittle finish.)
FOR EXPLANATION OF TABLES SEE NOTE 5.1

BS 5950-1: 2000
BS 4-1: 1993

BENDING

UC SECTIONS
SUBJECT TO BENDING

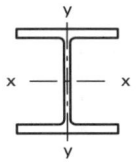

RESTRAINED BEAM CAPACITIES FOR S355

Section Designation	Section Classification	Shear Capacity P_v kN	Moment Capacity M_{cx} kNm	Ultimate U.D.L. Capacity (kN) for Restrained Beams for Lengths, L (m)											
				2.0	3.0	4.0	5.0	6.0	7.0	8.0	9.0	10.0	11.0	12.0	13.0	14.0
356 x 406 x 634 #	Plastic	4410	4520	8810	8810	8060	6940	6020	5160	4520	*4010*	*3610*	*3280*	*3010*	*2780*	*2580*
356 x 406 x 551 #	Plastic	3740	3890	7480	7480	6860	5900	5110	4440	3890	*3450*	*3110*	*2830*	*2590*	*2390*	*2220*
356 x 406 x 467 #	Plastic	3140	3350	6280	6280	5850	5030	4350	3800	3350	*2980*	*2680*	*2440*	*2230*	*2060*	*1910*
356 x 406 x 393 #	Plastic	2580	2750	5150	5150	4820	4140	3580	3120	2750	*2450*	*2200*	*2000*	*1840*	*1690*	*1570*
356 x 406 x 340 #	Plastic	2170	2340	4350	4350	4110	3520	3050	2660	*2340*	*2080*	*1880*	*1710*	*1560*	*1440*	*1340*
356 x 406 x 287 #	Plastic	1840	2010	3680	3680	3510	3010	2600	*2270*	*2010*	*1780*	*1600*	*1460*	*1340*	*1230*	*1150*
356 x 406 x 235 #	Plastic	1450	1620	2900	2900	2820	2420	2100	*1830*	*1620*	*1440*	*1290*	*1180*	*1080*	*995*	*924*
356 x 368 x 202 #	Plastic	1280	1370	2560	2560	2420	2070	1790	*1560*	*1370*	*1220*	*1100*	*997*	*914*	*843*	*783*
356 x 368 x 177 #	Plastic	1100	1190	2200	2200	2100	1800	1550	*1350*	*1190*	*1060*	*954*	*867*	*795*	*734*	*681*
356 x 368 x 153 #	Semi-compact	922	1020	1840	1840	1790	1540	1330	*1160*	*1020*	*909*	*818*	*743*	*681*	*629*	*584*
356 x 368 x 129 #	Semi-compact	766	821	1530	1530	1460	1240	1070	*933*	*821*	*730*	*657*	*597*	*548*	*505*	*469*
305 x 305 x 283	Plastic	1970	1710	3940	3840	3190	2670	2280	*1950*	*1710*	*1520*	*1370*	*1240*	*1140*	*1050*	*977*
305 x 305 x 240	Plastic	1680	1470	3360	3300	2730	2290	1950	*1670*	*1470*	*1300*	*1170*	*1070*	*977*	*902*	*837*
305 x 305 x 198	Plastic	1340	1190	2690	2670	2210	1850	1580	*1360*	*1190*	*1050*	*949*	*863*	*791*	*730*	*678*
305 x 305 x 158	Plastic	1070	925	2140	2100	1740	1450	1230	*1060*	*925*	*822*	*740*	*672*	*616*	*569*	*528*
305 x 305 x 137	Plastic	916	792	1830	1810	1490	1240	*1060*	*906*	*792*	*704*	*634*	*576*	*528*	*488*	*453*
305 x 305 x 118	Compact	781	676	1560	1540	1270	1060	*901*	*772*	*676*	*600*	*540*	*491*	*450*	*416*	*386*
305 x 305 x 97	Semi-compact	649	548	1300	1260	1040	*862*	*730*	*626*	*548*	*487*	*438*	*398*	*365*	*337*	*313*
254 x 254 x 167	Plastic	1150	836	2300	2020	1630	*1340*	*1120*	*956*	*836*	*743*	*669*	*608*	*558*	*515*	*478*
254 x 254 x 132	Plastic	875	645	1750	1560	1250	*1030*	*860*	*737*	*645*	*573*	*516*	*469*	*430*	*397*	*368*
254 x 254 x 107	Plastic	707	512	1410	1240	997	*819*	*683*	*585*	*512*	*455*	*410*	*372*	*341*	*315*	*293*
254 x 254 x 89	Plastic	555	422	1110	1020	818	*671*	*563*	*483*	*422*	*375*	*338*	*307*	*282*	*260*	*241*
254 x 254 x 73	Semi-compact	465	350	*931*	*846*	*680*	*557*	*467*	*400*	*350*	*311*	*280*	*255*	*234*	*216*	*200*
203 x 203 x 86	Plastic	584	337	1150	*865*	*674*	*539*	*449*	*385*	*337*	*300*	*270*	*245*	*225*	*207*	*193*
203 x 203 x 71	Plastic	447	276	**893**	*701*	*547*	*441*	*368*	*315*	*276*	*245*	*221*	*200*	*184*	*170*	*158*
203 x 203 x 60	Plastic	420	233	*807*	*602*	*466*	*373*	*311*	*266*	*233*	*207*	*186*	*169*	*155*	*143*	*133*
203 x 203 x 52	Compact	347	201	*689*	*518*	*403*	*322*	*268*	*230*	*201*	*179*	*161*	*146*	*134*	*124*	*115*
203 x 203 x 46	Semi-compact	312	174	*603*	*450*	*348*	*278*	*232*	*199*	*174*	*155*	*139*	*127*	*116*	*107*	*99.4*
152 x 152 x 37	Plastic	276	110	*419*	*293*	*219*	*176*	*146*	*125*	*110*	*97.5*	*87.8*	*79.8*	*73.1*	*67.5*	*62.7*
152 x 152 x 30	Compact	218	88.0	*336*	*235*	*176*	*141*	*117*	*101*	*88.0*	*78.5*	*70.4*	*64.0*	*58.7*	*54.2*	*50.3*
152 x 152 x 23	Semi-compact	188	60.5	*239*	*161*	*121*	*96.8*	*80.7*	*69.2*	*60.5*	*53.8*	*48.4*	*44.0*	*40.3*	*37.2*	*34.6*

\# Check availability.
Section classification given applies to members subject to bending about the x-x axis only.
M_{cx} values in *italic type* are governed by $1.2p_yZ$ and a higher value may be used in some circumstances.
Loads given are the total Ultimate factored uniformaly distributed load supported over a beam span L assuming full lateral restraint to the compression flange. Self weight of the section has not been allowed for.
UDL values in **bold type** are governed by the shear capacity.
The unfactored imposed load is assumed to be 40% of the ultimate load given. Unless otherwise indicated, it is only necessary to perform deflection checks if the ratio of unfactored imposed load to dead load is greater than 1.556.
UDL values to the right of the zigzag line in *italic type* may be susceptible to serviceability deflections > L/360 and so should be checked under imposed loading. (This deflection limit is for beams which carry a brittle finish.)
Shaded UDL values may be susceptible to serviceability deflections > L/200 and so should be checked under imposed loading. (This deflection limit is for beams which do not carry a brittle finish.)
FOR EXPLANATION OF TABLES SEE NOTE 5.1

BS 5950-1: 2000
BS 4-1: 1993

BENDING

JOISTS SUBJECT TO BENDING

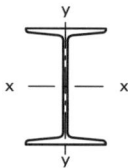

RESTRAINED BEAM CAPACITIES FOR S355

Section Designation	Section Classification	Shear Capacity P_v kN	Moment Capacity M_{cx} kNm	Ultimate U.D.L. Capacity (kN) for Restrained Beams for Lengths, L (m)											
				1.0	1.5	2.0	2.5	3.0	3.5	4.0	4.5	5.0	5.5	6.0	6.5	7.0
254 x 203 x 82 #	Plastic	536	372	**1070**	**1070**	**1070**	1030	915	814	729	656	595	540	495	457	425
254 x 114 x 37 #	Plastic	411	163	**822**	730	611	513	435	372	326	290	261	237	217	201	186
203 x 152 x 52 #	Plastic	374	187	**749**	**749**	668	570	490	427	373	332	299	271	249	230	213
152 x 127 x 37 #	Plastic	338	99.0	648	501	396	317	264	226	198	176	158	144	132	122	113
127 x 114 x 29 #	Plastic	276	64.3	461	339	257	206	171	147	129	114	103	93.5	85.7	79.1	73.4
127 x 114 x 27 #	Plastic	200	61.1	**400**	309	242	195	163	140	122	109	97.7	88.8	81.4	75.2	69.8
127 x 76 x 16 #	Plastic	151	36.9	262	194	148	118	98.5	84.4	73.8	65.6	59.1	53.7	49.2	45.4	42.2
114 x 114 x 27 #	Plastic	231	53.6	388	283	214	172	143	123	107	95.3	85.8	78.0	71.5	66.0	61.3
102 x 102 x 23 #	Plastic	206	40.1	305	214	160	128	107	91.7	80.2	71.3	64.2	58.3	53.5	49.4	45.8
102 x 44 x 7 #	Plastic	93.1	12.6	101	67.0	50.3	40.2	33.5	28.7	25.1	22.3	20.1	18.3	16.8	15.5	14.4
89 x 89 x 19 #	Plastic	180	29.4	231	157	117	93.9	78.3	67.1	58.7	52.2	47.0	42.7	39.1	36.1	33.6
76 x 76 x 15 #	Plastic	144	19.2	154	103	77.0	61.6	51.3	44.0	38.5	34.2	30.8	28.0	25.7	23.7	22.0
76 x 76 x 13 #	Plastic	82.8	17.3	131	92.2	69.2	55.3	46.1	39.5	34.6	30.7	27.7	25.1	23.1	21.3	19.8

Check availability.
Section classification given applies to members subject to bending about the x-x axis only.
Loads given are the total Ultimate factored uniformaly distributed load supported over a beam span L assuming full lateral restraint to the compression flange. Self weight of the section has not been allowed for.
UDL values in **bold type** are governed by the shear capacity.
The unfactored imposed load is assumed to be 40% of the ultimate load given. Unless otherwise indicated, it is only necessary to perform deflection checks if the ratio of unfactored imposed load to dead load is greater than 1.556.
UDL values to the right of the zigzag line in *italic type* may be susceptible to serviceability deflections > L/360 and so should be checked under imposed loading. (This deflection limit is for beams which carry a brittle finish.)
Shaded UDL values may be susceptible to serviceability deflections > L/200 and so should be checked under imposed loading. (This deflection limit is for beams which do not carry a brittle finish.)
FOR EXPLANATION OF TABLES SEE NOTE 5.1

BS 5950-1: 2000
BS 4-1: 1993

BENDING

PARALLEL FLANGE CHANNELS
SUBJECT TO BENDING

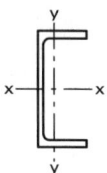

RESTRAINED BEAM CAPACITIES FOR S355

Section Designation	Section Classification	Shear Capacity P_v kN	Moment Capacity M_{cx} kNm	Ultimate U.D.L. Capacity (kN) for Restrained Beams for Lengths, L (m)											
				1.0	1.5	2.0	2.5	3.0	3.5	4.0	4.5	5.0	5.5	6.0	6.5	7.0
430 x 100 x 64	Plastic	979	422	**1960**	1720	1490	1290	1120	964	843	749	675	613	562	519	482
380 x 100 x 54	Plastic	747	322	**1490**	1330	1150	990	858	736	644	572	515	468	429	*396*	*368*
300 x 100 x 46	Plastic	559	221	**1120**	974	822	695	590	505	442	393	354	*322*	*295*	*272*	*253*
300 x 90 x 41	Plastic	575	202	1110	930	773	645	538	461	403	358	*323*	*293*	*269*	*248*	*230*
260 x 90 x 35	Plastic	443	151	863	711	584	483	402	345	302	268	*241*	*219*	*201*	*186*	*172*
260 x 75 x 28	Plastic	388	116	706	571	460	373	311	266	233	*207*	*186*	*169*	*155*	*143*	*133*
230 x 90 x 32	Plastic	367	126	732	598	489	403	336	288	252	224	*202*	*183*	*168*	*155*	*144*
230 x 75 x 26	Plastic	318	98.7	599	483	389	316	263	226	*197*	*175*	*158*	*144*	*132*	*121*	*113*
200 x 90 x 30	Plastic	298	103	**596**	494	401	331	275	236	*207*	*184*	*165*	*150*	*138*	*127*	*118*
200 x 75 x 23	Plastic	256	80.6	495	396	317	258	215	*184*	*161*	*143*	*129*	*117*	*107*	*99.2*	*92.1*
180 x 90 x 26	Plastic	249	82.4	**498**	402	323	264	220	*188*	*165*	*146*	*132*	*120*	*110*	*101*	*94.1*
180 x 75 x 20	Plastic	230	62.5	413	319	250	200	*167*	*143*	*125*	*111*	*100*	*90.9*	*83.3*	*76.9*	*71.4*
150 x 90 x 24	Plastic	208	63.5	410	319	252	*203*	*169*	*145*	*127*	*113*	*102*	*92.4*	*84.7*	*78.2*	*72.6*
150 x 75 x 18	Plastic	176	46.9	318	242	187	*150*	*125*	*107*	*93.7*	*83.3*	*75.0*	*68.2*	*62.5*	*57.7*	*53.6*
125 x 65 x 15 #	Plastic	146	31.9	234	170	128	*102*	*85.1*	*72.9*	*63.8*	*56.7*	*51.1*	*46.4*	*42.6*	*39.3*	*36.5*
100 x 50 x 10 #	Plastic	107	17.4	136	92.6	69.4	*55.6*	*46.3*	*39.7*	*34.7*	*30.9*	*27.8*	*25.3*	*23.1*	*21.4*	*19.8*

\# Check availability.
Section classification given applies to members subject to bending about the x-x axis only.
Loads given are the total Ultimate factored uniformaly distributed load supported over a beam span L assuming full lateral restraint to the compression flange. Self weight of the section has not been allowed for.
UDL values in **bold type** are governed by the shear capacity.
The unfactored imposed load is assumed to be 40% of the ultimate load given. Unless otherwise indicated, it is only necessary to perform deflection checks if the ratio of unfactored imposed load to dead load is greater than 1.556.
UDL values to the right of the zigzag line in *italic type* may be susceptible to serviceability deflections > L/360 and so should be checked under imposed loading. (This deflection limit is for beams which carry a brittle finish.)
Shaded UDL values may be susceptible to serviceability deflections > L/200 and so should be checked under imposed loading. (This deflection limit is for beams which do not carry a brittle finish.)
FOR EXPLANATION OF TABLES SEE NOTE 5.1

BS 5950-1: 2000
BS 4-1: 1993

WEB BEARING AND BUCKLING

UB SECTIONS

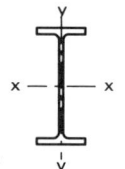

BEARING AND BUCKLING VALUES FOR UNSTIFFENED WEBS FOR S355

Section Designation	Web Thickness	Depth Between Fillets	Bearing				Buckling			Shear Capacity
			End Bearing		Continuous Over Bearing		End Bearing		Continuous Over Bearing	
			Beam Factor	Stiff Bearing Factor	Beam Factor	Stiff Bearing Factor	Buckling Factor	1.4d	Buckling Factor	
	t	d	C1	C2	C1	C2	C4		C4	P_v
	mm	mm	kN	kN/mm	kN	kN/mm	kN	mm	kN	kN
1016x305x487 # +	30.0	868	1690	10.1	4230	10.1	5350	1220	5350	6250
1016x305x437 # +	26.9	868	1420	9.01	3560	9.01	3850	1220	3850	5550
1016x305x393 # +	24.4	868	1210	8.17	3020	8.17	2880	1220	2880	4980
1016x305x349 # +	21.1	868	1020	7.28	2550	7.28	1860	1220	1860	4400
1016x305x314 # +	19.1	868	868	6.59	2170	6.59	1380	1220	1380	3950
1016x305x272 # +	16.5	868	694	5.69	1740	5.69	889	1220	889	3380
1016x305x249 # +	16.5	868	638	5.69	1590	5.69	889	1220	889	3350
1016x305x222 # +	16.0	868	564	5.52	1410	5.52	811	1220	811	3210
914 x 419 x 388 #	21.4	800	896	7.38	2240	7.38	2110	1120	2110	4080
914 x 419 x 343 #	19.4	800	751	6.69	1880	6.69	1570	1120	1570	3660
914 x 305 x 289 #	19.5	824	688	6.73	1720	6.73	1550	1150	1550	3740
914 x 305 x 253 #	17.3	824	561	5.97	1400	5.97	1080	1150	1080	3290
914 x 305 x 224 #	15.9	824	472	5.49	1180	5.49	838	1150	838	3000
914 x 305 x 201 #	15.1	824	409	5.21	1020	5.21	718	1150	718	2820
838 x 292 x 226 #	16.1	762	495	5.55	1240	5.55	942	1070	942	2840
838 x 292 x 194 #	14.7	762	401	5.07	1000	5.07	717	1070	717	2560
838 x 292 x 176 #	14.0	762	354	4.83	884	4.83	619	1070	619	2420
762 x 267 x 197	15.6	686	451	5.38	1130	5.38	951	960	951	2490
762 x 267 x 173	14.3	686	376	4.93	940	4.93	733	960	733	2260
762 x 267 x 147	12.8	686	300	4.42	751	4.42	525	960	525	2000
762 x 267 x 134	12.0	686	273	4.26	682	4.26	433	960	433	1920
686 x 254 x 170	14.5	615	389	5.00	973	5.00	852	861	852	2080
686 x 254 x 152	13.2	615	330	4.55	824	4.55	643	861	643	1880
686 x 254 x 140	12.4	615	293	4.28	732	4.28	533	861	533	1750
686 x 254 x 125	11.7	615	253	4.04	634	4.04	448	861	448	1640
610 x 305 x 238	18.4	540	608	6.35	1520	6.35	1980	756	1980	2420
610 x 305 x 179	14.1	540	390	4.86	975	4.86	892	756	892	1810
610 x 305 x 149	11.8	540	295	4.07	737	4.07	523	756	523	1500
610 x 229 x 140	13.1	548	315	4.52	786	4.52	706	767	706	1670
610 x 229 x 125	11.9	548	265	4.11	663	4.11	529	767	529	1510
610 x 229 x 113	11.1	548	230	3.83	574	3.83	429	767	429	1400
610 x 229 x 101	10.5	548	205	3.73	513	3.73	363	767	363	1350

+ Section is not given in BS 4-1: 1993.
# Check availability.
Web bearing capacity, $P_w = C1 + b_1 \, C2$
Web buckling resistance, $P_x = K \, (C4 \, P_w)^{0.5}$
For Continuous over bearing: K = 1.0
For End bearing: K = Min. { $(a_e / 1.4d) + 0.5$, 1.0 }
Where b_1 is the stiff bearing length and a_e is the distance from the load or reaction to the nearer end of the member.
FOR EXPLANATION OF TABLES SEE NOTE 6.1

BS 5950-1: 2000
BS 4-1: 1993

WEB BEARING AND BUCKLING

UB SECTIONS

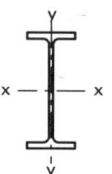

BEARING AND BUCKLING VALUES FOR UNSTIFFENED WEBS FOR S355

Section Designation	Web Thickness	Depth Between Fillets	Bearing				Buckling			Shear Capacity
			End Bearing		Continuous Over Bearing		End Bearing		Continuous Over Bearing	
			Beam Factor	Stiff Bearing Factor	Beam Factor	Stiff Bearing Factor	Buckling Factor	1.4d	Buckling Factor	
	t	d	C1	C2	C1	C2	C4		C4	P_v
	mm	mm	kN	kN/mm	kN	kN/mm	kN	mm	kN	kN
533 x 210 x 122	12.7	477	298	4.38	745	4.38	739	667	739	1430
533 x 210 x 109	11.6	477	252	4.00	630	4.00	563	667	563	1300
533 x 210 x 101	10.8	477	224	3.73	561	3.73	454	667	454	1200
533 x 210 x 92	10.1	477	203	3.59	507	3.59	372	667	372	1150
533 x 210 x 82	9.6	477	177	3.41	441	3.41	319	667	319	1080
457 x 191 x 98	11.4	408	234	3.93	586	3.93	625	571	625	1100
457 x 191 x 89	10.5	408	202	3.62	505	3.62	488	571	488	1010
457 x 191 x 82	9.9	408	184	3.51	460	3.51	409	571	409	970
457 x 191 x 74	9.0	408	158	3.20	395	3.20	307	571	307	876
457 x 191 x 67	8.5	408	138	3.02	346	3.02	259	571	259	821
457 x 152 x 82	10.5	408	211	3.62	527	3.62	488	571	488	1010
457 x 152 x 74	9.6	408	180	3.31	450	3.31	373	571	373	918
457 x 152 x 67	9.0	408	161	3.20	403	3.20	307	571	307	878
457 x 152 x 60	8.1	408	135	2.88	338	2.88	224	571	224	784
457 x 152 x 52	7.6	408	114	2.70	285	2.70	185	571	185	728
406 x 178 x 74	9.5	360	177	3.37	442	3.37	409	505	409	835
406 x 178 x 67	8.8	360	153	3.12	383	3.12	325	505	325	767
406 x 178 x 60	7.9	360	129	2.80	323	2.80	235	505	235	684
406 x 178 x 54	7.7	360	115	2.73	288	2.73	218	505	218	660
406 x 140 x 46	6.8	360	103	2.41	258	2.41	150	505	150	584
406 x 140 x 39	6.4	360	85.4	2.27	214	2.27	125	505	125	543
356 x 171 x 67	9.1	312	167	3.23	418	3.23	416	436	416	704
356 x 171 x 57	8.1	312	133	2.88	334	2.88	293	436	293	618
356 x 171 x 51	7.4	312	114	2.63	285	2.63	224	436	224	560
356 x 171 x 45	7.0	312	98.9	2.49	247	2.49	189	436	189	524
356 x 127 x 39	6.6	312	97.9	2.34	245	2.34	159	436	159	497
356 x 127 x 33	6.0	312	79.7	2.13	199	2.13	119	436	119	446
305 x 165 x 54	7.9	265	127	2.80	317	2.80	320	371	320	522
305 x 165 x 46	6.7	265	90.5	2.38	246	2.38	195	371	195	438
305 x 165 x 40	6.0	265	81.4	2.13	203	2.13	140	371	140	388

Web bearing capacity, $P_w = C1 + b_1 C2$
Web buckling resistance, $P_x = K (C4 \, P_w)^{0.5}$
For Continuous over bearing: $K = 1.0$
For End bearing: $K = \text{Min.} \{ (a_e / 1.4d) + 0.5 , 1.0 \}$
Where b_1 is the stiff bearing length and a_e is the distance from the load or reaction to the nearer end of the member.
FOR EXPLANATION OF TABLES SEE NOTE 6.1

WEB BEARING AND BUCKLING

UB SECTIONS

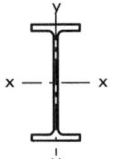

BEARING AND BUCKLING VALUES FOR UNSTIFFENED WEBS FOR S355

Section Designation	Web Thickness	Depth Between Fillets	Bearing				Buckling			Shear Capacity
			End Bearing		Continuous Over Bearing		End Bearing		Continuous Over Bearing	
			Beam Factor	Stiff Bearing Factor	Beam Factor	Stiff Bearing Factor	Buckling Factor	1.4d	Buckling Factor	
	t	d	C1	C2	C1	C2	C4		C4	P_v
	mm	mm	kN	kN/mm	kN	kN/mm	kN	mm	kN	kN
305 x 127 x 48	9.0	265	146	3.20	366	3.20	472	371	472	596
305 x 127 x 42	8.0	265	119	2.84	298	2.84	332	371	332	523
305 x 127 x 37	7.1	265	98.8	2.52	247	2.52	232	371	232	460
305 x 102 x 33	6.6	276	86.2	2.34	216	2.34	179	386	179	440
305 x 102 x 28	6.0	276	69.9	2.13	175	2.13	135	386	135	395
305 x 102 x 25	5.8	276	60.1	2.06	150	2.06	122	386	122	377
254 x 146 x 43	7.2	219	104	2.56	259	2.56	293	307	293	398
254 x 146 x 37	6.3	219	82.8	2.24	207	2.24	196	307	196	344
254 x 146 x 31	6.0	219	69.0	2.13	173	2.13	170	307	170	321
254 x 102 x 28	6.3	225	78.7	2.24	197	2.24	191	315	191	349
254 x 102 x 25	6.0	225	68.2	2.13	170	2.13	165	315	165	329
254 x 102 x 22	5.7	225	58.3	2.02	146	2.02	141	315	141	308
203 x 133 x 30	6.4	172	78.2	2.27	195	2.27	261	241	261	282
203 x 133 x 25	5.7	172	62.3	2.02	156	2.02	185	241	185	247
203 x 102 x 23	5.4	169	64.8	1.92	162	1.92	160	237	160	234
178 x 102 x 19	4.8	147	52.8	1.70	132	1.70	129	206	129	182
152 x 89 x 16	4.5	122	48.9	1.60	122	1.60	129	171	129	146
127 x 76 x 13	4.0	96.6	43.2	1.42	108	1.42	114	135	114	108

Web bearing capacity, $P_w = C1 + b_1 C2$
Web buckling resistance, $P_x = K (C4 \, P_w)^{0.5}$
For Continuous over bearing: K = 1.0
For End bearing: K = Min. { $(a_e / 1.4d) + 0.5$, 1.0 }
Where b_1 is the stiff bearing length and a_e is the distance from the load or reaction to the nearer end of the member.
FOR EXPLANATION OF TABLES SEE NOTE 6.1

BS 5950-1: 2000
BS 4-1: 1993

WEB BEARING AND BUCKLING

UC SECTIONS

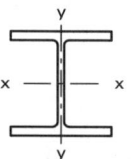

BEARING AND BUCKLING VALUES FOR UNSTIFFENED WEBS FOR S355

Section Designation	Web Thickness	Depth Between Fillets	Bearing				Buckling				Shear Capacity
			End Bearing		Continuous Over Bearing		End Bearing		Continuous Over Bearing		
			Beam Factor	Stiff Bearing Factor	Beam Factor	Stiff Bearing Factor	Buckling Factor	1.4d	Buckling Factor		
	t	d	C1	C2	C1	C2	C4		C4		P_v
	mm	mm	kN	kN/mm	kN	kN/mm	kN	mm	kN		kN
356 x 406 x 634 #	47.6	290	2850	15.5	7130	15.5	63900	406	63900		4410
356 x 406 x 551 #	42.1	290	2260	13.7	5660	13.7	44200	406	44200		3740
356 x 406 x 467 #	35.8	290	1760	12.0	4390	12.0	27200	406	27200		3140
356 x 406 x 393 #	30.6	290	1320	10.3	3300	10.3	17000	406	17000		2580
356 x 406 x 340 #	26.6	290	1040	8.91	2590	8.91	11100	406	11100		2170
356 x 406 x 287 #	22.6	290	806	7.80	2020	7.80	6840	406	6840		1840
356 x 406 x 235 #	18.4	290	576	6.35	1440	6.35	3690	406	3690		1450
356 x 368 x 202 #	16.5	290	480	5.69	1200	5.69	2660	406	2660		1280
356 x 368 x 177 #	14.4	290	388	4.97	969	4.97	1770	406	1770		1100
356 x 368 x 153 #	12.3	290	305	4.24	762	4.24	1100	406	1100		922
356 x 368 x 129 #	10.4	290	235	3.59	587	3.59	666	406	666		766
305 x 305 x 283	26.8	247	1060	8.98	2660	8.98	13400	345	13400		1970
305 x 305 x 240	23.0	247	840	7.94	2100	7.94	8480	345	8480		1680
305 x 305 x 198	19.1	247	614	6.59	1540	6.59	4850	345	4850		1340
305 x 305 x 158	15.8	247	438	5.45	1100	5.45	2750	345	2750		1070
305 x 305 x 137	13.8	247	351	4.76	878	4.76	1830	345	1830		916
305 x 305 x 118	12.0	247	281	4.14	702	4.14	1200	345	1200		781
305 x 305 x 97	9.9	247	215	3.51	538	3.51	676	345	676		649
254 x 254 x 167	19.2	200	588	6.62	1470	6.62	6070	280	6070		1150
254 x 254 x 132	15.3	200	401	5.28	1000	5.28	3070	280	3070		875
254 x 254 x 107	12.8	200	293	4.42	733	4.42	1800	280	1800		707
254 x 254 x 89	10.3	200	213	3.55	533	3.55	938	280	938		555
254 x 254 x 73	8.6	200	164	3.05	411	3.05	546	280	546		465
203 x 203 x 86	12.7	161	269	4.38	673	4.38	2190	225	2190		584
203 x 203 x 71	10.0	161	190	3.45	474	3.45	1070	225	1070		447
203 x 203 x 60	9.4	161	163	3.34	407	3.34	888	225	888		420
203 x 203 x 52	7.9	161	127	2.80	318	2.80	527	225	527		347
203 x 203 x 46	7.2	161	108	2.56	271	2.56	399	225	399		312
152 x 152 x 37	8.0	124	108	2.84	271	2.84	712	173	712		276
152 x 152 x 30	6.5	124	78.5	2.31	196	2.31	382	173	382		218
152 x 152 x 23	5.8	124	59.3	2.06	148	2.06	271	173	271		188

\# Check availability.
Web bearing capacity, $P_w = C1 + b_1 C2$
Web buckling resistance, $P_x = K (C4\, P_w)^{0.5}$
For Continuous over bearing: K = 1.0
For End bearing: K = Min. { $(a_e / 1.4d) + 0.5$, 1.0 }
Where b_1 is the stiff bearing length and a_e is the distance from the load or reaction to the nearer end of the member.
FOR EXPLANATION OF TABLES SEE NOTE 6.1

BS 5950-1: 2000
BS 4-1: 1993

WEB BEARING AND BUCKLING

JOISTS

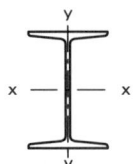

BEARING AND BUCKLING VALUES FOR UNSTIFFENED WEBS FOR S355

Section Designation	Web Thickness	Depth Between Fillets	Bearing				Buckling			Shear Capacity
			End Bearing		Continuous Over Bearing		End Bearing		Continuous Over Bearing	
			Beam Factor	Stiff Bearing Factor	Beam Factor	Stiff Bearing Factor	Buckling Factor	1.4d	Buckling Factor	
	t	d	C1	C2	C1	C2	C4		C4	P_v
	mm	mm	kN	kN/mm	kN	kN/mm	kN	mm	kN	kN
254 x 203 x 82 #	10.2	167	278	3.52	695	3.52	1090	233	1090	536
254 x 114 x 37 ‡	7.6	199	136	2.70	340	2.70	379	279	379	411
203 x 152 x 52 #	8.9	133	197	3.07	491	3.07	910	186	910	374
152 x 127 x 37 #	10.4	94.3	197	3.69	493	3.69	2050	132	2050	338
127 x 114 x 29 #	10.2	79.5	155	3.62	387	3.62	2290	111	2290	276
127 x 114 x 27 #	7.4	79.5	112	2.63	280	2.63	876	111	876	200
127 x 76 x 16 ‡	5.6	86.5	75.5	1.99	189	1.99	349	121	349	151
114 x 114 x 27 ‡	9.5	60.8	168	3.37	420	3.37	2420	85.1	2420	231
102 x 102 x 23 #	9.5	55.2	144	3.37	361	3.37	2670	77.3	2670	206
102 x 44 x 7 #	4.3	74.6	39.7	1.53	99.2	1.53	183	104	183	93.1
89 x 89 x 19 #	9.5	44.2	142	3.37	354	3.37	3330	61.9	3330	180
76 x 76 x 15 ‡	8.9	38.1	112	3.16	281	3.16	3180	53.3	3180	144
76 x 76 x 13 #	5.1	38.1	64.5	1.81	161	1.81	598	53.3	598	82.8

‡ Not available from some leading producers. Check availability.
# Check availability.
Web bearing capacity, $P_w = C1 + b_1 C2$
Web buckling resistance, $P_x = K (C4 P_w)^{0.5}$
For Continuous over bearing: $K = 1.0$
For End bearing: $K = \text{Min.} \{ (a_e / 1.4d) + 0.5 , 1.0 \}$
Where b_1 is the stiff bearing length and a_e is the distance from the load or reaction to the nearer end of the member.
FOR EXPLANATION OF TABLES SEE NOTE 6.1

BS 5950-1: 2000
BS 4-1: 1993

WEB BEARING AND BUCKLING

PARALLEL FLANGE CHANNELS

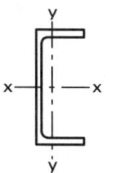

BEARING AND BUCKLING VALUES FOR UNSTIFFENED WEBS FOR S355

Section Designation	Web Thickness	Depth Between Fillets	Bearing					Buckling				Shear Capacity
			End Bearing			Continuous Over Bearing		End Bearing			Continuous Over Bearing	
			Beam Factor	Stiff Bearing Factor	Reduc. Factor	Beam Factor	Stiff Bearing Factor	Buckling Factor	1.4d	Reduc. Factor	Buckling Factor	
	t	d	C1	C2	K_b	C1	C2	C4		K_w	C4	P_v
	mm	mm	kN	kN/mm		kN	kN/mm	kN	mm		kN	kN
430 x 100 x 64	11.0	362	258	3.80	0.748	645	3.80	632	507	0.785	632	979
380 x 100 x 54	9.5	315	213	3.28	0.713	533	3.28	468	441	0.759	468	747
300 x 100 x 46	9.0	237	196	3.11	0.699	489	3.11	529	332	0.719	529	559
300 x 90 x 41	9.0	245	176	3.20	0.699	439	3.20	511	343	0.724	511	575
260 x 90 x 35	8.0	208	148	2.84	0.667	369	2.84	423	291	0.693	423	443
260 x 75 x 28	7.0	212	119	2.49	0.627	298	2.49	278	297	0.682	278	388
230 x 90 x 32	7.5	178	138	2.66	0.648	346	2.66	407	249	0.668	407	367
230 x 75 x 26	6.5	181	113	2.31	0.604	283	2.31	261	253	0.652	261	318
200 x 90 x 30	7.0	148	129	2.49	0.627	323	2.49	398	207	0.640	398	298
200 x 75 x 23	6.0	151	104	2.13	0.577	261	2.13	246	211	0.618	246	256
180 x 90 x 26	6.5	131	113	2.31	0.604	283	2.31	360	183	0.617	360	249
180 x 75 x 20	6.0	135	95.9	2.13	0.577	240	2.13	275	189	0.605	275	230
150 x 90 x 24	6.5	102	111	2.31	0.604	277	2.31	463	143	0.601	463	208
150 x 75 x 18	5.5	106	85.9	1.95	0.548	215	1.95	270	148	0.568	270	176
125 x 65 x 15 #	5.5	82.0	84.0	1.95	0.565	210	1.95	349	115	0.567	349	146
100 x 50 x 10 #	5.0	65.0	62.1	1.78	0.571	155	1.78	331	91.0	0.566	331	107

\# Check availability.
Web bearing capacity, $P_w = K_b (C1 + b_1 C2)$
Web buckling resistance, $P_x = K_w K (C4 P_w)^{0.5}$
For Continuous over bearing: K = 1.0
For End bearing: K = Min. { $(a_e / 1.4d) + 0.5$, 1.0 }
Where b_1 is the stiff bearing length and a_e is the distance from the load or reaction to the nearer end of the member.
Reduction factors K_b and K_w should be applied where appropriate to take account of the eccentricity due to the heel radius.
FOR EXPLANATION OF TABLES SEE NOTE 6.2

BS 5950-1: 2000
BS EN 10056-1: 1999

TIES

EQUAL ANGLES
SUBJECT TO AXIAL TENSION

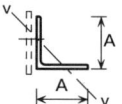

TENSION CAPACITY FOR S355

Section Designation		Mass per Metre	Radius of Gyration Axis	Gross Area	Weld or Bolt Size	Holes Deducted From Angle		Equivalent Tension Area	Tension Capacity
A x A	t		v-v			No.	Diameter		P_t
mm	mm	kg	cm	cm²			mm	cm²	kN
200 x 200	24 #	71.1	3.90	90.6	Weld	0	-	77.8	2680
					M24	1	26	67.2	2320
					M24	2	26	60.4	2080
					M20	3	22	56.7	1960
	20	59.9	3.92	76.3	Weld	0	-	65.4	2260
					M24	1	26	56.4	1950
					M24	2	26	50.7	1750
					M20	3	22	47.6	1640
	18	54.3	3.90	69.1	Weld	0	-	59.2	2040
					M24	1	26	51.0	1760
					M24	2	26	45.9	1580
					M20	3	22	43.1	1490
	16	48.5	3.94	61.8	Weld	0	-	52.9	1880
					M24	1	26	45.5	1620
					M24	2	26	40.9	1450
					M20	3	22	38.5	1370
150 x 150	18 #	40.1	2.93	51.2	Weld	0	-	43.9	1520
					M24	1	26	36.7	1260
					M20	2	22	33.1	1140
	15	33.8	2.93	43.0	Weld	0	-	36.9	1310
					M24	1	26	30.7	1090
					M20	2	22	27.7	985
	12	27.3	2.95	34.8	Weld	0	-	29.8	1060
					M24	1	26	24.8	879
					M20	2	22	22.4	795
	10	23.0	2.97	29.3	Weld	0	-	25.0	888
					M24	1	26	20.8	738
					M20	2	22	18.8	668
120 x 120	15 #	26.6	2.34	34.0	Weld	0	-	29.2	1040
					M24	1	26	23.5	835
					M20	1	22	24.2	858
	12	21.6	2.35	27.5	Weld	0	-	23.6	837
					M24	1	26	19.0	673
					M20	1	22	19.5	692
	10	18.2	2.36	23.2	Weld	0	-	19.8	704
					M24	1	26	15.9	566
					M20	1	22	16.4	581
	8 #	14.7	2.38	18.8	Weld	0	-	16.0	569
					M24	1	26	12.9	457
					M20	1	22	13.2	469

\# Check availability.
FOR EXPLANATION OF TABLES SEE NOTE 7.1

BS 5950-1: 2000
BS EN 10056-1: 1999

TIES

EQUAL ANGLES
SUBJECT TO AXIAL TENSION

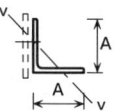

TENSION CAPACITY FOR S355

Section Designation		Mass per Metre	Radius of Gyration Axis	Gross Area	Weld or Bolt Size	Holes Deducted From Angle		Equivalent Tension Area	Tension Capacity
A x A	t		v-v			No.	Diameter		P_t
mm	mm	kg	cm	cm²			mm	cm²	kN
100 x 100	15 #	21.9	1.94	28.0	Weld	0	-	24.1	856
					M24	1	26	18.7	664
					M20	1	22	19.4	688
	12	17.8	1.94	22.7	Weld	0	-	19.5	692
					M24	1	26	15.1	537
					M20	1	22	15.6	555
	10	15.0	1.95	19.2	Weld	0	-	16.4	584
					M24	1	26	12.7	452
					M20	1	22	13.2	468
	8	12.2	1.96	15.5	Weld	0	-	13.3	470
					M24	1	26	10.3	364
					M20	1	22	10.6	377
90 x 90	12 #	15.9	1.75	20.3	Weld	0	-	17.5	619
					M20	1	22	13.7	487
					M16	1	18	14.3	506
	10	13.4	1.75	17.1	Weld	0	-	14.7	521
					M20	1	22	11.5	409
					M16	1	18	12.0	425
	8	10.9	1.76	13.9	Weld	0	-	11.9	422
					M20	1	22	9.33	331
					M16	1	18	9.69	344
	7 #	9.61	1.77	12.2	Weld	0	-	10.4	370
					M20	1	22	8.19	291
					M16	1	18	8.49	302
80 x 80	10 ‡	11.9	1.55	15.1	Weld	0	-	13.0	460
					M20	1	22	9.93	353
					M16	1	18	10.4	368
	8 ‡	9.63	1.56	12.3	Weld	0	-	10.5	374
					M20	1	22	8.05	286
					M16	1	18	8.41	298
75 x 75	8 ‡	8.99	1.46	11.4	Weld	0	-	9.78	347
					M20	1	22	7.30	261
					M16	1	18	7.72	274
	6 ‡	6.85	1.47	8.73	Weld	0	-	7.46	265
					M20	1	22	5.61	199
					M16	1	18	5.88	209

‡ Not available from some leading producers. Check availability.
\# Check availability.
FOR EXPLANATION OF TABLES SEE NOTE 7.1

BS 5950-1: 2000
BS EN 10056-1: 1999

TIES

EQUAL ANGLES
SUBJECT TO AXIAL TENSION

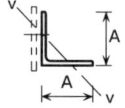

TENSION CAPACITY FOR S355

Section Designation		Mass per Metre	Radius of Gyration Axis	Gross Area	Weld or Bolt Size	Holes Deducted From Angle		Equivalent Tension Area	Tension Capacity
A x A	t		v-v			No.	Diameter		P_t
mm	mm	kg	cm	cm²			mm	cm²	kN
70 x 70	7 ‡	7.38	1.36	9.40	Weld	0	-	8.05	286
					M20	1	22	5.95	211
					M16	1	18	6.25	222
	6 ‡	6.38	1.37	8.13	Weld	0	-	6.95	247
					M20	1	22	5.13	182
					M16	1	18	5.40	192
65 x 65	7 ‡	6.83	1.26	8.73	Weld	0	-	7.48	265
					M20	1	22	5.40	192
					M16	1	18	5.71	203
60 x 60	8 ‡	7.09	1.16	9.03	Weld	0	-	7.76	276
					M16	1	18	5.81	206
	6 ‡	5.42	1.17	6.91	Weld	0	-	5.92	210
					M16	1	18	4.43	157
	5 ‡	4.57	1.17	5.82	Weld	0	-	4.97	177
					M16	1	18	3.72	132
50 x 50	6 ‡	4.47	0.968	5.69	Weld	0	-	4.88	173
					M12	1	14	3.72	132
	5 ‡	3.77	0.973	4.80	Weld	0	-	4.11	146
					M12	1	14	3.13	111
	4 ‡	3.06	0.979	3.89	Weld	0	-	3.32	118
					M12	1	14	2.53	89.8
45 x 45	4.5 ‡	3.06	0.870	3.90	Weld	0	-	3.34	118
					M12	1	14	2.47	87.8
40 x 40	5 ‡	2.97	0.773	3.79	Weld	0	-	3.25	115
					M12	1	14	2.33	82.5
	4 ‡	2.42	0.777	3.08	Weld	0	-	2.64	93.6
					M12	1	14	1.88	66.9
35 x 35	4 ‡	2.09	0.678	2.67	Weld	0	-	2.29	81.3
					M12	1	14	1.56	55.3
30 x 30	4 ‡	1.78	0.577	2.27	Weld	0	-	1.95	69.2
					M12	1	14	1.24	44.0
	3 ‡	1.36	0.581	1.74	Weld	0	-	1.49	52.8
					M12	1	14	0.948	33.7
25 x 25	4 ‡	1.45	0.482	1.85	Weld	0	-	1.60	56.6
					M12	1	14	0.909	32.3
	3 ‡	1.12	0.484	1.42	Weld	0	-	1.22	43.3
					M12	1	14	0.698	24.8
20 x 20	3 ‡	0.882	0.383	1.12	Weld	0	-	0.964	34.2
					M12	1	14	0.458	16.3

‡ Not available from some leading producers. Check availability.
FOR EXPLANATION OF TABLES SEE NOTE 7.1

BS 5950-1: 2000
BS EN 10056-1: 1999

TIES

TWO EQUAL ANGLES
BACK TO BACK
SUBJECT TO AXIAL TENSION

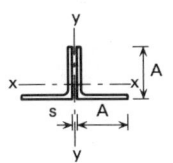

TENSION CAPACITY FOR S355

Composed Of Two Angles		Mass per Metre	Space Between Angles	Radius of Gyration		Gross Area	Weld or Bolt Size	Holes Deducted From Angle		Gusset Between Angles		Gusset On Back of Angles	
				Axis x-x	Axis y-y			No.	Diameter	Equivalent Tension Area	Tension Capacity	Equivalent Tension Area	Tension Capacity
A x A mm	t mm	kg	mm	cm	cm	cm²			mm	cm²	kN	cm²	kN
200 x 200	24 #	142	15	6.06	8.95	181	Weld	0	-	168	5810	156	5370
							M24	1	26.0	156	5370	134	4640
							M24	2	26.0	142	4900	121	4170
							M20	3	22.0	135	4650	113	3910
	20	120	15	6.11	8.87	153	Weld	0	-	142	4890	131	4510
							M24	1	26.0	131	4520	113	3890
							M24	2	26.0	120	4130	101	3500
							M20	3	22.0	113	3910	95.3	3290
	18	109	15	6.13	8.83	138	Weld	0	-	128	4430	118	4080
							M24	1	26.0	119	4090	102	3520
							M24	2	26.0	108	3730	91.7	3160
							M20	3	22.0	103	3540	86.2	2970
	16	97.0	15	6.16	8.79	124	Weld	0	-	115	4070	106	3750
							M24	1	26.0	106	3760	91.0	3230
							M24	2	26.0	96.8	3440	81.9	2910
							M20	3	22.0	91.9	3260	77.0	2730
150 x 150	18 #	80.2	12	4.55	6.75	102	Weld	0	-	95.1	3280	87.9	3030
							M24	1	26.0	85.4	2950	73.3	2530
							M20	2	22.0	78.3	2700	66.2	2280
	15	67.6	12	4.57	6.66	86.0	Weld	0	-	79.9	2830	73.7	2620
							M24	1	26.0	71.7	2540	61.4	2180
							M20	2	22.0	65.7	2330	55.5	1970
	12	54.6	12	4.60	6.59	69.6	Weld	0	-	64.6	2290	59.5	2110
							M24	1	26.0	57.9	2060	49.5	1760
							M20	2	22.0	53.2	1890	44.8	1590
	10	46.0	12	4.62	6.54	58.6	Weld	0	-	54.3	1930	50.0	1780
							M24	1	26.0	48.7	1730	41.6	1480
							M20	2	22.0	44.8	1590	37.6	1340
120 x 120	15 #	53.2	12	3.63	5.49	68.0	Weld	0	-	63.2	2240	58.4	2070
							M24	1	26.0	55.0	1950	47.0	1670
							M20	1	22.0	56.3	2000	48.3	1720
	12	43.2	12	3.65	5.42	55.0	Weld	0	-	51.1	1810	47.1	1670
							M24	1	26.0	44.5	1580	37.9	1350
							M20	1	22.0	45.5	1620	39.0	1380
	10	36.4	12	3.67	5.36	46.4	Weld	0	-	43.0	1530	39.7	1410
							M24	1	26.0	37.5	1330	31.9	1130
							M20	1	22.0	38.4	1360	32.8	1160
	8 #	29.4	12	3.71	5.34	37.6	Weld	0	-	34.8	1240	32.1	1140
							M24	1	26.0	30.3	1080	25.7	914
							M20	1	22.0	31.0	1100	26.4	939

Check availability.
FOR EXPLANATION OF TABLES SEE NOTE 7.2

BS 5950-1: 2000
BS EN 10056-1: 1999

TIES

TWO EQUAL ANGLES BACK TO BACK SUBJECT TO AXIAL TENSION

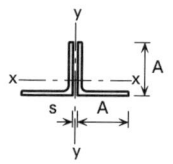

TENSION CAPACITY FOR S355

Composed Of Two Angles		Mass per Metre	Space Between Angles s	Radius of Gyration		Gross Area	Weld or Bolt Size	Holes Deducted From Angle		Gusset Between Angles		Gusset On Back of Angles	
A x A mm	t mm	kg	mm	Axis x-x cm	Axis y-y cm	cm²		No.	Diameter mm	Equivalent Tension Area cm²	Tension Capacity kN	Equivalent Tension Area cm²	Tension Capacity kN
100 x 100	15 #	43.8	10	2.99	4.62	56.0	Weld	0	-	52.1	1850	48.2	1710
							M24	1	26.0	43.9	1560	37.4	1330
							M20	1	22.0	45.2	1610	38.7	1380
	12	35.6	10	3.02	4.55	45.4	Weld	0	-	42.2	1500	39.0	1380
							M24	1	26.0	35.6	1260	30.2	1070
							M20	1	22.0	36.6	1300	31.3	1110
	10	30.0	10	3.04	4.50	38.4	Weld	0	-	35.6	1270	32.9	1170
							M24	1	26.0	30.1	1070	25.5	905
							M20	1	22.0	31.0	1100	26.4	936
	8	24.4	10	3.06	4.46	31.0	Weld	0	-	28.8	1020	26.5	941
							M24	1	26.0	24.3	862	20.5	729
							M20	1	22.0	25.0	887	21.2	754
90 x 90	12 #	31.8	10	2.71	4.16	40.6	Weld	0	-	37.8	1340	34.9	1240
							M20	1	22.0	32.2	1140	27.5	975
							M16	1	18.0	33.3	1180	28.5	1010
	10	26.8	10	2.72	4.11	34.2	Weld	0	-	31.8	1130	29.3	1040
							M20	1	22.0	27.1	962	23.1	819
							M16	1	18.0	28.0	994	23.9	850
	8	21.8	10	2.74	4.06	27.8	Weld	0	-	25.8	916	23.8	844
							M20	1	22.0	22.0	782	18.7	663
							M16	1	18.0	22.7	807	19.4	688
	7 #	19.2	10	2.75	4.04	24.4	Weld	0	-	22.6	803	20.9	741
							M20	1	22.0	19.3	686	16.4	581
							M16	1	18.0	19.9	708	17.0	603
80 x 80	10 ‡	23.8	8	2.41	3.65	30.2	Weld	0	-	28.1	996	25.9	921
							M20	1	22.0	23.4	831	19.9	705
							M16	1	18.0	24.3	862	20.7	736
	8 ‡	19.3	8	2.43	3.60	24.6	Weld	0	-	22.8	810	21.1	748
							M20	1	22.0	19.1	677	16.1	572
							M16	1	18.0	19.8	702	16.8	597
75 x 75	8 ‡	18.0	8	2.27	3.41	22.8	Weld	0	-	21.2	752	19.6	694
							M20	1	22.0	17.4	619	14.7	523
							M16	1	18.0	18.1	644	15.4	548
	6 ‡	13.7	8	2.29	3.35	17.4	Weld	0	-	16.2	575	14.9	530
							M20	1	22.0	13.3	474	11.2	399
							M16	1	18.0	13.9	492	11.8	417

‡ Not available from some leading producers. Check availability.
# Check availability.
FOR EXPLANATION OF TABLES SEE NOTE 7.2

BS 5950-1: 2000
BS EN 10056-1: 1999

TIES

TWO EQUAL ANGLES
BACK TO BACK
SUBJECT TO AXIAL TENSION

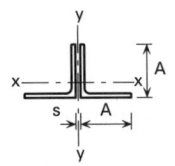

TENSION CAPACITY FOR S355

Composed Of Two Angles	Mass per Metre	Space Between Angles	Radius of Gyration		Gross Area	Weld or Bolt Size	Holes Deducted From Angle		Gusset Between Angles		Gusset On Back of Angles	
A x A	t			Axis x-x	Axis y-y			No.	Diameter	Equivalent Tension Area	Tension Capacity	Equivalent Tension Area	Tension Capacity
mm	mm	kg	mm	cm	cm	cm²			mm	cm²	kN	cm²	kN
70 x 70	7 ‡	14.8	8	2.12	3.18	18.8	Weld	0	-	17.5	619	16.1	572
							M20	1	22.0	14.1	502	11.9	422
							M16	1	18.0	14.8	524	12.5	444
	6 ‡	12.8	8	2.13	3.16	16.3	Weld	0	-	15.1	535	13.9	494
							M20	1	22.0	12.2	434	10.3	364
							M16	1	18.0	12.8	453	10.8	383
65 x 65	7 ‡	13.7	8	1.96	3.14	17.5	Weld	0	-	16.2	575	15.0	531
							M20	1	22.0	12.9	458	10.8	383
							M16	1	18.0	13.5	480	11.4	405
60 x 60	8 ‡	14.2	8	1.80	2.82	18.1	Weld	0	-	16.8	596	15.5	551
							M16	1	18.0	13.7	488	11.6	413
	6 ‡	10.8	8	1.82	2.77	13.8	Weld	0	-	12.8	455	11.8	420
							M16	1	18.0	10.5	373	8.85	314
	5 ‡	9.14	8	1.82	2.74	11.6	Weld	0	-	10.8	383	9.95	353
							M16	1	18.0	8.85	314	7.44	264
50 x 50	6 ‡	8.94	8	1.50	2.38	11.4	Weld	0	-	10.6	375	9.77	347
							M12	1	14.0	8.79	312	7.44	264
	5 ‡	7.54	8	1.51	2.35	9.60	Weld	0	-	8.91	316	8.22	292
							M12	1	14.0	7.41	263	6.26	222
	4 ‡	6.12	8	1.52	2.32	7.78	Weld	0	-	7.21	256	6.65	236
							M12	1	14.0	6.00	213	5.06	180

‡ Not available from some leading producers. Check availability.
FOR EXPLANATION OF TABLES SEE NOTE 7.2

BS 5950-1: 2000
BS EN 10056-1: 1999

TIES

UNEQUAL ANGLES
LONG LEG ATTACHED
SUBJECT TO AXIAL TENSION

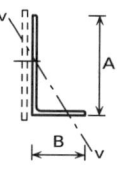

TENSION CAPACITY FOR S355

Section Designation A x B mm	t mm	Mass per Metre kg	Radius of Gyration Axis v-v cm	Gross Area cm²	Weld or Bolt Size	Holes Deducted From Angle		Equivalent Tension Area cm²	Tension Capacity P_t kN
						No.	Diameter mm		
200 x 150	18 #	47.1	3.22	60.1	Weld	0	-	52.9	1820
					M24	1	26	46.5	1600
					M24	2	26	41.4	1430
					M20	3	22	38.6	1330
	15	39.6	3.23	50.5	Weld	0	-	44.4	1570
					M24	1	26	39.0	1380
					M24	2	26	34.7	1230
					M20	3	22	32.4	1150
	12	32.0	3.25	40.8	Weld	0	-	35.8	1270
					M24	1	26	31.4	1110
					M24	2	26	27.9	992
					M20	3	22	26.1	926
200 x 100	15	33.8	2.12	43.0	Weld	0	-	39.1	1390
					M24	1	26	35.2	1250
					M24	2	26	30.9	1100
					M20	3	22	28.6	1020
	12	27.3	2.14	34.8	Weld	0	-	31.6	1120
					M24	1	26	28.4	1010
					M24	2	26	24.9	885
					M20	3	22	23.1	820
	10	23.0	2.15	29.2	Weld	0	-	26.4	939
					M24	1	26	23.7	843
					M24	2	26	20.9	741
					M20	3	22	19.3	687
150 x 90	15	26.6	1.93	33.9	Weld	0	-	30.5	1080
					M24	1	26	26.2	929
					M20	2	22	23.2	823
	12	21.6	1.94	27.5	Weld	0	-	24.7	875
					M24	1	26	21.1	750
					M20	2	22	18.7	665
	10	18.2	1.95	23.2	Weld	0	-	20.7	736
					M24	1	26	17.7	630
					M20	2	22	15.8	559
150 x 75	15	24.8	1.58	31.7	Weld	0	-	28.9	1030
					M24	1	26	25.1	890
					M20	2	22	22.1	784
	12	20.2	1.59	25.7	Weld	0	-	23.4	830
					M24	1	26	20.2	718
					M20	2	22	17.8	633
	10	17.0	1.60	21.7	Weld	0	-	19.7	699
					M24	1	26	17.0	603
					M20	2	22	15.0	533

\# Check availability.
FOR EXPLANATION OF TABLES SEE NOTE 7.1

BS 5950-1: 2000
BS EN 10056-1: 1999

TIES

UNEQUAL ANGLES
LONG LEG ATTACHED
SUBJECT TO AXIAL TENSION

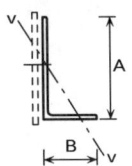

TENSION CAPACITY FOR S355

Section Designation A x B mm	t mm	Mass per Metre kg	Radius of Gyration Axis v-v cm	Gross Area cm^2	Weld or Bolt Size	Holes Deducted From Angle No.	Diameter mm	Equivalent Tension Area cm^2	Tension Capacity P_t kN
125 x 75	12	17.8	1.61	22.7	Weld	0	-	20.4	724
					M24	1	26	16.9	601
					M20	2	22	14.5	516
	10	15.0	1.61	19.1	Weld	0	-	17.1	608
					M24	1	26	14.2	504
					M20	2	22	12.2	433
	8	12.2	1.63	15.5	Weld	0	-	13.9	492
					M24	1	26	11.5	407
					M20	2	22	9.88	351
100 x 75	12	15.4	1.59	19.7	Weld	0	-	17.4	617
					M24	1	26	13.6	483
					M20	1	22	14.1	502
	10	13.0	1.59	16.6	Weld	0	-	14.6	519
					M24	1	26	11.4	406
					M20	1	22	11.9	422
	8	10.6	1.60	13.5	Weld	0	-	11.9	421
					M24	1	26	9.26	329
					M20	1	22	9.61	341
100 x 65	10 #	12.3	1.39	15.6	Weld	0	-	13.9	494
					M24	1	26	10.9	388
					M20	1	22	11.4	404
	8 #	9.94	1.40	12.7	Weld	0	-	11.3	401
					M24	1	26	8.86	315
					M20	1	22	9.21	327
	7 #	8.77	1.40	11.2	Weld	0	-	9.94	353
					M24	1	26	7.80	277
					M20	1	22	8.11	288
100 x 50	8 ‡	8.97	1.06	11.4	Weld	0	-	10.4	368
					M24	1	26	8.21	292
					M20	1	22	8.56	304
	6 ‡	6.84	1.07	8.71	Weld	0	-	7.90	280
					M24	1	26	6.24	221
					M20	1	22	6.50	231
80 x 60	7 ‡	7.36	1.28	9.38	Weld	0	-	8.25	293
					M20	1	22	6.36	226
					M16	1	18	6.66	237

‡ Not available from some leading producers. Check availability.
# Check availability.
FOR EXPLANATION OF TABLES SEE NOTE 7.1

BS 5950-1: 2000
BS EN 10056-1: 1999

TIES

UNEQUAL ANGLES
LONG LEG ATTACHED
SUBJECT TO AXIAL TENSION

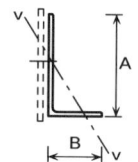

TENSION CAPACITY FOR S355

Section Designation A x B mm	t mm	Mass per Metre kg	Radius of Gyration Axis v-v cm	Gross Area cm^2	Weld or Bolt Size	Holes Deducted From Angle		Equivalent Tension Area cm^2	Tension Capacity P_t kN
						No.	Diameter mm		
80 x 40	8 ‡	7.07	0.838	9.01	Weld	0	-	8.23	292
					M20	1	22	6.41	228
					M16	1	18	6.76	240
	6 ‡	5.41	0.845	6.89	Weld	0	-	6.26	222
					M20	1	22	4.87	173
					M16	1	18	5.14	182
75 x 50	8 ‡	7.39	1.07	9.41	Weld	0	-	8.39	298
					M20	1	22	6.37	226
					M16	1	18	6.72	239
	6 ‡	5.65	1.08	7.19	Weld	0	-	6.38	227
					M20	1	22	4.84	172
					M16	1	18	5.11	181
70 x 50	6 ‡	5.41	1.07	6.89	Weld	0	-	6.08	216
					M20	1	22	4.51	160
					M16	1	18	4.78	170
65 x 50	5 ‡	4.35	1.07	5.54	Weld	0	-	4.85	172
					M20	1	22	3.51	125
					M16	1	18	3.73	132
60 x 40	6 ‡	4.46	0.855	5.68	Weld	0	-	5.06	179
					M16	1	18	3.81	135
	5 ‡	3.76	0.860	4.79	Weld	0	-	4.25	151
					M16	1	18	3.21	114
60 x 30	5 ‡	3.36	0.633	4.28	Weld	0	-	3.90	138
					M16	1	18	2.95	105
50 x 30	5 ‡	2.96	0.639	3.78	Weld	0	-	3.40	121
					M12	1	14	2.62	93.0
45 x 30	4 ‡	2.25	0.640	2.87	Weld	0	-	2.55	90.5
					M12	1	14	1.90	67.4
40 x 25	4 ‡	1.93	0.534	2.46	Weld	0	-	2.20	78.2
					M12	1	14	1.57	55.9
40 x 20	4 ‡	1.77	0.417	2.26	Weld	0	-	2.06	73.2
					M12	1	14	1.47	52.3
30 x 20	4 ‡	1.46	0.421	1.86	Weld	0	-	1.66	59.0
					M12	1	14	1.03	36.7
	3 ‡	1.12	0.424	1.43	Weld	0	-	1.27	45.1
					M12	1	14	0.793	28.2

‡ Not available from some leading producers. Check availability.
FOR EXPLANATION OF TABLES SEE NOTE 7.1

BS 5950-1: 2000
BS EN 10056-1: 1999

TIES

TWO UNEQUAL ANGLES
LONG LEGS BACK TO BACK
SUBJECT TO AXIAL TENSION

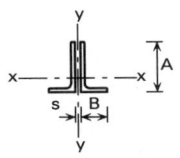

TENSION CAPACITY FOR S355

Composed Of Two Angles A x B mm	t mm	Mass per Metre kg	Space Between Angles s mm	Radius of Gyration Axis x-x cm	Radius of Gyration Axis y-y cm	Gross Area cm^2	Weld or Bolt Size	Holes Deducted From Angle No.	Holes Deducted From Angle Diameter mm	Gusset Between Angles Equivalent Tension Area cm^2	Gusset Between Angles Tension Capacity kN	Gusset On Back of Angles Equivalent Tension Area cm^2	Gusset On Back of Angles Tension Capacity kN
200x150	18 #	94.2	15	6.30	6.36	120	Weld	0	-	113	3900	100	3460
							M24	1	26	105	3620	82.2	2840
							M20	2	22	97.9	3380	75.1	2590
	15	79.2	15	6.33	6.28	101	Weld	0	-	94.9	3370	84.2	2990
							M24	1	26	88.2	3130	68.9	2450
							M20	2	22	82.2	2920	63.0	2240
	12	64.0	15	6.36	6.22	81.6	Weld	0	-	76.6	2720	67.9	2410
							M24	1	26	71.1	2530	55.5	1970
							M20	2	22	66.4	2360	50.8	1800
200x100	15	67.5	15	6.40	3.97	86.0	Weld	0	-	82.1	2910	69.2	2460
							M24	1	26	76.9	2730	52.4	1860
							M20	1	22	78.2	2780	53.7	1910
	12	54.6	15	6.43	3.90	69.6	Weld	0	-	66.4	2360	55.9	1990
							M24	1	26	62.1	2210	42.3	1500
							M20	1	22	63.2	2240	43.4	1540
	10	46.0	15	6.46	3.85	58.4	Weld	0	-	55.6	1980	46.9	1660
							M24	1	26	52.1	1850	35.5	1260
							M20	1	22	53.0	1880	36.4	1290
150x90	15	53.2	12	4.74	3.749	67.8	Weld	0	-	64.4	2290	55.6	1970
							M20	1	22	59.3	2110	42.8	1520
							M16	1	18	60.7	2150	44.2	1570
	12	43.2	12	4.77	3.69	55.0	Weld	0	-	52.2	1850	45.0	1600
							M20	1	22	48.0	1710	34.7	1230
							M16	1	18	49.1	1740	35.7	1270
	10	36.4	12	4.80	3.64	46.4	Weld	0	-	43.9	1560	37.9	1340
							M20	1	22	40.5	1440	29.2	1040
							M16	1	18	41.3	1470	30.0	1070
150x75	15	49.6	12	4.75	3.09	63.4	Weld	0	-	60.6	2150	51.1	1820
							M20	1	22	56.0	1990	37.9	1350
							M16	1	18	57.4	2040	39.3	1390
	12	40.4	12	4.78	3.02	51.4	Weld	0	-	49.1	1740	41.4	1470
							M20	1	22	45.3	1610	30.7	1090
							M16	1	18	46.4	1650	31.7	1130
	10	34.0	12	4.81	2.97	43.4	Weld	0	-	41.4	1470	34.9	1240
							M20	1	22	38.2	1360	25.9	918
							M16	1	18	39.1	1390	26.7	949

Check availability.
FOR EXPLANATION OF TABLES SEE NOTE 7.2

BS 5950-1: 2000
BS EN 10056-1: 1999

TIES

TWO UNEQUAL ANGLES
LONG LEGS BACK TO BACK
SUBJECT TO AXIAL TENSION

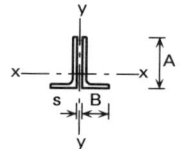

TENSION CAPACITY FOR S355

Composed Of Two Angles		Mass per Metre	Space Between Angles	Radius of Gyration		Gross Area	Weld or Bolt Size	Holes Deducted From Angle		Gusset Between Angles		Gusset On Back of Angles	
A x B	t			Axis x-x	Axis y-y			No.	Diameter	Equivalent Tension Area	Tension Capacity	Equivalent Tension Area	Tension Capacity
mm	mm	kg	mm	cm	cm	cm²			mm	cm²	kN	cm²	kN
125x75	12	35.6	12	3.95	3.19	45.4	Weld	0	-	43.1	1530	37.2	1320
							M20	1	22	38.7	1380	27.7	983
							M16	1	18	39.8	1410	28.7	1020
	10	30.0	12	3.97	3.14	38.2	Weld	0	-	36.2	1290	31.2	1110
							M20	1	22	32.6	1160	23.3	826
							M16	1	18	33.4	1190	24.1	857
	8	24.4	12	4.00	3.09	31.0	Weld	0	-	29.4	1040	25.3	898
							M20	1	22	26.4	936	18.8	668
							M16	1	18	27.1	961	19.5	693
100x75	12	30.8	10	3.10	3.31	39.4	Weld	0	-	37.1	1320	33.0	1170
							M20	1	22	32.1	1140	24.7	877
							M16	1	18	33.2	1180	25.7	914
	10	26.0	10	3.12	3.27	33.2	Weld	0	-	31.2	1110	27.7	985
							M20	1	22	27.1	961	20.8	737
							M16	1	18	27.9	992	21.6	768
	8	21.2	10	3.14	3.22	27.0	Weld	0	-	25.3	900	22.5	799
							M20	1	22	22.0	780	16.8	597
							M16	1	18	22.7	805	17.5	622
100x65	10 #	24.6	10	3.14	2.79	31.2	Weld	0	-	29.5	1050	25.7	914
							M20	1	22	25.6	907	18.6	659
							M16	1	18	26.4	939	19.4	690
	8 #	19.9	10	3.16	2.74	25.4	Weld	0	-	24.0	852	20.9	742
							M20	1	22	20.8	738	15.1	535
							M16	1	18	21.5	763	15.8	560
	7 #	17.5	10	3.17	2.72	22.4	Weld	0	-	21.1	750	18.4	654
							M20	1	22	18.3	650	13.3	471
							M16	1	18	18.9	672	13.9	493
100x50	8 ‡	17.9	10	3.19	2.09	22.8	Weld	0	-	21.8	773	18.4	652
							M12	1	14	20.2	718	13.7	488
	6 ‡	13.7	10	3.21	2.04	17.4	Weld	0	-	16.6	590	14.0	497
							M12	1	14	15.4	547	10.5	371

‡ Not available from some leading producers. Check availability.
# Check availability.
FOR EXPLANATION OF TABLES SEE NOTE 7.2

BS 5950-1: 2000
BS EN 10056-1: 1999

TIES

TWO UNEQUAL ANGLES
LONG LEGS BACK TO BACK
SUBJECT TO AXIAL TENSION

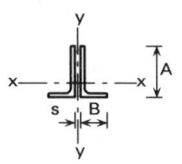

TENSION CAPACITY FOR S355

Composed Of Two Angles		Mass per Metre	Space Between Angles	Radius of Gyration		Gross Area	Weld or Bolt Size	Holes Deducted From Angle		Gusset Between Angles		Gusset On Back of Angles	
A x B	t		s	Axis x-x	Axis y-y			No.	Diameter	Equivalent Tension Area	Tension Capacity	Equivalent Tension Area	Tension Capacity
mm	mm	kg	mm	cm	cm	cm²			mm	cm²	kN	cm²	kN
80x60	7 ‡	14.7	8	2.51	2.59	18.8	Weld	0	-	17.6	626	15.7	556
							M16	1	18	15.2	540	11.6	414
80x40	8 ‡	14.1	8	2.53	1.71	18.0	Weld	0	-	17.2	612	14.5	516
							M12	1	14	15.5	551	10.4	369
	6 ‡	10.8	8	2.55	1.66	13.8	Weld	0	-	13.2	467	11.1	394
							M12	1	14	11.8	421	7.92	281
75x50	8 ‡	14.8	8	2.35	2.19	18.8	Weld	0	-	17.8	632	15.6	553
							M12	1	14	15.9	563	11.7	417
	6 ‡	11.3	8	2.37	2.14	14.4	Weld	0	-	13.6	482	11.9	421
							M12	1	14	12.1	429	8.94	317
70x50	6 ‡	10.8	8	2.20	2.19	13.8	Weld	0	-	13.0	461	11.4	406
							M12	1	14	11.4	406	8.64	307
65x50	5 ‡	8.70	8	2.05	2.21	11.1	Weld	0	-	10.4	369	9.26	329
							M12	1	14	9.05	321	7.00	249
60x40	6 ‡	8.92	8	1.88	1.80	11.4	Weld	0	-	10.7	381	9.39	333
							M12	1	14	9.19	326	6.71	238
	5 ‡	7.52	8	1.89	1.78	9.58	Weld	0	-	9.04	321	7.91	281
							M12	1	14	7.75	275	5.65	201

‡ Not available from some leading producers. Check availability.
FOR EXPLANATION OF TABLES SEE NOTE 7.2

BS 5950-1: 2000
BS 4-1: 1993

STRUTS

UB SECTIONS
SUBJECT TO AXIAL COMPRESSION

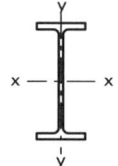

COMPRESSION RESISTANCE FOR S355

Section Designation	Axis	Compression resistance, P_{cx}, P_{cy} (kN) for Effective lengths, L_E (m)												
		2.0	3.0	4.0	5.0	6.0	7.0	8.0	9.0	10.0	11.0	12.0	13.0	14.0
1016x305x487 # +	P_{cx}	20800	20800	20800	20800	20800	20600	20500	20300	20100	19900	19700	19400	19200
	P_{cy}	19000	16900	14500	11900	9630	7750	6310	5200	4340	3670	3140	2720	2380
1016x305x437 # +	P_{cx}	18700	18700	18700	18700	18700	18600	18400	18300	18200	18000	17900	17700	17500
	P_{cy}	17300	15600	13600	11200	9010	7200	5810	4760	3960	3340	2850	2460	2140
1016x305x393 # +	P_{cx}	16800	16800	16800	16800	16800	16700	16500	16400	16300	16200	16000	15900	15700
	P_{cy}	15500	14000	12100	9920	7930	6320	5100	4170	3470	2920	2490	2150	1870
* 1016x305x349 # +	P_{cx}	*14500*	*14500*	*14500*	*14500*	*14500*	*14500*	*14400*	*14300*	*14200*	*14200*	*14100*	*14000*	*13900*
	P_{cy}	*13700*	*12600*	*11600*	*9560*	*7600*	*6020*	*4820*	*3920*	*3240*	*2720*	*2320*	*1990*	*1730*
* 1016x305x314 # +	P_{cx}	*12600*	*12600*	*12600*	*12600*	*12600*	*12500*	*12500*	*12400*	*12300*	*12300*	*12200*	*12100*	*12000*
	P_{cy}	*11900*	*11000*	*9700*	*8490*	*6730*	*5310*	*4250*	*3460*	*2860*	*2400*	*2040*	*1760*	*1530*
* 1016x305x272 # +	P_{cx}	*10400*	*10400*	*10400*	*10400*	*10400*	*10400*	*10300*	*10300*	*10200*	*10100*	*10100*	*10000*	*9970*
	P_{cy}	*9820*	*9100*	*8110*	*6860*	*5810*	*4590*	*3670*	*2980*	*2470*	*2070*	*1760*	*1510*	*1320*
* 1016x305x249 # +	P_{cx}	*9350*	*9350*	*9350*	*9350*	*9350*	*9320*	*9280*	*9230*	*9180*	*9130*	*9070*	*9020*	*8960*
	P_{cy}	*8800*	*8110*	*7160*	*5960*	*4990*	*3910*	*3120*	*2530*	*2090*	*1750*	*1490*	*1280*	*1110*
* 1016x305x222 # +	P_{cx}	*8130*	*8130*	*8130*	*8130*	*8130*	*8100*	*8060*	*8020*	*7970*	*7930*	*7880*	*7830*	*7780*
	P_{cy}	*7620*	*6980*	*6090*	*5000*	*4150*	*3230*	*2560*	*2070*	*1710*	*1430*	*1210*	*1040*	*906*
* 914 x 419 x 388 #	P_{cx}	16800	16800	16800	16800	16800	16700	16600	16500	16400	16300	16200	16100	15900
	P_{cy}	16400	15900	15100	14100	12800	11300	9770	8350	7130	6120	5280	4590	4030
* 914 x 419 x 343 #	P_{cx}	14400	14400	14400	14400	14400	14300	14200	14100	14000	13900	13800	13700	13600
	P_{cy}	14100	13500	12800	12100	11200	9870	8500	7240	6180	5290	4560	3970	3480
* 914 x 305 x 289 #	P_{cx}	11800	11800	11800	11800	11800	11800	11700	11600	11600	11500	11400	11300	11200
	P_{cy}	11200	10300	9650	8000	6380	5060	4060	3310	2740	2300	1960	1680	1460
* 914 x 305 x 253 #	P_{cx}	9910	9910	9910	9910	9910	9860	9800	9740	9690	9630	9560	9490	9420
	P_{cy}	9370	8680	7730	6920	5490	4340	3480	2830	2340	1970	1670	1440	1250
* 914 x 305 x 224 #	P_{cx}	8460	8460	8460	8460	8460	8410	8370	8320	8270	8220	8160	8110	8050
	P_{cy}	*8000*	*7400*	*6580*	*5550*	*4700*	*3700*	*2960*	*2400*	*1990*	*1670*	*1420*	*1220*	*1060*
* 914 x 305 x 201 #	P_{cx}	7350	7350	7350	7350	7350	7310	7270	7220	7180	7140	7090	7040	6990
	P_{cy}	6930	6390	5660	4740	4010	3140	2500	2030	1670	1400	1190	1030	891
* 838 x 292 x 226 #	P_{cx}	8930	8930	8930	8930	8910	8860	8800	8750	8690	8630	8560	8490	8420
	P_{cy}	8420	7760	6860	6030	4750	3740	2990	2430	2010	1680	1430	1230	1070
* 838 x 292 x 194 #	P_{cx}	7320	7320	7320	7320	7300	7260	7210	7170	7120	7070	7020	6960	6900
	P_{cy}	6890	6340	5580	4630	3860	3020	2410	1950	1610	1350	1150	987	857
* 838 x 292 x 176 #	P_{cx}	6460	6460	6460	6460	6450	6410	6370	6330	6290	6240	6190	6140	6090
	P_{cy}	*6070*	*5570*	*4880*	*4030*	*3360*	*2620*	*2080*	*1690*	*1390*	*1170*	*990*	*851*	*739*
* 762 x 267 x 197	P_{cx}	7970	7970	7970	7960	7910	7860	7800	7740	7680	7610	7540	7460	7370
	P_{cy}	7390	6680	5960	4670	3580	2780	2200	1780	1470	1230	1040	896	778
* 762 x 267 x 173	P_{cx}	6730	6730	6730	6720	6680	6630	6590	6540	6490	6430	6370	6300	6230
	P_{cy}	6230	5630	4780	3970	3030	2340	1850	1500	1230	1030	875	751	652
* 762 x 267 x 147	P_{cx}	5440	5440	5440	5440	5410	5370	5330	5290	5250	5210	5160	5110	5050
	P_{cy}	*5040*	*4540*	*3840*	*3220*	*2440*	*1880*	*1480*	*1200*	*985*	*824*	*699*	*600*	*521*
* 762 x 267 x 134	P_{cx}	4950	4950	4950	4950	4920	4880	4850	4810	4770	4730	4690	4640	4590
	P_{cy}	*4570*	*4110*	*3460*	*2720*	*2180*	*1670*	*1320*	*1060*	*873*	*729*	*618*	*531*	*460*

Check availability.
+ Section is not given in BS 4-1: 1993.
* Section may be slender under axial compression.
Values in *italic type* indicate that the section is slender when loaded to capacity and allowance has been made in calculating the capacity.
FOR EXPLANATION OF TABLES SEE NOTE 8.1

BS 5950-1: 2000
BS 4-1: 1993

STRUTS

UB SECTIONS
SUBJECT TO AXIAL COMPRESSION

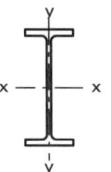

COMPRESSION RESISTANCE FOR S355

Section Designation	Axis	Compression resistance, P_{cx}, P_{cy} (kN) for Effective lengths, L_E (m)												
		2.0	3.0	4.0	5.0	6.0	7.0	8.0	9.0	10.0	11.0	12.0	13.0	14.0
* 686 x 254 x 170	P_{cx}	7000	7000	7000	6970	6920	6870	6810	6750	6680	6610	6530	6440	6340
	P_{cy}	6450	5780	5010	3860	2940	2270	1800	1450	1190	1000	848	729	632
* 686 x 254 x 152	P_{cx}	6040	6040	6040	6020	5970	5930	5880	5830	5770	5710	5640	5570	5490
	P_{cy}	5570	4990	4350	3390	2580	1990	1570	1270	1040	873	740	636	552
* 686 x 254 x 140	P_{cx}	5400	5400	5400	5390	5350	5300	5260	5220	5170	5110	5050	4990	4920
	P_{cy}	4980	4470	3740	3060	2320	1780	1410	1140	934	781	663	569	494
* 686 x 254 x 125	P_{cx}	4690	4690	4690	4670	4640	4600	4570	4530	4480	4440	4390	4330	4260
	P_{cy}	4310	3850	3210	2620	1970	1520	1200	963	792	662	561	482	418
610 x 305 x 238	P_{cx}	10500	10500	10500	10400	10300	10200	10100	9990	9870	9740	9590	9420	9230
	P_{cy}	9960	9300	8420	7280	6030	4900	3980	3270	2720	2290	1960	1690	1470
* 610 x 305 x 179	P_{cx}	7690	7690	7690	7630	7560	7490	7420	7340	7260	7160	7070	7060	6910
	P_{cy}	7320	6960	6270	5370	4410	3560	2890	2370	1970	1660	1410	1220	1060
* 610 x 305 x 149	P_{cx}	6070	6070	6070	6030	5980	5930	5870	5810	5740	5670	5590	5490	5390
	P_{cy}	5790	5410	4900	4400	3630	2930	2370	1940	1610	1360	1160	996	866
* 610 x 229 x 140	P_{cx}	5780	5780	5780	5730	5680	5630	5570	5510	5440	5370	5280	5180	5070
	P_{cy}	5230	4750	3720	2770	2060	1580	1240	1000	821	686	581	499	432
* 610 x 229 x 125	P_{cx}	4980	4980	4980	4940	4900	4860	4810	4760	4700	4640	4570	4480	4390
	P_{cy}	4510	3930	3280	2430	1810	1380	1090	873	717	599	507	435	377
* 610 x 229 x 113	P_{cx}	4390	4390	4390	4350	4320	4280	4240	4190	4140	4090	4020	3950	3870
	P_{cy}	3970	3460	2910	2140	1590	1210	951	764	627	524	444	381	330
* 610 x 229 x 101	P_{cx}	3920	3920	3910	3880	3850	3810	3780	3740	3690	3640	3580	3520	3440
	P_{cy}	3530	3050	2400	1850	1370	1040	813	653	536	447	378	324	281
* 533 x 210 x 122	P_{cx}	5250	5250	5220	5170	5110	5050	4990	4960	4910	4810	4690	4550	4390
	P_{cy}	4720	3940	2970	2150	1580	1200	944	758	621	518	439	376	326
* 533 x 210 x 109	P_{cx}	4550	4550	4520	4480	4430	4380	4330	4260	4190	4110	4010	3900	3860
	P_{cy}	4020	3500	2610	1880	1380	1050	823	661	542	452	382	328	284
* 533 x 210 x 101	P_{cx}	4110	4110	4090	4050	4010	3970	3920	3860	3800	3730	3640	3540	3430
	P_{cy}	3650	3180	2400	1730	1270	964	755	606	496	414	351	301	260
* 533 x 210 x 92	P_{cx}	3720	3720	3700	3670	3630	3590	3540	3490	3440	3370	3290	3200	3100
	P_{cy}	3290	2760	2170	1540	1130	857	670	537	440	367	311	266	231
* 533 x 210 x 82	P_{cx}	3260	3260	3240	3210	3170	3140	3100	3050	3000	2950	2880	2800	2700
	P_{cy}	2870	2390	1870	1320	964	729	569	456	374	311	264	226	196
* 457 x 191 x 98	P_{cx}	4310	4310	4260	4210	4160	4000	4020	3930	3830	3700	3550	3380	3180
	P_{cy}	3730	3000	2160	1530	1120	847	661	530	434	362	306	262	227
* 457 x 191 x 89	P_{cx}	3810	3810	3770	3730	3680	3620	3560	3480	3460	3370	3230	3070	2880
	P_{cy}	3390	2710	1950	1370	1000	759	593	475	389	324	274	235	204
* 457 x 191 x 82	P_{cx}	3480	3480	3440	3400	3360	3310	3250	3180	3100	3000	2890	2830	2650
	P_{cy}	3010	2490	1760	1230	897	677	528	423	346	289	244	209	181
* 457 x 191 x 74	P_{cx}	3070	3070	3030	3000	2960	2920	2870	2810	2740	2660	2550	2430	2300
	P_{cy}	2660	2240	1580	1110	806	608	474	380	311	259	219	188	163
* 457 x 191 x 67	P_{cx}	2710	2710	2680	2650	2610	2570	2530	2480	2410	2340	2250	2140	2020
	P_{cy}	2340	1880	1390	970	703	530	413	331	271	226	191	163	141

* Section may be slender under axial compression.
Values in *italic type* indicate that the section is slender when loaded to capacity and allowance has been made in calculating the capacity.
FOR EXPLANATION OF TABLES SEE NOTE 8.1

BS 5950-1: 2000
BS 4-1: 1993

STRUTS

UB SECTIONS
SUBJECT TO AXIAL COMPRESSION

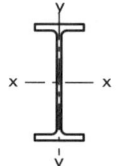

COMPRESSION RESISTANCE FOR S355

Section Designation	Axis	Compression resistance, P_{cx}, P_{cy} (kN) for Effective lengths, L_E (m)												
		1.0	1.5	2.0	2.5	3.0	3.5	4.0	4.5	5.0	5.5	6.0	6.5	7.0
* 457 x 152 x 82	P_{cx}	3500	3500	3500	3500	3500	3480	3460	3440	3420	3400	3370	3350	3320
	P_{cy}	3320	3160	2810	2360	1910	1530	1230	1000	831	699	595	513	446
* 457 x 152 x 74	P_{cx}	3050	3050	3050	3050	3040	3030	3010	2990	2970	2960	2930	2910	2890
	P_{cy}	2890	2680	2480	2100	1690	1350	1080	884	732	615	524	451	392
* 457 x 152 x 67	P_{cx}	2750	2750	2750	2750	2750	2730	2720	2700	2680	2670	2650	2630	2610
	P_{cy}	2600	2410	2150	1900	1510	1200	958	779	645	541	460	396	344
* 457 x 152 x 60	P_{cx}	2350	2350	2350	2350	2350	2340	2330	2310	2300	2280	2270	2250	2240
	P_{cy}	2230	2070	1840	1560	1320	1040	835	679	561	471	400	345	299
* 457 x 152 x 52	P_{cx}	1990	1990	1990	1990	1990	1970	1960	1950	1940	1930	1920	1900	1890
	P_{cy}	1880	1730	1540	1290	1090	857	683	554	458	384	326	280	244
* 406 x 178 x 74	P_{cx}	3270	3270	3270	3270	3250	3230	3210	3190	3160	3140	3110	3080	3050
	P_{cy}	3150	2990	2820	2520	2160	1800	1490	1240	1040	877	750	648	565
* 406 x 178 x 67	P_{cx}	2880	2880	2880	2880	2860	2850	2830	2810	2790	2770	2740	2720	2690
	P_{cy}	2780	2630	2430	2260	1930	1610	1320	1100	917	776	664	573	500
* 406 x 178 x 60	P_{cx}	2490	2490	2490	2490	2470	2460	2440	2430	2410	2390	2370	2350	2330
	P_{cy}	2400	2270	2110	1900	1720	1430	1180	973	814	688	588	508	443
* 406 x 178 x 54	P_{cx}	2210	2210	2210	2210	2200	2180	2170	2150	2140	2120	2100	2080	2060
	P_{cy}	2130	2010	1860	1660	1500	1230	1010	834	696	588	502	433	377
* 406 x 140 x 46	P_{cx}	1790	1790	1790	1790	1780	1770	1760	1750	1740	1720	1710	1690	1680
	P_{cy}	1680	1540	1350	1120	925	722	574	465	384	322	273	235	204
* 406 x 140 x 39	P_{cx}	1460	1460	1460	1460	1450	1440	1430	1420	1410	1400	1390	1380	1370
	P_{cy}	1360	1240	1080	877	721	559	442	357	294	246	209	180	156
356 x 171 x 67	P_{cx}	3040	3040	3040	3030	3000	2980	2960	2930	2910	2880	2850	2810	2770
	P_{cy}	2920	2760	2540	2260	1930	1610	1320	1100	917	776	664	573	500
* 356 x 171 x 57	P_{cx}	2500	2500	2500	2490	2480	2460	2440	2420	2400	2370	2350	2320	2280
	P_{cy}	2410	2270	2140	1890	1610	1330	1090	901	752	636	543	469	409
* 356 x 171 x 51	P_{cx}	2170	2170	2170	2160	2150	2130	2120	2100	2080	2060	2040	2010	1990
	P_{cy}	2090	1970	1810	1680	1420	1160	955	788	657	555	474	409	357
* 356 x 171 x 45	P_{cx}	1870	1870	1870	1870	1850	1840	1830	1810	1790	1780	1760	1730	1710
	P_{cy}	1800	1690	1560	1400	1210	991	808	665	554	468	399	344	300
* 356 x 127 x 39	P_{cx}	1580	1580	1580	1580	1570	1550	1540	1530	1520	1500	1480	1470	1440
	P_{cy}	1450	1300	1090	859	647	497	392	316	260	217	184	158	137
* 356 x 127 x 33	P_{cx}	1280	1280	1280	1280	1270	1260	1250	1240	1230	1210	1200	1190	1170
	P_{cy}	1170	1040	854	687	513	393	309	249	204	171	145	124	108
305 x 165 x 54	P_{cx}	2440	2440	2440	2420	2400	2380	2350	2330	2300	2270	2230	2190	2140
	P_{cy}	2350	2210	2030	1800	1530	1270	1040	861	719	608	520	449	391
* 305 x 165 x 46	P_{cx}	2010	2010	2010	2000	1980	1960	1940	1920	1900	1880	1850	1810	1780
	P_{cy}	1940	1830	1730	1530	1290	1070	878	725	605	512	437	377	329
* 305 x 165 x 40	P_{cx}	1710	1710	1710	1690	1680	1660	1650	1630	1610	1590	1570	1540	1510
	P_{cy}	1640	1550	1430	1270	1120	921	755	623	520	439	375	324	282

* Section may be slender under axial compression.
Values in *italic type* indicate that the section is slender when loaded to capacity and allowance has been made in calculating the capacity.
FOR EXPLANATION OF TABLES SEE NOTE 8.1

BS 5950-1: 2000
BS 4-1: 1993

STRUTS

UB SECTIONS
SUBJECT TO AXIAL COMPRESSION

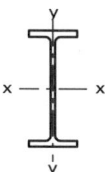

COMPRESSION RESISTANCE FOR S355

Section Designation	Axis	Compression resistance, P_{cx}, P_{cy} (kN) for Effective lengths, L_E (m)												
		1.0	1.5	2.0	2.5	3.0	3.5	4.0	4.5	5.0	5.5	6.0	6.5	7.0
305 x 127 x 48	P_{cx}	2170	2170	2170	2150	2130	2110	2090	2060	2040	2000	1970	1930	1880
	P_{cy}	1980	1750	1430	1090	824	635	501	404	332	278	236	202	176
305 x 127 x 42	P_{cx}	1900	1900	1890	1880	1860	1840	1820	1800	1770	1750	1710	1680	1630
	P_{cy}	1730	1520	1230	931	702	540	426	343	282	236	200	172	149
* 305 x 127 x 37	P_{cx}	*1640*	*1640*	*1630*	*1620*	*1610*	*1590*	*1570*	*1550*	*1530*	*1530*	*1510*	*1480*	*1440*
	P_{cy}	*1520*	*1330*	*1070*	*809*	*609*	*468*	*369*	*297*	*244*	*204*	*173*	*149*	*129*
* 305 x 102 x 33	P_{cx}	*1380*	*1380*	*1380*	*1370*	*1360*	*1350*	*1330*	*1320*	*1300*	*1280*	*1260*	*1240*	*1210*
	P_{cy}	*1200*	*1020*	*724*	*510*	*371*	*281*	*219*	*175*	*144*	*120*	*101*	*86.8*	*75.2*
* 305 x 102 x 28	P_{cx}	*1140*	*1140*	*1140*	*1130*	*1120*	*1110*	*1100*	*1080*	*1070*	*1050*	*1040*	*1020*	*995*
	P_{cy}	*985*	*826*	*592*	*414*	*300*	*227*	*177*	*142*	*116*	*96.5*	*81.6*	*69.9*	*60.5*
* 305 x 102 x 25	P_{cx}	*974*	*974*	*973*	*965*	*956*	*947*	*938*	*927*	*915*	*902*	*887*	*869*	*849*
	P_{cy}	*833*	*684*	*480*	*332*	*240*	*180*	*140*	*112*	*91.7*	*76.5*	*64.7*	*55.4*	*47.9*
254 x 146 x 43	P_{cx}	1950	1950	1930	1910	1890	1870	1840	1810	1780	1740	1690	1640	1570
	P_{cy}	1850	1720	1540	1310	1070	859	694	568	471	397	338	291	253
254 x 146 x 37	P_{cx}	1680	1680	1660	1650	1630	1610	1590	1560	1530	1500	1450	1400	1350
	P_{cy}	1590	1470	1320	1120	907	727	587	480	398	335	285	245	214
* 254 x 146 x 31	P_{cx}	*1390*	*1390*	*1380*	*1370*	*1350*	*1330*	*1330*	*1310*	*1280*	*1250*	*1210*	*1160*	*1110*
	P_{cy}	*1330*	*1230*	*1090*	*907*	*728*	*579*	*465*	*379*	*314*	*264*	*225*	*193*	*168*
* 254 x 102 x 28	P_{cx}	*1270*	*1270*	*1260*	*1250*	*1240*	*1230*	*1210*	*1190*	*1160*	*1130*	*1100*	*1060*	*1010*
	P_{cy}	*1110*	*902*	*654*	*464*	*339*	*257*	*201*	*161*	*132*	*110*	*93.0*	*79.7*	*69.1*
* 254 x 102 x 25	P_{cx}	*1110*	*1110*	*1100*	*1080*	*1070*	*1060*	*1040*	*1030*	*1030*	*999*	*967*	*928*	*883*
	P_{cy}	*977*	*778*	*554*	*390*	*284*	*215*	*168*	*134*	*110*	*91.6*	*77.5*	*66.4*	*57.5*
* 254 x 102 x 22	P_{cx}	*944*	*944*	*936*	*926*	*915*	*903*	*889*	*873*	*854*	*832*	*832*	*803*	*761*
	P_{cy}	*832*	*654*	*455*	*318*	*230*	*174*	*135*	*108*	*88.7*	*73.8*	*62.5*	*53.5*	*46.3*
203 x 133 x 30	P_{cx}	1360	1350	1330	1310	1290	1270	1240	1210	1160	1110	1040	971	894
	P_{cy}	1270	1160	1010	819	645	508	405	329	272	228	194	167	145
203 x 133 x 25	P_{cx}	1140	1130	1120	1100	1080	1060	1040	1000	966	918	862	799	733
	P_{cy}	1060	964	829	669	523	410	326	265	219	183	156	134	116
203 x 102 x 23	P_{cx}	1040	1040	1020	1010	993	973	949	919	882	837	784	725	664
	P_{cy}	922	770	578	418	308	234	183	147	121	101	85.2	73.0	63.3
178 x 102 x 19	P_{cx}	863	854	840	825	808	786	758	723	679	627	570	514	460
	P_{cy}	763	638	480	348	256	195	153	123	100	83.8	71.0	60.9	52.8
152 x 89 x 16	P_{cx}	720	708	694	677	657	620	580	540	492	437	385	338	299
	P_{cy}	615	483	340	238	173	131	102	81.5	66.7	55.6	47.0	40.3	34.9
127 x 76 x 13	P_{cx}	581	569	554	534	506	468	419	365	313	269	232	201	176
	P_{cy}	474	340	225	154	110	82.9	64.5	51.5	42.1	35.0	29.6	25.3	21.9

* Section may be slender under axial compression.
Values in *italic type* indicate that the section is slender when loaded to capacity and allowance has been made in calculating the capacity.
FOR EXPLANATION OF TABLES SEE NOTE 8.1

BS 5950-1: 2000
BS 4-1: 1993

STRUTS

UC SECTIONS
SUBJECT TO AXIAL COMPRESSION

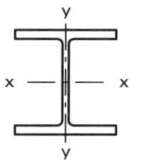

COMPRESSION RESISTANCE FOR S355

Section Designation	Axis	Compression resistance, P_{cx}, P_{cy} (kN) for Effective lengths, L_E (m)												
		2.0	3.0	4.0	5.0	6.0	7.0	8.0	9.0	10.0	11.0	12.0	13.0	14.0
356 x 406 x 634 #	P_{cx}	26300	26200	25400	24500	23700	22800	21800	20800	19700	18600	17500	16400	15300
	P_{cy}	25700	23800	21900	19900	18000	16100	14200	12600	11100	9800	8680	7720	6890
356 x 406 x 551 #	P_{cx}	22800	22700	22000	21200	20500	19700	18800	17900	16900	16000	15000	13900	13000
	P_{cy}	22300	20600	18900	17200	15500	13800	12300	10800	9530	8410	7440	6610	5900
356 x 406 x 467 #	P_{cx}	19900	19700	19100	18400	17700	17000	16200	15400	14500	13600	12600	11700	10800
	P_{cy}	19400	17900	16400	14900	13300	11800	10400	9130	8010	7040	6220	5510	4910
356 x 406 x 393 #	P_{cx}	16800	16600	16200	15700	15200	14600	14000	13300	12600	11800	11000	10200	9440
	P_{cy}	16400	15300	14100	12900	11600	10300	9110	7970	6970	6100	5360	4730	4200
356 x 406 x 340 #	P_{cx}	14500	14300	13900	13500	13100	12600	12000	11400	10800	10100	9390	8680	7990
	P_{cy}	14100	13200	12200	11100	9980	8870	7790	6810	5940	5200	4570	4030	3580
356 x 406 x 287 #	P_{cx}	12600	12500	12200	11900	11600	11200	10700	10200	9680	9060	8400	7740	7090
	P_{cy}	12300	11600	10800	9960	9010	8000	7010	6100	5300	4610	4030	3540	3130
356 x 406 x 235 #	P_{cx}	10300	10200	9960	9700	9420	9100	8730	8320	7850	7330	6780	6230	5700
	P_{cy}	10100	9470	8820	8100	7310	6480	5670	4920	4270	3710	3240	2850	2520
356 x 368 x 202 #	P_{cx}	8870	8760	8550	8320	8080	7790	7470	7110	6690	6240	5760	5280	4820
	P_{cy}	8590	8040	7430	6760	6020	5260	4540	3910	3360	2910	2530	2210	1950
356 x 368 x 177 #	P_{cx}	7800	7700	7510	7310	7080	6830	6540	6210	5840	5430	5000	4580	4170
	P_{cy}	7540	7060	6520	5920	5270	4600	3970	3410	2930	2530	2200	1930	1700
356 x 368 x 153 #	P_{cx}	6730	6640	6470	6300	6100	5880	5630	5340	5010	4660	4290	3920	3570
	P_{cy}	6510	6100	5620	5100	4530	3950	3400	2920	2510	2170	1880	1650	1450
356 x 368 x 129 #	P_{cx}	5660	5580	5440	5290	5120	4930	4710	4460	4180	3870	3560	3250	2950
	P_{cy}	5470	5110	4710	4270	3790	3300	2840	2430	2090	1800	1570	1370	1210
305 x 305 x 283	P_{cx}	12100	11800	11400	11000	10500	10000	9460	8840	8190	7520	6860	6230	5640
	P_{cy}	11300	10300	9200	8030	6880	5820	4900	4150	3530	3030	2630	2290	2020
305 x 305 x 240	P_{cx}	10600	10300	10100	9750	9400	8990	8510	7970	7370	6740	6110	5510	4960
	P_{cy}	10000	9200	8290	7260	6200	5210	4360	3660	3100	2650	2280	1990	1740
305 x 305 x 198	P_{cx}	8690	8510	8270	8000	7700	7350	6940	6470	5960	5430	4900	4410	3960
	P_{cy}	8210	7540	6780	5920	5040	4220	3520	2960	2500	2130	1840	1600	1400
305 x 305 x 158	P_{cx}	6930	6770	6580	6360	6110	5820	5470	5090	4660	4230	3810	3410	3060
	P_{cy}	6530	5990	5360	4660	3950	3290	2740	2290	1940	1650	1420	1230	1080
305 x 305 x 137	P_{cx}	6000	5860	5680	5490	5270	5010	4700	4360	3980	3600	3230	2900	2590
	P_{cy}	5650	5170	4620	4010	3380	2820	2340	1960	1650	1410	1210	1050	922
305 x 305 x 118	P_{cx}	5180	5050	4890	4730	4530	4300	4040	3730	3410	3080	2760	2470	2210
	P_{cy}	4860	4450	3970	3430	2890	2400	2000	1670	1410	1200	1030	895	784
305 x 305 x 97	P_{cx}	4370	4250	4120	3970	3800	3600	3360	3090	2800	2520	2250	2000	1790
	P_{cy}	4090	3730	3310	2850	2390	1970	1630	1360	1140	971	835	724	634

Check availability.
FOR EXPLANATION OF TABLES SEE NOTE 8.1

BS 5950-1: 2000
BS 4-1: 1993

STRUTS

UC SECTIONS
SUBJECT TO AXIAL COMPRESSION

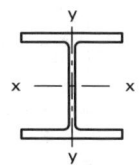

COMPRESSION RESISTANCE FOR S355

Section Designation	Axis	Compression resistance, P_{cx}, P_{cy} (kN) for Effective lengths, L_E (m)												
		1.0	1.5	2.0	2.5	3.0	3.5	4.0	4.5	5.0	5.5	6.0	6.5	7.0
254 x 254 x 167	P_{cx}	7350	7350	7310	7190	7070	6950	6820	6670	6520	6350	6160	5950	5730
	P_{cy}	7350	7070	6750	6410	6040	5640	5210	4770	4330	3910	3510	3160	2840
254 x 254 x 132	P_{cx}	5800	5800	5760	5660	5570	5460	5350	5240	5110	4970	4810	4640	4450
	P_{cy}	5800	5560	5300	5030	4730	4410	4060	3710	3350	3020	2710	2430	2180
254 x 254 x 107	P_{cx}	4690	4690	4650	4570	4490	4410	4310	4210	4100	3980	3850	3700	3550
	P_{cy}	4690	4490	4280	4050	3810	3540	3250	2960	2670	2400	2150	1930	1730
254 x 254 x 89	P_{cx}	3900	3900	3860	3800	3730	3660	3580	3490	3400	3300	3190	3060	2930
	P_{cy}	3900	3730	3550	3360	3150	2930	2690	2450	2210	1980	1770	1590	1420
254 x 254 x 73	P_{cx}	3310	3310	3270	3210	3150	3090	3020	2950	2870	2780	2680	2570	2450
	P_{cy}	3300	3150	3000	2830	2650	2460	2250	2040	1830	1630	1460	1300	1170
203 x 203 x 86	P_{cx}	3800	3780	3710	3630	3540	3450	3340	3230	3090	2950	2790	2610	2440
	P_{cy}	3720	3510	3290	3040	2770	2480	2190	1920	1670	1460	1280	1130	1000
203 x 203 x 71	P_{cx}	3120	3110	3040	2980	2910	2830	2740	2640	2530	2410	2270	2130	1980
	P_{cy}	3060	2880	2700	2490	2270	2030	1790	1560	1360	1190	1040	917	812
203 x 203 x 60	P_{cx}	2710	2700	2640	2580	2510	2440	2360	2270	2170	2050	1920	1790	1660
	P_{cy}	2650	2490	2330	2140	1940	1720	1500	1310	1140	987	862	758	669
203 x 203 x 52	P_{cx}	2350	2340	2290	2240	2180	2120	2050	1960	1870	1770	1660	1550	1430
	P_{cy}	2300	2160	2020	1850	1670	1490	1300	1130	980	852	744	653	577
203 x 203 x 46	P_{cx}	2080	2070	2020	1980	1930	1870	1810	1730	1650	1560	1460	1350	1250
	P_{cy}	2030	1910	1780	1630	1470	1300	1140	988	856	743	648	569	503
152 x 152 x 37	P_{cx}	1670	1630	1580	1530	1460	1390	1300	1200	1100	989	886	792	708
	P_{cy}	1570	1430	1270	1100	920	761	630	525	442	376	323	281	246
152 x 152 x 30	P_{cx}	1360	1320	1280	1240	1190	1120	1050	969	880	792	708	632	564
	P_{cy}	1270	1160	1030	884	739	611	504	420	353	300	258	224	196
152 x 152 x 23	P_{cx}	1040	1010	974	938	895	844	785	718	648	580	516	458	408
	P_{cy}	965	875	770	654	542	444	365	303	254	216	185	160	140

FOR EXPLANATION OF TABLES SEE NOTE 8.1

BS 5950-1: 2000
BS EN 10056-1: 1999

STRUTS

EQUAL ANGLES
SUBJECT TO AXIAL COMPRESSION

TWO OR MORE BOLTS IN LINE
OR EQUIVALENT WELDED AT EACH END

COMPRESSION RESISTANCE FOR STEEL S355

Section Designation		Area	Radius of Gyration			Compression resistance, P_c (kN) for Length between Intersections (m)												
			Axis a-a	Axis b-b	Axis v-v													
A x A	t					1.0	1.5	2.0	2.5	3.0	3.5	4.0	4.5	5.0	5.5	6.0	6.5	7.0
mm	mm	cm²	cm	cm	cm													
200 x 200	24.0 #	90.6	6.06	6.06	3.90	2620	2490	2360	2190	1960	1730	1520	1290	1110	954	827	723	636
	20.0	76.3	6.11	6.11	3.92	2210	2100	1990	1850	1650	1470	1280	1100	940	810	702	614	540
	* 18.0	69.1	6.13	6.13	3.90	*1870*	*1790*	*1700*	*1580*	*1420*	*1260*	*1110*	*956*	*822*	*710*	*618*	*541*	*477*
	* 16.0	61.8	6.16	6.16	3.94	*1350*	*1300*	*1240*	*1170*	*1070*	*974*	*876*	*769*	*673*	*590*	*519*	*459*	*407*
150 x 150	18.0 #	51.2	4.55	4.55	2.93	1430	1330	1190	1020	858	695	568	469	392	332	285	246	215
	15.0	43.0	4.57	4.57	2.93	1240	1150	1020	874	731	591	481	397	332	281	240	208	182
	* 12.0	34.8	4.60	4.60	2.95	*740*	*697*	*639*	*566*	*493*	*413*	*346*	*292*	*247*	*212*	*183*	*159*	*140*
	* 10.0	29.3	4.62	4.62	2.97	*447*	*424*	*396*	*360*	*324*	*282*	*244*	*211*	*183*	*159*	*139*	*123*	*109*
120 x 120	15.0 #	34.0	3.63	3.63	2.34	942	840	690	547	420	328	261	213	176	148	126	109	94.5
	12.0	27.5	3.65	3.65	2.35	762	681	560	445	342	267	213	173	143	121	103	88.5	77.1
	* 10.0	23.2	3.67	3.67	2.36	*515*	*469*	*400*	*330*	*262*	*209*	*169*	*139*	*116*	*98.0*	*83.9*	*72.5*	*63.3*
	* 8.0 #	18.8	3.71	3.71	2.38	*280*	*260*	*231*	*201*	*169*	*141*	*118*	*99.1*	*84.1*	*72.1*	*62.4*	*54.5*	*47.9*
100 x 100	15.0 #	28.0	2.99	2.99	1.94	744	615	472	343	255	196	155	125	103	86.1	73.1	62.9	54.6
	12.0	22.7	3.02	3.02	1.94	604	498	383	278	207	159	125	101	83.3	69.8	59.3	51.0	44.3
	10.0	19.2	3.04	3.04	1.95	512	423	326	237	177	136	107	86.4	71.1	59.6	50.6	43.5	37.8
	* 8.0	15.5	3.06	3.06	1.96	*310*	*268*	*219*	*168*	*129*	*101*	*81.0*	*66.1*	*54.9*	*46.2*	*39.4*	*34.0*	*29.7*
90 x 90	12.0 #	20.3	2.71	2.71	1.75	525	411	297	211	155	118	92.9	74.8	61.5	51.4	43.6	37.5	32.5
	10.0	17.1	2.72	2.72	1.75	443	346	250	178	131	99.7	78.3	63.0	51.8	43.3	36.7	31.6	27.4
	* 8.0	13.9	2.74	2.74	1.76	*325*	*261*	*194*	*140*	*104*	*79.8*	*62.9*	*50.8*	*41.9*	*35.1*	*29.8*	*25.6*	*22.3*
	* 7.0 #	12.2	2.75	2.75	1.77	*228*	*190*	*149*	*112*	*85.1*	*66.3*	*52.8*	*43.0*	*35.6*	*29.9*	*25.5*	*22.0*	*19.2*
80 x 80	10.0 ‡	15.1	2.41	2.41	1.55	372	273	185	128	93.3	70.7	55.3	44.4	36.4	30.4	25.8	22.1	19.2
	8.0 ‡	12.3	2.43	2.43	1.56	304	224	152	106	76.9	58.3	45.6	36.6	30.0	25.1	21.2	18.2	15.8
75 x 75	8.0 ‡	11.4	2.27	2.27	1.46	271	193	127	87.4	63.3	47.8	37.4	30.0	24.6	20.5	17.3	14.9	12.9
	* 6.0 ‡	8.73	2.29	2.29	1.47	*160*	*123*	*86.5*	*61.7*	*45.6*	*34.9*	*27.5*	*22.2*	*18.3*	*15.3*	*13.0*	*11.2*	*9.73*
70 x 70	7.0 ‡	9.40	2.12	2.12	1.36	214	145	93.2	63.6	45.9	34.6	27.0	21.6	17.7	14.7	12.5	10.7	9.28
	* 6.0 ‡	8.13	2.13	2.13	1.37	*161*	*115*	*76.5*	*53.2*	*38.8*	*29.4*	*23.0*	*18.5*	*15.2*	*12.7*	*10.8*	*9.24*	*8.02*
65 x 65	7.0 ‡	8.73	1.96	1.96	1.26	188	121	76.2	51.6	37.1	27.9	21.7	17.4	14.2	11.8	10.0	8.58	7.43
60 x 60	8.0 ‡	9.03	1.80	1.80	1.16	182	110	68.4	46.0	32.9	24.7	19.2	15.3	12.5	10.4	8.83	7.56	6.55
	6.0 ‡	6.91	1.82	1.82	1.17	140	85.3	53.1	35.8	25.6	19.2	14.9	11.9	9.76	8.12	6.87	5.88	5.09
	* 5.0 ‡	5.82	1.82	1.82	1.17	*99.8*	*65.0*	*41.9*	*28.7*	*20.7*	*15.6*	*12.2*	*9.79*	*8.03*	*6.70*	*5.67*	*4.87*	*4.22*
50 x 50	6.0 ‡	5.69	1.50	1.50	0.968	95.6	51.7	31.3	20.8	14.8	11.1	8.57	6.83	5.58	4.64	3.92	3.35	2.90
	5.0 ‡	4.80	1.51	1.51	0.973	81.2	44.0	26.6	17.7	12.6	9.42	7.30	5.82	4.75	3.95	3.34	2.86	2.47
	* 4.0 ‡	3.89	1.52	1.52	0.979	*54.8*	*32.4*	*20.3*	*13.7*	*9.88*	*7.43*	*5.79*	*4.63*	*3.79*	*3.16*	*2.68*	*2.29*	*1.99*

‡ Not available from some leading producers. Check availability.
# Check availability.
* Section is slender under axial compression and allowance has been made in calculating the compression resistance which is given in *italic type*.
FOR EXPLANATION OF TABLES SEE NOTE 8.2

BS 5950-1: 2000
BS EN 10056-1: 1999

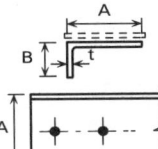

STRUTS

UNEQUAL ANGLES
LONG LEG ATTACHED
SUBJECT TO AXIAL COMPRESSION

TWO OR MORE BOLTS IN LINE
OR EQUIVALENT WELDED AT EACH END

COMPRESSION RESISTANCE FOR STEEL S355

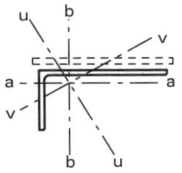

Section Designation		Area	Radius of Gyration			Compression resistance, P_c (kN) for Length between Intersections, L (m)												
			Axis a-a	Axis b-b	Axis v-v	1.0	1.5	2.0	2.5	3.0	3.5	4.0	4.5	5.0	5.5	6.0	6.5	7.0
A x B	t																	
mm	mm	cm²	cm	cm	cm													
200 x 150	18.0 #	60.1	4.38	6.30	3.22	1670	1550	1420	1280	1110	932	770	642	540	460	395	343	300
	*15.0	50.5	4.40	6.33	3.23	*1210*	*1130*	*1040*	*956*	*844*	*722*	*606*	*511*	*434*	*371*	*321*	*280*	*245*
	*12.0	40.8	4.44	6.36	3.25	*653*	*618*	*582*	*546*	*499*	*445*	*389*	*339*	*296*	*259*	*228*	*201*	*179*
200 x 100	*15.0	43.0	2.64	6.40	2.12	*1090*	*924*	*767*	*597*	*451*	*349*	*277*	*224*	*185*	*155*	*132*	*114*	*99.0*
	*12.0	34.8	2.67	6.43	2.14	*605*	*539*	*472*	*392*	*312*	*250*	*203*	*167*	*140*	*118*	*101*	*87.9*	*76.8*
	*10.0	29.2	2.68	6.46	2.15	*368*	*335*	*301*	*260*	*216*	*179*	*149*	*125*	*106*	*91.1*	*78.8*	*68.8*	*60.5*
150 x 90	15.0	33.9	2.46	4.74	1.93	851	709	567	412	306	235	185	150	123	103	87.7	75.4	65.5
	12.0	27.5	2.49	4.77	1.94	693	579	463	337	251	192	152	123	101	84.5	71.8	61.7	53.6
	*10.0	23.2	2.51	4.80	1.95	*476*	*412*	*342*	*259*	*198*	*154*	*123*	*99.7*	*82.6*	*69.5*	*59.2*	*51.0*	*44.4*
150 x 75	15.0	31.7	1.94	4.75	1.58	724	562	399	278	203	154	120	96.6	79.3	66.2	56.1	48.1	41.8
	12.0	25.7	1.97	4.78	1.59	592	462	327	228	166	126	98.6	79.3	65.0	54.3	46.0	39.5	34.3
	*10.0	21.7	1.99	4.81	1.60	*414*	*338*	*251*	*180*	*134*	*103*	*80.9*	*65.3*	*53.8*	*45.1*	*38.3*	*32.9*	*28.6*
125 x 75	12.0	22.7	2.05	3.95	1.61	532	420	294	206	150	114	89.2	71.7	58.8	49.1	41.6	35.7	31.0
	10.0	19.1	2.07	3.97	1.61	449	356	247	173	126	95.8	75.0	60.3	49.5	41.3	35.0	30.1	26.1
	*8.0	15.5	2.09	4.00	1.63	*282*	*236*	*176*	*129*	*96.4*	*74.3*	*58.8*	*47.6*	*39.3*	*33.0*	*28.0*	*24.1*	*21.0*
100 x 75	12.0	19.7	2.14	3.10	1.59	470	366	250	175	127	96.6	75.6	60.7	49.8	41.6	35.3	30.3	26.3
	10.0	16.6	2.16	3.12	1.59	397	308	211	147	107	81.4	63.7	51.2	42.0	35.1	29.7	25.5	22.1
	*8.0	13.5	2.18	3.14	1.60	*308*	*242*	*169*	*119*	*87.0*	*66.1*	*51.9*	*41.7*	*34.3*	*28.7*	*24.3*	*20.9*	*18.1*
100 x 65	10.0 #	15.6	1.81	3.14	1.39	345	247	160	110	79.3	59.8	46.6	37.4	30.6	25.5	21.6	18.5	16.1
	8.0 ‡	12.7	1.83	3.16	1.40	282	203	132	90.5	65.4	49.3	38.5	30.8	25.3	21.1	17.8	15.3	13.3
	*7.0 #	11.2	1.83	3.17	1.40	*213*	*160*	*108*	*75.7*	*55.3*	*42.0*	*32.9*	*26.5*	*21.8*	*18.2*	*15.4*	*13.2*	*11.5*
100 x 50	8.0 ‡	11.4	1.31	3.19	1.06	204	120	73.7	49.3	35.2	26.3	20.4	16.3	13.3	11.1	9.36	8.01	6.94
	*6.0 ‡	8.71	1.33	3.21	1.07	*118*	*78.1*	*50.8*	*35.0*	*25.4*	*19.2*	*15.0*	*12.1*	*9.92*	*8.29*	*7.02*	*6.03*	*5.23*
80 x 60	7.0 ‡	9.38	1.74	2.51	1.28	203	133	86.1	57.0	41.0	30.8	24.0	19.2	15.7	13.1	11.1	9.50	8.23
80 x 40	8.0 ‡	9.01	1.03	2.53	0.838	124	63.9	38.1	25.2	17.8	13.3	10.3	8.19	6.68	5.55	4.68	4.01	3.46
	*6.0 ‡	6.89	1.05	2.55	0.845	*95.2*	*49.4*	*29.5*	*19.5*	*13.8*	*10.3*	*7.98*	*6.36*	*5.18*	*4.31*	*3.64*	*3.11*	*2.69*
75 x 50	8.0 ‡	9.41	1.40	2.35	1.07	176	101	61.9	41.4	29.5	22.1	17.2	13.7	11.2	9.31	7.87	6.74	5.83
	6.0 ‡	7.19	1.42	2.37	1.08	136	78.2	48.1	32.2	23.0	17.2	13.3	10.7	8.70	7.24	6.12	5.24	4.54
70 x 50	6.0 ‡	6.89	1.43	2.20	1.07	129	73.9	45.3	30.3	21.6	16.2	12.6	10.0	8.19	6.82	5.76	4.93	4.27
65 x 50	*5.0 ‡	5.54	1.47	2.05	1.07	*93.9*	*56.1*	*35.1*	*23.7*	*17.0*	*12.8*	*9.93*	*7.94*	*6.50*	*5.41*	*4.58*	*3.92*	*3.40*
60 x 40	6.0 ‡	5.68	1.12	1.00	0.855	80.5	41.7	24.9	16.5	11.7	8.71	6.74	5.37	4.38	3.64	3.07	2.63	2.27
	5.0 ‡	4.79	1.13	1.89	0.860	68.4	35.6	21.2	14.0	9.96	7.43	5.75	4.58	3.73	3.10	2.62	2.24	1.94

‡ Not available from some leading producers. Check availability.
# Check availability.
* Section is slender under axial compression and allowance has been made in calculating the compression resistance which is given in *italic type*.
FOR EXPLANATION OF TABLES SEE NOTE 8.2

BS 5950-1: 2000
BS EN 10056-1: 1999

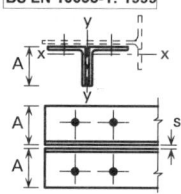

STRUTS

TWO EQUAL ANGLES BACK TO BACK CONNECTED TO ONE SIDE OF GUSSET OR MEMBER SUBJECT TO AXIAL COMPRESSION

TWO OR MORE BOLTS IN LINE ALONG EACH ANGLE OR EQUIVALENT WELDED AT EACH END

COMPRESSION RESISTANCE FOR STEEL S355

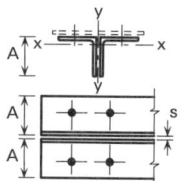

Section Designation	t	Space	Total Area	Radius of Gyration		Elastic Mod.	Compression resistance, P_c (kN) for Length between Intersections, L (m)												
A x A		s		Axis x-x	Axis y-y	Axis x-x													
mm	mm	mm	cm²	cm	cm	cm³	1.0	1.5	2.0	2.5	3.0	3.5	4.0	4.5	5.0	5.5	6.0	6.5	7.0
200 x 200	24.0 #	15	181	6.06	8.95	470	5250	4990	4710	4430	4130	3840	3540	3270	3000	2760	2540	2270	2020
	20.0	15	153	6.11	8.87	398	4420	4210	3980	3740	3490	3240	3000	2770	2550	2340	2160	1930	1730
	* 18.0	15	138	6.13	8.83	362	*3750*	*3570*	*3390*	*3200*	*3000*	*2800*	*2600*	*2410*	*2230*	*2060*	*1900*	*1710*	*1530*
	* 16.0	15	124	6.16	8.79	324	*2700*	*2590*	*2480*	*2360*	*2230*	*2110*	*1990*	*1860*	*1740*	*1630*	*1520*	*1390*	*1250*
150 x 150	18.0 #	12	102	4.55	6.75	200	2870	2660	2450	2220	2000	1800	1610	1440	1240	1060	922	805	708
	15.0	12	86.0	4.57	6.66	167	2470	2290	2100	1910	1720	1540	1370	1230	1060	907	786	686	603
	* 12.0	12	69.6	4.60	6.59	135	*1480*	*1390*	*1300*	*1210*	*1120*	*1020*	934	851	751	658	578	510	**453**
	* 10.0	12	58.6	4.62	6.54	114	*894*	*849*	*803*	*757*	*711*	*665*	*619*	*575*	*520*	*466*	*418*	*375*	*337*
120 x 120	15.0 #	12	68.0	3.63	5.49	106	1880	1700	1500	1320	1140	991	825	684	573	**486**	**417**	**361**	**316**
	12.0	12	55.0	3.65	5.42	85.4	1520	1380	1220	1070	928	806	673	558	468	**397**	**340**	**295**	**258**
	* 10.0	12	46.4	3.67	5.36	72.0	*1030*	*946*	*858*	*769*	*684*	*606*	*518*	*436*	*370*	*317*	**274**	**239**	**210**
	* 8.0 #	12	37.6	3.71	5.34	59.0	*559*	*523*	*486*	*449*	*413*	*377*	*335*	*292*	*255*	*223*	**196**	**173**	**154**
100 x 100	15.0 #	10	56.0	2.99	4.62	71.6	1490	1300	1100	931	783	618	496	**405**	**336**	**283**	**242**	**208**	**182**
	12.0	10	45.4	3.02	4.55	58.2	1210	1060	902	761	642	509	409	**334**	**277**	**234**	**200**	**172**	**150**
	10.0	10	38.4	3.04	4.50	49.2	1020	895	766	648	547	435	350	**286**	**237**	**200**	**171**	**147**	**128**
	* 8.0	10	31.0	3.06	4.46	39.8	*621*	*559*	*496*	*435*	*379*	*312*	*257*	*213*	**179**	**153**	**131**	**114**	**99.7**
90 x 90	12.0 #	10	40.6	2.71	4.16	47.0	1050	896	747	617	489	381	304	**247**	**204**	**172**	**146**	**126**	**110**
	10.0	10	34.2	2.72	4.11	39.6	886	756	631	522	414	323	257	**209**	**173**	**146**	**124**	**107**	**92.9**
	* 8.0	10	27.8	2.74	4.06	32.2	*650*	*565*	*480*	*403*	*325*	*256*	*206*	**168**	**139**	**117**	**100**	**86.5**	**75.4**
	* 7.0 #	10	24.4	2.75	4.04	28.2	*455*	*405*	*354*	*306*	*255*	*206*	*168*	**139**	**116**	**98.4**	**84.5**	**73.2**	**64.0**
80 x 80	10.0 ‡	8	30.2	2.41	3.65	30.8	753	624	505	402	302	232	**184**	**149**	**123**	**103**	**87.3**	**75.1**	**65.3**
	8.0 ‡	8	24.6	2.43	3.60	25.2	615	511	415	331	249	192	**152**	**123**	**101**	**85.0**	**72.3**	**62.2**	**54.0**
75 x 75	8.0 ‡	8	22.8	2.27	3.41	22.0	556	453	362	277	206	**158**	**125**	**101**	**82.8**	**69.4**	**58.9**	**50.7**	**44.0**
	* 6.0 ‡	8	17.5	2.29	3.35	16.8	*326*	*279*	*234*	*187*	*144*	**113**	**90.2**	**73.6**	**61.0**	**51.4**	**43.9**	**37.8**	**33.0**
70 x 70	7.0 ‡	8	18.8	2.12	3.18	16.8	447	357	280	205	151	**115**	**90.8**	**73.2**	**60.2**	**50.3**	**42.7**	**36.7**	**31.9**
	* 6.0 ‡	8	16.3	2.13	3.16	14.5	*332*	*274*	*222*	*167*	*125*	**96.5**	**76.4**	**61.8**	**51.0**	**42.8**	**36.4**	**31.3**	**27.3**
65 x 65	7.0 ‡	8	17.5	1.96	3.14	14.4	401	312	238	167	**123**	**93.2**	**73.0**	**58.7**	**48.2**	**40.3**	**34.2**	**29.3**	**25.5**
60 x 60	8.0 ‡	8	18.1	1.80	2.82	13.8	398	301	216	150	**109**	**82.6**	**64.6**	**51.8**	**42.5**	**35.5**	**30.1**	**25.8**	**22.4**
	6.0 ‡	8	13.8	1.82	2.77	10.6	306	233	168	117	**85.1**	**64.5**	**50.4**	**40.5**	**33.2**	**27.7**	**23.5**	**20.2**	**17.5**
	* 5.0 ‡	8	11.6	1.82	2.74	8.90	*214*	*171*	*128*	*91.7*	**67.8**	**51.8**	**40.8**	**32.9**	**27.1**	**22.7**	**19.3**	**16.6**	**14.4**
50 x 50	6.0 ‡	8	11.4	1.50	2.38	7.22	225	160	101	**68.7**	**49.4**	**37.1**	**28.9**	**23.1**	**18.9**	**15.8**	**13.3**	**11.4**	**9.91**
	5.0 ‡	8	9.60	1.51	2.35	6.10	191	136	86.4	**58.7**	**42.2**	**31.7**	**24.7**	**19.8**	**16.2**	**13.5**	**11.4**	**9.78**	**8.47**
	* 4.0 ‡	8	7.78	1.52	2.32	4.92	*124*	*94.5*	*63.8*	**44.5**	**32.5**	**24.7**	**19.4**	**15.6**	**12.8**	**10.7**	**9.09**	**7.81**	**6.78**

‡ Not rolled by certain leading producers, check availability.
# Check availability.
* Section is slender under axial compression and allowance has been made in calculating the compression resistance which is given in *italic type*.
Values in **bold type** indicate that the section has been divided into more than 3 bays until $\lambda_c \leq 50$.
FOR EXPLANATION OF TABLES SEE NOTE 8.3

| BS 5950-1: 2000 |
| BS EN 10056-1: 1999 |

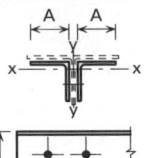

STRUTS

TWO EQUAL ANGLES BACK TO BACK
CONNECTED TO BOTH SIDES OF GUSSET OR MEMBER
SUBJECT TO AXIAL COMPRESSION

TWO OR MORE BOLTS IN LINE ALONG EACH ANGLE
AT EACH END

COMPRESSION RESISTANCE FOR STEEL S355

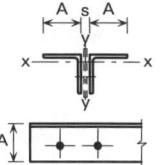

Section Designation		Space	Total Area	Radius of Gyration		Elastic Mod.	Compression resistance, P_c (kN) for Length between Intersections, L (m)												
				Axis x-x	Axis y-y	Axis x-x	1.0	1.5	2.0	2.5	3.0	3.5	4.0	4.5	5.0	5.5	6.0	6.5	7.0
A x A mm	t mm	s mm	cm²	cm	cm	cm³													
200 x 200	24.0 #	15	181	6.06	8.95	470	5250	4990	4710	4430	4130	3840	3540	3270	3000	2760	**2540**	**2340**	**2160**
	20.0	15	153	6.11	8.87	398	4420	4210	3980	3740	3490	3240	3000	2770	2550	2340	**2160**	**1990**	**1830**
	* 18.0	15	138	6.13	8.83	362	*3750*	*3570*	*3390*	*3200*	*3000*	*2800*	*2600*	*2410*	*2230*	*2060*	***1900***	***1750***	***1620***
	* 16.0	15	124	6.16	8.79	324	*2700*	*2590*	*2480*	*2360*	*2230*	*2110*	*1990*	*1860*	*1740*	*1630*	***1520***	***1410***	***1320***
150 x 150	18.0 #	12	102	4.55	6.75	200	2870	2660	2450	2220	2000	1800	1610	**1440**	**1290**	**1160**	**1040**	**944**	**857**
	15.0	12	86.0	4.57	6.66	167	2470	2290	2100	1910	1720	1540	1370	**1230**	**1100**	**986**	**889**	**803**	**729**
	* 12.0	12	69.6	4.60	6.59	135	*1480*	*1390*	*1300*	*1210*	*1120*	*1020*	*934*	***851***	***775***	***706***	***644***	***588***	***539***
	* 10.0	12	58.6	4.62	6.54	114	*894*	*849*	*803*	*757*	*711*	*665*	*619*	***575***	***533***	***494***	***457***	***424***	***392***
120 x 120	15.0 #	12	68.0	3.63	5.49	106	1880	1700	1500	1320	1140	991	**862**	**753**	**662**	**585**	**520**	**465**	**418**
	12.0	12	55.0	3.65	5.42	85.4	1520	1380	1220	1070	928	806	**701**	**613**	**539**	**477**	**424**	**379**	**341**
	* 10.0	12	46.4	3.67	5.36	72.0	*1030*	*946*	*858*	*769*	*684*	*606*	***536***	***475***	***422***	***376***	***337***	***303***	***274***
	* 8.0 #	12	37.6	3.71	5.34	59.0	*559*	*523*	*486*	*449*	*413*	*377*	***343***	***312***	***283***	***258***	***235***	***214***	***196***
100 x 100	15.0 #	10	56.0	2.99	4.62	71.6	1490	1300	1100	931	**784**	**663**	**565**	**486**	**422**	**369**	**324**	**281**	**245**
	12.0	10	45.4	3.02	4.55	58.2	1210	1060	902	761	**642**	**544**	**464**	**399**	**347**	**303**	**267**	**232**	**203**
	10.0	10	38.4	3.04	4.50	49.2	1020	895	766	648	**547**	**463**	**396**	**341**	**296**	**259**	**228**	**198**	**173**
	* 8.0	10	31.0	3.06	4.46	39.8	*621*	*569*	*496*	*435*	***379***	***329***	***286***	***250***	***220***	***194***	***172***	***151***	***133***
90 x 90	12.0 #	10	40.6	2.71	4.16	47.0	1050	896	747	**617**	**512**	**428**	**362**	**309**	**266**	**231**	**197**	**170**	**148**
	10.0	10	34.2	2.72	4.11	39.6	886	756	631	**522**	**433**	**362**	**306**	**261**	**225**	**195**	**167**	**144**	**126**
	* 8.0	10	27.8	2.74	4.06	32.2	*650*	*565*	*480*	***403***	***338***	***285***	***242***	***208***	***180***	***157***	***135***	***117***	***102***
	* 7.0 #	10	24.4	2.75	4.04	28.2	*455*	*405*	*354*	***306***	***263***	***226***	***195***	***169***	***148***	***130***	***112***	***97.7***	***85.7***
80 x 80	10.0 ‡	8	30.2	2.41	3.65	30.8	753	624	505	**408**	**333**	**275**	**230**	**195**	**165**	**139**	**118**	**102**	**88.7**
	8.0 ‡	8	24.6	2.43	3.60	25.2	615	511	415	**336**	**274**	**226**	**189**	**160**	**136**	**115**	**97.7**	**84.3**	**73.4**
75 x 75	8.0 ‡	8	22.8	2.27	3.41	22.0	556	453	362	**289**	**234**	**192**	**160**	**135**	**112**	**93.8**	**79.9**	**68.8**	**59.9**
	* 6.0 ‡	8	17.5	2.29	3.35	16.8	*326*	*279*	*234*	***194***	***161***	***134***	***114***	***96.8***	***81.3***	***68.8***	***58.9***	***51.0***	***44.5***
70 x 70	7.0 ‡	8	18.8	2.12	3.18	16.8	447	357	280	**221**	**177**	**144**	**119**	**98.5**	**81.3**	**68.2**	**58.0**	**49.9**	**43.4**
	* 6.0 ‡	8	16.3	2.13	3.16	14.5	*332*	*274*	*222*	***178***	***145***	***119***	***99.5***	***82.7***	***68.6***	***57.7***	***49.2***	***42.5***	***37.0***
65 x 65	7.0 ‡	8	17.5	1.96	3.14	14.4	401	312	**240**	**187**	**148**	**120**	**98.2**	**79.3**	**65.3**	**54.7**	**46.5**	**40.0**	**34.7**
60 x 60	8.0 ‡	8	18.1	1.80	2.82	13.8	398	301	**227**	**174**	**137**	**110**	**87.0**	**70.1**	**57.7**	**48.2**	**40.9**	**35.2**	**30.5**
	6.0 ‡	8	13.8	1.82	2.77	10.6	306	233	**176**	**135**	**106**	**85.3**	**68.0**	**54.8**	**45.0**	**37.7**	**32.0**	**27.5**	**23.9**
	* 5.0 ‡	8	11.6	1.82	2.74	8.90	*214*	*171*	***133***	***105***	***83.6***	***67.9***	***54.5***	***44.2***	***36.6***	***30.7***	***26.1***	***22.5***	***19.6***
50 x 50	6.0 ‡	8	11.4	1.50	2.38	7.22	225	**160**	**115**	**86.1**	**66.3**	**50.2**	**39.2**	**31.5**	**25.8**	**21.5**	**18.2**	**15.6**	**13.6**
	5.0 ‡	8	9.60	1.51	2.35	6.10	191	**136**	**98.1**	**73.3**	**56.5**	**42.8**	**33.5**	**26.9**	**22.0**	**18.4**	**15.6**	**13.4**	**11.6**
	* 4.0 ‡	8	7.78	1.52	2.32	4.92	*124*	***94.5***	***71.3***	***54.7***	***42.9***	***33.0***	***26.0***	***21.0***	***17.3***	***14.5***	***12.4***	***10.6***	***9.24***

‡ Not rolled by certain leading producers, check availability.
# Check availability.
* Section is slender under axial compression and allowance has been made in calculating the compression resistance which is given in *italic type*.
Values in **bold type** indicate that the section has been divided into more than 3 bays until $\lambda_c \leq 50$.
FOR EXPLANATION OF TABLES SEE NOTE 8.3

BS 5950-1: 2000
BS EN 10056-1: 1999

STRUTS

TWO UNEQUAL ANGLES LONG LEGS BACK TO BACK CONNECTED TO ONE SIDE OF GUSSET OR MEMBER SUBJECT TO AXIAL COMPRESSION

TWO OR MORE BOLTS IN LINE ALONG EACH ANGLE OR EQUIVALENT WELDED AT EACH END

COMPRESSION RESISTANCE FOR STEEL S355

Section Designation A x A mm	t mm	Space s mm	Total Area cm²	Radius of Gyration Axis x-x cm	Radius of Gyration Axis y-y cm	Elastic Mod. Axis x-x cm³	Compression resistance, P_c (kN) for Length between Intersections, L (m) 1.0	1.5	2.0	2.5	3.0	3.5	4.0	4.5	5.0	5.5	6.0	6.5	7.0
200 x 150	18.0 #	15	120	6.30	6.36	350	3490	3330	3160	2970	2790	2600	2410	2230	**2060**	**1900**	**1750**	**1590**	1430
	*15.0	15	101	6.33	6.28	294	2510	2410	2290	2180	2060	1940	1810	1690	*1580*	*1470*	*1360*	*1250*	*1130*
	*12.0	15	81.6	6.36	6.22	238	*1350*	*1300*	*1250*	*1200*	*1150*	*1100*	*1050*	*998*	*948*	*898*	*850*	*798*	*735*
200 x 100	*15.0	15	86.0	6.40	3.97	274	2520	2410	2250	1930	1610	**1480**	**1240**	**1110**	**944**	**844**	**729**	**654**	**575**
	*12.0	15	69.6	6.43	3.90	222	*1360*	*1310*	*1240*	*1100*	*966*	*904*	*787*	*720*	*628*	*569*	*501*	*453*	*403*
	*10.0	15	58.4	6.46	3.85	186	*813*	*786*	*747*	*678*	*609*	*577*	*516*	*479*	*428*	*393*	*352*	*322*	*290*
150 x 90	15.0	12	67.8	4.74	3.75	155	1960	1820	1680	1450	**1310**	**1080**	956	800	706	603	535	465	417
	12.0	12	55.0	4.77	3.69	127	1590	1480	1370	1160	**1050**	864	762	637	561	478	424	368	329
	*10.0	12	46.4	4.80	3.64	107	*1070*	*1010*	*941*	*819*	*749*	*634*	*566*	*481*	*427*	*368*	*328*	*287*	*258*
150 x 75	15.0	12	63.4	4.75	3.09	150	1830	1710	1390	**1210**	956	818	693	571	491	416	363	319	279
	12.0	12	51.4	4.78	3.02	123	1490	1380	1110	960	757	644	544	447	384	325	283	249	217
	*10.0	12	43.4	4.81	2.97	103	*999*	*931*	*780*	*687*	*559*	*483*	*397*	*343*	*297*	*253*	*221*	*195*	*171*
125 x 75	12.0	12	45.4	3.95	3.19	86.4	1280	1160	1020	**891**	711	611	497	429	370	313	274	241	211
	10.0	12	38.2	3.97	3.14	73.0	1080	981	850	741	589	505	410	353	304	258	225	198	173
	*8.0	12	31.0	4.00	3.09	59.2	*648*	*603*	*536*	*480*	*398*	*348*	*290*	*253*	*220*	*188*	*166*	*144*	*129*
100 x 75	12.0	10	39.4	3.10	3.31	56.0	1060	926	796	**676**	572	460	371	303	252	213	182	157	137
	10.0	10	33.2	3.12	3.27	47.6	891	783	673	572	485	392	316	258	215	181	155	134	117
	*8.0	10	27.0	3.14	3.22	38.6	*684*	*606*	*526*	*451*	*385*	*314*	*254*	*209*	*174*	*147*	*126*	*109*	*95.0*
100 x 65	10.0 #	10	31.2	3.14	2.79	46.4	838	738	610	**520**	432	357	287	242	204	172	147	127	111
	8.0 #	10	25.4	3.16	2.74	37.8	683	602	492	417	345	271	228	192	163	137	119	104	90.5
	*7.0 #	10	22.4	3.17	2.72	33.2	*500*	*449*	*376*	*326*	*274*	*220*	*186*	*158*	*135*	*114*	*99.4*	*87.0*	*76.1*
100 x 50	8.0 ‡	10	22.8	3.19	2.09	36.4	615	449	**350**	270	210	166	134	110	91.8	77.6	66.4	57.4	50.1
	*6.0 ‡	10	17.4	3.21	2.04	27.6	*313*	*248*	*205*	*166*	*133*	*108*	*88.4*	*73.5*	*61.8*	*52.6*	*45.3*	*39.3*	*34.5*
80 x 60	7.0 ‡	8	18.8	2.51	2.59	21.4	474	397	**325**	265	200	155	123	99.4	82.0	68.8	58.5	50.4	43.8
80 x 40	8.0 ‡	8	18.0	2.53	1.71	22.8	425	**307**	218	159	123	94.7	75.1	60.9	50.4	42.5	36.2	31.2	27.1
	*6.0 ‡	8	13.8	2.55	1.66	17.5	*316*	*226*	*160*	*116*	*89.1*	*68.6*	*54.3*	*44.0*	*36.4*	*30.7*	*26.1*	*22.5*	*19.5*
75 x 50	8.0 ‡	8	18.8	2.35	2.19	20.8	465	383	**304**	237	180	139	109	88.5	72.9	61.1	51.9	44.6	38.8
	6.0 ‡	8	14.4	2.37	2.14	16.0	356	290	**228**	176	138	107	84.9	68.6	56.6	47.4	40.3	34.7	30.1
70 x 50	6.0 ‡	8	13.8	2.20	2.19	14.0	332	268	**212**	159	118	90.4	71.2	57.4	47.2	39.5	33.6	28.9	25.1
65 x 50	*5.0 ‡	8	11.1	2.05	2.21	10.3	*229*	*186*	*148*	*108*	*80.7*	*61.9*	*48.9*	*39.5*	*32.6*	*27.3*	*23.2*	*19.9*	*17.3*
60 x 40	6.0 ‡	8	11.4	1.88	1.80	10.1	256	**197**	146	102	74.1	56.2	44.0	35.4	29.0	24.2	20.5	17.6	15.3
	5.0 ‡	8	9.58	1.89	1.78	8.50	216	**166**	123	86.4	63.1	47.9	37.5	30.1	24.7	20.6	17.5	15.0	13.0

‡ Not rolled by certain leading producers, check availability.
\# Check availability.
* Section is slender under axial compression and allowance has been made in calculating the compression resistance which is given in *italic type*.
Values in **bold type** indicate that the section has been divided into more than 3 bays until $\lambda_c \leq 50$.
FOR EXPLANATION OF TABLES SEE NOTE 8.3

BS 5950-1: 2000
BS EN 10056-1: 1999

STRUTS

TWO UNEQUAL ANGLES LONG LEGS BACK TO BACK CONNECTED TO BOTH SIDES OF GUSSET OR MEMBER SUBJECT TO AXIAL COMPRESSION

TWO OR MORE BOLTS IN LINE ALONG EACH ANGLE AT EACH END

COMPRESSION RESISTANCE FOR STEEL S355

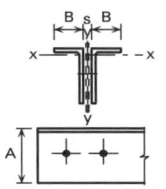

Section Designation	t	Space s	Total Area	Radius of Gyration Axis x-x	Radius of Gyration Axis y-y	Elastic Mod. Axis x-x	Compression resistance, P_c (kN) for Length between Intersections, L (m)												
A x A mm	mm	mm	cm²	cm	cm	cm³	1.0	1.5	2.0	2.5	3.0	3.5	4.0	4.5	5.0	5.5	6.0	6.5	7.0
200 x 150	18.0 #	15	120	6.30	6.36	350	3490	3330	3160	2970	2790	2600	2380	2080	**2000**	**1760**	**1550**	**1450**	**1290**
	* 15.0	15	101	6.33	6.28	294	*2510*	*2410*	*2290*	*2180*	*2060*	*1940*	*1780*	*1580*	*1510*	*1350*	*1200*	*1120*	*1010*
	* 12.0	15	81.6	6.36	6.22	238	*1350*	*1300*	*1250*	*1200*	*1150*	*1100*	*1030*	*941*	*911*	*833*	*760*	*692*	*659*
200 x 100	* 15.0	15	86.0	6.40	3.97	274	2520	2410	2100	1740	1410	**1250**	**1020**	**894**	**751**	**660**	**567**	**503**	**440**
	* 12.0	15	69.6	6.43	3.90	222	*1360*	*1310*	*1170*	*1020*	*871*	*790*	*672*	*599*	*514*	*457*	*398*	*356*	*314*
	* 10.0	15	58.4	6.46	3.85	186	*813*	*786*	*714*	*636*	*560*	*517*	*453*	*410*	*360*	*324*	*287*	*259*	*231*
150 x 90	15.0	12	67.8	4.74	3.75	155	1960	1820	1600	1290	**1120**	**902**	**775**	**640**	**554**	**470**	**412**	**357**	**317**
	12.0	12	55.0	4.77	3.69	127	1590	1480	1290	1030	**895**	**718**	**616**	**508**	**439**	**372**	**326**	**282**	**250**
	* 10.0	12	46.4	4.80	3.64	107	*1070*	*1010*	*890*	*740*	*653*	*537*	*466*	*390*	*339*	*290*	*255*	*222*	*197*
150 x 75	15.0	12	63.4	4.75	3.09	150	1830	1600	1240	**1030**	**794**	**658**	**545**	**445**	**378**	**318**	**276**	**241**	**210**
	12.0	12	51.4	4.78	3.02	123	1490	1290	992	**816**	**626**	**517**	**427**	**348**	**295**	**248**	**215**	**187**	**163**
	* 10.0	12	43.4	4.81	2.97	103	*999*	*878*	*707*	*596*	*470*	*393*	*318*	*270*	*230*	*195*	*169*	*148*	*129*
125 x 75	12.0	12	45.4	3.95	3.19	86.4	1280	1160	917	**766**	**593**	**494**	**396**	**335**	**285**	**240**	**209**	**182**	**159**
	10.0	12	38.2	3.97	3.14	73.0	1080	976	763	**635**	**490**	**407**	**326**	**276**	**234**	**197**	**171**	**149**	**130**
	* 8.0	12	31.0	4.00	3.09	59.2	*648*	*597*	*492*	*423*	*340*	*288*	*235*	*201*	*172*	*146*	*127*	*110*	*97.6*
100 x 75	12.0	10	39.4	3.10	3.31	56.0	1060	926	796	**676**	**537**	**451**	**377**	**308**	**263**	**222**	**193**	**169**	**147**
	10.0	10	33.2	3.12	3.27	47.6	891	783	673	**572**	**445**	**373**	**311**	**254**	**217**	**183**	**159**	**139**	**121**
	* 8.0	10	27.0	3.14	3.22	38.6	*684*	*606*	*525*	*444*	*347*	*291*	*234*	*199*	*170*	*143*	*125*	*109*	*95.1*
100 x 65	10.0 #	10	31.2	3.14	2.79	46.4	838	727	537	**437**	**351**	**282**	**224**	**187**	**157**	**132**	**114**	**98.7**	**86.5**
	8.0 #	10	25.4	3.16	2.74	37.8	683	586	431	**349**	**279**	**216**	**178**	**148**	**124**	**104**	**89.6**	**77.8**	**67.7**
	* 7.0 #	10	22.4	3.17	2.72	33.2	*500*	*437*	*336*	*277*	*225*	*177*	*147*	*123*	*103*	*87.2*	*75.3*	*65.6*	*57.2*
100 x 50	8.0 ‡	10	22.8	3.19	2.09	36.4	581	395	291	**216**	**164**	**128**	**102**	**83.0**	**68.9**	**57.9**	**49.3**	**42.5**	**37.0**
	* 6.0 ‡	10	17.4	3.21	2.04	27.6	*299*	*225*	*176*	*137*	*107*	*84.6*	*68.3*	*56.1*	*46.9*	*39.7*	*34.0*	*29.4*	*25.7*
80 x 60	7.0 ‡	8	18.8	2.51	2.59	21.4	474	397	**321**	**234**	**186**	**149**	**121**	**99.5**	**82.2**	**69.7**	**59.8**	**51.8**	**45.1**
80 x 40	8.0 ‡	8	18.0	2.53	1.71	22.8	387	**259**	**176**	**125**	**93.8**	**71.6**	**56.4**	**45.5**	**37.5**	**31.5**	**26.7**	**23.0**	**19.9**
	* 6.0 ‡	8	13.8	2.55	1.66	17.5	*287*	*190*	*128*	*90.5*	*67.9*	*51.8*	*40.7*	*32.8*	*27.0*	*22.7*	*19.3*	*16.5*	*14.4*
75 x 50	8.0 ‡	8	18.8	2.35	2.19	20.8	465	340	**255**	**191**	**146**	**114**	**91.3**	**74.4**	**61.8**	**52.0**	**44.4**	**38.3**	**33.4**
	6.0 ‡	8	14.4	2.37	2.14	16.0	356	256	**190**	**142**	**108**	**84.1**	**67.1**	**54.7**	**45.3**	**38.1**	**32.5**	**28.0**	**24.3**
70 x 50	6.0 ‡	8	13.8	2.20	2.19	14.0	332	249	**186**	**139**	**106**	**83.2**	**66.5**	**54.2**	**45.0**	**37.9**	**32.3**	**27.9**	**24.3**
65 x 50	* 5.0 ‡	8	11.1	2.05	2.21	10.3	*229*	*183*	*141*	*108*	*83.1*	*65.5*	*52.7*	*43.1*	*35.9*	*30.3*	*25.9*	*22.4*	*19.5*
60 x 40	6.0 ‡	8	11.4	1.88	1.80	10.1	254	**174**	**120**	**85.4**	**64.6**	**49.5**	**39.0**	**31.5**	**26.0**	**21.8**	**18.6**	**16.0**	**13.9**
	5.0 ‡	8	9.58	1.89	1.78	8.50	*213*	*145*	*99.1*	*70.7*	*52.5*	*40.2*	*32.2*	*26.0*	*21.4*	*17.9*	*15.2*	*13.1*	*11.4*

‡ Not rolled by certain leading producers, check availability.
# Check availability.
* Section is slender under axial compression and allowance has been made in calculating the compression resistance which is given in *italic type*.
Values in **bold type** indicate that the section has been divided into more than 3 bays until $\lambda_c \leq 50$.
FOR EXPLANATION OF TABLES SEE NOTE 8.3

BS 5950-1: 2000
BS 4-1: 1993

AXIAL LOAD & BENDING

UB SECTIONS SUBJECT TO AXIAL LOAD (COMPRESSION OR TENSION) AND BENDING

CROSS-SECTION CAPACITY CHECK

CAPACITIES FOR S355

Section Designation and Axial load Capacity P_z (kN)	F/P_z Limit Semi-Compact Compact		Moment Capacity M_{cx}, M_{cy} (kNm) and Reduced Moment Capacity M_{rx}, M_{ry} (kNm) for Ratios of Axial Load to Axial Load Capacity F/P_z										
		F/P_z	0.0	0.1	0.2	0.3	0.4	0.5	0.6	0.7	0.8	0.9	1.0
1016x305x487 # + $P_z = A_g p_y = 20800$	n/a 1.00	M_{cx}	7770	7770	7770	7770	7770	7770	7770	7770	7770	7770	7770
		M_{cy}	696	696	696	696	696	696	696	696	696	696	696
		M_{rx}	7770	7660	7340	6810	6060	5120	4140	3130	2110	1070	0
		M_{ry}	696	696	696	696	696	696	696	696	693	522	291	0
1016x305x437 # + $P_z = A_g p_y = 18700$	n/a 1.00	M_{cx}	6950	6950	6950	6950	6950	6950	6950	6950	6950	6950	6950
		M_{cy}	617	617	617	617	617	617	617	617	617	617	617
		M_{rx}	6950	6860	6570	6090	5410	4570	3690	2790	1880	949	0
		M_{ry}	617	617	617	617	617	617	617	616	464	258	0
1016x305x393 # + $P_z = A_g p_y = 16800$	n/a 1.00	M_{cx}	6210	6210	6210	6210	6210	6210	6210	6210	6210	6210	6210
		M_{cy}	544	544	544	544	544	544	544	544	544	544	544
		M_{rx}	6210	6130	5870	5440	4840	4080	3290	2490	1670	844	0
		M_{ry}	544	544	544	544	544	544	544	544	412	230	0
1016x305x349 # + $P_z = A_g p_y = 15400$	0.802 0.321	M_{cx}	5720	5720	5720	5720	5450	5380	5310	5190	4960	3730	$
		M_{cy}	506	506	506	506	506	506	506	506	506	318	$
		M_{rx}	5720	5640	5400	5000	-	-	-	-	-	-	-
		M_{ry}	506	506	506	506	-	-	-	-	-	-	-
1016x305x314 # + $P_z = A_g p_y = 13800$	0.678 0.266	M_{cx}	5120	5120	5120	4990	4740	4660	4560	2740	2740	2740	$
		M_{cy}	448	448	448	448	448	448	448	230	230	230	$
		M_{rx}	5120	5050	4830	-	-	-	-	-	-	-	-
		M_{ry}	448	448	448	-	-	-	-	-	-	-	-
1016x305x272 # + $P_z = A_g p_y = 12000$	0.518 0.192	M_{cx}	4430	4430	4390	4100	3960	3880	1780	1780	1780	$	$
		M_{cy}	387	387	387	387	387	387	148	148	148	$	$
		M_{rx}	4430	4360	-	-	-	-	-	-	-	-	-
		M_{ry}	387	387	-	-	-	-	-	-	-	-	-
1016x305x249 # + $P_z = A_g p_y = 10900$	0.518 0.210	M_{cx}	3920	3920	3920	3650	3490	3400	1560	1560	1560	$	$
		M_{cy}	325	325	325	325	325	325	125	125	125	$	$
		M_{rx}	3920	3860	3710	-	-	-	-	-	-	-	-
		M_{ry}	325	325	325	-	-	-	-	-	-	-	-

+ Section is not given in BS 4-1: 1993.
# Check availability.
F = Factored axial load.
$ For these values of F/P_z the section would be overloaded due to F alone even when M is zero, because F would exceed the local buckling resistance of the section.
- Not applicable for semi-compact and slender sections.
The values in this table are conservative for tension as the more onerous compression section classification limits have been used.
FOR EXPLANATION OF TABLES SEE NOTE 9.1

| BS 5950-1: 2000 |
| BS 4-1: 1993 |

AXIAL LOAD & BENDING

UB SECTIONS SUBJECT TO AXIAL COMPRESSION AND BENDING

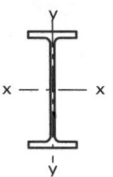

MEMBER BUCKLING CHECK

RESISTANCES AND CAPACITIES FOR S355

Section Designation and Capacities (kN, kNm)	F/P_z Limit	Compression Resistance P_{cx}, P_{cy} (kN) and Buckling Resistance Moment M_b, M_{bs} (kNm) for Varying effective lengths L_E (m) within the limiting value of F/P_z												
		L_E (m)	2.0	3.0	4.0	5.0	6.0	7.0	8.0	9.0	10.0	11.0	12.0	13.0	14.0
1016x305x487 # +	1.00	P_{cx}	20800	20800	20800	20800	20800	20600	20500	20300	20100	19900	19700	19400	19200
$P_z = A_g p_y = 20800$		P_{cy}	19000	16900	14500	11900	9630	7750	6310	5200	4340	3670	3140	2720	2380
$p_y Z_x = 6610$	1.00	M_b	7770	7340	6580	5850	5180	4590	4100	3690	3360	3070	2830	2630	2450
$p_y Z_y = 580$		M_{bs}	7770	7770	7770	7310	6780	6220	5650	5080	4530	4030	3590	3200	2860
1016x305x437 # +	1.00	P_{cx}	18700	18700	18700	18700	18700	18600	18400	18300	18200	18000	17900	17700	17500
$P_z = A_g p_y = 18700$		P_{cy}	17300	15600	13600	11200	9010	7200	5810	4760	3960	3340	2850	2460	2140
$p_y Z_x = 5940$	1.00	M_b	6950	6530	5810	5120	4490	3950	3500	3130	2830	2580	2370	2190	2040
$p_y Z_y = 514$		M_{bs}	6950	6950	6950	6510	6030	5520	5000	4490	4000	3550	3160	2810	2510
1016x305x393 # +	1.00	P_{cx}	16800	16800	16800	16800	16800	16700	16500	16400	16300	16200	16000	15900	15700
$P_z = A_g p_y = 16800$		P_{cy}	15500	14000	12100	9920	7930	6320	5100	4170	3470	2920	2490	2150	1870
$p_y Z_x = 5330$	1.00	M_b	6210	5780	5120	4470	3880	3380	2970	2640	2370	2150	1960	1810	1680
$p_y Z_y = 453$		M_{bs}	6210	6210	6200	5790	5350	4890	4410	3950	3510	3120	2770	2460	2200
* 1016x305x349 # +	0.802	P_{cx}	15400	15400	15400	15400	15400	15300	15200	15100	15000	14900	14800	14800	14600
$P_z = A_g p_y = 15400$		P_{cy}	14400	13200	11600	9560	7600	6020	4820	3920	3240	2720	2320	1990	1730
$p_y Z_x = 4950$	0.802	M_b	4950	4700	4200	3710	3240	2830	2490	2210	1980	1790	1640	1510	1390
$p_y Z_y = 422$		M_{bs}	4950	4950	4930	4600	4250	3880	3510	3130	2780	2470	2190	1940	1730
	0.321	M_b	5720	5300	4660	4040	3470	2990	2610	2300	2050	1850	1680	1540	1430
		M_{bs}	5720	5720	5700	5320	4920	4490	4050	3620	3220	2850	2530	2250	2000
* 1016x305x314 # +	0.678	P_{cx}	13800	13800	13800	13800	13800	13700	13700	13600	13500	13400	13300	13300	13200
$P_z = A_g p_y = 13800$		P_{cy}	12900	11800	10300	8490	6730	5310	4250	3460	2860	2400	2040	1760	1530
$p_y Z_x = 4440$	0.678	M_b	4440	4200	3730	3270	2830	2450	2140	1890	1680	1510	1380	1260	1160
$p_y Z_y = 373$		M_{bs}	4440	4440	4420	4120	3800	3460	3120	2780	2460	2180	1930	1720	1530
	0.266	M_b	5120	4710	4130	3540	3020	2580	2230	1950	1730	1560	1410	1290	1190
		M_{bs}	5120	5120	5090	4750	4380	3990	3590	3200	2840	2510	2230	1980	1760
* 1016x305x272 # +	0.518	P_{cx}	12000	12000	12000	12000	12000	11900	11900	11800	11700	11600	11600	11500	11400
$P_z = A_g p_y = 12000$		P_{cy}	11200	10300	8950	7340	5810	4590	3670	2980	2470	2070	1760	1510	1320
$p_y Z_x = 3860$	0.518	M_b	3860	3630	3210	2790	2390	2050	1770	1550	1370	1220	1110	1010	926
$p_y Z_y = 322$		M_{bs}	3860	3860	3830	3570	3290	3000	2700	2410	2130	1890	1670	1480	1320
	0.192	M_b	4430	4060	3530	3010	2540	2150	1840	1600	1410	1260	1130	1030	944
		M_{bs}	4430	4430	4390	4090	3780	3440	3100	2760	2450	2160	1920	1700	1520
* 1016x305x249 # +	0.518	P_{cx}	10900	10900	10900	10900	10900	10900	10800	10800	10700	10600	10600	10500	10400
$P_z = A_g p_y = 10900$		P_{cy}	10200	9250	7940	6390	4990	3910	3120	2530	2090	1750	1490	1280	1110
$p_y Z_x = 3390$	0.518	M_b	3390	3150	2770	2380	2020	1710	1470	1270	1120	995	894	812	743
$p_y Z_y = 270$		M_{bs}	3390	3390	3330	3080	2830	2550	2280	2020	1780	1560	1380	1220	1090
	0.210	M_b	3920	3540	3060	2570	2140	1800	1520	1310	1150	1020	914	828	757
		M_{bs}	3920	3920	3840	3560	3270	2950	2640	2330	2050	1810	1590	1410	1250

+ Section is not given in BS 4-1: 1993.
# Check availability.
* The section can become slender under axial compression only. Under combined axial compression and bending the section becomes slender when the semi-compact F/P$_z$ limit is exceeded.
Under combined axial compression and bending the capacities are only valid up to the given F/P$_z$ limit. For higher values F/P$_z$ the section would be overloaded due to F alone even when M is zero, because F would exceed the local buckling resistance of the section.
M_b is obtained using an equivalent slenderness = u.v.L$_E$/r$_y$.β$_w$$^{0.5}$
M_{bs} is obtained using an equivalent slenderness = 0.5 L/r$_y$. Effective length L$_E$ = L.
FOR EXPLANATION OF TABLES SEE NOTE 9.1

BS 5950-1: 2000
BS 4-1: 1993

AXIAL LOAD & BENDING

UB SECTIONS SUBJECT TO AXIAL LOAD (COMPRESSION OR TENSION) AND BENDING

CROSS-SECTION CAPACITY CHECK

CAPACITIES FOR S355

Section Designation and Axial load Capacity P_z (kN)	F/P_z Limit Semi-Compact / Compact		Moment Capacity M_{cx}, M_{cy} (kNm) and Reduced Moment Capacity M_{rx}, M_{ry} (kNm) for Ratios of Axial Load to Axial Load Capacity F/P_z										
		F/P_z	0.0	0.1	0.2	0.3	0.4	0.5	0.6	0.7	0.8	0.9	1.0
1016x305x222 # + $P_z = A_g p_y = 9760$	0.487 / 0.211	M_{cx}	3380	3380	3380	3140	2980	1260	1260	1260	1260	$	$
		M_{cy}	263	263	263	263	263	95.1	95.1	95.1	95.1	$	$
		M_{rx}	3380	3340	3210	-	-	-	-	-	-	-	-
		M_{ry}	263	263	263	-	-	-	-	-	-	-	-
914 x 419 x 388 # $P_z = A_g p_y = 17000$	0.934 / 0.321	M_{cx}	6100	6100	6100	6100	6000	5990	5960	5930	5850	5620	$
		M_{cy}	895	895	895	895	895	895	895	895	895	895	$
		M_{rx}	6100	6000	5700	5210	-	-	-	-	-	-	-
		M_{ry}	895	895	895	895	-	-	-	-	-	-	-
914 x 419 x 343 # $P_z = A_g p_y = 15100$	0.800 / 0.276	M_{cx}	5340	5340	5340	5260	5110	5070	5020	4920	3560	3560	$
		M_{cy}	775	775	775	775	775	775	775	775	485	485	$
		M_{rx}	5340	5260	5000	-	-	-	-	-	-	-	-
		M_{ry}	775	775	775	-	-	-	-	-	-	-	-
914 x 305 x 289 # $P_z = A_g p_y = 12700$	0.767 / 0.324	M_{cx}	4340	4340	4340	4340	4130	4040	3980	3870	2680	2680	$
		M_{cy}	420	420	420	420	420	420	420	420	250	250	$
		M_{rx}	4340	4280	4100	3800	-	-	-	-	-	-	-
		M_{ry}	420	420	420	420	-	-	-	-	-	-	-
914 x 305 x 253 # $P_z = A_g p_y = 11100$	0.624 / 0.257	M_{cx}	3770	3770	3770	3650	3470	3380	3300	1840	1840	$	$
		M_{cy}	361	361	361	361	361	361	361	169	169	$	$
		M_{rx}	3770	3720	3570	-	-	-	-	-	-	-	-
		M_{ry}	361	361	361	-	-	-	-	-	-	-	-
914 x 305 x 224 # $P_z = A_g p_y = 9870$	0.533 / 0.221	M_{cx}	3290	3290	3290	3090	2950	2880	1350	1350	1350	$	$
		M_{cy}	306	306	306	306	306	306	121	121	121	$	$
		M_{rx}	3290	3250	3110	-	-	-	-	-	-	-	-
		M_{ry}	306	306	306	-	-	-	-	-	-	-	-

+ Section is not given in BS 4-1: 1993.
# Check availability.
F = Factored axial load.
$ For these values of F/P_z the section would be overloaded due to F alone even when M is zero, because F would exceed the local buckling resistance of the section.
- Not applicable for semi-compact and slender sections.
The values in this table are conservative for tension as the more onerous compression section classification limits have been used.
FOR EXPLANATION OF TABLES SEE NOTE 9.1

BS 5950-1: 2000
BS 4-1: 1993

AXIAL LOAD & BENDING

UB SECTIONS SUBJECT TO AXIAL COMPRESSION AND BENDING

MEMBER BUCKLING CHECK

RESISTANCES AND CAPACITIES FOR S355

Section Designation and Capacities (kN, kNm)	F/P_z Limit		Compression Resistance P_{cx}, P_{cy} (kN) and Buckling Resistance Moment M_b, M_{bs} (kNm) for Varying effective lengths L_E (m) within the limiting value of F/P_z											
			L_E (m)	2.0	3.0	4.0	5.0	6.0	7.0	8.0	9.0	10.0	11.0	12.0	13.0	14.0
* 1016x305x222 # + $P_z = A_g p_y = 9760$ $p_y Z_x = 2900$ $p_y Z_y = 219$	0.487	P_{cx}	9760	9760	9760	9760	9750	9700	9640	9590	9530	9460	9400	9330	9250
		P_{cy}	9020	8110	6820	5380	4150	3230	2560	2070	1710	1430	1210	1040	906
	0.487	M_b	2900	2670	2320	1970	1650	1390	1180	1010	885	782	700	632	576
		M_{bs}	2900	2900	2810	2590	2350	2110	1860	1640	1430	1250	1100	971	861
	0.211	M_b	3380	3010	2570	2130	1750	1450	1220	1050	909	801	715	645	587
		M_{bs}	3380	3380	3280	3020	2740	2460	2170	1910	1670	1460	1280	1130	1000
* 914 x 419 x 388 # $P_z = A_g p_y = 17000$ $p_y Z_x = 5390$ $p_y Z_y = 746$	0.934	P_{cx}	17000	17000	17000	17000	17000	16900	16800	16700	16600	16500	16400	16300	16200
		P_{cy}	16700	16000	15100	14100	12800	11300	9770	8350	7130	6120	5280	4590	4030
	0.934	M_b	5390	5390	5250	4880	4510	4140	3780	3440	3140	2870	2640	2430	2260
		M_{bs}	5390	5390	5390	5390	5360	5120	4880	4620	4350	4070	3790	3520	3260
	0.321	M_b	6100	6100	5830	5380	4920	4460	4030	3640	3300	3000	2740	2520	2330
		M_{bs}	6100	6100	6100	6100	6060	5790	5510	5220	4910	4600	4290	3980	3680
* 914 x 419 x 343 # $P_z = A_g p_y = 15100$ $p_y Z_x = 4740$ $p_y Z_y = 645$	0.800	P_{cx}	15100	15100	15100	15100	15100	15000	14900	14800	14700	14600	14500	14400	14300
		P_{cy}	14800	14100	13400	12400	11200	9870	8500	7240	6180	5290	4560	3970	3480
	0.800	M_b	4740	4740	4590	4250	3910	3560	3230	2930	2650	2410	2200	2020	1860
		M_{bs}	4740	4740	4740	4740	4690	4480	4260	4030	3790	3540	3290	3050	2820
	0.276	M_b	5340	5340	5080	4670	4250	3830	3440	3080	2770	2500	2270	2080	1910
		M_{bs}	5340	5340	5340	5340	5290	5050	4800	4540	4270	3990	3710	3440	3170
* 914 x 305 x 289 # $P_z = A_g p_y = 12700$ $p_y Z_x = 3750$ $p_y Z_y = 350$	0.767	P_{cx}	12700	12700	12700	12700	12700	12600	12500	12400	12400	12300	12200	12100	12000
		P_{cy}	11900	11000	9650	8000	6380	5060	4060	3310	2740	2300	1960	1680	1460
	0.767	M_b	3750	3570	3190	2810	2440	2110	1840	1620	1450	1300	1180	1080	995
		M_{bs}	3750	3750	3750	3500	3240	2970	2680	2400	2140	1900	1680	1500	1340
	0.324	M_b	4340	4030	3540	3050	2610	2230	1930	1690	1490	1340	1210	1110	1020
		M_{bs}	4340	4340	4330	4050	3750	3430	3100	2770	2470	2190	1950	1730	1540
* 914 x 305 x 253 # $P_z = A_g p_y = 11100$ $p_y Z_x = 3280$ $p_y Z_y = 300$	0.624	P_{cx}	11100	11100	11100	11100	11100	11100	11000	10900	10900	10800	10700	10600	10500
		P_{cy}	10500	9590	8390	6920	5490	4340	3480	2830	2340	1970	1670	1440	1250
	0.624	M_b	3280	3100	2750	2400	2060	1770	1530	1340	1180	1060	952	867	795
		M_{bs}	3280	3280	3260	3040	2810	2570	2310	2070	1840	1630	1440	1280	1140
	0.257	M_b	3770	3480	3040	2600	2200	1860	1590	1380	1220	1080	975	886	811
		M_{bs}	3770	3770	3760	3510	3240	2960	2660	2380	2110	1870	1660	1470	1320
* 914 x 305 x 224 # $P_z = A_g p_y = 9870$ $p_y Z_x = 2850$ $p_y Z_y = 255$	0.533	P_{cx}	9870	9870	9870	9870	9840	9780	9720	9660	9600	9530	9460	9380	9300
		P_{cy}	9230	8430	7310	5960	4700	3700	2960	2400	1990	1670	1420	1220	1060
	0.533	M_b	2850	2680	2370	2040	1740	1480	1270	1100	963	855	767	695	635
		M_{bs}	2850	2850	2820	2630	2420	2200	1970	1750	1550	1370	1210	1080	958
	0.221	M_b	3290	3010	2610	2210	1850	1550	1320	1130	990	876	784	709	647
		M_{bs}	3290	3290	3250	3030	2790	2530	2270	2020	1790	1580	1400	1240	1100

+ Section is not given in BS 4-1: 1993.
# Check availability.
* The section can become slender under axial compression only. Under combined axial compression and bending the section becomes slender when the semi-compact F/P_z limit is exceeded.
Under combined axial compression and bending the capacities are only valid up to the given F/P_z limit. For higher values F/P_z the section would be overloaded due to F alone even when M is zero, because F would exceed the local buckling resistance of the section.
M_b is obtained using an equivalent slenderness = $u.v.L_E/r_y.\beta_w^{0.5}$
M_{bs} is obtained using an equivalent slenderness = $0.5 L/r_y$. Effective length $L_E = L$.
FOR EXPLANATION OF TABLES SEE NOTE 9.1

		AXIAL LOAD & BENDING										

BS 5950-1: 2000
BS 4-1: 1993

AXIAL LOAD & BENDING

UB SECTIONS SUBJECT TO AXIAL LOAD (COMPRESSION OR TENSION) AND BENDING

CROSS-SECTION CAPACITY CHECK

CAPACITIES FOR S355

Section Designation and Axial load Capacity P_z (kN)	F/P_z Limit Semi-Compact Compact		Moment Capacity M_{cx}, M_{cy} (kNm) and Reduced Moment Capacity M_{rx}, M_{ry} (kNm) for Ratios of Axial Load to Axial Load Capacity F/P_z									
		F/P_z	0.0	0.1	0.2	0.3	0.4	0.5	0.6	0.7	0.8	0.9	1.0
914 x 305 x 201 # $P_z = A_g p_y = 8830$	0.481 0.206	M_{cx} M_{cy} M_{rx} M_{ry}	2880 257 2880 257	2880 257 2840 257	2880 257 2730 257	2670 257 - -	2550 257 - -	1060 91.7 - -	1060 91.7 - -	1060 91.7 - -	1060 91.7 - -	$ $ - -	$ $ - -
838 x 292 x 226 # $P_z = A_g p_y = 9970$	0.632 0.251	M_{cx} M_{cy} M_{rx} M_{ry}	3160 320 3160 320	3160 320 3110 320	3160 320 2980 320	3050 320 - -	2900 320 - -	2850 320 - -	2780 320 - -	1570 152 - -	1570 152 - -	$ $ - -	$ $ - -
838 x 292 x 194 # $P_z = A_g p_y = 8520$	0.534 0.219	M_{cx} M_{cy} M_{rx} M_{ry}	2640 257 2640 257	2640 257 2600 257	2640 257 2490 257	2480 257 - -	2370 257 - -	2310 257 - -	1090 102 - -	1090 102 - -	1090 102 - -	$ $ - -	$ $ - -
838 x 292 x 176 # $P_z = A_g p_y = 7730$	0.485 0.203	M_{cx} M_{cy} M_{rx} M_{ry}	2350 221 2350 221	2350 221 2320 221	2350 221 2230 221	2180 221 - -	2080 221 - -	876 79.5 - -	876 79.5 - -	876 79.5 - -	876 79.5 - -	$ $ - -	$ $ - -
762 x 267 x 197 $P_z = A_g p_y = 8660$	0.718 0.293	M_{cx} M_{cy} M_{rx} M_{ry}	2470 253 2470 253	2470 253 2440 253	2470 253 2330 253	2460 253 - -	2320 253 - -	2280 253 - -	2240 253 - -	2170 253 - -	1420 139 - -	1420 139 - -	$ - -

\# Check availability.
F = Factored axial load.
$ For these values of F/P_z the section would be overloaded due to F alone even when M is zero, because F would exceed the local buckling resistance of the section.
- Not applicable for semi-compact and slender sections.
The values in this table are conservative for tension as the more onerous compression section classification limits have been used.
FOR EXPLANATION OF TABLES SEE NOTE 9.1

BS 5950-1: 2000
BS 4-1: 1993

AXIAL LOAD & BENDING

UB SECTIONS SUBJECT TO AXIAL COMPRESSION AND BENDING

MEMBER BUCKLING CHECK

RESISTANCES AND CAPACITIES FOR S355

Section Designation and Capacities (kN, kNm)	F/P_z Limit		Compression Resistance P_{cx}, P_{cy} (kN) and Buckling Resistance Moment M_b, M_{bs} (kNm) for Varying effective lengths L_E (m) within the limiting value of F/P_z											
		L_E (m)	2.0	3.0	4.0	5.0	6.0	7.0	8.0	9.0	10.0	11.0	12.0	13.0	14.0
* 914 x 305 x 201 #	0.481	P_{cx}	8830	8830	8830	8830	8800	8750	8700	8640	8580	8520	8450	8380	8310
$P_z = A_g p_y$ = 8830		P_{cy}	8220	7460	6390	5140	4010	3140	2500	2030	1670	1400	1190	1030	891
$p_y Z_x$ = 2490	0.481	M_b	2490	2310	2030	1740	1460	1230	1050	902	786	694	620	559	509
$p_y Z_y$ = 214		M_{bs}	2490	2490	2440	2260	2070	1870	1670	1470	1300	1140	1010	891	792
	0.206	M_b	2880	2610	2240	1880	1550	1290	1090	930	808	711	633	570	518
		M_{bs}	2880	2880	2830	2620	2400	2170	1930	1710	1500	1320	1170	1030	918
* 838 x 292 x 226 #	0.632	P_{cx}	9970	9970	9970	9970	9920	9860	9800	9730	9660	9580	9500	9420	9320
$P_z = A_g p_y$ = 9970		P_{cy}	9320	8520	7390	6030	4750	3740	2990	2430	2010	1680	1430	1230	1070
$p_y Z_x$ = 2750	0.632	M_b	2750	2580	2280	1980	1690	1450	1250	1090	968	866	782	713	655
$p_y Z_y$ = 267		M_{bs}	2750	2750	2730	2540	2330	2120	1900	1690	1500	1320	1170	1040	925
	0.251	M_b	3160	2880	2510	2130	1800	1520	1300	1130	995	888	800	728	668
		M_{bs}	3160	3160	3130	2910	2680	2430	2180	1940	1720	1520	1340	1190	1060
* 838 x 292 x 194 #	0.534	P_{cx}	8520	8520	8520	8520	8480	8420	8360	8310	8240	8180	8100	8030	7940
$P_z = A_g p_y$ = 8520		P_{cy}	7930	7190	6160	4950	3860	3020	2410	1950	1610	1350	1150	987	857
$p_y Z_x$ = 2290	0.534	M_b	2290	2120	1860	1590	1350	1140	972	840	736	653	586	531	485
$p_y Z_y$ = 214		M_{bs}	2290	2290	2250	2080	1910	1720	1530	1360	1190	1050	926	819	728
	0.219	M_b	2640	2380	2050	1710	1420	1190	1010	866	756	669	599	541	494
		M_{bs}	2640	2640	2580	2390	2190	1980	1770	1560	1370	1210	1070	942	838
* 838 x 292 x 176 #	0.485	P_{cx}	7730	7730	7730	7730	7680	7630	7580	7520	7470	7400	7340	7270	7190
$P_z = A_g p_y$ = 7730		P_{cy}	7160	6460	5470	4340	3360	2620	2080	1690	1390	1170	990	851	739
$p_y Z_x$ = 2030	0.485	M_b	2030	1870	1630	1380	1160	972	824	708	617	545	487	439	400
$p_y Z_y$ = 185		M_{bs}	2030	2030	1980	1830	1660	1500	1330	1170	1020	897	789	697	619
	0.203	M_b	2350	2100	1790	1490	1220	1010	853	729	633	557	497	447	407
		M_{bs}	2350	2350	2280	2110	1920	1730	1530	1350	1180	1040	912	805	715
* 762 x 267 x 197	0.718	P_{cx}	8660	8660	8660	8640	8580	8520	8460	8390	8320	8240	8150	8060	7960
$P_z = A_g p_y$ = 8660		P_{cy}	7980	7140	5960	4670	3580	2780	2200	1780	1470	1230	1040	896	778
$p_y Z_x$ = 2150	0.718	M_b	2150	1950	1700	1440	1220	1040	898	786	697	625	567	519	478
$p_y Z_y$ = 210		M_{bs}	2150	2150	2070	1900	1730	1540	1360	1190	1040	905	794	700	620
	0.293	M_b	2470	2180	1850	1550	1290	1090	930	809	715	640	579	529	487
		M_{bs}	2470	2470	2380	2190	1980	1770	1560	1360	1190	1040	913	805	713

\# Check availability.
* The section can become slender under axial compression only. Under combined axial compression and bending the section becomes slender when the semi-compact F/P_z limit is exceeded.
Under combined axial compression and bending the capacities are only valid up to the given F/P_z limit. For higher values F/P_z the section would be overloaded due to F alone even when M is zero, because F would exceed the local buckling resistance of the section.
M_b is obtained using an equivalent slenderness = $u.v.L_E/r_y.\beta_w^{0.5}$
M_{bs} is obtained using an equivalent slenderness = $0.5 L/r_y$. Effective length L_E = L.
FOR EXPLANATION OF TABLES SEE NOTE 9.1

BS 5950-1: 2000
BS 4-1: 1993

AXIAL LOAD & BENDING

UB SECTIONS SUBJECT TO AXIAL LOAD (COMPRESSION OR TENSION) AND BENDING

CROSS-SECTION CAPACITY CHECK

CAPACITIES FOR S355

Section Designation and Axial load Capacity P_z (kN)	F/P_z Limit Semi-Compact Compact		Moment Capacity M_{cx}, M_{cy} (kNm) and Reduced Moment Capacity M_{rx}, M_{ry} (kNm) for Ratios of Axial Load to Axial Load Capacity F/P_z									
		F/P_z	0.0	0.1	0.2	0.3	0.4	0.5	0.6	0.7	0.8	0.9	1.0
762 x 267 x 173 $P_z = A_g p_y = 7590$	0.617 **0.256**	M_{cx} M_{cy} M_{rx} M_{ry}	2140 213 2140 213	2140 213 2110 213	2140 213 2020 213	2070 213 - -	1960 213 - -	1910 213 - -	1870 213 - -	1030 98.3 - -	1030 98.3 - -	$ $ - -	$ $ - -
762 x 267 x 147 $P_z = A_g p_y = 6450$	0.500 **0.208**	M_{cx} M_{cy} M_{rx} M_{ry}	1780 170 1780 170	1780 170 1760 170	1780 170 1680 170	1660 170 - -	1580 170 - -	685 63.0 - -	685 63.0 - -	685 63.0 - -	685 63.0 - -	$ $ - -	$ $ - -
762 x 267 x 134 $P_z = A_g p_y = 6070$	0.424 **0.173**	M_{cx} M_{cy} M_{rx} M_{ry}	1650 154 1650 154	1650 154 1630 154	1600 154 - -	1490 154 - -	1440 154 - -	541 48.7 - -	541 48.7 - -	541 48.7 - -	541 48.7 - -	$ $ - -	$ $ - -
686 x 254 x 170 $P_z = A_g p_y = 7490$	0.763 **0.303**	M_{cx} M_{cy} M_{rx} M_{ry}	1940 214 1940 214	1940 214 1910 214	1940 214 1830 214	1940 214 1690 214	1840 214 - -	1810 214 - -	1790 214 - -	1740 214 - -	1200 127 - -	1200 127 - -	$ $ - -
686 x 254 x 152 $P_z = A_g p_y = 6690$	0.650 **0.256**	M_{cx} M_{cy} M_{rx} M_{ry}	1730 188 1730 188	1730 188 1700 188	1730 188 1630 188	1670 188 - -	1590 188 - -	1560 188 - -	1530 188 - -	886 92.2 - -	886 92.2 - -	886 92.2 - -	$ - - -
686 x 254 x 140 $P_z = A_g p_y = 6140$	0.580 **0.228**	M_{cx} M_{cy} M_{rx} M_{ry}	1570 169 1570 169	1570 169 1550 169	1570 169 1480 169	1490 169 - -	1430 169 - -	1400 169 - -	713 73.1 - -	713 73.1 - -	713 73.1 - -	$ $ - -	$ $ - -

F = Factored axial load.
$ For these values of F/P_z the section would be overloaded due to F alone even when M is zero, because F would exceed the local buckling resistance of the section.
- Not applicable for semi-compact and slender sections.
The values in this table are conservative for tension as the more onerous compression section classification limits have been used.
FOR EXPLANATION OF TABLES SEE NOTE 9.1

| BS 5950-1: 2000 |
| BS 4-1: 1993 |

AXIAL LOAD & BENDING

UB SECTIONS SUBJECT TO AXIAL COMPRESSION AND BENDING

MEMBER BUCKLING CHECK

RESISTANCES AND CAPACITIES FOR S355

Section Designation and Capacities (kN, kNm)	F/P_z Limit	Compression Resistance P_{cx}, P_{cy} (kN) and Buckling Resistance Moment M_b, M_{bs} (kNm) for Varying effective lengths L_E (m) within the limiting value of F/P_z												
		L_E (m)	2.0	3.0	4.0	5.0	6.0	7.0	8.0	9.0	10.0	11.0	12.0	13.0	14.0
* 762 x 267 x 173 $P_z = A_g p_y = 7590$ $p_y Z_x = 1860$ $p_y Z_y = 177$	0.617	P_{cx}	7590	7590	7590	7570	7520	7460	7410	7350	7280	7210	7130	7050	6960
		P_{cy}	6960	6200	5120	3970	3030	2340	1850	1500	1230	1030	875	751	652
	0.617	M_b	1860	1670	1440	1210	1010	853	729	632	557	496	447	407	374
		M_{bs}	1860	1860	1780	1630	1470	1300	1140	997	867	756	662	583	516
	0.256	M_b	2140	1860	1570	1290	1060	887	753	650	570	507	456	415	380
		M_{bs}	2140	2140	2040	1870	1690	1500	1320	1150	998	870	762	670	594
* 762 x 267 x 147 $P_z = A_g p_y = 6450$ $p_y Z_x = 1540$ $p_y Z_y = 142$	0.500	P_{cx}	6450	6450	6450	6430	6390	6340	6290	6240	6180	6120	6050	5980	5890
		P_{cy}	5880	5190	4220	3220	2440	1880	1480	1200	985	824	699	600	521
	0.500	M_b	1540	1370	1160	966	796	662	559	480	419	371	332	301	274
		M_{bs}	1540	1540	1460	1330	1190	1050	916	793	687	597	522	458	405
	0.208	M_b	1770	1520	1270	1030	835	687	576	493	429	378	338	306	279
		M_{bs}	1780	1780	1680	1530	1370	1210	1060	915	793	689	602	529	467
* 762 x 267 x 134 $P_z = A_g p_y = 6070$ $p_y Z_x = 1430$ $p_y Z_y = 129$	0.424	P_{cx}	6070	6070	6070	6050	6000	5960	5910	5860	5800	5740	5680	5600	5520
		P_{cy}	5510	4810	3850	2900	2180	1670	1320	1060	873	729	618	531	460
	0.424	M_b	1430	1250	1050	865	705	581	487	416	361	318	283	256	232
		M_{bs}	1430	1430	1330	1210	1080	945	818	705	608	527	459	403	356
	0.173	M_b	1620	1390	1150	919	737	602	501	426	369	324	288	260	236
		M_{bs}	1650	1650	1540	1400	1250	1090	946	815	703	609	531	466	411
* 686 x 254 x 170 $P_z = A_g p_y = 7490$ $p_y Z_x = 1700$ $p_y Z_y = 179$	0.763	P_{cx}	7490	7490	7490	7450	7390	7330	7260	7200	7120	7040	6950	6840	6730
		P_{cy}	6860	6090	5010	3860	2940	2270	1800	1450	1190	1000	848	729	632
	0.763	M_b	1700	1520	1310	1110	936	797	689	604	537	483	438	402	371
		M_{bs}	1700	1700	1620	1480	1330	1180	1030	900	782	681	596	524	464
	0.303	M_b	1940	1680	1420	1180	984	829	711	621	550	493	447	409	378
		M_{bs}	1940	1940	1850	1690	1530	1350	1180	1030	895	780	682	600	531
* 686 x 254 x 152 $P_z = A_g p_y = 6690$ $p_y Z_x = 1510$ $p_y Z_y = 157$	0.650	P_{cx}	6690	6690	6690	6660	6600	6550	6490	6430	6360	6290	6200	6110	6010
		P_{cy}	6120	5410	4420	3390	2580	1990	1570	1270	1040	873	740	636	552
	0.650	M_b	1510	1340	1150	961	803	677	580	505	446	399	361	330	303
		M_{bs}	1510	1510	1430	1310	1180	1040	907	787	683	594	519	457	404
	0.256	M_b	1710	1480	1240	1020	841	702	598	518	456	407	368	335	308
		M_{bs}	1730	1730	1640	1500	1340	1190	1040	900	781	679	594	522	462
* 686 x 254 x 140 $P_z = A_g p_y = 6140$ $p_y Z_x = 1380$ $p_y Z_y = 141$	0.580	P_{cx}	6140	6140	6140	6100	6060	6010	5950	5890	5830	5760	5680	5600	5500
		P_{cy}	5600	4930	4010	3060	2320	1780	1410	1140	934	781	663	569	494
	0.580	M_b	1380	1210	1030	860	713	597	509	440	387	345	311	283	260
		M_{bs}	1380	1380	1300	1180	1060	936	815	706	611	531	464	408	360
	0.228	M_b	1560	1340	1120	912	745	618	523	451	396	352	317	288	264
		M_{bs}	1570	1570	1490	1350	1210	1070	932	807	699	607	530	466	412

* The section can become slender under axial compression only. Under combined axial compression and bending the section becomes slender when the semi-compact F/P_z limit is exceeded.
Under combined axial compression and bending the capacities are only valid up to the given F/P_z limit. For higher values F/P_z the section would be overloaded due to F alone even when M is zero, because F would exceed the local buckling resistance of the section.
M_b is obtained using an equivalent slenderness = $u.v.L_E/r_y.\beta_w^{0.5}$
M_{bs} is obtained using an equivalent slenderness = $0.5 L/r_y$. Effective length $L_E = L$.
FOR EXPLANATION OF TABLES SEE NOTE 9.1

BS 5950-1: 2000
BS 4-1: 1993

AXIAL LOAD & BENDING

UB SECTIONS SUBJECT TO AXIAL LOAD (COMPRESSION OR TENSION) AND BENDING

CROSS-SECTION CAPACITY CHECK

CAPACITIES FOR S355

Section Designation and Axial load Capacity P_z (kN)	F/P_z Limit Semi-Compact **Compact**		Moment Capacity M_{cx}, M_{cy} (kNm) and Reduced Moment Capacity M_{rx}, M_{ry} (kNm) for Ratios of Axial Load to Axial Load Capacity F/P_z										
		F/P_z	0.0	0.1	0.2	0.3	0.4	0.5	0.6	0.7	0.8	0.9	1.0
686 x 254 x 125 $P_z = A_g p_y = 5490$	0.519 **0.211**	M_{cx}	1380	1380	1380	1290	1240	1210	554	554	554	$	$
		M_{cy}	143	143	143	143	143	143	55.1	55.1	55.1	$	$
		M_{rx}	1380	1360	1300	-	-	-	-	-	-	-	-
		M_{ry}	143	143	143	-	-	-	-	-	-	-	-
610 x 305 x 238 $P_z = A_g p_y = 10500$	n/a **1.00**	M_{cx}	2580	2580	2580	2580	2580	2580	2580	2580	2580	2580	2580
		M_{cy}	421	421	421	421	421	421	421	421	421	421	421
		M_{rx}	2580	2540	2410	2200	1900	1600	1290	974	654	330	0
		M_{ry}	421	421	421	421	421	421	421	421	375	275	150	0
610 x 305 x 179 $P_z = A_g p_y = 7870$	0.899 **0.296**	M_{cx}	1910	1910	1910	1910	1870	1860	1850	1840	1800	1480	$
		M_{cy}	308	308	308	308	308	308	308	308	308	223	$
		M_{rx}	1910	1880	1790	-	-	-	-	-	-	-	-
		M_{ry}	308	308	308	-	-	-	-	-	-	-	-
610 x 305 x 149 $P_z = A_g p_y = 6560$	0.671 **0.213**	M_{cx}	1580	1580	1580	1510	1480	1470	1440	864	864	864	$
		M_{cy}	253	253	253	253	253	253	253	128	128	128	$
		M_{rx}	1580	1560	1480	-	-	-	-	-	-	-	-
		M_{ry}	253	253	253	-	-	-	-	-	-	-	-
610 x 229 x 140 $P_z = A_g p_y = 6140$	0.781 **0.305**	M_{cx}	1430	1430	1430	1430	1350	1340	1320	1290	912	912	$
		M_{cy}	162	162	162	162	162	162	162	162	98.5	98.5	$
		M_{rx}	1430	1410	1350	1240	-	-	-	-	-	-	-
		M_{ry}	162	162	162	162	-	-	-	-	-	-	-
610 x 229 x 125 $P_z = A_g p_y = 5490$	0.664 **0.257**	M_{cx}	1270	1270	1270	1230	1170	1160	1130	669	669	669	$
		M_{cy}	142	142	142	142	142	142	142	71.3	71.3	71.3	$
		M_{rx}	1270	1250	1190	-	-	-	-	-	-	-	-
		M_{ry}	142	142	142	-	-	-	-	-	-	-	-

F = Factored axial load.
$ For these values of F/P_z the section would be overloaded due to F alone even when M is zero, because F would exceed the local buckling resistance of the section.
- Not applicable for semi-compact and slender sections.
The values in this table are conservative for tension as the more onerous compression section classification limits have been used.
FOR EXPLANATION OF TABLES SEE NOTE 9.1

BS 5950-1: 2000
BS 4-1: 1993

AXIAL LOAD & BENDING

UB SECTIONS SUBJECT TO AXIAL COMPRESSION AND BENDING

MEMBER BUCKLING CHECK

RESISTANCES AND CAPACITIES FOR S355

Section Designation and Capacities (kN, kNm)	F/P_z Limit	Compression Resistance P_{cx}, P_{cy} (kN) and Buckling Resistance Moment M_b, M_{bs} (kNm) for Varying effective lengths L_E (m) within the limiting value of F/P_z												
		L_E (m)	2.0	3.0	4.0	5.0	6.0	7.0	8.0	9.0	10.0	11.0	12.0	13.0	14.0
* 686 x 254 x 125	0.519	P_{cx}	5490	5490	5490	5450	5410	5360	5310	5260	5200	5140	5060	4980	4890
$P_z = A_g p_y$ = 5490		P_{cy}	4970	4340	3480	2620	1970	1520	1200	963	792	662	561	482	418
$p_y Z_x$ = 1200	0.519	M_b	1200	1050	884	727	596	495	418	359	314	278	250	226	207
$p_y Z_y$ = 119		M_{bs}	1200	1200	1120	1020	908	795	689	594	512	444	387	340	300
	0.211	M_b	1360	1160	956	770	622	511	429	368	320	283	254	230	210
		M_{bs}	1380	1380	1290	1170	1040	912	790	681	588	510	444	390	344
610 x 305 x 238	1.00	P_{cx}	10500	10500	10500	10400	10300	10200	10100	9990	9870	9740	9590	9420	9230
$P_z = A_g p_y$ = 10500		P_{cy}	9960	9300	8420	7280	6030	4900	3980	3270	2720	2290	1960	1690	1470
$p_y Z_x$ = 2270	1.00	M_b	2580	2480	2240	2010	1780	1590	1420	1270	1160	1060	972	900	839
$p_y Z_y$ = 351		M_{bs}	2580	2580	2580	2500	2340	2170	2000	1830	1650	1490	1340	1200	1080
* 610 x 305 x 179	0.899	P_{cx}	7870	7870	7860	7800	7730	7660	7590	7500	7410	7310	7200	7060	6910
$P_z = A_g p_y$ = 7870		P_{cy}	7480	6960	6270	5370	4410	3560	2890	2370	1970	1660	1410	1220	1060
$p_y Z_x$ = 1700	0.899	M_b	1700	1650	1490	1320	1170	1030	905	804	721	652	594	546	505
$p_y Z_y$ = 256		M_{bs}	1700	1700	1700	1630	1530	1420	1300	1180	1060	955	855	766	688
	0.296	M_b	1910	1820	1620	1430	1240	1080	942	832	742	669	609	558	515
		M_{bs}	1910	1910	1910	1840	1720	1590	1460	1330	1200	1070	961	861	773
* 610 x 305 x 149	0.671	P_{cx}	6560	6560	6550	6500	6440	6380	6320	6250	6170	6080	5990	5870	5750
$P_z = A_g p_y$ = 6560		P_{cy}	6220	5780	5190	4440	3630	2930	2370	1940	1610	1360	1160	996	866
$p_y Z_x$ = 1420	0.671	M_b	1420	1360	1220	1080	940	816	711	624	554	497	449	410	377
$p_y Z_y$ = 211		M_{bs}	1420	1420	1420	1360	1270	1170	1070	974	877	786	703	629	565
	0.213	M_b	1580	1500	1330	1160	993	852	736	643	568	508	459	418	383
		M_{bs}	1580	1580	1580	1520	1420	1310	1200	1090	980	878	786	703	631
* 610 x 229 x 140	0.781	P_{cx}	6140	6140	6130	6080	6030	5970	5900	5840	5760	5670	5570	5460	5330
$P_z = A_g p_y$ = 6140		P_{cy}	5520	4750	3720	2770	2060	1580	1240	1000	821	686	581	499	432
$p_y Z_x$ = 1250	0.781	M_b	1240	1070	905	752	629	534	461	405	361	325	296	272	252
$p_y Z_y$ = 135		M_{bs}	1250	1250	1150	1040	915	794	682	584	502	434	377	330	291
	0.305	M_b	1390	1180	975	796	656	552	475	415	369	332	302	277	256
		M_{bs}	1430	1430	1320	1190	1050	908	780	668	574	496	431	378	333
* 610 x 229 x 125	0.664	P_{cx}	5490	5490	5480	5430	5380	5330	5270	5210	5140	5060	4980	4870	4760
$P_z = A_g p_y$ = 5490		P_{cy}	4920	4210	3280	2430	1810	1380	1090	873	717	599	507	435	377
$p_y Z_x$ = 1110	0.664	M_b	1100	945	790	649	537	451	387	338	299	268	244	223	206
$p_y Z_y$ = 118		M_{bs}	1110	1110	1020	916	806	697	598	511	438	378	329	288	254
	0.257	M_b	1230	1040	849	685	559	466	397	345	305	273	248	227	209
		M_{bs}	1270	1270	1160	1040	920	796	682	583	500	432	375	328	289

* The section can become slender under axial compression only. Under combined axial compression and bending the section becomes slender when the semi-compact F/P_z limit is exceeded.
Under combined axial compression and bending the capacities are only valid up to the given F/P_z limit. For higher values F/P_z the section would be overloaded due to F alone even when M is zero, because F would exceed the local buckling resistance of the section.
M_b is obtained using an equivalent slenderness = $u.v.L_E/r_y.\beta_w^{0.5}$
M_{bs} is obtained using an equivalent slenderness = $0.5 L/r_y$. Effective length $L_E = L$.
FOR EXPLANATION OF TABLES SEE NOTE 9.1

BS 5950-1: 2000
BS 4-1: 1993

AXIAL LOAD & BENDING

UB SECTIONS SUBJECT TO AXIAL LOAD (COMPRESSION OR TENSION) AND BENDING

CROSS-SECTION CAPACITY CHECK

CAPACITIES FOR S355

Section Designation and Axial load Capacity P_z (kN)	F/P_z Limit Semi-Compact Compact		Moment Capacity M_{cx}, M_{cy} (kNm) and Reduced Moment Capacity M_{rx}, M_{ry} (kNm) for Ratios of Axial Load to Axial Load Capacity F/P_z										
		F/P_z	0.0	0.1	0.2	0.3	0.4	0.5	0.6	0.7	0.8	0.9	1.0
610 x 229 x 113 $P_z = A_g p_y = 4970$	0.586 **0.228**	M_{cx}	1130	1130	1130	1080	1030	1010	520	520	520	$	$
		M_{cy}	125	125	125	125	125	125	54.4	54.4	54.4	$	$
		M_{rx}	1130	1120	1070	-	-	-	-	-	-	-	-
		M_{ry}	125	125	125	-	-	-	-	-	-	-	-
610 x 229 x 101 $P_z = A_g p_y = 4580$	0.513 **0.204**	M_{cx}	1020	1020	1020	954	917	895	407	407	407	$	$
		M_{cy}	109	109	109	109	109	109	41.4	41.4	41.4	$	$
		M_{rx}	1020	1010	966	-	-	-	-	-	-	-	-
		M_{ry}	109	109	109	-	-	-	-	-	-	-	-
533 x 210 x 122 $P_z = A_g p_y = 5350$	0.928 **0.359**	M_{cx}	1100	1100	1100	1100	1080	1080	1070	1070	1050	1000	$
		M_{cy}	132	132	132	132	132	132	132	132	132	132	$
		M_{rx}	1100	1090	1040	956	-	-	-	-	-	-	-
		M_{ry}	132	132	132	132	-	-	-	-	-	-	-
533 x 210 x 109 $P_z = A_g p_y = 4800$	0.804 **0.311**	M_{cx}	976	976	976	976	929	922	912	894	857	646	$
		M_{cy}	116	116	116	116	116	116	116	116	116	72.8	$
		M_{rx}	976	961	918	846	-	-	-	-	-	-	-
		M_{ry}	116	116	116	116	-	-	-	-	-	-	-
533 x 210 x 101 $P_z = A_g p_y = 4450$	0.714 **0.272**	M_{cx}	901	901	901	884	842	833	819	795	518	518	$
		M_{cy}	106	106	106	106	106	106	106	106	106	57.9	$
		M_{rx}	901	888	848	-	-	-	-	-	-	-	-
		M_{ry}	106	106	106	-	-	-	-	-	-	-	-
533 x 210 x 92 $P_z = A_g p_y = 4150$	0.619 **0.237**	M_{cx}	838	838	838	802	769	756	740	410	410	$	$
		M_{cy}	97.1	97.1	97.1	97.1	97.1	97.1	97.1	45.1	45.1	$	$
		M_{rx}	838	826	790	-	-	-	-	-	-	-	-
		M_{ry}	97.1	97.1	97.1	-	-	-	-	-	-	-	-

F = Factored axial load.
$ For these values of F/P_z the section would be overloaded due to F alone even when M is zero, because F would exceed the local buckling resistance of the section.
- Not applicable for semi-compact and slender sections.
The values in this table are conservative for tension as the more onerous compression section classification limits have been used.
FOR EXPLANATION OF TABLES SEE NOTE 9.1

BS 5950-1: 2000
BS 4-1: 1993

AXIAL LOAD & BENDING

UB SECTIONS SUBJECT TO AXIAL COMPRESSION AND BENDING

MEMBER BUCKLING CHECK

RESISTANCES AND CAPACITIES FOR S355

Section Designation and Capacities (kN, kNm)	F/P_z Limit		Compression Resistance P_{cx}, P_{cy} (kN) and Buckling Resistance Moment M_b, M_{bs} (kNm) for Varying effective lengths L_E (m) within the limiting value of F/P_z											
		L_E (m)	2.0	3.0	4.0	5.0	6.0	7.0	8.0	9.0	10.0	11.0	12.0	13.0	14.0
* 610 x 229 x 113	0.586	P_{cx}	4970	4970	4960	4920	4870	4820	4770	4710	4650	4580	4490	4400	4290
$P_z = A_g p_y = 4970$		P_{cy}	4430	3770	2910	2140	1590	1210	951	764	627	524	444	381	330
$p_y Z_x = 992$	0.586	M_b	973	834	690	561	459	383	325	282	249	222	201	183	168
$p_y Z_y = 104$		M_{bs}	992	991	903	808	708	610	521	444	380	328	285	249	219
	0.228	M_b	1090	916	740	590	477	394	334	288	253	226	204	186	171
		M_{bs}	1130	1130	1030	923	809	697	595	507	434	374	325	284	250
* 610 x 229 x 101	0.513	P_{cx}	4580	4580	4570	4530	4480	4440	4390	4330	4270	4200	4120	4020	3910
$P_z = A_g p_y = 4580$		P_{cy}	4050	3390	2560	1850	1370	1040	813	653	536	447	378	324	281
$p_y Z_x = 893$	0.513	M_b	867	736	600	480	387	319	269	231	202	180	162	147	134
$p_y Z_y = 90.9$		M_{bs}	893	883	800	711	617	527	446	378	323	277	240	210	184
	0.204	M_b	973	809	642	503	401	328	275	236	206	183	164	149	136
		M_{bs}	1020	1010	917	814	707	603	511	434	370	317	275	240	211
* 533 x 210 x 122	0.928	P_{cx}	5350	5350	5320	5260	5210	5150	5080	5000	4910	4810	4690	4550	4390
$P_z = A_g p_y = 5350$		P_{cy}	4720	3940	2970	2150	1580	1200	944	758	621	518	439	376	326
$p_y Z_x = 964$	0.928	M_b	935	800	668	553	463	395	344	304	272	247	226	208	193
$p_y Z_y = 110$		M_{bs}	964	952	862	764	662	565	478	405	346	297	257	224	197
	0.359	M_b	1050	879	716	583	482	409	354	312	278	252	230	212	197
		M_{bs}	1100	1090	986	874	758	646	547	464	395	340	294	257	226
* 533 x 210 x 109	0.804	P_{cx}	4800	4800	4770	4720	4670	4610	4550	4480	4400	4310	4200	4070	3910
$P_z = A_g p_y = 4800$		P_{cy}	4220	3500	2610	1880	1380	1050	823	661	542	452	382	328	284
$p_y Z_x = 855$	0.804	M_b	825	700	577	471	390	329	284	249	222	201	183	168	156
$p_y Z_y = 96.3$		M_{bs}	855	840	759	671	579	492	415	352	299	257	222	194	171
	0.311	M_b	922	766	616	494	404	339	291	255	227	204	186	171	158
		M_{bs}	976	960	866	766	661	562	474	401	342	293	254	221	195
* 533 x 210 x 101	0.714	P_{cx}	4450	4450	4420	4380	4330	4280	4220	4160	4080	4000	3890	3770	3630
$P_z = A_g p_y = 4450$		P_{cy}	3910	3230	2400	1730	1270	964	755	606	496	414	351	301	260
$p_y Z_x = 791$	0.714	M_b	761	643	526	426	350	294	252	221	196	176	160	147	136
$p_y Z_y = 88.3$		M_{bs}	791	776	700	618	532	452	381	322	274	235	203	177	156
	0.272	M_b	849	702	561	446	362	302	258	225	200	179	163	149	138
		M_{bs}	901	885	798	704	607	515	434	367	312	268	232	202	178
* 533 x 210 x 92	0.619	P_{cx}	4150	4150	4120	4080	4040	3990	3930	3870	3800	3710	3610	3490	3350
$P_z = A_g p_y = 4150$		P_{cy}	3620	2960	2170	1540	1130	857	670	537	440	367	311	266	231
$p_y Z_x = 736$	0.619	M_b	702	590	474	378	307	255	218	189	167	150	136	124	114
$p_y Z_y = 80.9$		M_{bs}	736	717	644	565	484	408	343	289	245	210	181	158	139
	0.237	M_b	782	640	503	395	317	262	222	193	170	152	138	126	116
		M_{bs}	838	816	734	644	551	465	390	329	279	239	206	180	158

* The section can become slender under axial compression only. Under combined axial compression and bending the section becomes slender when the semi-compact F/P_z limit is exceeded.
Under combined axial compression and bending the capacities are only valid up to the given F/P_z limit. For higher values F/P_z the section would be overloaded due to F alone even when M is zero, because F would exceed the local buckling resistance of the section.
M_b is obtained using an equivalent slenderness = $u.v.L_E/r_y.\beta_w^{0.5}$
M_{bs} is obtained using an equivalent slenderness = $0.5 L/r_y$. Effective length $L_E = L$.
FOR EXPLANATION OF TABLES SEE NOTE 9.1

| BS 5950-1: 2000 |
| BS 4-1: 1993 |

AXIAL LOAD & BENDING

UB SECTIONS SUBJECT TO AXIAL LOAD (COMPRESSION OR TENSION) AND BENDING

CROSS-SECTION CAPACITY CHECK

CAPACITIES FOR S355

Section Designation and Axial load Capacity P_z (kN)	F/P_z Limit Semi-Compact Compact		Moment Capacity M_{cx}, M_{cy} (kNm) and Reduced Moment Capacity M_{rx}, M_{ry} (kNm) for Ratios of Axial Load to Axial Load Capacity F/P_z									
		F/P_z	0.0	0.1	0.2	0.3	0.4	0.5	0.6	0.7	0.8	0.9	1.0
533 x 210 x 82 $P_z = A_g p_y = 3730$	0.564 0.225	M_{cx} M_{cy} M_{rx} M_{ry}	731 81.8 731 81.8	731 81.8 721 81.8	731 81.8 690 81.8	692 81.8 - -	663 81.8 - -	649 81.8 - -	321 34.3 - -	321 34.3 - -	321 34.3 - -	$ - - -	$ - - -
457 x 191 x 98 $P_z = A_g p_y = 4310$	0.998 0.371	M_{cx} M_{cy} M_{rx} M_{ry}	770 101 770 101	770 101 758 101	770 101 723 101	770 101 664 -	770 101 - -	770 101 - -	770 101 - -	769 101 - -	769 101 - -	768 101 - -	$ $ - -
457 x 191 x 89 $P_z = A_g p_y = 3930$	0.880 0.325	M_{cx} M_{cy} M_{rx} M_{ry}	695 90.3 695 90.3	695 90.3 684 90.3	695 90.3 652 90.3	695 90.3 599 -	675 90.3 - -	672 90.3 - -	667 90.3 - -	659 90.3 - -	643 90.3 - -	517 63.7 - -	$ $ - -
457 x 191 x 82 $P_z = A_g p_y = 3690$	0.783 0.294	M_{cx} M_{cy} M_{rx} M_{ry}	650 83.5 650 83.5	650 83.5 640 83.5	650 83.5 611 83.5	647 83.5 - -	617 83.5 - -	612 83.5 - -	605 83.5 - -	592 83.5 - -	418 50.9 - -	418 50.9 - -	$ $ - -
457 x 191 x 74 $P_z = A_g p_y = 3360$	0.666 0.244	M_{cx} M_{cy} M_{rx} M_{ry}	587 75.0 587 75.0	587 75.0 578 75.0	587 75.0 552 75.0	565 75.0 - -	544 75.0 - -	538 75.0 - -	528 75.0 - -	313 37.8 - -	313 37.8 - -	313 37.8 - -	$ $ - -
457 x 191 x 67 $P_z = A_g p_y = 3040$	0.601 0.226	M_{cx} M_{cy} M_{rx} M_{ry}	522 65.2 522 65.2	522 65.2 515 65.2	522 65.2 492 65.2	497 65.2 - -	478 65.2 - -	471 65.2 - -	460 65.2 - -	248 29.3 - -	248 29.3 - -	$ - $ -	$ $ - -

F = Factored axial load.
$ For these values of F/P_z the section would be overloaded due to F alone even when M is zero, because F would exceed the local buckling resistance of the section.
- Not applicable for semi-compact and slender sections.
The values in this table are conservative for tension as the more onerous compression section classification limits have been used.
FOR EXPLANATION OF TABLES SEE NOTE 9.1

BS 5950-1: 2000
BS 4-1: 1993

AXIAL LOAD & BENDING

UB SECTIONS SUBJECT TO AXIAL COMPRESSION AND BENDING

MEMBER BUCKLING CHECK

RESISTANCES AND CAPACITIES FOR S355

Section Designation and Capacities (kN, kNm)	F/P_z Limit		Compression Resistance P_{cx}, P_{cy} (kN) and Buckling Resistance Moment M_b, M_{bs} (kNm) for Varying effective lengths L_E (m) within the limiting value of F/P_z											
			L_E (m)	2.0	3.0	4.0	5.0	6.0	7.0	8.0	9.0	10.0	11.0	12.0	13.0	14.0
*533 x 210 x 82 $P_z = A_g p_y = 3730$ $p_y Z_x = 639$ $p_y Z_y = 68.2$	0.564	P_{cx}	3730	3730	3700	3660	3620	3570	3520	3460	3390	3310	3220	3100	2970
		P_{cy}	3220	2590	1870	1320	964	729	569	456	374	311	264	226	196
	0.564	M_b	606	503	400	315	253	208	176	152	133	119	107	97.5	89.6
		M_{bs}	639	617	552	480	408	342	286	240	203	173	150	130	114
	0.225	M_b	676	548	424	328	261	213	179	154	135	120	109	98.8	90.7
		M_{bs}	731	706	631	550	467	391	327	274	232	198	171	149	131
*457 x 191 x 98 $P_z = A_g p_y = 4310$ $p_y Z_x = 675$ $p_y Z_y = 83.8$	0.998	P_{cx}	4310	4310	4260	4210	4150	4090	4020	3930	3830	3700	3550	3380	3180
		P_{cy}	3730	3000	2160	1530	1120	847	661	530	434	362	306	262	227
	0.998	M_b	640	539	444	366	307	263	229	204	183	167	153	141	131
		M_{bs}	675	652	583	507	431	361	302	254	215	184	158	138	121
	0.371	M_b	714	588	473	383	318	271	235	208	187	170	156	144	133
		M_{bs}	770	744	665	578	492	412	344	289	245	209	181	157	138
*457 x 191 x 89 $P_z = A_g p_y = 3930$ $p_y Z_x = 611$ $p_y Z_y = 75.2$	0.880	P_{cx}	3930	3930	3880	3840	3790	3730	3660	3580	3490	3370	3230	3070	2880
		P_{cy}	3390	2710	1950	1370	1000	759	593	475	389	324	274	235	204
	0.880	M_b	576	481	391	318	264	224	195	172	154	140	128	118	109
		M_{bs}	611	588	525	455	386	323	269	226	191	163	141	123	108
	0.325	M_b	641	523	415	333	273	231	199	176	157	142	130	120	111
		M_{bs}	695	669	597	518	439	367	307	257	218	186	160	140	123
*457 x 191 x 82 $P_z = A_g p_y = 3690$ $p_y Z_x = 572$ $p_y Z_y = 69.6$	0.783	P_{cx}	3690	3690	3640	3600	3550	3490	3430	3350	3250	3140	3000	2830	2650
		P_{cy}	3160	2490	1760	1230	897	677	528	423	346	289	244	209	181
	0.783	M_b	535	442	353	283	232	196	169	148	132	119	109	100	92.7
		M_{bs}	572	546	485	418	352	293	243	203	172	146	126	110	96.2
	0.294	M_b	593	478	374	295	240	201	172	151	134	121	110	101	93.9
		M_{bs}	650	621	551	476	400	333	276	231	195	166	143	125	109
*457 x 191 x 74 $P_z = A_g p_y = 3360$ $p_y Z_x = 518$ $p_y Z_y = 62.5$	0.666	P_{cx}	3360	3350	3310	3270	3230	3180	3120	3040	2960	2850	2730	2580	2410
		P_{cy}	2860	2250	1580	1110	806	608	474	380	311	259	219	188	163
	0.666	M_b	482	395	313	248	201	168	144	126	112	100	91.3	83.8	77.5
		M_{bs}	518	493	437	376	316	262	218	182	154	131	113	98.1	86.0
	0.244	M_b	533	427	329	257	207	172	147	128	113	102	92.6	84.9	78.4
		M_{bs}	587	559	496	427	358	297	247	206	174	148	128	111	97.5
*457 x 191 x 67 $P_z = A_g p_y = 3040$ $p_y Z_x = 460$ $p_y Z_y = 54.3$	0.601	P_{cx}	3040	3030	2990	2950	2910	2860	2810	2740	2660	2560	2440	2300	2140
		P_{cy}	2570	2000	1390	970	703	530	413	331	271	226	191	163	141
	0.601	M_b	425	346	270	211	170	141	120	104	91.7	82.2	74.5	68.2	62.9
		M_{bs}	460	436	385	329	275	227	188	157	132	113	97.0	84.3	73.9
	0.226	M_b	471	373	284	219	175	144	122	106	93.1	83.3	75.5	69.0	63.6
		M_{bs}	522	495	437	374	312	258	214	178	150	128	110	95.6	83.8

* The section can become slender under axial compression only. Under combined axial compression and bending the section becomes slender when the semi-compact F/P_z limit is exceeded.
Under combined axial compression and bending the capacities are only valid up to the given F/P_z limit. For higher values F/P_z the section would be overloaded due to F alone even when M is zero, because F would exceed the local buckling resistance of the section.
M_b is obtained using an equivalent slenderness = $u.v.L_E/r_y.\beta_w^{0.5}$
M_{bs} is obtained using an equivalent slenderness = $0.5 L/r_y$. Effective length $L_E = L$.
FOR EXPLANATION OF TABLES SEE NOTE 9.1

BS 5950-1: 2000
BS 4-1: 1993

AXIAL LOAD & BENDING

UB SECTIONS SUBJECT TO AXIAL LOAD (COMPRESSION OR TENSION) AND BENDING

CROSS-SECTION CAPACITY CHECK

CAPACITIES FOR S355

Section Designation and Axial load Capacity P_z (kN)	F/P_z Limit Semi-Compact Compact		Moment Capacity M_{cx}, M_{cy} (kNm) and Reduced Moment Capacity M_{rx}, M_{ry} (kNm) for Ratios of Axial Load to Axial Load Capacity F/P_z										
		F/P_z	0.0	0.1	0.2	0.3	0.4	0.5	0.6	0.7	0.8	0.9	1.0
457 x 152 x 82 $P_z = A_g p_y = 3620$	0.880 **0.353**	M_{cx}	625	625	625	625	607	602	597	589	574	459	$
		M_{cy}	63.3	63.3	63.3	63.3	63.3	63.3	63.3	63.3	63.3	44.7	$
		M_{rx}	625	616	589	543	-	-	-	-	-	-	-
		M_{ry}	63.3	63.3	63.3	63.3	-	-	-	-	-	-	-
457 x 152 x 74 $P_z = A_g p_y = 3260$	0.762 **0.304**	M_{cx}	561	561	561	561	530	523	515	502	345	345	$
		M_{cy}	56.3	56.3	56.3	56.3	56.3	56.3	56.3	56.3	33.2	33.2	$
		M_{rx}	561	553	529	489	-	-	-	-	-	-	-
		M_{ry}	56.3	56.3	56.3	56.3	-	-	-	-	-	-	-
457 x 152 x 67 $P_z = A_g p_y = 3040$	0.666 **0.270**	M_{cx}	516	516	516	504	478	468	458	271	271	271	$
		M_{cy}	50.7	50.7	50.7	50.7	50.7	50.7	50.7	25.5	25.5	25.5	$
		M_{rx}	516	509	487	-	-	-	-	-	-	-	-
		M_{ry}	50.7	50.7	50.7	-	-	-	-	-	-	-	-
457 x 152 x 60 $P_z = A_g p_y = 2710$	0.549 **0.216**	M_{cx}	457	457	457	430	412	403	195	195	195	$	$
		M_{cy}	44.3	44.3	44.3	44.3	44.3	44.3	18.1	18.1	18.1	$	$
		M_{rx}	457	451	431	-	-	-	-	-	-	-	-
		M_{ry}	44.3	44.3	44.3	-	-	-	-	-	-	-	-
457 x 152 x 52 $P_z = A_g p_y = 2360$	0.485 **0.199**	M_{cx}	389	389	389	360	345	145	145	145	145	$	$
		M_{cy}	36.0	36.0	36.0	36.0	36.0	12.9	12.9	12.9	12.9	$	$
		M_{rx}	389	384	-	-	-	-	-	-	-	-	-
		M_{ry}	36.0	36.0	-	-	-	-	-	-	-	-	-
406 x 178 x 74 $P_z = A_g p_y = 3350$	0.892 **0.319**	M_{cx}	533	533	533	533	519	517	514	508	498	404	$
		M_{cy}	73.3	73.3	73.3	73.3	73.3	73.3	73.3	73.3	73.3	52.6	$
		M_{rx}	533	525	499	458	-	-	-	-	-	-	-
		M_{ry}	73.3	73.3	73.3	73.3	-	-	-	-	-	-	-

F = Factored axial load.
$ For these values of F/P_z the section would be overloaded due to F alone even when M is zero, because F would exceed the local buckling resistance of the section.
- Not applicable for semi-compact and slender sections.
The values in this table are conservative for tension as the more onerous compression section classification limits have been used.
FOR EXPLANATION OF TABLES SEE NOTE 9.1

BS 5950-1: 2000
BS 4-1: 1993

AXIAL LOAD & BENDING

UB SECTIONS SUBJECT TO AXIAL COMPRESSION AND BENDING

MEMBER BUCKLING CHECK

RESISTANCES AND CAPACITIES FOR S355

Compression Resistance P_{cx}, P_{cy} (kN) and Buckling Resistance Moment M_b, M_{bs} (kNm) for Varying effective lengths L_E (m) within the limiting value of F/P_z

Section Designation and Capacities (kN, kNm)	F/P_z Limit		L_E (m)	1.0	1.50	2.00	2.50	3.00	3.50	4.00	4.50	5.00	5.50	6.00	6.50	7.00
* 457 x 152 x 82	0.880		P_{cx}	3620	3620	3620	3620	3620	3600	3580	3550	3530	3510	3480	3460	3430
$P_z = A_g p_y = 3620$			P_{cy}	3420	3160	2810	2360	1910	1530	1230	1000	831	699	595	513	446
$p_y Z_x = 542$	0.880		M_b	542	522	470	418	368	323	286	254	228	207	189	174	161
$p_y Z_y = 52.8$			M_{bs}	542	542	542	512	476	438	399	359	321	286	255	228	204
	0.353		M_b	625	587	522	456	395	343	300	265	237	213	194	178	165
			M_{bs}	625	625	625	590	549	505	459	414	370	330	294	262	235
* 457 x 152 x 74	0.762		P_{cx}	3260	3260	3260	3260	3250	3240	3220	3200	3180	3160	3130	3110	3080
$P_z = A_g p_y = 3260$			P_{cy}	3080	2840	2510	2100	1690	1350	1080	884	732	615	524	451	392
$p_y Z_x = 488$	0.762		M_b	488	467	419	370	323	281	247	218	195	175	160	146	135
$p_y Z_y = 46.9$			M_{bs}	488	488	488	459	426	391	355	319	285	254	226	201	180
	0.304		M_b	561	525	464	402	346	297	258	226	201	181	164	150	138
			M_{bs}	561	561	561	528	490	450	409	368	328	292	260	232	207
* 457 x 152 x 67	0.666		P_{cx}	3040	3040	3040	3040	3030	3010	3000	2980	2960	2940	2910	2890	2870
$P_z = A_g p_y = 3040$			P_{cy}	2860	2620	2300	1900	1510	1200	958	779	645	541	460	396	344
$p_y Z_x = 448$	0.666		M_b	448	425	378	331	286	246	214	187	166	149	134	123	113
$p_y Z_y = 42.2$			M_{bs}	448	448	447	418	386	353	319	285	253	225	199	177	158
	0.270		M_b	516	477	418	359	304	259	223	194	171	152	138	125	115
			M_{bs}	516	516	514	480	444	406	367	328	291	258	229	204	182
* 457 x 152 x 60	0.549		P_{cx}	2710	2710	2710	2710	2700	2680	2670	2650	2630	2610	2590	2570	2550
$P_z = A_g p_y = 2710$			P_{cy}	2540	2330	2030	1670	1320	1040	835	679	561	471	400	345	299
$p_y Z_x = 398$	0.549		M_b	398	375	333	289	247	211	182	158	139	124	112	101	92.8
$p_y Z_y = 36.9$			M_{bs}	398	398	396	369	341	311	280	250	222	196	174	154	138
	0.216		M_b	457	420	366	312	262	221	189	163	143	127	114	103	94.6
			M_{bs}	457	457	454	424	391	357	321	287	254	225	199	177	158
* 457 x 152 x 52	0.485		P_{cx}	2360	2360	2360	2360	2360	2340	2330	2310	2300	2280	2260	2240	2220
$P_z = A_g p_y = 2360$			P_{cy}	2210	2010	1730	1400	1090	857	683	554	458	384	326	280	244
$p_y Z_x = 337$	0.485		M_b	337	315	277	237	201	170	144	125	109	96.2	86.0	77.7	70.8
$p_y Z_y = 30$			M_{bs}	337	337	332	308	283	256	229	203	179	158	139	123	110
	0.199		M_b	389	353	305	256	213	177	150	128	112	98.4	87.8	79.2	72.1
			M_{bs}	389	389	383	356	326	296	264	234	207	182	161	142	126
* 406 x 178 x 74	0.892		P_{cx}	3350	3350	3350	3350	3340	3320	3290	3270	3250	3220	3190	3160	3130
$P_z = A_g p_y = 3350$			P_{cy}	3230	3050	2820	2520	2160	1800	1490	1240	1040	877	750	648	565
$p_y Z_x = 470$	0.892		M_b	470	470	432	393	354	316	283	253	228	206	188	173	160
$p_y Z_y = 61.1$			M_{bs}	470	470	470	467	442	416	388	360	331	302	275	249	226
	0.319		M_b	533	525	478	430	382	337	298	265	237	213	194	178	164
			M_{bs}	533	533	533	530	501	472	440	408	375	343	312	283	256

* The section can become slender under axial compression only. Under combined axial compression and bending the section becomes slender when the semi-compact F/P_z limit is exceeded.
Under combined axial compression and bending the capacities are only valid up to the given F/P_z limit. For higher values F/P_z the section would be overloaded due to F alone even when M is zero, because F would exceed the local buckling resistance of the section.
M_b is obtained using an equivalent slenderness = $u.v.L_E/r_y.\beta_w^{0.5}$
M_{bs} is obtained using an equivalent slenderness = $0.5 L/r_y$. Effective length $L_E = L$.
FOR EXPLANATION OF TABLES SEE NOTE 9.1

BS 5950-1: 2000
BS 4-1: 1993

AXIAL LOAD & BENDING

UB SECTIONS SUBJECT TO AXIAL LOAD (COMPRESSION OR TENSION) AND BENDING

CROSS-SECTION CAPACITY CHECK

CAPACITIES FOR S355

Section Designation and Axial load Capacity P_z (kN)	F/P_z Limit Semi-Compact Compact		Moment Capacity M_{cx}, M_{cy} (kNm) and Reduced Moment Capacity M_{rx}, M_{ry} (kNm) for Ratios of Axial Load to Axial Load Capacity F/P_z										
		F/P_z	0.0	0.1	0.2	0.3	0.4	0.5	0.6	0.7	0.8	0.9	1.0
406 x 178 x 67 $P_z = A_g p_y = 3040$	0.789 0.284	M_{cx} M_{cy} M_{rx} M_{ry}	478 65.2 478 65.2	478 65.2 470 65.2	478 65.2 448 65.2	473 65.2 - -	455 65.2 - -	452 65.2 - -	446 65.2 - -	437 65.2 - -	312 40.1 - -	312 40.1 - -	$ $ - -
406 x 178 x 60 $P_z = A_g p_y = 2720$	0.658 0.231	M_{cx} M_{cy} M_{rx} M_{ry}	426 57.5 426 57.5	426 57.5 419 57.5	426 57.5 399 57.5	408 57.5 - -	395 57.5 - -	390 57.5 - -	383 57.5 - -	225 28.5 - -	225 28.5 - -	225 28.5 - -	$ $ - -
406 x 178 x 54 $P_z = A_g p_y = 2450$	0.628 0.236	M_{cx} M_{cy} M_{rx} M_{ry}	375 49.0 375 49.0	375 49.0 369 49.0	375 49.0 353 49.0	359 49.0 - -	345 49.0 - -	340 49.0 - -	333 49.0 - -	187 23.1 - -	187 23.1 - -	187 23.1 - -	$ $ - -
406 x 140 x 46 $P_z = A_g p_y = 2080$	0.496 0.184	M_{cx} M_{cy} M_{rx} M_{ry}	315 32.2 315 32.2	315 32.2 311 32.2	311 32.2 - -	291 32.2 - -	282 32.2 - -	122 11.9 - -	122 11.9 - -	122 11.9 - -	122 11.9 - -	$ $ - -	$ $ - -
406 x 140 x 39 $P_z = A_g p_y = 1760$	0.438 0.174	M_{cx} M_{cy} M_{rx} M_{ry}	257 24.6 257 24.6	257 24.6 254 24.6	250 24.6 - -	234 24.6 - -	225 24.6 - -	87.3 8.02 - -	87.3 8.02 - -	87.3 8.02 - -	87.3 8.02 - -	$ $ - -	$ $ - -
356 x 171 x 67 $P_z = A_g p_y = 3040$	n/a 1.00	M_{cx} M_{cy} M_{rx} M_{ry}	430 66.9 430 66.9	430 66.9 423 66.9	430 66.9 401 66.9	430 66.9 366 66.9	430 66.9 317 66.9	430 66.9 266 66.9	430 66.9 215 66.9	430 66.9 162 66.9	430 66.9 109 60.3	430 66.9 54.8 44.3	430 66.9 0 24.2
356 x 171 x 57 $P_z = A_g p_y = 2580$	0.873 0.298	M_{cx} M_{cy} M_{rx} M_{ry}	359 55.0 359 55.0	359 55.0 353 55.0	359 55.0 335 55.0	358 55.0 - -	348 55.0 - -	347 55.0 - -	344 55.0 - -	340 55.0 - -	332 55.0 - -	266 38.4 - -	$ $ - -

F = Factored axial load.
$ For these values of F/P_z the section would be overloaded due to F alone even when M is zero, because F would exceed the local buckling resistance of the section.
- Not applicable for semi-compact and slender sections.
The values in this table are conservative for tension as the more onerous compression section classification limits have been used.
FOR EXPLANATION OF TABLES SEE NOTE 9.1

BS 5950-1: 2000
BS 4-1: 1993

AXIAL LOAD & BENDING

UB SECTIONS SUBJECT TO AXIAL COMPRESSION AND BENDING

MEMBER BUCKLING CHECK

RESISTANCES AND CAPACITIES FOR S355

Section Designation and Capacities (kN, kNm)	F/P_z Limit		Compression Resistance P_{cx}, P_{cy} (kN) and Buckling Resistance Moment M_b, M_{bs} (kNm) for Varying effective lengths L_E (m) within the limiting value of F/P_z											
		L_E (m)	1.0	1.50	2.00	2.50	3.00	3.50	4.00	4.50	5.00	5.50	6.00	6.50	7.00
* 406 x 178 x 67	0.789	P_{cx}	3040	3040	3040	3040	3020	3000	2980	2960	2940	2910	2890	2860	2830
$P_z = A_g p_y = 3040$		P_{cy}	2920	2760	2540	2260	1930	1610	1320	1100	917	776	664	573	500
$p_y Z_x = 422$	0.789	M_b	422	421	386	349	313	278	246	219	196	177	160	147	135
$p_y Z_y = 54.3$		M_{bs}	422	422	422	418	395	372	346	320	294	268	243	220	200
	0.284	M_b	478	469	426	381	336	295	259	228	203	182	165	150	138
		M_{bs}	478	478	478	473	448	421	392	363	333	303	275	250	226
* 406 x 178 x 60	0.658	P_{cx}	2720	2720	2720	2720	2700	2680	2660	2650	2630	2600	2580	2560	2530
$P_z = A_g p_y = 2720$		P_{cy}	2610	2460	2270	2020	1720	1430	1180	973	814	688	588	508	443
$p_y Z_x = 377$	0.658	M_b	377	375	343	310	276	243	214	189	168	151	136	124	114
$p_y Z_y = 47.9$		M_{bs}	377	377	377	373	353	331	309	285	262	238	216	196	177
	0.231	M_b	426	417	377	336	295	257	224	197	174	155	140	127	116
		M_{bs}	426	426	426	421	398	374	348	322	295	269	244	221	200
* 406 x 178 x 54	0.628	P_{cx}	2450	2450	2450	2450	2430	2420	2400	2380	2360	2340	2320	2300	2270
$P_z = A_g p_y = 2450$		P_{cy}	2350	2210	2020	1780	1500	1230	1010	834	696	588	502	433	377
$p_y Z_x = 330$	0.628	M_b	330	327	298	267	236	207	180	158	140	125	112	101	92.6
$p_y Z_y = 40.8$		M_{bs}	330	330	330	324	305	286	265	244	222	202	182	165	149
	0.236	M_b	375	364	328	291	253	218	189	164	144	128	115	104	94.5
		M_{bs}	375	375	375	367	346	324	301	276	252	229	207	187	169
* 406 x 140 x 46	0.496	P_{cx}	2080	2080	2080	2080	2070	2050	2040	2020	2010	1990	1970	1950	1930
$P_z = A_g p_y = 2080$		P_{cy}	1930	1750	1490	1190	925	722	574	465	384	322	273	235	204
$p_y Z_x = 276$	0.496	M_b	276	254	221	189	159	134	115	99.3	87.2	77.5	69.7	63.3	57.9
$p_y Z_y = 26.9$		M_{bs}	276	276	270	250	228	206	183	161	142	125	110	96.9	86.1
	0.184	M_b	315	282	242	202	167	140	118	102	89.3	79.2	71.0	64.4	58.8
		M_{bs}	315	315	308	285	261	235	209	184	162	142	125	111	98.2
* 406 x 140 x 39	0.438	P_{cx}	1760	1760	1760	1760	1750	1740	1730	1710	1700	1680	1670	1650	1630
$P_z = A_g p_y = 1760$		P_{cy}	1620	1450	1210	942	721	559	442	357	294	246	209	180	156
$p_y Z_x = 223$	0.438	M_b	223	202	174	146	121	101	84.8	72.7	63.2	55.7	49.7	44.8	40.8
$p_y Z_y = 20.5$		M_{bs}	223	223	215	197	179	159	140	122	107	93.1	81.6	71.9	63.7
	0.174	M_b	257	225	190	156	127	105	87.5	74.6	64.6	56.8	50.6	45.5	41.4
		M_{bs}	257	257	247	227	206	183	161	141	123	107	93.9	82.7	73.3
356 x 171 x 67	1.00	P_{cx}	3040	3040	3040	3030	3000	2980	2960	2930	2910	2880	2850	2810	2770
$P_z = A_g p_y = 3040$		P_{cy}	2920	2760	2540	2260	1930	1610	1320	1100	917	776	664	573	500
$p_y Z_x = 380$	1.00	M_b	430	422	384	346	309	274	241	218	196	178	163	150	139
$p_y Z_y = 55.7$		M_{bs}	430	430	430	426	403	378	353	326	299	273	248	224	203
* 356 x 171 x 57	0.873	P_{cx}	2580	2580	2580	2570	2550	2530	2510	2490	2470	2440	2410	2380	2350
$P_z = A_g p_y = 2580$		P_{cy}	2480	2330	2140	1890	1610	1330	1090	901	752	636	543	469	409
$p_y Z_x = 318$	0.873	M_b	318	315	288	261	233	207	183	163	146	132	120	110	102
$p_y Z_y = 45.8$		M_{bs}	318	318	318	313	296	277	258	238	217	198	179	162	146
	0.298	M_b	359	350	317	283	249	219	192	170	151	136	124	113	104
		M_{bs}	359	359	359	353	333	313	291	268	245	223	202	182	165

* The section can become slender under axial compression only. Under combined axial compression and bending the section becomes slender when the semi-compact F/P_z limit is exceeded.
Under combined axial compression and bending the capacities are only valid up to the given F/P_z limit. For higher values F/P_z the section would be overloaded due to F alone even when M is zero, because F would exceed the local buckling resistance of the section.
M_b is obtained using an equivalent slenderness = $u.v.L_E/r_y.\beta_w^{0.5}$
M_{bs} is obtained using an equivalent slenderness = $0.5 L/r_y$. Effective length $L_E = L$.
FOR EXPLANATION OF TABLES SEE NOTE 9.1

BS 5950-1: 2000
BS 4-1: 1993

AXIAL LOAD & BENDING

UB SECTIONS SUBJECT TO AXIAL LOAD (COMPRESSION OR TENSION) AND BENDING

CROSS-SECTION CAPACITY CHECK

CAPACITIES FOR S355

Section Designation and Axial load Capacity P_z (kN)	F/P_z Limit Semi-Compact Compact		Moment Capacity M_{cx}, M_{cy} (kNm) and Reduced Moment Capacity M_{rx}, M_{ry} (kNm) for Ratios of Axial Load to Axial Load Capacity F/P_z										
		F/P_z	0.0	0.1	0.2	0.3	0.4	0.5	0.6	0.7	0.8	0.9	1.0
356 x 171 x 51 $P_z = A_g p_y = 2300$	0.754 0.258	M_{cx}	318	318	318	310	301	299	295	288	198	198	$
		M_{cy}	48.1	48.1	48.1	48.1	48.1	48.1	48.1	48.1	28.0	28.0	$
		M_{rx}	318	313	298	-	-	-	-	-	-	-	-
		M_{ry}	48.1	48.1	48.1	-	-	-	-	-	-	-	-
356 x 171 x 45 $P_z = A_g p_y = 2030$	0.686 0.00	M_{cx}	275	275	275	266	257	254	250	153	153	153	$
		M_{cy}	40.4	40.4	40.4	40.4	40.4	40.4	40.4	21.1	21.1	21.1	$
		M_{rx}	-	-	-	-	-	-	-	-	-	-	-
		M_{ry}	-	-	-	-	-	-	-	-	-	-	-
356 x 127 x 39 $P_z = A_g p_y = 1770$	0.619 0.238	M_{cx}	234	234	234	224	214	210	206	114	114	$	$
		M_{cy}	24.2	24.2	24.2	24.2	24.2	24.2	24.2	11.2	11.2	$	$
		M_{rx}	234	231	221	-	-	-	-	-	-	-	-
		M_{ry}	24.2	24.2	24.2	-	-	-	-	-	-	-	-
356 x 127 x 33 $P_z = A_g p_y = 1490$	0.517 0.206	M_{cx}	193	193	193	180	173	169	77.2	77.2	77.2	$	$
		M_{cy}	19.0	19.0	19.0	19.0	19.0	19.0	7.29	7.29	7.29	$	$
		M_{rx}	193	190	182	-	-	-	-	-	-	-	-
		M_{ry}	19.0	19.0	19.0	-	-	-	-	-	-	-	-
305 x 165 x 54 $P_z = A_g p_y = 2440$	n/a 1.00	M_{cx}	300	300	300	300	300	300	300	300	300	300	300
		M_{cy}	54.1	54.1	54.1	54.1	54.1	54.1	54.1	54.1	54.1	54.1	54.1
		M_{rx}	300	295	279	252	218	183	148	111	74.8	37.7	0
		M_{ry}	54.1	54.1	54.1	54.1	54.1	54.1	54.1	47.3	34.6	18.8	0
305 x 165 x 46 $P_z = A_g p_y = 2080$	0.834 0.247	M_{cx}	256	256	256	248	247	246	244	240	234	181	$
		M_{cy}	46.0	46.0	46.0	46.0	46.0	46.0	46.0	46.0	46.0	30.3	$
		M_{rx}	256	251	237	-	-	-	-	-	-	-	-
		M_{ry}	46.0	46.0	46.0	-	-	-	-	-	-	-	-
305 x 165 x 40 $P_z = A_g p_y = 1820$	0.695 0.205	M_{cx}	221	221	221	210	208	206	203	126	126	126	$
		M_{cy}	39.4	39.4	39.4	39.4	39.4	39.4	39.4	20.9	20.9	20.9	$
		M_{rx}	221	217	206	-	-	-	-	-	-	-	-
		M_{ry}	39.4	39.4	39.4	-	-	-	-	-	-	-	-

F = Factored axial load.
$ For these values of F/P_z the section would be overloaded due to F alone even when M is zero, because F would exceed the local buckling resistance of the section.
- Not applicable for semi-compact and slender sections.
The values in this table are conservative for tension as the more onerous compression section classification limits have been used.
FOR EXPLANATION OF TABLES SEE NOTE 9.1

BS 5950-1: 2000
BS 4-1: 1993

AXIAL LOAD & BENDING

UB SECTIONS SUBJECT TO AXIAL COMPRESSION AND BENDING

MEMBER BUCKLING CHECK

RESISTANCES AND CAPACITIES FOR S355

Section Designation and Capacities (kN, kNm)	F/P_z Limit	Compression Resistance P_{cx}, P_{cy} (kN) and Buckling Resistance Moment M_b, M_{bs} (kNm) for Varying effective lengths L_E (m) within the limiting value of F/P_z											
		L_E (m)	1.0	1.50	2.00	2.50	3.00	3.50	4.00	4.50	5.00	5.50	6.00	6.50	7.00
* 356 x 171 x 51	0.754	P_{cx}	2300	2300	2300	2300	2280	2260	2240	2220	2200	2180	2150	2130	2090
$P_z = A_g p_y = 2300$		P_{cy}	2210	2080	1900	1680	1420	1160	955	788	657	555	474	409	357
$p_y Z_x = 283$	0.754	M_b	283	279	254	229	203	179	157	139	124	111	101	91.7	84.3
$p_y Z_y = 40.1$		M_{bs}	283	283	283	277	262	245	227	209	191	173	157	141	128
	0.258	M_b	318	309	279	247	217	188	164	144	128	114	103	93.8	86.0
		M_{bs}	318	318	318	312	294	276	256	235	215	195	176	159	144
* 356 x 171 x 45	0.686	P_{cx}	2030	2030	2030	2020	2010	1990	1980	1960	1940	1920	1900	1870	1840
$P_z = A_g p_y = 2030$		P_{cy}	1950	1820	1660	1450	1210	991	808	665	554	468	399	344	300
$p_y Z_x = 244$	0.686	M_b	244	239	217	194	171	149	130	114	101	89.7	80.7	73.3	67.0
$p_y Z_y = 33.7$		M_{bs}	244	244	244	238	223	209	193	177	161	145	131	118	106
	0.00	M_b	275	265	238	210	182	157	135	118	104	92.1	82.6	74.8	68.3
		M_{bs}	275	275	275	268	252	235	218	199	181	164	147	133	120
* 356 x 127 x 39	0.619	P_{cx}	1770	1770	1770	1760	1750	1730	1720	1700	1680	1660	1640	1620	1590
$P_z = A_g p_y = 1770$		P_{cy}	1610	1410	1130	859	647	497	392	316	260	217	184	158	137
$p_y Z_x = 204$	0.619	M_b	204	179	152	127	105	88.6	75.8	66.0	58.3	52.1	47.1	43.0	39.6
$p_y Z_y = 20.2$		M_{bs}	204	204	192	175	156	137	119	103	88.7	76.9	67.0	58.8	52.0
	0.238	M_b	230	198	165	134	110	91.8	78.0	67.6	59.5	53.1	48.0	43.8	40.2
		M_{bs}	234	234	220	200	178	157	136	117	101	88.0	76.7	67.3	59.5
* 356 x 127 x 33	0.517	P_{cx}	1490	1490	1490	1490	1470	1460	1450	1440	1420	1400	1380	1360	1340
$P_z = A_g p_y = 1490$		P_{cy}	1350	1170	921	687	513	393	309	249	204	171	145	124	108
$p_y Z_x = 168$	0.517	M_b	167	145	121	99.2	81.1	67.2	56.8	48.9	42.8	38.0	34.1	31.0	28.4
$p_y Z_y = 15.9$		M_{bs}	168	168	156	141	124	108	93.3	80.1	68.8	59.5	51.8	45.3	40.0
	0.206	M_b	188	160	131	105	84.5	69.4	58.3	50.0	43.7	38.7	34.7	31.5	28.8
		M_{bs}	193	193	179	161	143	124	107	91.9	79.0	68.3	59.4	52.1	45.9
305 x 165 x 54	1.00	P_{cx}	2440	2440	2440	2420	2400	2380	2350	2330	2300	2270	2230	2190	2140
$P_z = A_g p_y = 2440$		P_{cy}	2350	2210	2030	1800	1530	1270	1040	861	719	608	520	449	391
$p_y Z_x = 268$	1.00	M_b	300	294	267	240	214	190	169	151	136	124	113	104	96.9
$p_y Z_y = 45.1$		M_{bs}	300	300	300	296	280	262	244	225	206	188	170	154	139
* 305 x 165 x 46	0.834	P_{cx}	2080	2080	2080	2070	2050	2030	2010	1990	1960	1940	1900	1870	1830
$P_z = A_g p_y = 2080$		P_{cy}	2000	1880	1730	1530	1290	1070	878	725	605	512	437	377	329
$p_y Z_x = 229$	0.834	M_b	229	226	207	187	167	148	131	117	105	95.2	86.8	79.8	73.7
$p_y Z_y = 38.3$		M_{bs}	229	229	229	226	213	200	186	171	156	142	129	116	105
	0.247	M_b	250	249	225	201	177	156	137	121	108	97.8	88.9	81.5	75.2
		M_{bs}	256	256	256	252	237	223	207	191	174	158	143	130	117
* 305 x 165 x 40	0.695	P_{cx}	1820	1820	1820	1800	1790	1770	1750	1730	1710	1690	1660	1630	1590
$P_z = A_g p_y = 1820$		P_{cy}	1750	1640	1500	1330	1120	921	755	623	520	439	375	324	282
$p_y Z_x = 199$	0.695	M_b	199	196	178	160	141	124	109	96.7	86.1	77.4	70.1	64.1	58.9
$p_y Z_y = 32.9$		M_{bs}	199	199	199	195	184	172	160	147	134	122	110	99.4	89.8
	0.205	M_b	221	214	193	171	150	130	114	99.9	88.6	79.3	71.7	65.3	60.0
		M_{bs}	221	221	221	217	205	192	178	164	149	135	122	111	99.9

* The section can become slender under axial compression only. Under combined axial compression and bending the section becomes slender when the semi-compact F/P_z limit is exceeded.
Under combined axial compression and bending the capacities are only valid up to the given F/P_z limit. For higher values F/P_z the section would be overloaded due to F alone even when M is zero, because F would exceed the local buckling resistance of the section.
M_b is obtained using an equivalent slenderness = $u.v.L_E/r_y.\beta_w^{0.5}$
M_{bs} is obtained using an equivalent slenderness = $0.5 L/r_y$. Effective length $L_E = L$.
FOR EXPLANATION OF TABLES SEE NOTE 9.1

BS 5950-1: 2000
BS 4-1: 1993

AXIAL LOAD & BENDING

UB SECTIONS SUBJECT TO AXIAL LOAD (COMPRESSION OR TENSION) AND BENDING

CROSS-SECTION CAPACITY CHECK

CAPACITIES FOR S355

Section Designation and Axial load Capacity P_z (kN)	F/P_z Limit Semi-Compact Compact		Moment Capacity M_{cx}, M_{cy} (kNm) and Reduced Moment Capacity M_{rx}, M_{ry} (kNm) for Ratios of Axial Load to Axial Load Capacity F/P_z										
		F/P_z	0.0	0.1	0.2	0.3	0.4	0.5	0.6	0.7	0.8	0.9	1.0
305 x 127 x 48 $P_z = A_g p_y = 2170$	n/a 1.00	M_{cx} M_{cy} M_{rx} M_{ry}	252 31.4 252 31.4	252 31.4 249 31.4	252 31.4 238 31.4	252 31.4 219 31.4	252 31.4 193 31.4	252 31.4 162 31.4	252 31.4 131 31.4	252 31.4 99.0 30.1	252 31.4 66.5 22.5	252 31.4 33.5 12.4	252 31.4 0 0
305 x 127 x 42 $P_z = A_g p_y = 1900$	n/a 1.00	M_{cx} M_{cy} M_{rx} M_{ry}	218 26.7 218 26.7	218 26.7 215 26.7	218 26.7 205 26.7	218 26.7 189 26.7	218 26.7 167 26.7	218 26.7 140 26.7	218 26.7 113 26.7	218 26.7 85.5 25.9	218 26.7 57.4 19.4	218 26.7 28.9 10.7	218 26.7 0 0
305 x 127 x 37 $P_z = A_g p_y = 1680$	0.914 0.361	M_{cx} M_{cy} M_{rx} M_{ry}	191 23.2 191 23.2	191 23.2 189 23.2	191 23.2 180 23.2	191 23.2 166 23.2	187 23.2 - -	186 23.2 - -	186 23.2 - -	184 23.2 - -	181 23.2 - -	170 23.2 - -	$ $ - -
305 x 102 x 33 $P_z = A_g p_y = 1480$	0.763 0.321	M_{cx} M_{cy} M_{rx} M_{ry}	171 16.1 171 16.1	171 16.1 168 16.1	171 16.1 161 16.1	171 16.1 150 16.1	162 16.1 - -	159 16.1 - -	156 16.1 - -	152 16.1 - -	105 9.54 - -	105 9.54 - -	$ $ - -
305 x 102 x 28 $P_z = A_g p_y = 1270$	0.648 0.281	M_{cx} M_{cy} M_{rx} M_{ry}	143 13.0 143 13.0	143 13.0 141 13.0	143 13.0 135 13.0	141 13.0 - -	133 13.0 - -	129 13.0 - -	126 13.0 - -	72.4 6.35 - -	72.4 6.35 - -	$ $ - -	$ $ - -
305 x 102 x 25 $P_z = A_g p_y = 1120$	0.610 0.287	M_{cx} M_{cy} M_{rx} M_{ry}	121 10.3 121 10.3	121 10.3 120 10.3	121 10.3 115 10.3	120 10.3 - -	112 10.3 - -	107 10.3 - -	104 10.3 - -	56.8 4.71 - -	56.8 4.71 - -	$ $ $ -	$ $ - -
254 x 146 x 43 $P_z = A_g p_y = 1950$	n/a 1.00	M_{cx} M_{cy} M_{rx} M_{ry}	201 39.2 201 39.2	201 39.2 197 39.2	201 39.2 186 39.2	201 39.2 168 39.2	201 39.2 145 39.2	201 39.2 122 39.2	201 39.2 98.1 39.2	201 39.2 74.1 33.5	201 39.2 49.8 24.5	201 39.2 25.1 13.3	201 39.2 0 0
254 x 146 x 37 $P_z = A_g p_y = 1680$	n/a 1.00	M_{cx} M_{cy} M_{rx} M_{ry}	171 33.2 171 33.2	171 33.2 168 33.2	171 33.2 159 33.2	171 33.2 143 33.2	171 33.2 124 33.2	171 33.2 104 33.2	171 33.2 83.6 33.2	171 33.2 63.1 28.6	171 33.2 42.4 20.9	171 33.2 21.3 11.4	171 33.2 0 0

F = Factored axial load.
$ For these values of F/P_z the section would be overloaded due to F alone even when M is zero, because F would exceed the local buckling resistance of the section.
- Not applicable for semi-compact and slender sections.
The values in this table are conservative for tension as the more onerous compression section classification limits have been used.
FOR EXPLANATION OF TABLES SEE NOTE 9.1

BS 5950-1: 2000
BS 4-1: 1993

AXIAL LOAD & BENDING

UB SECTIONS SUBJECT TO AXIAL COMPRESSION AND BENDING

MEMBER BUCKLING CHECK

RESISTANCES AND CAPACITIES FOR S355

Section Designation and Capacities (kN, kNm)	F/P_z Limit		Compression Resistance P_{cx}, P_{cy} (kN) and Buckling Resistance Moment M_b, M_{bs} (kNm) for Varying effective lengths L_E (m) within the limiting value of F/P_z											
			L_E (m)	1.0	1.50	2.00	2.50	3.00	3.50	4.00	4.50	5.00	5.50	6.00	6.50	7.00
305 x 127 x 48	1.00	P_{cx}	2170	2170	2170	2150	2130	2110	2090	2060	2040	2000	1970	1930	1880
$P_z = A_g p_y = 2170$		P_{cy}	1980	1750	1430	1090	824	635	501	404	332	278	236	202	176
$p_y Z_x = 219$	1.00	M_b	251	220	188	160	136	118	103	91.7	82.5	75.0	68.8	63.5	59.1
$p_y Z_y = 26.1$		M_{bs}	252	252	239	218	196	173	151	130	113	98.2	85.7	75.3	66.6
305 x 127 x 42	1.00	P_{cx}	1900	1900	1890	1880	1860	1840	1820	1800	1770	1750	1710	1680	1630
$P_z = A_g p_y = 1900$		P_{cy}	1730	1520	1230	931	702	540	426	343	282	236	200	172	149
$p_y Z_x = 190$	1.00	M_b	215	187	159	133	112	95.1	82.6	72.8	65.1	58.9	53.8	49.5	45.9
$p_y Z_y = 22.2$		M_{bs}	218	218	205	187	167	147	128	110	95.5	82.9	72.3	63.5	56.1
* 305 x 127 x 37	0.914	P_{cx}	1680	1680	1670	1660	1640	1630	1610	1590	1570	1540	1510	1480	1440
$P_z = A_g p_y = 1680$		P_{cy}	1520	1330	1070	809	609	468	369	297	244	204	173	149	129
$p_y Z_x = 167$	0.914	M_b	167	147	126	106	89.3	76.1	65.9	58.0	51.7	46.6	42.5	39.0	36.1
$p_y Z_y = 19.3$		M_{bs}	167	167	157	143	127	112	96.8	83.5	72.1	62.5	54.5	47.8	42.2
	0.361	M_b	188	163	136	113	93.6	79.0	68.0	59.5	52.9	47.6	43.3	39.7	36.7
		M_{bs}	191	191	179	163	146	128	111	95.6	82.5	71.5	62.4	54.7	48.3
* 305 x 102 x 33	0.763	P_{cx}	1480	1480	1480	1470	1460	1440	1430	1410	1390	1370	1340	1320	1280
$P_z = A_g p_y = 1480$		P_{cy}	1280	1020	724	510	371	281	219	175	144	120	101	86.8	75.2
$p_y Z_x = 148$	0.763	M_b	140	117	94.4	76.1	62.5	52.7	45.4	39.8	35.5	32.0	29.2	26.8	24.9
$p_y Z_y = 13.5$		M_{bs}	148	142	126	109	92.5	77.2	64.3	53.9	45.5	38.9	33.5	29.2	25.6
	0.321	M_b	157	128	101	79.8	64.9	54.3	46.5	40.7	36.2	32.6	29.7	27.3	25.2
		M_{bs}	171	164	146	127	107	89.2	74.3	62.3	52.6	44.9	38.8	33.7	29.6
* 305 x 102 x 28	0.648	P_{cx}	1270	1270	1270	1260	1250	1240	1220	1210	1190	1170	1150	1120	1090
$P_z = A_g p_y = 1270$		P_{cy}	1080	846	592	414	300	227	177	142	116	96.5	81.6	69.9	60.5
$p_y Z_x = 124$	0.648	M_b	116	95.2	75.4	59.6	48.2	40.0	34.1	29.7	26.2	23.5	21.3	19.5	18.0
$p_y Z_y = 10.8$		M_{bs}	124	117	104	89.2	74.7	61.8	51.2	42.8	36.1	30.7	26.5	23.0	20.2
	0.281	M_b	130	104	80.3	62.3	49.8	41.1	34.9	30.3	26.7	23.9	21.7	19.8	18.3
		M_{bs}	143	136	120	103	86.5	71.6	59.3	49.5	41.8	35.6	30.7	26.7	23.4
* 305 x 102 x 25	0.610	P_{cx}	1120	1120	1120	1110	1100	1090	1070	1060	1040	1020	1000	979	951
$P_z = A_g p_y = 1120$		P_{cy}	935	705	480	332	240	180	140	112	91.9	76.5	64.7	55.4	47.9
$p_y Z_x = 104$	0.610	M_b	95.8	77.6	60.2	46.7	37.3	30.7	25.9	22.4	19.7	17.6	15.9	14.5	13.3
$p_y Z_y = 8.59$		M_{bs}	104	96.7	84.4	71.3	58.9	48.2	39.6	32.9	27.7	23.5	20.2	17.5	15.4
	0.287	M_b	109	85.1	64.0	48.8	38.5	31.5	26.5	22.8	20.0	17.8	16.1	14.7	13.5
		M_{bs}	121	113	98.9	83.6	68.9	56.4	46.4	38.5	32.4	27.5	23.7	20.5	18.0
254 x 146 x 43	1.00	P_{cx}	1950	1950	1930	1910	1890	1870	1840	1810	1780	1740	1690	1640	1570
$P_z = A_g p_y = 1950$		P_{cy}	1850	1720	1540	1310	1070	859	694	568	471	397	338	291	253
$p_y Z_x = 179$	1.00	M_b	201	191	171	152	134	119	106	94.8	85.7	78.3	72.0	66.6	62.0
$p_y Z_y = 32.7$		M_{bs}	201	201	201	192	179	166	152	137	124	111	98.8	88.4	79.3
254 x 146 x 37	1.00	P_{cx}	1680	1680	1660	1650	1630	1610	1590	1560	1530	1500	1450	1400	1350
$P_z = A_g p_y = 1680$		P_{cy}	1590	1470	1320	1120	907	727	587	480	398	335	285	245	214
$p_y Z_x = 154$	1.00	M_b	171	162	144	127	111	96.5	84.9	75.5	67.8	61.4	56.1	51.7	47.9
$p_y Z_y = 27.7$		M_{bs}	171	171	171	163	152	140	128	116	104	93.0	83.1	74.2	66.5

* The section can become slender under axial compression only. Under combined axial compression and bending the section becomes slender when the semi-compact F/P_z limit is exceeded.
Under combined axial compression and bending the capacities are only valid up to the given F/P_z limit. For higher values F/P_z the section would be overloaded due to F alone even when M is zero, because F would exceed the local buckling resistance of the section.
M_b is obtained using an equivalent slenderness = $u.v.L_E/r_y.\beta_w^{0.5}$
M_{bs} is obtained using an equivalent slenderness = $0.5 L/r_y$. Effective length $L_E = L$.
FOR EXPLANATION OF TABLES SEE NOTE 9.1

BS 5950-1: 2000
BS 4-1: 1993

AXIAL LOAD & BENDING

UB SECTIONS SUBJECT TO AXIAL LOAD (COMPRESSION OR TENSION) AND BENDING

CROSS-SECTION CAPACITY CHECK

CAPACITIES FOR S355

Section Designation and Axial load Capacity P_z (kN)	F/P_z Limit Semi-Compact Compact		Moment Capacity M_{cx}, M_{cy} (kNm) and Reduced Moment Capacity M_{rx}, M_{ry} (kNm) for Ratios of Axial Load to Axial Load Capacity F/P_z										
		F/P_z	0.0	0.1	0.2	0.3	0.4	0.5	0.6	0.7	0.8	0.9	1.0
254 x 146 x 31 $P_z = A_g p_y = 1410$	0.947 **0.311**	M_{cx} M_{cy} M_{rx} M_{ry}	140 26.1 140 26.1	140 26.1 137 26.1	140 26.1 130 26.1	140 26.1 119 26.1	138 26.1 - -	138 26.1 - -	137 26.1 - -	137 26.1 - -	135 26.1 - -	131 26.1 - -	$ $ - -
254 x 102 x 28 $P_z = A_g p_y = 1280$	0.977 **0.383**	M_{cx} M_{cy} M_{rx} M_{ry}	125 14.9 125 14.9	125 14.9 123 14.9	125 14.9 118 14.9	125 14.9 109 14.9	125 14.9 - -	124 14.9 - -	124 14.9 - -	124 14.9 - -	123 14.9 - -	122 14.9 - -	$ $ - -
254 x 102 x 25 $P_z = A_g p_y = 1140$	0.907 **0.379**	M_{cx} M_{cy} M_{rx} M_{ry}	109 12.4 109 12.4	109 12.4 107 12.4	109 12.4 103 12.4	109 12.4 95.0 12.4	107 12.4 - -	106 12.4 - -	105 12.4 - -	104 12.4 - -	102 12.4 - -	95.4 12.4 - -	$ $ - -
254 x 102 x 22 $P_z = A_g p_y = 994$	0.837 **0.375**	M_{cx} M_{cy} M_{rx} M_{ry}	91.9 10.0 91.9 10.0	91.9 10.0 90.7 10.0	91.9 10.0 87.1 10.0	91.9 10.0 81.0 10.0	90.5 10.0 - -	87.3 10.0 - -	86.4 10.0 - -	84.8 10.0 - -	81.7 10.0 - -	63.1 6.62 - -	$ $ - -
203 x 133 x 30 $P_z = A_g p_y = 1360$	n/a **1.00**	M_{cx} M_{cy} M_{rx} M_{ry}	111 24.5 111 24.5	111 24.5 109 24.5	111 24.5 103 24.5	111 24.5 93.3 24.5	111 24.5 80.7 24.5	111 24.5 67.7 24.5	111 24.5 54.5 24.5	111 24.5 41.2 21.2	111 24.5 27.7 15.5	111 24.5 13.9 8.40	111 24.5 0 0
203 x 133 x 25 $P_z = A_g p_y = 1140$	n/a **1.00**	M_{cx} M_{cy} M_{rx} M_{ry}	91.6 19.7 91.6 19.7	91.6 19.7 90.0 19.7	91.6 19.7 85.2 19.7	91.6 19.7 77.2 19.7	91.6 19.7 66.8 19.7	91.6 19.7 56.0 19.7	91.6 19.7 45.1 19.7	91.6 19.7 34.0 17.5	91.6 19.7 22.8 12.8	91.6 19.7 11.5 6.98	91.6 19.7 0 0
203 x 102 x 23 $P_z = A_g p_y = 1040$	n/a **1.00**	M_{cx} M_{cy} M_{rx} M_{ry}	83.1 13.7 83.1 13.7	83.1 13.7 81.6 13.7	83.1 13.7 77.4 13.7	83.1 13.7 70.3 13.7	83.1 13.7 60.9 13.7	83.1 13.7 51.1 13.7	83.1 13.7 41.2 13.7	83.1 13.7 31.1 12.2	83.1 13.7 20.9 8.98	83.1 13.7 10.5 4.90	83.1 13.7 0 0

F = Factored axial load.
$ For these values of F/P_z the section would be overloaded due to F alone even when M is zero, because F would exceed the local buckling resistance of the section.
- Not applicable for semi-compact and slender sections.
The values in this table are conservative for tension as the more onerous compression section classification limits have been used.
FOR EXPLANATION OF TABLES SEE NOTE 9.1

BS 5950-1: 2000
BS 4-1: 1993

AXIAL LOAD & BENDING

UB SECTIONS SUBJECT TO AXIAL COMPRESSION AND BENDING

MEMBER BUCKLING CHECK

RESISTANCES AND CAPACITIES FOR S355

Section Designation and Capacities (kN, kNm)	F/P_z Limit	Compression Resistance P_{cx}, P_{cy} (kN) and Buckling Resistance Moment M_b, M_{bs} (kNm) for Varying effective lengths L_E (m) within the limiting value of F/P_z												
		L_E (m)	1.0	1.50	2.00	2.50	3.00	3.50	4.00	4.50	5.00	5.50	6.00	6.50	7.00
* 254 x 146 x 31	0.947	P_{cx}	1410	1410	1400	1380	1370	1350	1330	1310	1280	1250	1210	1160	1110
$P_z = A_g p_y = 1410$		P_{cy}	1330	1230	1090	907	728	579	465	379	314	264	225	193	168
$p_y Z_x = 125$	0.947	M_b	125	118	106	92.8	80.7	70.0	61.2	53.9	48.1	43.3	39.3	36.0	33.2
$p_y Z_y = 21.8$		M_{bs}	125	125	125	117	109	99.9	90.6	81.4	72.7	64.7	57.5	51.2	45.8
	0.311	M_b	140	130	115	99.2	85.1	73.1	63.3	55.6	49.3	44.3	40.1	36.7	33.8
		M_{bs}	140	140	140	131	122	112	101	91.2	81.4	72.4	64.4	57.4	51.2
* 254 x 102 x 28	0.977	P_{cx}	1280	1280	1270	1260	1240	1230	1210	1190	1160	1130	1100	1060	1010
$P_z = A_g p_y = 1280$		P_{cy}	1110	902	654	464	339	257	201	161	132	110	93.0	79.7	69.1
$p_y Z_x = 109$	0.977	M_b	104	88.1	72.4	59.4	49.5	42.2	36.6	32.4	29.0	26.3	24.1	22.2	20.6
$p_y Z_y = 12.4$		M_{bs}	109	106	95.0	83.1	70.8	59.5	49.8	41.9	35.5	30.4	26.2	22.8	20.0
	0.383	M_b	117	96.5	77.4	62.3	51.4	43.5	37.6	33.1	29.6	26.8	24.5	22.6	20.9
		M_{bs}	125	122	109	95.2	81.2	68.2	57.1	48.0	40.7	34.8	30.0	26.2	23.0
* 254 x 102 x 25	0.907	P_{cx}	1140	1140	1130	1110	1100	1090	1070	1050	1030	999	967	928	883
$P_z = A_g p_y = 1140$		P_{cy}	977	778	554	390	284	215	168	134	110	91.6	77.5	66.4	57.5
$p_y Z_x = 94.4$	0.907	M_b	89.3	74.5	60.2	48.5	39.9	33.6	28.9	25.4	22.6	20.4	18.6	17.1	15.9
$p_y Z_y = 10.4$		M_{bs}	94.4	90.7	80.8	70.0	59.2	49.4	41.1	34.4	29.1	24.9	21.4	18.6	16.4
	0.379	M_b	100	81.6	64.2	50.8	41.3	34.6	29.6	25.9	23.1	20.8	18.9	17.4	16.1
		M_{bs}	109	104	92.9	80.5	68.1	56.8	47.3	39.6	33.5	28.6	24.7	21.4	18.8
* 254 x 102 x 22	0.837	P_{cx}	994	994	984	973	961	948	932	915	894	869	838	803	761
$P_z = A_g p_y = 994$		P_{cy}	842	654	455	318	230	174	135	108	88.7	73.8	62.5	53.5	46.3
$p_y Z_x = 79.5$	0.837	M_b	74.3	61.1	48.4	38.3	31.1	25.9	22.1	19.2	17.0	15.3	13.9	12.7	11.8
$p_y Z_y = 8.34$		M_{bs}	79.5	75.3	66.5	56.9	47.6	39.3	32.5	27.1	22.9	19.5	16.8	14.6	12.8
	0.375	M_b	83.6	66.9	51.5	40.0	32.1	26.6	22.6	19.6	17.4	15.6	14.1	12.9	11.9
		M_{bs}	91.9	87.1	76.9	65.8	55.0	45.4	37.6	31.4	26.4	22.5	19.4	16.8	14.8
203 x 133 x 30	1.00	P_{cx}	1360	1350	1330	1310	1290	1270	1240	1210	1160	1110	1040	971	894
$P_z = A_g p_y = 1360$		P_{cy}	1270	1160	1010	819	645	508	405	329	272	228	194	167	145
$p_y Z_x = 99.4$	1.00	M_b	111	103	90.7	79.3	69.2	60.7	53.7	48.1	43.4	39.6	36.4	33.7	31.4
$p_y Z_y = 20.4$		M_{bs}	111	111	110	103	94.5	85.9	77.1	68.6	60.6	53.5	47.3	42.0	37.4
203 x 133 x 25	1.00	P_{cx}	1140	1130	1120	1100	1080	1060	1040	1000	966	918	862	799	733
$P_z = A_g p_y = 1140$		P_{cy}	1060	964	829	669	523	410	326	265	219	183	156	134	116
$p_y Z_x = 81.7$	1.00	M_b	91.6	83.2	72.8	62.6	53.6	46.3	40.4	35.7	32.0	28.9	26.4	24.3	22.5
$p_y Z_y = 16.4$		M_{bs}	91.6	91.6	90.0	83.6	76.7	69.4	62.1	55.0	48.4	42.7	37.6	33.3	29.6
203 x 102 x 23	1.00	P_{cx}	1040	1040	1020	1010	993	973	949	919	882	837	784	725	664
$P_z = A_g p_y = 1040$		P_{cy}	922	770	578	418	308	234	183	147	121	101	85.2	73.0	63.3
$p_y Z_x = 73.5$	1.00	M_b	78.9	66.7	55.2	45.8	38.6	33.1	29.0	25.8	23.3	21.2	19.5	18.0	16.8
$p_y Z_y = 11.4$		M_{bs}	83.1	82.0	74.3	65.9	57.1	48.7	41.2	34.9	29.7	25.5	22.1	19.3	17.0

* The section can become slender under axial compression only. Under combined axial compression and bending the section becomes slender when the semi-compact F/P_z limit is exceeded.
Under combined axial compression and bending the capacities are only valid up to the given F/P_z limit. For higher values F/P_z the section would be overloaded due to F alone even when M is zero, because F would exceed the local buckling resistance of the section.
M_b is obtained using an equivalent slenderness = $u.v.L_E/r_y.\beta_w^{0.5}$
M_{bs} is obtained using an equivalent slenderness = $0.5 L/r_y$. Effective length $L_E = L$.
FOR EXPLANATION OF TABLES SEE NOTE 9.1

BS 5950-1: 2000
BS 4-1: 1993

AXIAL LOAD & BENDING

UB SECTIONS SUBJECT TO AXIAL LOAD (COMPRESSION OR TENSION) AND BENDING

CROSS-SECTION CAPACITY CHECK

CAPACITIES FOR S355

Section Designation and Axial load Capacity P_z (kN)	F/P_z Limit Semi-Compact Compact		Moment Capacity M_{cx}, M_{cy} (kNm) and Reduced Moment Capacity M_{rx}, M_{ry} (kNm) for Ratios of Axial Load to Axial Load Capacity F/P_z										
		F/P_z	0.0	0.1	0.2	0.3	0.4	0.5	0.6	0.7	0.8	0.9	1.0
178 x 102 x 19 $P_z = A_g p_y = 863$	n/a 1.00	M_{cx}	60.7	60.7	60.7	60.7	60.7	60.7	60.7	60.7	60.7	60.7	60.7
		M_{cy}	11.5	11.5	11.5	11.5	11.5	11.5	11.5	11.5	11.5	11.5	11.5
		M_{rx}	60.7	59.6	56.3	50.9	44.1	37.1	29.8	22.5	15.1	7.62	0
		M_{ry}	11.5	11.5	11.5	11.5	11.5	11.5	11.5	11.5	10.1	7.40	4.03
152 x 89 x 16 $P_z = A_g p_y = 721$	n/a 1.00	M_{cx}	43.7	43.7	43.7	43.7	43.7	43.7	43.7	43.7	43.7	43.7	43.7
		M_{cy}	8.61	8.61	8.61	8.61	8.61	8.61	8.61	8.61	8.61	8.61	8.61
		M_{rx}	43.7	42.9	40.4	36.4	31.5	26.4	21.3	16.1	10.8	5.45	0
		M_{ry}	8.61	8.61	8.61	8.61	8.61	8.61	8.61	8.61	7.45	5.44	2.96
127 x 76 x 13 $P_z = A_g p_y = 586$	n/a 1.00	M_{cx}	29.9	29.9	29.9	29.9	29.9	29.9	29.9	29.9	29.9	29.9	29.9
		M_{cy}	6.26	6.26	6.26	6.26	6.26	6.26	6.26	6.26	6.26	6.26	6.26
		M_{rx}	29.9	29.3	27.5	24.5	21.2	17.8	14.4	10.9	7.31	3.69	0
		M_{ry}	6.26	6.26	6.26	6.26	6.26	6.26	6.26	6.26	5.25	3.82	2.07

F = Factored axial load.
$ For these values of F/P_z the section would be overloaded due to F alone even when M is zero, because F would exceed the local buckling resistance of the section.
- Not applicable for semi-compact and slender sections.
The values in this table are conservative for tension as the more onerous compression section classification limits have been used.
FOR EXPLANATION OF TABLES SEE NOTE 9.1

BS 5950-1: 2000
BS 4-1: 1993

AXIAL LOAD & BENDING

UB SECTIONS SUBJECT TO AXIAL COMPRESSION AND BENDING

MEMBER BUCKLING CHECK

RESISTANCES AND CAPACITIES FOR S355

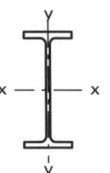

Section Designation and Capacities (kN, kNm)	F/P_z Limit		Compression Resistance P_{cx}, P_{cy} (kN) and Buckling Resistance Moment M_b, M_{bs} (kNm) for Varying effective lengths L_E (m) within the limiting value of F/P_z											
		L_E (m)	1.0	1.50	2.00	2.50	3.00	3.50	4.00	4.50	5.00	5.50	6.00	6.50	7.00
178 x 102 x 19	1.00	P_{cx}	863	854	840	825	808	786	758	723	679	627	570	514	460
$P_z = A_g p_y = 863$		P_{cy}	763	638	480	348	256	195	153	123	100	83.8	71.0	60.9	52.8
$p_y Z_x = 54.3$	1.00	M_b	57.7	48.9	40.5	33.5	28.2	24.3	21.3	18.9	17.0	15.5	14.2	13.2	12.3
$p_y Z_y = 9.59$		M_{bs}	60.7	60.0	54.4	48.3	41.9	35.7	30.3	25.7	21.9	18.8	16.3	14.2	12.5
152 x 89 x 16	1.00	P_{cx}	720	708	694	677	657	629	593	546	492	437	385	338	299
$P_z = A_g p_y = 721$		P_{cy}	615	483	340	238	173	131	102	81.5	66.7	55.6	47.0	40.3	34.9
$p_y Z_x = 38.7$	1.00	M_b	40.1	33.3	27.4	22.8	19.3	16.8	14.8	13.3	12.0	11.0	10.1	9.41	8.78
$p_y Z_y = 7.17$		M_{bs}	43.7	41.6	36.9	31.8	26.7	22.1	18.4	15.3	13.0	11.0	9.51	8.27	7.25
127 x 76 x 13	1.00	P_{cx}	581	569	554	534	506	468	419	365	313	269	232	201	176
$P_z = A_g p_y = 586$		P_{cy}	474	340	225	154	110	82.9	64.5	51.7	42.1	35.0	29.6	25.3	21.9
$p_y Z_x = 26.5$	1.00	M_b	26.4	21.7	17.9	15.1	13.0	11.4	10.1	9.13	8.31	7.63	7.06	6.57	6.15
$p_y Z_y = 5.22$		M_{bs}	29.9	27.2	23.3	19.3	15.6	12.6	10.3	8.48	7.10	6.02	5.16	4.48	3.91

* The section can become slender under axial compression only. Under combined axial compression and bending the section becomes slender when the semi-compact F/P_z limit is exceeded.
Under combined axial compression and bending the capacities are only valid up to the given F/P_z limit. For higher values F/P_z the section would be overloaded due to F alone even when M is zero, because F would exceed the local buckling resistance of the section.
M_b is obtained using an equivalent slenderness = $u.v.L_E/r_y.\beta_w^{0.5}$
M_{bs} is obtained using an equivalent slenderness = $0.5 L/r_y$. Effective length $L_E = L$.
FOR EXPLANATION OF TABLES SEE NOTE 9.1

BS 5950-1: 2000
BS 4-1: 1993

AXIAL LOAD & BENDING

UC SECTIONS SUBJECT TO AXIAL LOAD (COMPRESSION OR TENSION) AND BENDING

CROSS-SECTION CAPACITY CHECK

CAPACITIES FOR S355

Section Designation and Axial load Capacity P_z (kN)	F/P_z Limit Semi-Compact Compact		Moment Capacity M_{cx}, M_{cy} (kNm) and Reduced Moment Capacity M_{rx}, M_{ry} (kNm) for Ratios of Axial Load to Axial Load Capacity F/P_z										
		F/P_z	0.0	0.1	0.2	0.3	0.4	0.5	0.6	0.7	0.8	0.9	1.0
356 x 406 x 634 # $P_z = A_g p_y = 26300$	n/a 1.00	M_{cx}	4520	4520	4520	4520	4520	4520	4520	4520	4520	4520	4520
		M_{cy}	1810	1810	1810	1810	1810	1810	1810	1810	1810	1810	1810
		M_{rx}	4520	4520	4180	3750	3290	2800	2290	1760	1200	611	0
		M_{ry}	1810	1810	1810	1810	1810	1810	1680	1360	976	522	0
356 x 406 x 551 # $P_z = A_g p_y = 22800$	n/a 1.00	M_{cx}	3890	3890	3890	3890	3890	3890	3890	3890	3890	3890	3890
		M_{cy}	1540	1540	1540	1540	1540	1540	1540	1540	1540	1540	1540
		M_{rx}	3890	3830	3550	3170	2770	2360	1930	1470	1000	510	0
		M_{ry}	1540	1540	1540	1540	1540	1540	1440	1170	836	448	0
356 x 406 x 467 # $P_z = A_g p_y = 19900$	n/a 1.00	M_{cx}	3350	3350	3350	3350	3350	3350	3350	3350	3350	3350	3350
		M_{cy}	1320	1320	1320	1320	1320	1320	1320	1320	1320	1320	1320
		M_{rx}	3350	3270	3020	2690	2350	2000	1630	1240	841	428	0
		M_{ry}	1320	1320	1320	1320	1320	1320	1230	1000	719	385	0
356 x 406 x 393 # $P_z = A_g p_y = 16800$	n/a 1.00	M_{cx}	2750	2750	2750	2750	2750	2750	2750	2750	2750	2750	2750
		M_{cy}	1090	1090	1090	1090	1090	1090	1090	1090	1090	1090	1090
		M_{rx}	2750	2690	2480	2210	1920	1630	1320	1010	683	346	0
		M_{ry}	1090	1090	1090	1090	1090	1090	1020	832	598	320	0
356 x 406 x 340 # $P_z = A_g p_y = 14500$	n/a 1.00	M_{cx}	2340	2340	2340	2340	2340	2340	2340	2340	2340	2340	2340
		M_{cy}	935	935	935	935	935	935	935	935	935	935	935
		M_{rx}	2340	2290	2110	1870	1630	1380	1120	849	574	291	0
		M_{ry}	935	935	935	935	935	935	876	712	511	274	0
356 x 406 x 287 # $P_z = A_g p_y = 12600$	n/a 1.00	M_{cx}	2010	2010	2010	2010	2010	2010	2010	2010	2010	2010	2010
		M_{cy}	803	803	803	803	803	803	803	803	803	803	803
		M_{rx}	2010	1950	1800	1600	1390	1170	948	719	485	246	0
		M_{ry}	803	803	803	803	803	803	754	613	441	236	0
356 x 406 x 235 # $P_z = A_g p_y = 10300$	n/a 1.00	M_{cx}	1620	1620	1620	1620	1620	1620	1620	1620	1620	1620	1620
		M_{cy}	650	650	650	650	650	650	650	650	650	650	650
		M_{rx}	1620	1580	1450	1280	1110	934	755	572	385	195	0
		M_{ry}	650	650	650	650	650	650	610	496	356	191	0
356 x 368 x 202 # $P_z = A_g p_y = 8870$	n/a 1.00	M_{cx}	1370	1370	1370	1370	1370	1370	1370	1370	1370	1370	1370
		M_{cy}	523	523	523	523	523	523	523	523	523	523	523
		M_{rx}	1370	1340	1230	1090	942	792	640	485	326	165	0
		M_{ry}	523	523	523	523	523	523	496	403	290	156	0
356 x 368 x 177 # $P_z = A_g p_y = 7800$	n/a 1.00	M_{cx}	1190	1190	1190	1190	1190	1190	1190	1190	1190	1190	1190
		M_{cy}	456	456	456	456	456	456	456	456	456	456	456
		M_{rx}	1190	1160	1070	947	819	688	555	420	282	142	0
		M_{ry}	456	456	456	456	456	456	433	352	253	136	0

\# Check availability.
F = Factored axial load.
- Not applicable for semi-compact and slender sections.
The values in this table are conservative for tension as the more onerous compression section classification limits have been used.
FOR EXPLANATION OF TABLES SEE NOTE 9.1

BS 5950-1: 2000
BS 4-1: 1993

AXIAL LOAD & BENDING

UC SECTIONS SUBJECT TO AXIAL COMPRESSION AND BENDING

MEMBER BUCKLING CHECK

RESISTANCES AND CAPACITIES FOR S355

Section Designation and Capacities (kN, kNm)	F/P_z Limit		Compression Resistance P_{cx}, P_{cy} (kN) and Buckling Resistance Moment M_b, M_{bs} (kNm) for Varying effective lengths L_E (m) within the limiting value of F/P_z											
		L_E (m)	2.0	3.0	4.0	5.0	6.0	7.0	8.0	9.0	10.0	11.0	12.0	13.0	14.0
356 x 406 x 634 # $P_z = A_g p_y = 26300$ $p_y Z_x = 3760$ $p_y Z_y = 1500$	1.00 1.00	P_{cx} P_{cy} M_b M_{bs}	26300 25700 4520 4520	26200 23800 4520 4520	25400 21900 4520 4520	24500 19900 4520 4520	23700 18000 4520 4520	22800 16100 4520 4520	21800 14200 4510 4520	20800 12600 4420 4440	19700 11100 4330 4260	18600 9800 4250 4070	17500 8680 4170 3870	16400 7720 4100 3680	15300 6890 4020 3470 3270
356 x 406 x 551 # $P_z = A_g p_y = 22800$ $p_y Z_x = 3240$ $p_y Z_y = 1280$	1.00 1.00	P_{cx} P_{cy} M_b M_{bs}	22800 22300 3890 3890	22700 20600 3890 3890	22000 18900 3890 3890	21200 17200 3890 3890	20500 15500 3890 3890	19700 13800 3860 3890	18800 12300 3770 3760	17900 10800 3690 3600	16900 9530 3610 3440	16000 8410 3530 3270	15000 7440 3460 3100	13900 6610 3390 2930	13000 5900 3320 2750
356 x 406 x 467 # $P_z = A_g p_y = 19900$ $p_y Z_x = 2810$ $p_y Z_y = 1100$	1.00 1.00	P_{cx} P_{cy} M_b M_{bs}	19900 19300 3350 3350	19700 17900 3350 3350	19100 16400 3350 3350	18400 14900 3350 3350	17700 13300 3310 3350	17000 11800 3220 3300	16200 10400 3130 3170	15400 9130 3060 3030	14500 8010 2980 2890	13600 7040 2910 2740	12600 6220 2840 2590	11700 5510 2770 2440	10800 4910 2710 2290
356 x 406 x 393 # $P_z = A_g p_y = 16800$ $p_y Z_x = 2340$ $p_y Z_y = 912$	1.00 1.00	P_{cx} P_{cy} M_b M_{bs}	16800 16400 2750 2750	16600 15300 2750 2750	16200 14100 2750 2750	15700 12900 2750 2750	15200 11600 2670 2750	14600 10300 2590 2700	14000 9110 2510 2590	13300 7970 2440 2470	12600 6970 2370 2350	11800 6100 2310 2230	11000 5360 2240 2100	10200 4730 2180 1970	9440 4200 2120 1850
356 x 406 x 340 # $P_z = A_g p_y = 14500$ $p_y Z_x = 2020$ $p_y Z_y = 779$	1.00 1.00	P_{cx} P_{cy} M_b M_{bs}	14500 14100 2340 2340	14300 13200 2340 2340	13900 12200 2340 2340	13500 11100 2330 2340	13100 9980 2240 2340	12600 8870 2160 2290	12000 7790 2090 2200	11400 6810 2020 2100	10800 5940 1960 1990	10100 5200 1900 1890	9390 4570 1840 1780	8680 4030 1780 1670	7990 3580 1730 1560
356 x 406 x 287 # $P_z = A_g p_y = 12600$ $p_y Z_x = 1750$ $p_y Z_y = 669$	1.00 1.00	P_{cx} P_{cy} M_b M_{bs}	12600 12300 2010 2010	12500 11600 2010 2010	12200 10800 2010 2010	11900 9960 1960 2010	11600 9010 1880 2010	11200 8000 1800 1950	10700 7010 1730 1860	10200 6100 1670 1780	9680 5300 1600 1690	9060 4610 1540 1590	8400 4030 1490 1500	7740 3540 1430 1400	7090 3130 1380 1310
356 x 406 x 235 # $P_z = A_g p_y = 10300$ $p_y Z_x = 1430$ $p_y Z_y = 542$	1.00 1.00	P_{cx} P_{cy} M_b M_{bs}	10300 10100 1620 1620	10200 9470 1620 1620	9960 8820 1620 1620	9700 8100 1550 1620	9420 7310 1480 1620	9100 6480 1410 1570	8730 5670 1350 1500	8320 4920 1290 1430	7850 4270 1230 1350	7330 3710 1170 1280	6780 3240 1120 1200	6230 2850 1080 1120	5700 2520 1030 1040
356 x 368 x 202 # $P_z = A_g p_y = 8870$ $p_y Z_x = 1220$ $p_y Z_y = 436$	1.00 1.00	P_{cx} P_{cy} M_b M_{bs}	8870 8590 1370 1370	8760 8040 1370 1370	8550 7430 1370 1370	8320 6760 1360 1370	8080 6020 1280 1360	7790 5260 1210 1300	7470 4540 1150 1240	7110 3910 1080 1170	6690 3360 1020 1110	6240 2910 971 1040	5760 2530 920 965	5280 2210 874 895	4820 1950 831 828 791
356 x 368 x 177 # $P_z = A_g p_y = 7800$ $p_y Z_x = 1070$ $p_y Z_y = 380$	1.00 1.00	P_{cx} P_{cy} M_b M_{bs}	7800 7540 1190 1190	7700 7060 1190 1190	7510 6520 1170 1190	7310 5920 1100 1190	7080 5270 1040 1180	6830 4600 976 1130	6540 3970 917 1080	6210 3410 862 1020	5840 2930 811 958	5430 2530 764 896	5000 2200 721 835	4580 1930 683 774	4170 1700 647 715

Check availability.
Under combined axial compression and bending the capacities are only valid up to the given F/P_z limit. For higher values F/P_z the section would be overloaded due to F alone even when M is zero, because F would exceed the local buckling resistance of the section.
M_b is obtained using an equivalent slenderness = $u.v.L_E/r_y.\beta_w^{0.5}$
M_{bs} is obtained using an equivalent slenderness = $0.5 L/r_y$. Effective length $L_E = L$.
FOR EXPLANATION OF TABLES SEE NOTE 9.1

BS 5950-1: 2000
BS 4-1: 1993

AXIAL LOAD & BENDING

UC SECTIONS SUBJECT TO AXIAL LOAD (COMPRESSION OR TENSION) AND BENDING

CROSS-SECTION CAPACITY CHECK

CAPACITIES FOR S355

Section Designation and Axial load Capacity P_z (kN)	F/P_z Limit Semi-Compact Compact		Moment Capacity M_{cx}, M_{cy} (kNm) and Reduced Moment Capacity M_{rx}, M_{ry} (kNm) for Ratios of Axial Load to Axial Load Capacity F/P_z										
		F/P_z	0.0	0.1	0.2	0.3	0.4	0.5	0.6	0.7	0.8	0.9	1.0
356 x 368 x 153 # $P_z = A_g p_y = 6730$	1.00 0.00	M_{cx} M_{cy} M_{rx} M_{ry}	1020 392 - -	1020 392 - -	1020 392 - -	1020 392 - -	1020 392 - -	1020 392 - -	1020 392 - -	1020 392 - -	1020 392 - -	1020 392 - -	1020 392 - -
356 x 368 x 129 # $P_z = A_g p_y = 5660$	1.00 0.00	M_{cx} M_{cy} M_{rx} M_{ry}	821 328 - -	821 328 - -	821 328 - -	821 328 - -	821 328 - -	821 328 - -	821 328 - -	821 328 - -	821 328 - -	821 328 - -	821 328 - -
305 x 305 x 283 $P_z = A_g p_y = 12100$	n/a 1.00	M_{cx} M_{cy} M_{rx} M_{ry}	1710 615 1710 615	1710 615 1670 615	1710 615 1550 615	1710 615 1380 615	1710 615 1200 615	1710 615 1020 615	1710 615 827 580	1710 615 631 472	1710 615 427 339	1710 615 217 182	1710 615 0 0
305 x 305 x 240 $P_z = A_g p_y = 10600$	n/a 1.00	M_{cx} M_{cy} M_{rx} M_{ry}	1470 528 1470 528	1470 528 1430 528	1470 528 1320 528	1470 528 1180 528	1470 528 1030 528	1470 528 867 528	1470 528 704 501	1470 528 535 408	1470 528 362 293	1470 528 184 157	1470 528 0 0
305 x 305 x 198 $P_z = A_g p_y = 8690$	n/a 1.00	M_{cx} M_{cy} M_{rx} M_{ry}	1190 429 1190 429	1190 429 1160 429	1190 429 1070 429	1190 429 949 429	1190 429 824 429	1190 429 695 429	1190 429 563 407	1190 429 428 332	1190 429 289 239	1190 429 146 128	1190 429 0 0
305 x 305 x 158 $P_z = A_g p_y = 6930$	n/a 1.00	M_{cx} M_{cy} M_{rx} M_{ry}	925 335 925 335	925 335 903 335	925 335 836 335	925 335 739 335	925 335 640 335	925 335 539 335	925 335 436 320	925 335 330 261	925 335 222 188	925 335 112 101	925 335 0 0
305 x 305 x 137 $P_z = A_g p_y = 6000$	n/a 1.00	M_{cx} M_{cy} M_{rx} M_{ry}	792 286 792 286	792 286 774 286	792 286 717 286	792 286 632 286	792 286 547 286	792 286 460 286	792 286 371 275	792 286 281 224	792 286 189 162	792 286 95.4 86.8	792 286 0 0
305 x 305 x 118 $P_z = A_g p_y = 5180$	n/a 1.00	M_{cx} M_{cy} M_{rx} M_{ry}	676 244 676 244	676 244 659 244	676 244 611 244	676 244 539 244	676 244 466 244	676 244 391 244	676 244 315 235	676 244 238 192	676 244 160 138	676 244 80.7 74.4	676 244 0 0
305 x 305 x 97 $P_z = A_g p_y = 4370$	1.00 0.00	M_{cx} M_{cy} M_{rx} M_{ry}	548 204 - -	548 204 - -	548 204 - -	548 204 - -	548 204 - -	548 204 - -	548 204 - -	548 204 - -	548 204 - -	548 204 - -	548 204 - -

\# Check availability.
F = Factored axial load.
- Not applicable for semi-compact and slender sections.
The values in this table are conservative for tension as the more onerous compression section classification limits have been used.
FOR EXPLANATION OF TABLES SEE NOTE 9.1

BS 5950-1: 2000
BS 4-1: 1993

AXIAL LOAD & BENDING

UC SECTIONS SUBJECT TO AXIAL COMPRESSION AND BENDING

MEMBER BUCKLING CHECK

RESISTANCES AND CAPACITIES FOR S355

Section Designation and Capacities (kN, kNm)	F/P_z Limit	Compression Resistance P_{cx}, P_{cy} (kN) and Buckling Resistance Moment M_b, M_{bs} (kNm) for Varying effective lengths L_E (m) within the limiting value of F/P_z												
		L_E (m)	2.0	3.0	4.0	5.0	6.0	7.0	8.0	9.0	10.0	11.0	12.0	13.0	14.0
356 x 368 x 153 #	1.00	P_{cx}	6730	6640	6470	6300	6100	5880	5630	5340	5010	4660	4290	3920	3570
$P_z = A_g p_y = 6730$		P_{cy}	6510	6080	5620	5100	4530	3950	3400	2920	2510	2170	1880	1650	1450
$p_y Z_x = 926$	1.00	M_b	926	926	918	865	814	764	717	673	632	594	560	529	500
$p_y Z_y = 327$		M_{bs}	926	926	926	926	918	877	834	788	742	694	645	598	552
356 x 368 x 129 #	1.00	P_{cx}	5660	5580	5440	5290	5120	4930	4710	4460	4180	3870	3560	3250	2950
$P_z = A_g p_y = 5660$		P_{cy}	5470	5110	4710	4270	3790	3300	2840	2430	2090	1800	1570	1370	1210
$p_y Z_x = 781$	1.00	M_b	781	781	769	721	673	627	583	542	504	469	439	411	386
$p_y Z_y = 274$		M_{bs}	781	781	781	781	773	738	701	663	623	582	541	501	463
305 x 305 x 283	1.00	P_{cx}	12100	11800	11400	11000	10500	10000	9460	8840	8190	7520	6860	6230	5640
$P_z = A_g p_y = 12100$		P_{cy}	11300	10300	9200	8030	6880	5820	4900	4150	3530	3030	2630	2290	2020
$p_y Z_x = 1450$	1.00	M_b	1710	1710	1710	1640	1570	1510	1460	1410	1360	1310	1270	1220	1180
$p_y Z_y = 512$		M_{bs}	1710	1710	1710	1710	1640	1540	1450	1350	1250	1150	1050	960	875
305 x 305 x 240	1.00	P_{cx}	10600	10300	10100	9750	9400	8990	8510	7970	7370	6740	6110	5510	4960
$P_z = A_g p_y = 10600$		P_{cy}	10000	9200	8290	7260	6200	5210	4360	3660	3100	2650	2280	1990	1740
$p_y Z_x = 1260$	1.00	M_b	1470	1470	1440	1370	1310	1250	1200	1150	1100	1060	1010	975	938
$p_y Z_y = 440$		M_{bs}	1470	1470	1470	1460	1390	1310	1220	1140	1050	961	876	797	724
305 x 305 x 198	1.00	P_{cx}	8690	8510	8270	8000	7700	7350	6940	6470	5960	5430	4900	4410	3960
$P_z = A_g p_y = 8690$		P_{cy}	8210	7540	6780	5920	5040	4220	3520	2960	2500	2130	1840	1600	1400
$p_y Z_x = 1030$	1.00	M_b	1190	1190	1140	1080	1030	974	925	880	838	798	762	728	696
$p_y Z_y = 358$		M_{bs}	1190	1190	1190	1180	1120	1050	984	912	839	767	699	635	576
305 x 305 x 158	1.00	P_{cx}	6930	6770	6580	6360	6110	5820	5470	5090	4660	4230	3810	3410	3060
$P_z = A_g p_y = 6930$		P_{cy}	6530	5990	5360	4660	3950	3290	2740	2290	1940	1650	1420	1230	1080
$p_y Z_x = 817$	1.00	M_b	925	925	874	818	767	719	675	634	597	564	533	506	480
$p_y Z_y = 279$		M_{bs}	925	925	925	917	867	814	759	702	644	587	533	484	438
305 x 305 x 137	1.00	P_{cx}	6000	5860	5680	5490	5270	5010	4700	4360	3980	3600	3230	2900	2590
$P_z = A_g p_y = 6000$		P_{cy}	5650	5170	4620	4010	3380	2820	2340	1960	1650	1410	1210	1050	922
$p_y Z_x = 707$	1.00	M_b	792	792	740	688	639	594	553	515	482	452	425	401	379
$p_y Z_y = 239$		M_{bs}	792	792	792	784	740	695	647	597	547	499	452	410	371
305 x 305 x 118	1.00	P_{cx}	5180	5050	4890	4730	4530	4300	4040	3730	3410	3080	2760	2470	2210
$P_z = A_g p_y = 5180$		P_{cy}	4860	4450	3970	3430	2890	2400	2000	1670	1410	1200	1030	895	784
$p_y Z_x = 607$	1.00	M_b	676	674	624	576	530	488	450	416	386	359	336	315	297
$p_y Z_y = 203$		M_{bs}	676	676	676	667	629	590	549	506	463	422	382	346	313
305 x 305 x 97	1.00	P_{cx}	4370	4250	4120	3970	3800	3600	3360	3090	2800	2520	2250	2000	1790
$P_z = A_g p_y = 4370$		P_{cy}	4090	3730	3310	2850	2390	1970	1630	1360	1140	971	835	724	634
$p_y Z_x = 513$	1.00	M_b	513	513	474	436	399	365	334	306	281	260	242	226	212
$p_y Z_y = 170$		M_{bs}	513	513	513	503	474	444	411	378	345	313	283	255	230

Check availability.
Under combined axial compression and bending the capacities are only valid up to the given F/P_z limit. For higher values F/P_z the section would be overloaded due to F alone even when M is zero, because F would exceed the local buckling resistance of the section.
M_b is obtained using an equivalent slenderness = $u.v.L_E/r_y.\beta_w^{0.5}$
M_{bs} is obtained using an equivalent slenderness = $0.5 L/r_y$. Effective length $L_E = L$.
FOR EXPLANATION OF TABLES SEE NOTE 9.1

| BS 5950-1: 2000 |
| BS 4-1: 1993 |

AXIAL LOAD & BENDING

UC SECTIONS SUBJECT TO AXIAL LOAD (COMPRESSION OR TENSION) AND BENDING

CROSS-SECTION CAPACITY CHECK

CAPACITIES FOR S355

Section Designation and Axial load Capacity P_z (kN)	F/P_z Limit Semi-Compact Compact		Moment Capacity M_{cx}, M_{cy} (kNm) and Reduced Moment Capacity M_{rx}, M_{ry} (kNm) for Ratios of Axial Load to Axial Load Capacity F/P_z										
		F/P_z	0.0	0.1	0.2	0.3	0.4	0.5	0.6	0.7	0.8	0.9	1.0
254 x 254 x 167 $P_z = A_g p_y = 7350$	n/a 1.00	M_{cx} M_{cy} M_{rx} M_{ry}	836 308 836 308	836 308 816 308	836 308 755 308	836 308 671 308	836 308 584 308	836 308 494 308	836 308 401 291	836 308 305 237	836 308 207 170	836 308 105 91.3	836 308 0 0
254 x 254 x 132 $P_z = A_g p_y = 5800$	n/a 1.00	M_{cx} M_{cy} M_{rx} M_{ry}	645 238 645 238	645 238 629 238	645 238 581 238	645 238 515 238	645 238 447 238	645 238 377 238	645 238 305 226	645 238 232 184	645 238 156 132	645 238 79.1 70.9	645 238 0 0
254 x 254 x 107 $P_z = A_g p_y = 4690$	n/a 1.00	M_{cx} M_{cy} M_{rx} M_{ry}	512 190 512 190	512 190 500 190	512 190 462 190	512 190 408 190	512 190 353 190	512 190 297 190	512 190 240 181	512 190 182 147	512 190 123 106	512 190 62.0 56.8	512 190 0 0
254 x 254 x 89 $P_z = A_g p_y = 3900$	n/a 1.00	M_{cx} M_{cy} M_{rx} M_{ry}	422 157 422 157	422 157 412 157	422 157 380 157	422 157 334 157	422 157 289 157	422 157 243 157	422 157 196 149	422 157 148 121	422 157 99.8 87.2	422 157 50.3 46.8	422 157 0 0
254 x 254 x 73 $P_z = A_g p_y = 3310$	1.00 0.00	M_{cx} M_{cy} M_{rx} M_{ry}	350 131 - -	350 131 - -	350 131 - -	350 131 - -	350 131 - -	350 131 - -	350 131 - -	350 131 - -	350 131 - -	350 131 - -	350 131 - -
203 x 203 x 86 $P_z = A_g p_y = 3800$	n/a 1.00	M_{cx} M_{cy} M_{rx} M_{ry}	337 124 337 124	337 124 329 124	337 124 304 124	337 124 271 124	337 124 235 124	337 124 198 124	337 124 161 118	337 124 122 96.1	337 124 82.3 69.2	337 124 41.7 37.1	337 124 0 0
203 x 203 x 71 $P_z = A_g p_y = 3120$	n/a 1.00	M_{cx} M_{cy} M_{rx} M_{ry}	276 102 276 102	276 102 269 102	276 102 247 102	276 102 219 102	276 102 190 102	276 102 160 102	276 102 129 96.1	276 102 97.9 78.2	276 102 65.9 56.2	276 102 33.3 30.1	276 102 0 0
203 x 203 x 60 $P_z = A_g p_y = 2710$	n/a 1.00	M_{cx} M_{cy} M_{rx} M_{ry}	233 85.6 233 85.6	233 85.6 227 85.6	233 85.6 211 85.6	233 85.6 187 85.6	233 85.6 161 85.6	233 85.6 136 85.6	233 85.6 110 82.4	233 85.6 83.0 67.3	233 85.6 55.8 48.5	233 85.6 28.2 26.1	233 85.6 0 0

F = Factored axial load.
- Not applicable for semi-compact and slender sections.
The values in this table are conservative for tension as the more onerous compression section classification limits have been used.
FOR EXPLANATION OF TABLES SEE NOTE 9.1

| BS 5950-1: 2000 |
| BS 4-1: 1993 |

AXIAL LOAD & BENDING

UC SECTIONS SUBJECT TO AXIAL COMPRESSION AND BENDING

MEMBER BUCKLING CHECK

RESISTANCES AND CAPACITIES FOR S355

Section Designation and Capacities (kN, kNm)	F/P_z Limit	Compression Resistance P_{cx}, P_{cy} (kN) and Buckling Resistance Moment M_b, M_{bs} (kNm) for Varying effective lengths L_E (m) within the limiting value of F/P_z												
		L_E (m)	1.0	1.5	2.0	2.5	3.0	3.5	4.0	4.5	5.0	5.5	6.0	6.5	7.0
254 x 254 x 167	1.00	P_{cx}	7350	7350	7310	7190	7070	6950	6820	6670	6520	6350	6160	5950	5730
$P_z = A_g p_y = 7350$		P_{cy}	7350	7070	6750	6410	6040	5640	5210	4770	4330	3910	3510	3160	2840
$p_y Z_x = 716$	1.00	M_b	836	836	836	836	836	836	815	792	771	751	732	714	696	679
$p_y Z_y = 257$		M_{bs}	836	836	836	836	836	836	836	819	793	766	738	710	680
254 x 254 x 132	1.00	P_{cx}	5800	5800	5760	5660	5570	5460	5350	5240	5110	4970	4810	4640	4450
$P_z = A_g p_y = 5800$		P_{cy}	5800	5560	5300	5030	4730	4410	4060	3710	3350	3020	2710	2430	2180
$p_y Z_x = 563$	1.00	M_b	645	645	645	645	636	615	595	576	558	540	524	508	493
$p_y Z_y = 199$		M_{bs}	645	645	645	645	645	645	629	608	587	565	542	519
254 x 254 x 107	1.00	P_{cx}	4690	4690	4650	4570	4490	4410	4310	4210	4100	3980	3850	3700	3550
$P_z = A_g p_y = 4690$		P_{cy}	4690	4490	4280	4050	3810	3540	3250	2960	2670	2400	2150	1930	1730
$p_y Z_x = 453$	1.00	M_b	512	512	512	512	498	479	460	443	426	410	395	380	367
$p_y Z_y = 158$		M_{bs}	512	512	512	512	512	512	497	480	463	445	427	408
254 x 254 x 89	1.00	P_{cx}	3900	3900	3860	3800	3730	3660	3580	3490	3400	3300	3190	3060	2930
$P_z = A_g p_y = 3900$		P_{cy}	3900	3730	3550	3360	3150	2930	2690	2450	2210	1980	1770	1590	1420
$p_y Z_x = 378$	1.00	M_b	422	422	422	422	406	389	372	356	340	325	311	298	285
$p_y Z_y = 131$		M_{bs}	422	422	422	422	422	422	409	395	381	366	351	335
254 x 254 x 73	1.00	P_{cx}	3310	3310	3270	3210	3150	3090	3020	2950	2870	2780	2680	2570	2450
$P_z = A_g p_y = 3310$		P_{cy}	3300	3150	3000	2830	2650	2460	2250	2040	1830	1630	1460	1300	1170
$p_y Z_x = 319$	1.00	M_b	319	319	319	319	307	293	280	267	254	242	231	220	210
$p_y Z_y = 109$		M_{bs}	319	319	319	319	319	319	317	307	296	285	273	262	249
203 x 203 x 86	1.00	P_{cx}	3800	3780	3710	3630	3540	3450	3340	3230	3090	2950	2790	2610	2440
$P_z = A_g p_y = 3800$		P_{cy}	3720	3510	3290	3040	2770	2480	2190	1920	1670	1460	1280	1130	1000
$p_y Z_x = 293$	1.00	M_b	337	337	337	337	330	316	304	292	281	270	260	250	241	233
$p_y Z_y = 103$		M_{bs}	337	337	337	337	337	331	317	303	289	274	258	243	227
203 x 203 x 71	1.00	P_{cx}	3120	3110	3040	2980	2910	2830	2740	2640	2530	2410	2270	2130	1980
$P_z = A_g p_y = 3120$		P_{cy}	3060	2880	2700	2490	2270	2030	1790	1560	1360	1190	1040	917	812
$p_y Z_x = 244$	1.00	M_b	276	276	276	276	266	254	242	231	221	211	201	193	185	177
$p_y Z_y = 84.9$		M_{bs}	276	276	276	276	276	270	259	247	235	223	210	197	185
203 x 203 x 60	1.00	P_{cx}	2710	2700	2640	2580	2510	2440	2360	2270	2170	2050	1920	1790	1660
$P_z = A_g p_y = 2710$		P_{cy}	2650	2490	2330	2140	1940	1720	1500	1310	1140	987	862	758	669
$p_y Z_x = 207$	1.00	M_b	233	233	233	221	209	198	187	177	168	159	151	143	136
$p_y Z_y = 71.4$		M_{bs}	233	233	233	233	233	226	216	206	196	185	174	162	151

Under combined axial compression and bending the capacities are only valid up to the given F/P_z limit. For higher values F/P_z the section would be overloaded due to F alone even when M is zero, because F would exceed the local buckling resistance of the section.
M_b is obtained using an equivalent slenderness = $u.v.L_E/r_y.\beta_w^{0.5}$
M_{bs} is obtained using an equivalent slenderness = $0.5 L/r_y$. Effective length $L_E = L$.
FOR EXPLANATION OF TABLES SEE NOTE 9.1

BS 5950-1: 2000
BS 4-1: 1993

AXIAL LOAD & BENDING

UC SECTIONS SUBJECT TO AXIAL LOAD (COMPRESSION OR TENSION) AND BENDING

CROSS-SECTION CAPACITY CHECK

CAPACITIES FOR S355

Section Designation and Axial load Capacity P_z (kN)	F/P_z Limit Semi-Compact Compact		Moment Capacity M_{cx}, M_{cy} (kNm) and Reduced Moment Capacity M_{rx}, M_{ry} (kNm) for Ratios of Axial Load to Axial Load Capacity F/P_z										
		F/P_z	0.0	0.1	0.2	0.3	0.4	0.5	0.6	0.7	0.8	0.9	1.0
203 x 203 x 52 $P_z = A_g p_y = 2350$	n/a 1.00	M_{cx} M_{cy} M_{rx} M_{ry}	201 74.1 201 74.1	201 74.1 196 74.1	201 74.1 182 74.1	201 74.1 161 74.1	201 74.1 139 74.1	201 74.1 117 74.1	201 74.1 94.0 71.2	201 74.1 71.1 58.1	201 74.1 47.8 41.8	201 74.1 24.1 22.5	201 74.1 0 0
203 x 203 x 46 $P_z = A_g p_y = 2080$	1.00 0.00	M_{cx} M_{cy} M_{rx} M_{ry}	174 64.8 - -	174 64.8 - -	174 64.8 - -	174 64.8 - -	174 64.8 - -	174 64.8 - -	174 64.8 - -	174 64.8 - -	174 64.8 - -	174 64.8 - -	174 64.8 - -
152 x 152 x 37 $P_z = A_g p_y = 1670$	n/a 1.00	M_{cx} M_{cy} M_{rx} M_{ry}	110 39.0 110 39.0	110 39.0 107 39.0	110 39.0 99.9 39.0	110 39.0 88.4 39.0	110 39.0 76.6 39.0	110 39.0 64.4 39.0	110 39.0 52.1 37.9	110 39.0 39.4 31.0	110 39.0 26.5 22.4	110 39.0 13.4 12.1	110 39.0 0 0
152 x 152 x 30 $P_z = A_g p_y = 1360$	n/a 1.00	M_{cx} M_{cy} M_{rx} M_{ry}	88.0 31.2 88.0 31.2	88.0 31.2 86.0 31.2	88.0 31.2 80.0 31.2	88.0 31.2 70.8 31.2	88.0 31.2 61.2 31.2	88.0 31.2 51.4 31.2	88.0 31.2 41.5 30.5	88.0 31.2 31.4 25.0	88.0 31.2 21.1 18.0	88.0 31.2 10.6 9.70	88.0 31.2 0 0
152 x 152 x 23 $P_z = A_g p_y = 1040$	1.00 0.00	M_{cx} M_{cy} M_{rx} M_{ry}	60.5 22.2 - -	60.5 22.2 - -	60.5 22.2 - -	60.5 22.2 - -	60.5 22.2 - -	60.5 22.2 - -	60.5 22.2 - -	60.5 22.2 - -	60.5 22.2 - -	60.5 22.2 - -	60.5 22.2 - -

F = Factored axial load.
- Not applicable for semi-compact and slender sections.
The values in this table are conservative for tension as the more onerous compression section classification limits have been used.
FOR EXPLANATION OF TABLES SEE NOTE 9.1

BS 5950-1: 2000
BS 4-1: 1993

AXIAL LOAD & BENDING

UC SECTIONS SUBJECT TO AXIAL COMPRESSION AND BENDING

MEMBER BUCKLING CHECK

RESISTANCES AND CAPACITIES FOR S355

Section Designation and Capacities (kN, kNm)	F/P_z Limit	Compression Resistance P_{cx}, P_{cy} (kN) and Buckling Resistance Moment M_b, M_{bs} (kNm) for Varying effective lengths L_E (m) within the limiting value of F/P_z												
			L_E (m)	1.0	1.5	2.0	2.5	3.0	3.5	4.0	4.5	5.0	5.5	6.0	6.5	7.0
203 x 203 x 52	1.00	P_{cx}	2350	2340	2290	2240	2180	2120	2050	1960	1870	1770	1660	1550	1430
$P_z = A_g p_y = 2350$		P_{cy}	2300	2160	2020	1850	1670	1490	1300	1130	980	852	744	653	577
$p_y Z_x = 181$	1.00	M_b	201	201	200	189	178	168	158	148	140	131	124	117	111
$p_y Z_y = 61.8$		M_{bs}	201	201	201	201	201	195	187	178	169	159	150	140	130
203 x 203 x 46	1.00	P_{cx}	2080	2070	2020	1980	1930	1870	1810	1730	1650	1560	1460	1350	1250
$P_z = A_g p_y = 2080$		P_{cy}	2030	1910	1780	1630	1470	1300	1140	988	856	743	648	569	503
$p_y Z_x = 160$	1.00	M_b	160	160	160	152	143	135	127	119	112	105	98.9	93.3	88.2
$p_y Z_y = 54$		M_{bs}	160	160	160	160	160	154	148	141	133	126	118	110	102
152 x 152 x 37	1.00	P_{cx}	1670	1630	1580	1530	1460	1390	1300	1200	1100	989	886	792	708
$P_z = A_g p_y = 1670$		P_{cy}	1570	1430	1270	1100	920	761	630	525	442	376	323	281	246
$p_y Z_x = 96.9$	1.00	M_b	110	110	102	95.3	88.8	82.7	77.2	72.2	67.6	63.5	59.8	56.5	53.5
$p_y Z_y = 32.5$		M_{bs}	110	110	110	108	102	95.2	88.4	81.3	74.2	67.4	60.9	55.0	49.7
152 x 152 x 30	1.00	P_{cx}	1360	1320	1280	1240	1190	1120	1050	969	880	792	708	632	564
$P_z = A_g p_y = 1360$		P_{cy}	1270	1160	1030	884	739	611	504	420	353	300	258	224	196
$p_y Z_x = 78.8$	1.00	M_b	88.0	87.3	80.7	74.4	68.4	62.8	57.8	53.4	49.4	46.0	43.0	40.3	37.9
$p_y Z_y = 26$		M_{bs}	88.0	88.0	88.0	86.2	81.3	76.0	70.4	64.7	59.0	53.5	48.3	43.6	39.3
152 x 152 x 23	1.00	P_{cx}	1040	1010	974	938	895	844	785	718	648	580	516	458	408
$P_z = A_g p_y = 1040$		P_{cy}	965	875	770	654	542	444	365	303	254	216	185	160	140
$p_y Z_x = 58.2$	1.00	M_b	58.2	57.9	53.3	48.7	44.4	40.3	36.6	33.4	30.6	28.2	26.1	24.3	22.8
$p_y Z_y = 18.7$		M_{bs}	58.2	58.2	58.2	56.5	53.0	49.4	45.5	41.6	37.7	34.0	30.6	27.5	24.7

Under combined axial compression and bending the capacities are only valid up to the given F/P_z limit. For higher values F/P_z the section would be overloaded due to F alone even when M is zero, because F would exceed the local buckling resistance of the section.
M_b is obtained using an equivalent slenderness = $u.v.L_E/r_y.\beta_w^{0.5}$
M_{bs} is obtained using an equivalent slenderness = $0.5 L/r_y$. Effective length $L_E = L$.
FOR EXPLANATION OF TABLES SEE NOTE 9.1

BS 5950-1: 2000
BS 4190: 2001

BOLT CAPACITIES

NON-PRELOADED ORDINARY BOLTS

GRADE 4.6 BOLTS IN S355

Diameter of Bolt	Tensile Stress Area	Tension Capacity		Shear Capacity		Bearing Capacity in kN (Minimum of P_{bb} and P_{bs}) End distance equal to 2 x bolt diameter. Thickness in mm of ply passed through.								
		Nominal $0.8A_t p_t$ P_{nom}	Exact $A_t p_t$ P_t	Single Shear P_s	Double Shear $2P_s$									
mm	mm^2	kN	kN	kN	kN	5	6	7	8	9	10	12	15	20	25	30
12	84.3	16.2	20.2	13.5	27.0	*27.6*	*33.1*	*38.6*	*44.2*	*49.7*	*55.2*	*66.2*	*82.8*	*110*	*138*	*166*
16	157	30.1	37.7	25.1	50.2	36.8	44.2	*51.5*	*58.9*	*66.2*	*73.6*	*88.3*	*110*	*147*	*184*	*221*
20	245	47.0	58.8	39.2	78.4	46.0	55.2	64.4	73.6	82.8	92.0	*110*	*138*	*184*	*230*	*276*
22	303	58.2	72.7	48.5	97.0	50.6	60.7	70.8	81.0	91.1	*101*	*121*	*152*	*202*	*253*	*304*
24	353	67.8	84.7	56.5	113	**55.2**	66.2	77.3	88.3	99.4	110	*132*	*166*	*221*	*276*	*331*
27	459	88.1	110	73.4	147	**62.1**	74.5	86.9	99.4	112	124	149	*186*	*248*	*311*	*373*
30	561	108	135	89.8	180	**69.0**	**82.8**	96.6	110	124	138	166	207	*276*	*345*	*414*

Values in **bold** are less than the single shear capacity of the bolt.
Values in *italic* are greater than the double shear capacity of the bolt.
Bearing values assume standard clearance holes.
If oversize or short slotted holes are used, bearing values should be multiplied by 0.7.
If long slotted or kidney shaped holes are used, bearing values should be multiplied by 0.5.
If appropriate, shear capacity must be reduced for large packings, large grip lengths and long joints.
FOR EXPLANATION OF TABLES SEE NOTE 10.1

GRADE 8.8 BOLTS IN S355

Diameter of Bolt	Tensile Stress Area	Tension Capacity		Shear Capacity		Bearing Capacity in kN (Minimum of P_{bb} and P_{bs}) End distance equal to 2 x bolt diameter. Thickness in mm of ply passed through.								
		Nominal $0.8A_t p_t$ P_{nom}	Exact $A_t p_t$ P_t	Single Shear P_s	Double Shear $2P_s$									
mm	mm^2	kN	kN	kN	kN	5	6	7	8	9	10	12	15	20	25	30
12	84.3	37.8	47.2	31.6	63.2	33.0	39.6	46.2	52.8	59.4	66.0	79.2	*99.0*	*132*	*165*	*198*
16	157	70.3	87.9	58.9	118	**44.0**	52.8	61.6	70.4	79.2	88.0	106	*132*	*176*	*220*	*264*
20	245	110	137	91.9	184	**55.0**	66.0	77.0	88.0	99.0	110	132	165	*220*	*275*	*330*
22	303	136	170	114	227	**60.5**	72.6	84.7	96.8	**109**	121	145	182	*242*	*303*	*363*
24	353	158	198	132	265	**66.0**	**79.2**	92.4	**106**	**119**	**132**	158	198	*264*	*330*	*396*
27	459	206	257	172	344	**74.3**	**89.1**	**104**	**119**	**134**	**149**	178	223	297	*371*	*446*
30	561	251	314	210	421	**82.5**	**99.0**	**116**	**132**	**149**	**165**	**198**	248	330	413	*495*

Values in **bold** are less than the single shear capacity of the bolt.
Values in *italic* are greater than the double shear capacity of the bolt.
Bearing values assume standard clearance holes.
If oversize or short slotted holes are used, bearing values should be multiplied by 0.7.
If long slotted or kidney shaped holes are used, bearing values should be multiplied by 0.5.
If appropriate, shear capacity must be reduced for large packings, large grip lengths and long joints.
FOR EXPLANATION OF TABLES SEE NOTE 10.1

BS 5950-1: 2000
BS 4190: 2001

BOLT CAPACITIES

NON-PRELOADED ORDINARY BOLTS

GRADE 10.9 BOLTS IN S355

Diameter of Bolt	Tensile Stress Area	Tension Capacity		Shear Capacity		Bearing Capacity in kN (Minimum of P_{bb} and P_{bs}) End distance equal to 2 x bolt diameter. Thickness in mm of ply passed through.									
		Nominal $0.8A_t p_t$	Exact $A_t p_t$	Single Shear	Double Shear										
mm	A_t mm²	P_{nom} kN	P_t kN	P_s kN	$2P_s$ kN	5	6	7	8	9	10	12	15	20	25	30
12	84.3	47.2	59.0	33.7	67.4	**33.0**	39.6	46.2	52.8	59.4	66.0	79.2	99.0	*132*	*165*	*198*
16	157	87.9	110	62.8	126	**44.0**	**52.8**	61.6	70.4	79.2	88.0	106	*132*	*176*	*220*	*264*
20	245	137	172	98.0	196	**55.0**	**66.0**	**77.0**	**88.0**	99.0	110	132	165	*220*	*275*	*330*
22	303	170	212	121	242	**60.5**	**72.6**	**84.7**	**96.8**	**109**	**121**	145	182	242	*303*	*363*
24	353	198	247	141	282	**66.0**	**79.2**	**92.4**	**106**	**119**	**132**	158	198	264	*330*	*396*
27	459	257	321	184	367	**74.3**	**89.1**	**104**	**119**	**134**	**149**	**178**	223	297	*371*	*446*
30	561	314	393	224	449	**82.5**	**99.0**	**116**	**132**	**149**	**165**	**198**	248	330	413	*495*

Values in **bold** are less than the single shear capacity of the bolt.
Values in *italic* are greater than the double shear capacity of the bolt.
Bearing values assume standard clearance holes.
If oversize or short slotted holes are used, bearing values should be multiplied by 0.7.
If long slotted or kidney shaped holes are used, bearing values should be multiplied by 0.5.
If appropriate, shear capacity must be reduced for large packings, large grip lengths and long joints.
FOR EXPLANATION OF TABLES SEE NOTE 10.1

BS 5950-1: 2000
BS 4190: 2001
BS 4933: 1973

BOLT CAPACITIES

NON-PRELOADED COUNTERSUNK BOLTS

GRADE 4.6 COUNTERSUNK BOLTS IN S355

Diameter of Bolt	Tensile Stress Area	Tension Capacity		Shear Capacity		Bearing Capacity in kN (Minimum of P_{bb} and P_{bs}) End distance equal to 2 x bolt diameter. Thickness in mm of ply passed through.								
		Nominal $0.8A_t p_t$ P_{nom}	Exact $A_t p_t$ P_t	Single Shear P_s	Double Shear $2P_s$									
mm	mm²	kN	kN	kN	kN	5	6	7	8	9	10	12	15	20	25	30
12	84.3	16.2	20.2	13.5	27.0	**11.0**	16.6	22.1	27.6	33.1	38.6	49.7	66.2	*93.8*	*121*	*149*
16	157	30.1	37.7	25.1	50.2	**7.36**	**14.7**	22.1	29.4	36.8	44.2	58.9	81.0	*118*	*155*	*191*
20	245	47.0	58.8	39.2	78.4	0	**9.20**	18.4	27.6	36.8	46.0	64.4	92.0	*138*	*184*	*230*
22	303	58.2	72.7	48.5	97.0	0	**5.06**	15.2	25.3	35.4	45.5	65.8	96.1	*147*	*197*	*248*
24	353	67.8	84.7	56.5	113	0	0	**11.0**	22.1	33.1	44.2	66.2	99.4	*155*	*210*	*265*
27	459	88.1	110	73.4	147	0	0	**3.11**	**15.5**	27.9	40.4	65.2	102	*165*	*227*	*289*
30	561	108	135	89.8	180	0	0	0	**6.90**	20.7	34.5	62.1	104	*173*	*242*	*311*

Values in **bold** are less than the single shear capacity of the bolt.
Values in *italic* are greater than the double shear capacity of the bolt.
Bearing values assume standard clearance holes.
If oversize or short slotted holes are used, bearing values should be multiplied by 0.7.
If long slotted or kidney shaped holes are used, bearing values should be multiplied by 0.5.
Depth of countersink is taken as half the bolt diameter.
FOR EXPLANATION OF TABLES SEE NOTE 10.1

GRADE 8.8 COUNTERSUNK BOLTS IN S 355

Diameter of Bolt	Tensile Stress Area	Tension Capacity		Shear Capacity		Bearing Capacity in kN (Minimum of P_{bb} and P_{bs}) End distance equal to 2 x bolt diameter. Thickness in mm of ply passed through.								
		Nominal $0.8A_t p_t$ P_{nom}	Exact $A_t p_t$ P_t	Single Shear P_s	Double Shear $2P_s$									
mm	mm²	kN	kN	kN	kN	5	6	7	8	9	10	12	15	20	25	30
12	84.3	37.8	47.2	31.6	63.2	**13.2**	**19.8**	26.4	33.0	39.6	46.2	59.4	79.2	*112*	*145*	*178*
16	157	70.3	87.9	58.9	118	**8.80**	**17.6**	26.4	35.2	44.0	52.8	70.4	96.8	*141*	*185*	*229*
20	245	110	137	91.9	184	0	**11.0**	22.0	33.0	44.0	55.0	77.0	110	*165*	*220*	*275*
22	303	136	170	114	227	0	**6.05**	18.2	30.3	42.4	54.7	78.7	115	*175*	*236*	*296*
24	353	158	198	132	265	0	0	**13.2**	26.4	39.6	52.8	79.2	119	*185*	*251*	*317*
27	459	206	257	172	344	0	0	**3.71**	**18.6**	33.4	48.3	78.0	123	*197*	*271*	*345*
30	561	251	314	210	421	0	0	0	**8.25**	24.8	41.3	74.3	124	**206**	*289*	*371*

Values in **bold** are less than the single shear capacity of the bolt.
Values in *italic* are greater than the double shear capacity of the bolt.
Bearing values assume standard clearance holes.
If oversize or short slotted holes are used, bearing values should be multiplied by 0.7.
If long slotted or kidney shaped holes are used, bearing values should be multiplied by 0.5.
Depth of countersink is taken as half the bolt diameter.
FOR EXPLANATION OF TABLES SEE NOTE 10.1

BS 5950-1: 2000
BS 4190: 2001
BS 4933: 1973

BOLT CAPACITIES

NON-PRELOADED COUNTERSUNK BOLTS

GRADE 10.9 COUNTERSUNK BOLTS IN S355

Diameter of Bolt	Tensile Stress Area	Tension Capacity		Shear Capacity		Bearing Capacity in kN (Minimum of P_{bb} and P_{bs}) End distance equal to 2 x bolt diameter. Thickness in mm of ply passed through.								
		Nominal $0.8A_t p_t$	Exact $A_t p_t$	Single Shear	Double Shear									
	A_t	P_{nom}	P_t	P_s	$2P_s$									
mm	mm²	kN	kN	kN	kN	5	6	7	8	9	10	12	15	20	25	30
12	84.3	47.2	59.0	33.7	67.4	**13.2**	**19.8**	**26.4**	**33.0**	39.6	46.2	59.4	79.2	*112*	*145*	*178*
16	157	87.9	110	62.8	126	**8.80**	**17.6**	**26.4**	**35.2**	**44.0**	52.8	70.4	96.8	*141*	*185*	*229*
20	245	137	172	98.0	196	0	**11.0**	**22.0**	**33.0**	**44.0**	**55.0**	77.0	110	165	*220*	*275*
22	303	170	212	121	242	0	**6.05**	**18.2**	**30.3**	**42.4**	54.5	78.7	**115**	175	236	*296*
24	353	198	247	141	282	0	0	**13.2**	**26.4**	**39.6**	**52.8**	79.2	**119**	185	251	*317*
27	459	257	321	184	367	0	0	**3.71**	**18.6**	**33.4**	**48.3**	78.0	**123**	197	271	345
30	561	314	393	224	449	0	0	0	**8.25**	**24.8**	**41.3**	74.3	**124**	206	289	371

Values in **bold** are less than the single shear capacity of the bolt.
Values in *italic* are greater than the double shear capacity of the bolt.
Bearing values assume standard clearance holes.
If oversize or short slotted holes are used, bearing values should be multiplied by 0.7.
If long slotted or kidney shaped holes are used, bearing values should be multiplied by 0.5.
Depth of countersink is taken as half the bolt diameter.
FOR EXPLANATION OF TABLES SEE NOTE 10.1

BS 5950-1: 2000
BS 4395: 1969

BOLT CAPACITIES

NON-PRELOADED HSFG BOLTS

GENERAL GRADE HSFG BOLTS IN S355

Diameter of Bolt	Tensile Stress Area	Tension Capacity		Shear Capacity		Bearing Capacity in kN (Minimum of P_{bb} and P_{bs}) End distance equal to 2 x bolt diameter. Thickness in mm of ply passed through.								
		Nominal $0.8A_t p_t$ P_{nom}	Exact $A_t p_t$ P_t	Single Shear P_s	Double Shear $2P_s$									
mm	mm²	kN	kN	kN	kN	5	6	7	8	9	10	12	15	20	25	30
12	84.3	39.8	49.7	33.7	67.4	**33.0**	39.6	46.2	52.8	59.4	66.0	79.2	99.0	*132*	*165*	*198*
16	157	74.1	92.6	62.8	126	**44.0**	**52.8**	61.6	70.4	79.2	88.0	106	*132*	*176*	*220*	*264*
20	245	116	145	98.0	196	**55.0**	**66.0**	**77.0**	**88.0**	99.0	110	132	165	*220*	*275*	*330*
22	303	143	179	121	242	**60.5**	**72.6**	**84.7**	**96.8**	**109**	**121**	145	182	242	*303*	*363*
24	353	167	208	141	282	**66.0**	**79.2**	**92.4**	**106**	**119**	**132**	158	198	264	*330*	*396*
27	459	189	236	161	321	**74.3**	**89.1**	**104**	**119**	**134**	**149**	178	223	297	*371*	*446*
30	561	231	289	196	393	**82.5**	**99.0**	**116**	**132**	**149**	**165**	198	248	330	*413*	*495*

Values in **bold** are less than the single shear capacity of the bolt.
Values in *italic* are greater than the double shear capacity of the bolt.
Bearing values assume standard clearance holes.
If oversize or short slotted holes are used, bearing values should be multiplied by 0.7.
If long slotted or kidney shaped holes are used, bearing values should be multiplied by 0.5.
If appropriate, shear capacity must be reduced for large packings, large grip lengths and long joints.
FOR EXPLANATION OF TABLES SEE NOTE 10.1

HIGHER GRADE HSFG BOLTS IN S 355

Diameter of Bolt	Tensile Stress Area	Tension Capacity		Shear Capacity		Bearing Capacity in kN (Minimum of P_{bb} and P_{bs}) End distance equal to 2 x bolt diameter. Thickness in mm of ply passed through.								
		Nominal $0.8A_t p_t$ P_{nom}	Exact $A_t p_t$ P_t	Single Shear P_s	Double Shear $2P_s$									
mm	mm²	kN	kN	kN	kN	5	6	7	8	9	10	12	15	20	25	30
16	157	87.9	110	62.8	126	**44.0**	**52.8**	61.6	70.4	79.2	88.0	106	*132*	*176*	*220*	*264*
20	245	137	172	98.0	196	**55.0**	**66.0**	**77.0**	**88.0**	99.0	110	132	165	*220*	*275*	*330*
22	303	170	212	121	242	**60.5**	**72.6**	**84.7**	**96.8**	**109**	**121**	145	182	242	*303*	*363*
24	353	198	247	141	282	**66.0**	**79.2**	**92.4**	**106**	**119**	**132**	158	198	264	*330*	*396*
27	459	257	321	184	367	**74.3**	**89.1**	**104**	**119**	**134**	**149**	178	223	297	*371*	*446*
30	561	314	393	224	449	**82.5**	**99.0**	**116**	**132**	**149**	**165**	198	248	330	*413*	*495*

Values in **bold** are less than the single shear capacity of the bolt.
Values in *italic* are greater than the double shear capacity of the bolt.
Bearing values assume standard clearance holes.
If oversize or short slotted holes are used, bearing values should be multiplied by 0.7.
If long slotted or kidney shaped holes are used, bearing values should be multiplied by 0.5.
If appropriate, shear capacity must be reduced for large packings, large grip lengths and long joints.
FOR EXPLANATION OF TABLES SEE NOTE 10.1

BS 5950-1: 2000
BS 4395: 1969
BS 4604: 1970

BOLT CAPACITIES

PRELOADED HSFG BOLTS: NON-SLIP IN SERVICE

GENERAL GRADE HSFG BOLTS IN S355

Diameter of Bolt mm	Min. Shank Tension P_o kN	Tension 1.1P_o kN	Tension $A_t p_t$ kN	Shear Capacity Single Shear kN	Shear Capacity Double Shear kN	Slip Resistance for $\mu = 0.5$ Single Shear kN	Slip Resistance for $\mu = 0.5$ Double Shear kN	Bearing Capacity, P_{bg} in kN — End distance equal to 3 x bolt diameter. Thickness in mm of ply passed through.									
								5	6	7	8	9	10	12	15	20	25	30
12	49.4	54.3	49.7	33.7	67.4	27.2	54.3	49.5	59.4	69.3	79.2	89.1	99.0	119	149	198	248	297
16	92.1	101	92.6	62.8	126	50.7	101	66.0	79.2	92.4	106	119	132	158	198	264	330	396
20	144	158	145	98.0	196	79.2	158	82.5	99.0	116	132	149	165	198	248	330	413	495
22	177	195	179	121	242	97.4	195	**90.8**	**109**	127	145	163	182	218	272	363	454	545
24	207	228	208	141	282	114	228	**99.0**	**119**	**139**	158	178	198	238	297	396	495	594
27	234	257	236	161	321	129	257	**111**	**134**	**156**	178	200	223	267	334	446	557	668
30	286	315	289	196	393	157	315	**124**	**149**	**173**	**198**	223	248	297	371	495	619	743

Values in **bold** are less than the single shear capacity of the bolt.
Values in *italic* are greater than the double shear capacity of the bolt.
Shading indicates that the ply thickness is not suitable for an outer ply.
FOR EXPLANATION OF TABLES SEE NOTE 10.1

HIGHER GRADE HSFG BOLTS IN S355

Diameter of Bolt mm	Min. Shank Tension P_o kN	Tension 1.1P_o kN	Tension $A_t p_t$ kN	Shear Capacity Single Shear kN	Shear Capacity Double Shear kN	Slip Resistance for $\mu = 0.5$ Single Shear kN	Slip Resistance for $\mu = 0.5$ Double Shear kN	Bearing Capacity, P_{bg} in kN — End distance equal to 3 x bolt diameter. Thickness in mm of ply passed through.									
								5	6	7	8	9	10	12	15	20	25	30
16	104	114	110	62.8	126	57.1	114	66.0	79.2	92.4	106	119	132	158	198	264	330	396
20	162	178	172	98.0	196	89.0	178	82.5	99.0	116	132	149	165	198	248	330	413	495
22	200	220	212	121	242	110	220	**90.8**	**109**	127	145	163	182	218	272	363	454	545
24	233	257	247	141	282	128	257	**99.0**	**119**	**139**	158	178	198	238	297	396	495	594
27	303	333	321	184	367	167	333	**111**	**134**	**156**	178	200	223	267	334	446	557	668
30	370	407	393	224	449	204	407	**124**	**149**	**173**	**198**	223	248	297	371	495	619	743

Values in **bold** are less than the single shear capacity of the bolt.
Values in *italic* are greater than the double shear capacity of the bolt.
Shading indicates that the ply thickness is not suitable for an outer ply.
FOR EXPLANATION OF TABLES SEE NOTE 10.1

BS 5950-1: 2000
BS 4395: 1969
BS 4604: 1970

BOLT CAPACITIES

PRELOADED HSFG BOLTS: NON-SLIP UNDER FACTORED LOADS

GENERAL GRADE HSFG BOLTS IN S355

Diameter of Bolt	Min. Shank Tension	Bolt Tension Capacity	Slip Resistance P_{sL}							
			$\mu = 0.2$		$\mu = 0.3$		$\mu = 0.4$		$\mu = 0.5$	
			Single Shear	Double Shear	Single Shear	Double Shear	Single Shear	Double Shear	Single Shear	Double Shear
mm	P_o kN	$0.9P_o$ kN	kN	kN	kN	kN	kN	kN	kN	kN
12	49.4	44.5	8.89	17.8	13.3	26.7	17.8	35.6	22.2	44.5
16	92.1	82.9	16.6	33.2	24.9	49.7	33.2	66.3	41.4	82.9
20	144	130	25.9	51.8	38.9	77.8	51.8	104	64.8	130
22	177	159	31.9	63.7	47.8	95.6	63.7	127	79.7	159
24	207	186	37.3	74.5	55.9	112	74.5	149	93.2	186
27	234	211	42.1	84.2	63.2	126	84.2	168	105	211
30	286	257	51.5	103	77.2	154	103	206	129	257

FOR EXPLANATION OF TABLES SEE NOTE 10.1

HIGHER GRADE HSFG BOLTS IN S 355

Diameter of Bolt	Min. Shank Tension	Bolt Tension Capacity	Slip Resistance P_{sL}							
			$\mu = 0.2$		$\mu = 0.3$		$\mu = 0.4$		$\mu = 0.5$	
			Single Shear	Double Shear	Single Shear	Double Shear	Single Shear	Double Shear	Single Shear	Double Shear
mm	P_o kN	$0.9P_o$ kN	kN	kN	kN	kN	kN	kN	kN	kN
16	104	93.5	18.7	37.4	28.1	56.1	37.4	74.8	46.8	93.5
20	162	146	29.1	58.2	43.7	87.4	58.2	116	72.8	146
22	200	180	36.0	72.1	54.1	108	72.1	144	90.1	180
24	233	210	42.0	84.0	63.0	126	84.0	168	105	210
27	303	273	54.5	109	81.8	164	109	218	136	273
30	370	333	66.6	133	100	200	133	266	167	333

FOR EXPLANATION OF TABLES SEE NOTE 10.1

BS 5950-1: 2000
BS 4395: 1969
BS 4604: 1970
BS 4933: 1973

BOLT CAPACITIES

PRELOADED HSFG BOLTS: NON-SLIP IN SERVICE

GENERAL GRADE COUNTERSUNK HSFG BOLTS IN S355

Diameter of Bolt	Min. Shank Tension	Tension		Shear Capacity		Slip Resistance for $\mu = 0.5$		Bearing Capacity, P_{bg} in kN End distance equal to 3 x bolt diameter. Thickness in mm of ply passed through.									
		$1.1P_o$	$A_t p_t$	Single Shear	Double Shear	Single Shear	Double Shear										
mm	P_o kN	kN	kN	kN	kN	kN	kN	5	6	7	8	9	10	12	15	20	25	30
12	49.4	54.3	49.7	33.7	67.4	27.2	54.3	19.8	29.7	39.6	49.5	59.4	69.3	89.1	*119*	*168*	*218*	*267*
16	92.1	101	92.6	62.8	126	50.7	101	13.2	26.4	39.6	52.8	66.0	79.2	106	145	*211*	*277*	*343*
20	144	158	145	98.0	196	79.2	158	**0**	16.5	33.0	49.5	66.0	82.5	116	165	*248*	*330*	*413*
22	177	195	179	121	242	97.4	195	**0**	9.08	27.2	45.4	63.5	81.7	118	172	*263*	*354*	*445*
24	207	228	208	141	282	114	228	**0**	**0**	19.8	39.6	59.4	79.2	119	178	*277*	*376*	*475*
27	234	257	236	161	321	129	257	**0**	**0**	5.57	27.8	50.1	72.4	117	184	*295*	*407*	*518*
30	286	315	289	196	393	157	315	**0**	**0**	**0**	12.4	37.1	61.9	111	186	*309*	*433*	*557*

Values in **bold** are less than the single shear capacity of the bolt.
Values in *italic* are greater than the double shear capacity of the bolt.
Shading indicates that the ply thickness is not suitable for an outer ply.
FOR EXPLANATION OF TABLES SEE NOTE 10.1

HIGHER GRADE COUNTERSUNK HSFG BOLTS IN S355

Diameter of Bolt	Min. Shank Tension	Tension		Shear Capacity		Slip Resistance for $\mu = 0.5$		Bearing Capacity, P_{bg} in kN End distance equal to 3 x bolt diameter. Thickness in mm of ply passed through.									
		$1.1P_o$	$A_t p_t$	Single Shear	Double Shear	Single Shear	Double Shear										
mm	P_o kN	kN	kN	kN	kN	kN	kN	5	6	7	8	9	10	12	15	20	25	30
16	104	114	110	62.8	126	57.1	114	13.2	26.4	39.6	52.8	66.0	79.2	106	145	*211*	*277*	*343*
20	162	178	172	98.0	196	89.0	178	**0**	16.5	33.0	49.5	66.0	82.5	116	165	*248*	*330*	*413*
22	200	220	212	121	242	110	220	**0**	9.08	27.2	45.4	63.5	81.7	118	172	*263*	*354*	*445*
24	233	257	247	141	282	128	257	**0**	**0**	19.8	39.6	59.4	79.2	119	178	*277*	*376*	*475*
27	303	333	321	184	367	167	333	**0**	**0**	5.57	27.8	50.1	72.4	117	184	*295*	*407*	*518*
30	370	407	393	224	449	204	407	**0**	**0**	**0**	12.4	37.1	61.9	111	186	*309*	*433*	*557*

Values in **bold** are less than the single shear capacity of the bolt.
Values in *italic* are greater than the double shear capacity of the bolt.
Shading indicates that the ply thickness is not suitable for an outer ply.
FOR EXPLANATION OF TABLES SEE NOTE 10.1

BS 5950-1: 2000
BS 4395: 1969
BS 4604: 1970
BS 4933: 1973

BOLT CAPACITIES

PRELOADED HSFG BOLTS: NON-SLIP UNDER FACTORED LOADS

GENERAL GRADE COUNTERSUNK HSFG BOLTS IN S355

Diameter of Bolt	Min. Shank Tension	Bolt Tension Capacity	Slip Resistance P_{sL}							
			$\mu = 0.2$		$\mu = 0.3$		$\mu = 0.4$		$\mu = 0.5$	
			Single Shear	Double Shear	Single Shear	Double Shear	Single Shear	Double Shear	Single Shear	Double Shear
mm	P_o kN	$0.9P_o$ kN	kN	kN	kN	kN	kN	kN	kN	kN
12	49.4	44.5	8.89	17.8	13.3	26.7	17.8	35.6	22.2	44.5
16	92.1	82.9	16.6	33.2	24.9	49.7	33.2	66.3	41.4	82.9
20	144	130	25.9	51.8	38.9	77.8	51.8	104	64.8	130
22	177	159	31.9	63.7	47.8	95.6	63.7	127	79.7	159
24	207	186	37.3	74.5	55.9	112	74.5	149	93.2	186
27	234	211	42.1	84.2	63.2	126	84.2	168	105	211
30	286	257	51.5	103	77.2	154	103	206	129	257

FOR EXPLANATION OF TABLES SEE NOTE 10.1

HIGHER GRADE COUNTERSUNK HSFG BOLTS IN S 355

Diameter of Bolt	Min. Shank Tension	Bolt Tension Capacity	Slip Resistance P_{sL}							
			$\mu = 0.2$		$\mu = 0.3$		$\mu = 0.4$		$\mu = 0.5$	
			Single Shear	Double Shear	Single Shear	Double Shear	Single Shear	Double Shear	Single Shear	Double Shear
mm	P_o kN	$0.9P_o$ kN	kN	kN	kN	kN	kN	kN	kN	kN
16	104	93.5	18.7	37.4	28.1	56.1	37.4	74.8	46.8	93.5
20	162	146	29.1	58.2	43.7	87.4	58.2	116	72.8	146
22	200	180	36.0	72.1	54.1	108	72.1	144	90.1	180
24	233	210	42.0	84.0	63.0	126	84.0	168	105	210
27	303	273	54.5	109	81.8	164	109	218	136	273
30	370	333	66.6	133	100	200	133	266	167	333

FOR EXPLANATION OF TABLES SEE NOTE 10.1

BS 5950-1:2000
BS EN 440
BS EN 499
BS EN 756
BS EN 758
BS EN 1668

FILLET WELDS

WELD CAPACITIES WITH E42 ELECTRODE WITH S355

Leg Length	Throat Thickness	Longitudinal Capacity	Transverse Capacity
s mm	a mm	P_L kN/mm	P_T kN/mm
3.0	2.1	0.525	0.656
4.0	2.8	0.700	0.875
5.0	3.5	0.875	1.094
6.0	4.2	1.050	1.312
8.0	5.6	1.400	1.750
10.0	7.0	1.750	2.188
12.0	8.4	2.100	2.625
15.0	10.5	2.625	3.281
18.0	12.6	3.150	3.938
20.0	14.0	3.500	4.375
22.0	15.4	3.850	4.813
25.0	17.5	4.375	5.469

Welds are between two elements at 90° to each other.
$P_L = p_w a$
$P_T = K p_w a$
$p_w = 250 \text{ N/mm}^2$
$K = 1.25$ for elements at 90° to each other.
FOR EXPLANATION OF TABLES SEE NOTE 10.2

[BLANK PAGE]

[BLANK PAGE]

[BLANK PAGE]